ORGANIC CHEMISTRY AS A SECOND LANGUAGE, 6e

ORGANIC CHEMISTRY AS A SECOND LANGUAGE, 6e

Second Semester Topics

DAVID KLEIN
Johns Hopkins University

WILEY

DIRECTOR	Michelle Renda
EXECUTIVE EDITOR	Natalie Ruffatto
SENIOR MANAGING EDITOR	Judy Howarth
PRODUCTION EDITOR	Mahalakshmi babu
COVER PHOTO CREDIT	Abstract geometric texture © bgblue/Getty Images, Flask - Norm Christiansen, Colorful Paintbrushes © Maartje van Caspel/Getty Images, Honeycomb slice © eli_asenova/Getty Images, Color diffuse © Korolkoff/Getty Images, Rosemary © Tetiana Rostopira/Getty Images

This book was set in 9/11 Times LT Std Roman by Straive™.

Founded in 1807, John Wiley & Sons, Inc. has been a valued source of knowledge and understanding for more than 200 years, helping people around the world meet their needs and fulfill their aspirations. Our company is built on a foundation of principles that include responsibility to the communities we serve and where we live and work. In 2008, we launched a Corporate Citizenship Initiative, a global effort to address the environmental, social, economic, and ethical challenges we face in our business. Among the issues we are addressing are carbon impact, paper specifications and procurement, ethical conduct within our business and among our vendors, and community and charitable support. For more information, please visit our website: www.wiley.com/go/citizenship.

ISBN: 978-1-119-83705-3 (PBK)
ISBN: 978-1-119-83704-6 (EVALC)

Library of Congress Cataloging-in-Publication Data:

LCCN is 2024004372

The inside back cover will contain printing identification and country of origin if omitted from this page. In addition, if the ISBN on the back cover differs from the ISBN on this page, the one on the back cover is correct.

SKY10062818_020224

CONTENTS

CHAPTER 9 *AMINES* **303**

CHAPTER 10 *DIELS–ALDER REACTIONS* **325**

CHAPTER 1

AROMATICITY

If you are using this book, then you have likely begun the second half of your organic chemistry course. By now, you have certainly encountered aromatic rings, such as benzene. In this chapter, we will explore the criteria for aromaticity, and we will discover many compounds (other than benzene) that are also classified as aromatic.

1.1 INTRODUCTION TO AROMATIC COMPOUNDS

Consider the structure of benzene:

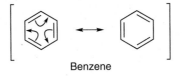

Benzene

Benzene is resonance-stabilized, as shown above, and is sometimes drawn in the following way:

This type of drawing (a hexagon with a circle in the center) is not suitable when drawing mechanisms of reactions, because mechanisms require that we keep track of electrons meticulously. But, it is helpful to see this type of drawing, even though we won't use it again in this book, because it represents all six π electrons of the ring as a single entity, rather than as three separate π bonds. Indeed, a benzene ring should be viewed as one functional group, rather than as three separate functional groups. This is perhaps most evident when we consider the special stability associated with a benzene ring. To illustrate this stability, we can compare the reactivity of cyclohexene and benzene:

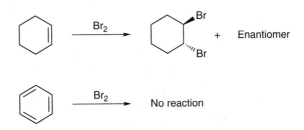

Cyclohexene is an alkene, and it will react with molecular bromine (Br_2) via an addition process, as expected for alkenes. In contrast, no reaction occurs when benzene is treated with Br_2, because the stability associated with the ring (of six π electrons) would be destroyed by an addition process. That is, the six π electrons of the ring represent a single functional group that does not react with Br_2, as alkenes do.

Understanding the source of the stability of benzene requires MO (molecular orbital) theory. You may or may not be responsible for MO theory in your course, so you should consult your textbook and/or lecture notes to see whether MO theory was covered.

Derivatives of benzene, called substituted benzenes, also exhibit the stability associated with a ring of six π electrons:

The ring can be monosubstituted, as shown above, or it can be disubstituted, or even polysubstituted (the ring can accommodate up to six different groups). Many derivatives of benzene were originally isolated from the fragrant extracts of trees and plants, so these compounds were described as being *aromatic,* in reference to their pleasant odors. Over time, it became apparent that many derivatives of benzene are, in fact, odorless. Nevertheless, the term *aromatic* is still currently used to describe derivatives of benzene, whether those compounds have odors or not.

1.2 NOMENCLATURE OF AROMATIC COMPOUNDS

As we have mentioned, an aromatic ring should be viewed as a single functional group. Compounds containing this functional group are generally referred to as *arenes.* In order to name arenes, recall that there are five parts of a systematic name, shown here (these five parts were discussed in Chapter 5 of the first volume of *Organic Chemistry as a Second Language: First Semester Topics*):

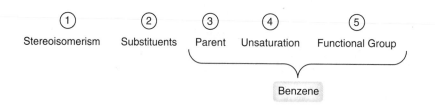

For benzene and its derivatives, the term *benzene* represents the parent, the unsaturation AND the suffix. Any other groups (connected to the ring) must be listed as substituents. For example, if a hydroxy (OH) group is connected to the ring, we do not refer to the compound as benzenol. We cannot add another suffix (-ol) to the term benzene, because that term is already a suffix itself. Therefore, the OH group is listed as a substituent, and the compound is called hydroxybenzene:

Hydroxybenzene

Similarly, other groups (connected to the ring) are also listed as substituents, as seen in the following examples:

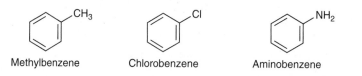

Methylbenzene   Chlorobenzene   Aminobenzene

Many monosubstituted derivatives of benzene (and even some disubstituted and polysubstituted derivatives) have common names. Several common names are shown here:

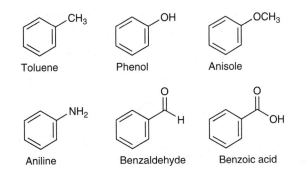

Toluene	Phenol	Anisole
Aniline	Benzaldehyde	Benzoic acid

So, for example, the first compound above can either be called methylbenzene (systematic name) or toluene (common name). Similarly, the second compound above can either be called hydroxybenzene (systematic name) or phenol (common name). When more than one substituent is present, common names are often used as parents. For example, the following compound exhibits a benzene ring with two substituents. But if we use the term *phenol* as the parent (rather than benzene), then we only have to list one substituent (bromo):

2-Bromophenol

The number (2) indicates the position of Br relative to the OH group. Notice that this compound can also be called 1-bromo-2-hydroxybenzene. Both names are acceptable, although it is generally more efficient to use the common name (2-bromophenol, rather than 1-bromo-2-hydroxybenzene).

When more than one group is present, numbers must be assigned, as seen in the previous example. When assigning numbers, the #1 position is determined by the parent. For example, the following compound will be named as a substituted toluene, so the methyl group is (by definition) at the #1 position:

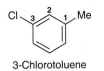

3-Chlorotoluene

The numbers are assigned in a manner that gives the lower possible number to the next substituent (Cl). So, in this case, we assign the numbers in a counterclockwise fashion, because that gives a lower number (3-chloro, rather than 5-chloro). The same method is applied for polysubstituted rings: first we choose the parent, and then we assign numbers in the direction that gives the lower possible number to the second substituent. Consider the following example:

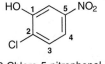

2-Chloro-5-nitrophenol

This compound can be named as a disubstituted phenol, rather than a trisubstituted benzene. So, the OH group is assigned the #1 position. Then, we continue assigning numbers in the direction that gives the lower possible number to the second substituent (2-chloro, rather than 3-nitro).

For disubstituted benzenes, there is special terminology that describes the proximity of the two groups:

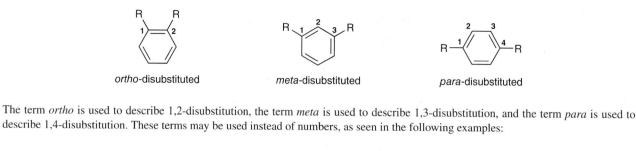

ortho-disubstituted   *meta*-disubstituted   *para*-disubstituted

The term *ortho* is used to describe 1,2-disubstitution, the term *meta* is used to describe 1,3-disubstitution, and the term *para* is used to describe 1,4-disubstitution. These terms may be used instead of numbers, as seen in the following examples:

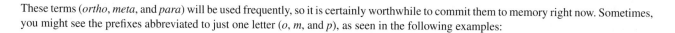

ortho-Chlorophenol   *meta*-Bromotoluene   *para*-Nitroanisole

These terms (*ortho*, *meta*, and *para*) will be used frequently, so it is certainly worthwhile to commit them to memory right now. Sometimes, you might see the prefixes abbreviated to just one letter (*o*, *m*, and *p*), as seen in the following examples:

o-Bromotoluene   *m*-Bromotoluene   *p*-Bromotoluene

WORKED PROBLEM 1.1   Provide a systematic name for the following compound:

Answer   This compound is a trisubstituted benzene, and it can be named as such, although it will be more efficient to name the compound as a disubstituted derivative of aniline. Since we are using a common name (aniline) as our parent, the position bearing the amino group is, by definition, position #1:

Then, we assign numbers in a clockwise fashion, rather than counterclockwise, so that the next substituent is at C2 rather than C3:

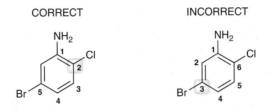

Therefore, the name is 5-bromo-2-chloroaniline. Notice that the substituents appear in the name in alphabetical order, in accordance with the general rules for IUPAC nomenclature.

PROBLEMS   Provide a systematic name for each of the following compounds:

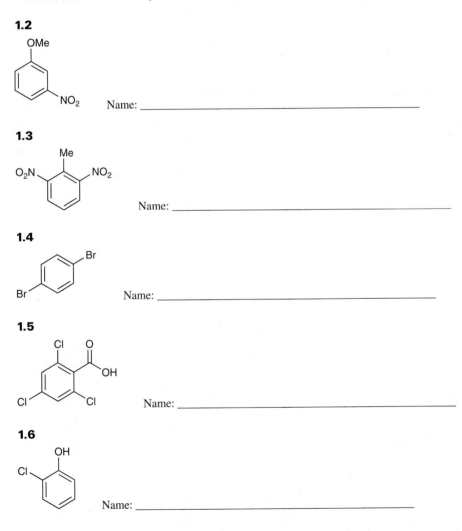

1.2

Name: _____

1.3

Name: _____

1.4

Name: _____

1.5

Name: _____

1.6

Name: _____

1.3 CRITERIA FOR AROMATICITY

In the previous section, we saw that benzene and its derivatives exhibit a special stability that is associated with the ring of six π electrons. We will now see that aromatic stabilization is not limited to benzene and its derivatives. Indeed, there are a large number of compounds and ions that exhibit aromatic stabilization. A few examples are shown here:

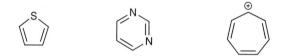

These structures are also said to be aromatic, illustrating that aromaticity is not strictly limited to six-membered rings, but indeed, even five-membered rings and seven-membered rings can be aromatic (if they meet certain criteria). In this section, we will explore the two criteria for aromaticity. Let's begin with the first criterion:

1) *There must be a ring comprised of continuously overlapping p orbitals.* The structures below do NOT satisfy this first criterion:

The first structure is not aromatic because it is not a ring, and the second structure is not aromatic because it lacks a continuous system of overlapping *p* orbitals. Six of the seven carbon atoms are *sp²* hybridized (and each of these carbon atoms does have a *p* orbital), but the seventh carbon atom (at the top of the ring) is *sp³* hybridized, thereby interrupting the overlap. The overlap of *p* orbitals must continue all the way around the ring, and it doesn't in this case because of the intervening *sp³* hybridized carbon atom.

Now let's explore the second criterion for aromaticity:

2) *The ring must contain an odd number of pairs of π electrons* (one pair of electrons, three pairs of electrons, five pairs of electrons, etc.). If we compare the following compounds, we find that only the middle compound (benzene) has an odd number of pairs of π electrons (three pairs of π electrons = 6 π electrons):

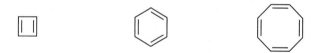

The first compound (cyclobutadiene) has two pairs of π electrons, and the last compound (cyclooctatetraene) has four pairs of π electrons. In order to understand the requirement for an odd number of pairs of π electrons, MO theory is required. Once again, consult your textbook and/or lecture notes for any coverage of MO theory. Another way of saying "an odd number of pairs of π electrons" is to say that the number of π electrons must be among the following series of numbers: 2, 6, 10, 14, 18, etc. These numbers are called Hückel numbers, and they can be summarized with the following formula: $4n + 2$, where *n* represents a series of integers (1, 2, 3, etc.). If we look closely at benzene (middle structure above), we find that there are six π electrons, which is a Hückel number. Therefore, benzene is aromatic. In contrast, each of the other two compounds above (cyclobutadiene and cyclooctatetraene) has an even number of pairs of π electrons. Cyclobutadiene has 4 π electrons, and cyclooctatetraene has 8 π electrons. These numbers (4 and 8) are NOT Hückel numbers. These numbers can be represented by the formula $4n$, where *n* represents a series of integers (1, 2, 3, etc.). These compounds are remarkably unstable (an observation that can be justified by MO theory). They are said to be *antiaromatic*. In order for a compound to be antiaromatic, it must satisfy the first criterion (it must possess a ring with a continuous system of overlapping *p* orbitals), but it must fail the second criterion (it must have $4n$, rather than $4n + 2$, π electrons).

Cyclooctatetraene can relieve much of the instability by puckering out of planarity, as shown:

In this way, the extent of continuous overlap (of p orbitals) is significantly reduced, so the first criterion (a ring of continuous overlapping p orbitals) is not fully satisfied. The compound therefore behaves as if it were nonaromatic, rather than antiaromatic. That is, it can be isolated, unlike antiaromatic compounds, which are generally too unstable to isolate. Moreover, it is observed to undergo addition reactions, unlike aromatic compounds which do not undergo addition reactions:

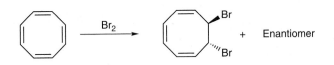

Now let's explore ions (structures that have a net charge). We will identify ions that are classified as aromatic, as well as examples of antiaromatic ions. Consider the structure of the following anion, which is resonance-stabilized. Draw all of the resonance structures in the space provided:

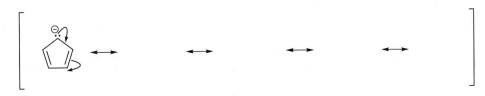

The remarkable stability of this anion cannot be explained with resonance alone. This anion is also aromatic, because it satisfies both criteria for aromaticity. To see how this is the case, we must recognize that the lone pair occupies a p orbital (because any lone pair that participates in resonance must occupy a p orbital), so this structure does indeed exhibit a ring with a continuous system of overlapping p orbitals, with a Hückel number of π electrons. The delocalized lone pair represents two π electrons, and the rest of the ring has another four π electrons, for a total of six π electrons. Indeed, the aromatic nature of this anion explains why cyclopentadiene is so acidic:

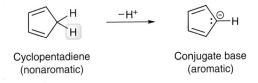

Cyclopentadiene is nonaromatic, but when it is deprotonated, the resulting conjugate base IS aromatic. Since the conjugate base is so stable, this renders cyclopentadiene fairly acidic (for a hydrocarbon). If we compare the pK_a values of cyclopentadiene and water, we find that they are similar in acidity:

This is truly remarkable, because it means that the following conjugate bases are similar in stability:

You often think of hydroxide as a strong base, but remember that basicity and acidity are relative concepts. Sure, hydroxide is a strong base when compared with certain weak bases, such as the acetate ion. But hydroxide is actually a relatively weak base when compared with other very strong bases, such as the amide ion (H_2N^-) or carbanions (C^-). Above, we see an example of a carbanion that is similar in stability to a hydroxide ion. You will not find many other examples of carbanions with such remarkable stability. And as we have seen, this stability is due to its aromatic nature.

The previous example was an anion. Let's now explore an example of a cation. The cation below, called a tropylium cation, is resonance-stabilized. Draw all of the resonance structures in the space provided.

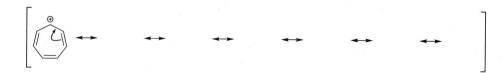

The remarkable stability of this cation is not fully explained by resonance. If we consider both criteria for aromaticity, we will find that this cation does indeed satisfy both criteria. Recall that a carbocation represents an empty p orbital, so we do have a ring with a continuous system of overlapping p orbitals, AND we have six π electrons (a Hückel number). Therefore, both criteria are satisfied, and this cation is aromatic.

Let's get some practice determining whether ions are aromatic, nonaromatic, or antiaromatic.

WORKED PROBLEM 1.7 Characterize the following ion as aromatic, nonaromatic, or antiaromatic:

Answer In order to be nonaromatic, it must fail the first criterion for aromaticity. In this case, it satisfies the first criterion, because the carbocation represents an empty p orbital, so we do have a ring with a continuous system of overlapping p orbitals. Since the first criterion is satisfied, we conclude that the structure will either be aromatic or antiaromatic, depending on whether the second criterion is met (a compound can only be nonaromatic if it fails the first criterion).

When we count the number of π electrons, we do NOT have a Hückel number in this case. With 4 π electrons, we expect this structure to be antiaromatic. That is, we expect this ion to be very unstable.

PROBLEMS Characterize each of the following structures as aromatic, nonaromatic, or antiaromatic:

1.8

Answer: _____

1.9

Answer: _____

1.10

Answer: _____

1.11

Answer: _____

1.12

Answer: _____

1.13

Answer: _____

1.4   LONE PAIRS

Compare the following two structures:

We saw in the previous section that the first structure is aromatic. The second structure, called pyrrole, is also aromatic, for the same reason. The nitrogen atom adopts an sp^2 hybridized state, which places the lone pair in a p orbital, thereby establishing a continuous system of overlapping p orbitals,

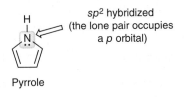

Pyrrole

and there are six π electrons (two from the lone pair + another two from each of the π bonds = 6). Both criteria for aromaticity have been satisfied, so the compound is aromatic. Notice that the lone pair is part of the aromatic system. As such, the lone pair is less available to function as a base:

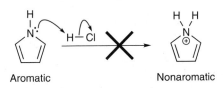

Aromatic   Nonaromatic

This doesn't occur, because the ring would lose aromaticity. If the nitrogen atom were to be protonated, the resulting nitrogen atom (with a positive charge) would be sp^3 hybridized. It would no longer have a p orbital, so the first criterion for aromaticity would not be satisfied (thus, nonaromatic). Protonation of the nitrogen atom would be extremely uphill in energy and is not observed.

In contrast, consider the nitrogen atom of pyridine:

Pyridine

In this case, the lone pair is NOT part of the aromatic system. This localized lone pair is not participating in resonance, and it does not occupy a p orbital. The nitrogen atom is sp^2 hybridized, and it does have a p orbital, but the p orbital is occupied by a π electron (as illustrated by the double bond that is drawn on the nitrogen atom). The lone pair actually occupies an sp^2 hybridized orbital, and it is therefore not contributing to the aromatic system. As such, it is available to function as a base, because protonation does not destroy aromaticity:

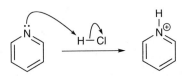

Aromatic   Still aromatic

Indeed, pyridine is often used as a mild base, especially in reactions where HCl is a byproduct. The presence of pyridine effectively neutralizes the acid as it is produced, and for this reason, pyridine is often referred to as "an acid sponge."

Let's explore one last example of an aromatic compound. This compound, called furan, is similar in structure to pyrrole:

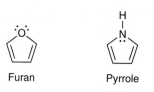

Furan   Pyrrole

The oxygen atom of furan is sp^2 hybridized (much like the nitrogen atom of pyrrole), which places one of the lone pairs in a p orbital, thereby establishing a continuous system of overlapping p orbitals, with a Hückel number of electrons (6π electrons). Therefore, furan is aromatic, for the same reason that pyrrole is aromatic. Notice that the oxygen atom in furan has two lone pairs, but only one of them occupies a p orbital. The other lone pair occupies an sp^2 hybridized orbital. As such, we only count one of the lone pairs of the oxygen atom (not both lone pairs) when we are counting to see if we have a Hückel number.

PROBLEMS

1.14 In the following compound, identify whether each lone pair is available to function as a base, and explain your choice:

1.15 Only one of the following compounds is aromatic. Identify the aromatic compound, and justify your choice:

END-OF-CHAPTER PROBLEMS

PRACTICE PROBLEMS *(Problems that involve only one skill)*

Identify whether each of the following structures is aromatic, antiaromatic, or nonaromatic.

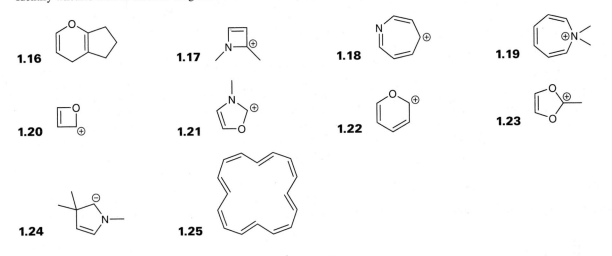

1.16   **1.17**   **1.18**   **1.19**

1.20   **1.21**   **1.22**   **1.23**

1.24   **1.25**

1.26   Below is the structure of Prilosec, a drug used for the treatment of ulcers and acid reflux. This structure has three rings. Determine whether each ring is aromatic or not.

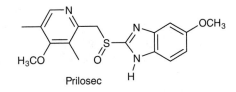

Prilosec

INTEGRATED PROBLEMS *(Problems that involve more than one skill)*

1.27   Cyclopentadiene and cycloheptatriene are both conjugated, cyclic hydrocarbons. Despite their similarity in structure, cyclopentadiene is relatively acidic (with a pK_a near that of H_2O), while cycloheptatriene does not function as an acid. Explain this significant difference in acidity.

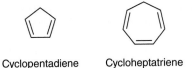

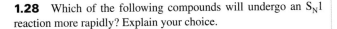

Cyclopentadiene   Cycloheptatriene

1.28   Which of the following compounds will undergo an S_N1 reaction more rapidly? Explain your choice.

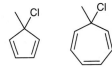

1.29   Calicene, shown below, has a measurable molecular dipole moment. Draw a resonance structure that justifies the observed dipole moment, and explain why that resonance structure contributes so much character to the overall resonance hybrid. (Note: when drawing resonance structures, we generally don't break C=C bonds to draw both C+ and C−, but this case represents an exception.)

Calicene

1.30   Which of the following compounds will undergo an S_N1 reaction more rapidly? Explain your choice.

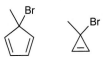

1.31 Which of the following compounds undergoes a rapid E2 reaction under weakly basic conditions? Explain your choice.

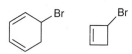

1.32 The following compound is aromatic. Draw resonance structures that justify this classification.

1.33 Explain why compound **A** is significantly more stable than compound **B**.

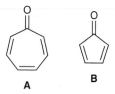

1.34 Compound **A** is insoluble in water, while compound **B** is highly water-soluble and functions much like a salt, by dissociating into a cation and an anion. Explain the observed behavior of compound **B**.

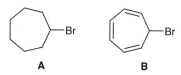

1.35 Which of the following is expected to be the stronger acid? Explain your choice.

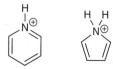

1.36 Which of the following compounds will undergo an S_N1 reaction more rapidly? Explain your choice.

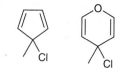

1.37 Which of the following compounds undergoes a rapid E2 reaction under weakly basic conditions? Explain your choice.

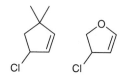

1.38 When treated with a strong acid, such as HBr, the following compound is protonated to give a highly stabilized cation. Draw the structure of this cation, and explain why it is so stable.

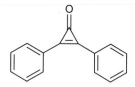

1.39 Borazine, shown below, is called "inorganic benzene" because of its remarkable stability. Identify the source of this stability.

Borazine

1.40 Identify which nitrogen atom in imidazole is more basic, and draw the conjugate acid that is formed when imidazole is protonated by a strong acid, such as HBr.

Imidazole

1.41 The following compound is too unstable to be made. Identify the source of its instability.

1.42 Draw the conjugate base that is formed when indene is deprotonated by a strong base, such as NaNH$_2$.

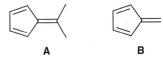

Indene (pK$_a$ = 20)

1.43 When treated with two equivalents of a strong base, such as NaH, the following compound is deprotonated to give a dianion that is particularly stable. Draw the structure of the dianion, and explain the source of its stability.

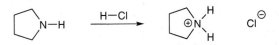

CHALLENGE PROBLEMS

1.44 Compound **A** is significantly more acidic than compound **B**. Explain the difference in acidity.

A   **B**

1.45 In general, a negative charge on a carbon atom is less stable than a negative charge on a nitrogen atom. As a result, alkanes are generally less acidic than amines. Nevertheless, cyclopentadiene is significantly more acidic than pyrrole. Explain.

Cyclopentadiene   Pyrrole

1.46 When treated with a strong acid, such as HCl, pyrrolidine is protonated to give a conjugate acid with a positive charge on the nitrogen atom.

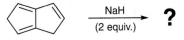

Pyrrolidine   Conjugate acid
of pyrrolidine

In contrast, when pyrrole is treated with a strong acid, the C2 position (indicated below with the number 2) is protonated, rather than the nitrogen atom. Draw the conjugate acid of pyrrole, and explain why protonation occurs at C2, rather than occurring at the nitrogen atom.

Pyrrole

CHAPTER 2

IR SPECTROSCOPY

Did you ever wonder how chemists are able to determine whether or not a reaction has produced the desired products? In your textbook, you will learn about many, many reactions. And an obvious question should be: "how do chemists *know* that those are the products of the reactions?"

Until about 50 years ago, it was actually VERY difficult to determine the structures of the products of a reaction. In fact, chemists would often spend many months, or even years to elucidate the structure of a single compound. But things got a lot simpler with the advent of spectroscopy. These days, the structure of a compound can be determined in minutes. Spectroscopy is, without a doubt, one of the most important tools available for determining the structure of a compound. Many Nobel prizes have been awarded to chemists who pioneered applications of spectroscopy.

The basic idea behind all forms of spectroscopy is that electromagnetic radiation (light) can interact with matter in predictable ways. Consider the following simple analogy: imagine that you have 10 friends, and you know what kind of bakery items they each like to eat every morning. John always has a brownie, Peter always has a French roll, Mary always has a blueberry muffin, etc. Now imagine that you walk into the bakery just after it opens, and you are told that some of your friends have already visited the bakery. By looking at what is missing from the bakery, you could figure out which of your friends had just been there. If you see that there is a brownie missing, then you deduce that John was in the bakery before you.

This simple analogy breaks down when you really get into the details of spectroscopy, but the basic idea is a good starting point. When electromagnetic radiation interacts with matter, certain frequencies are absorbed while other frequencies are not. By analyzing which frequencies were absorbed (which frequencies are missing once the light passes through a solution containing the unknown compound), we can glean useful information about the structure of the compound.

You may recall from your high school science classes that the range of all possible frequencies (of electromagnetic radiation) is known as the electromagnetic spectrum, which is divided into several regions (including X-rays, UV light, visible light, infrared radiation, microwaves, and radio waves). Different regions of the electromagnetic spectrum are used to probe different aspects of molecular structure, as seen in the table below:

Type of Spectroscopy	Region of Electromagnetic Spectrum	Information Obtained
NMR Spectroscopy	Radio Waves	The specific arrangement of all carbon and hydrogen atoms in the compound
IR Spectroscopy	Infrared (IR)	The functional groups present in the compound
UV-Vis Spectroscopy	Visible and Ultraviolet	Any conjugated π system present in the compound

We will not cover UV-Vis spectroscopy in this book. Your textbook may have a short section on that form of spectroscopy. In this chapter, we will focus on the information that can be obtained with IR spectroscopy. The next chapter will cover NMR spectroscopy.

2.1 VIBRATIONAL EXCITATION

Molecules can store energy in a variety of ways. They rotate in space, their bonds vibrate like springs, their electrons can occupy a number of possible molecular orbitals, etc. According to the principles of quantum mechanics, each of these forms of energy is quantized. For example, a bond in a molecule can only vibrate at specific energy levels:

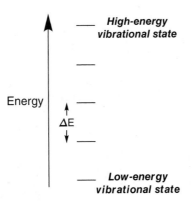

The horizontal lines in this diagram represent allowed vibrational energy levels for a particular bond. The bond is restricted to these energy levels, and cannot vibrate with an energy that is in between the allowed levels. The difference in energy (ΔE) between allowed energy levels is determined by the nature of the bond. If a photon of light possesses exactly this amount of energy, the bond (which was already vibrating) can absorb the photon to promote a *vibrational excitation*. That is, the bond will now vibrate more energetically. The energy of the photon is temporarily stored as vibrational energy, until that energy is released back into the environment, usually in the form of heat.

Bonds can store vibrational energy in a number of ways. They can *stretch*, very much the way a spring stretches, or they can *bend* in a number of ways. Your textbook will likely have images that illustrate these different kinds of vibrational excitation. In this chapter, we will devote most of our attention to stretching vibrations (as opposed to bending vibrations) because stretching vibrations generally provide the most useful information.

For each and every bond in a molecule, the energy gap between vibrational states is very much dependent on the nature of the bond. For example, the energy gap for a C—H bond is much larger than the energy gap for a C—O bond:

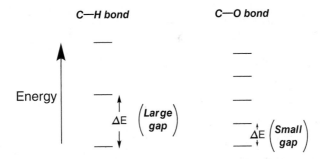

Both bonds will absorb infrared (IR) radiation, but the C—H bond will absorb a higher-energy photon. A similar analysis can be performed for other types of bonds as well, and we find that each type of bond will absorb a characteristic frequency, allowing us to determine which types of bonds are present in a compound. For example, a compound containing an O—H bond will absorb a frequency of IR radiation characteristic of O—H bonds. In this way, ***IR spectroscopy can be used to identify the presence of functional groups in a compound***. It is important to realize that IR spectroscopy does NOT reveal the entire structure of a compound. It can indicate that an unknown compound is an alcohol, but to determine the entire structure of the compound, we will need NMR spectroscopy (covered in the next chapter). For now, we are simply focusing on identifying which functional groups are present in an unknown compound. To get this information, we irradiate the compound with all frequencies of IR radiation, and then detect which frequencies were absorbed. This can be achieved with an IR spectrometer, which measures absorption as a function of frequency. The resulting plot is called an IR absorption spectrum (or IR spectrum, for short).

2.2   IR SPECTRA

An example of an IR spectrum is shown below:

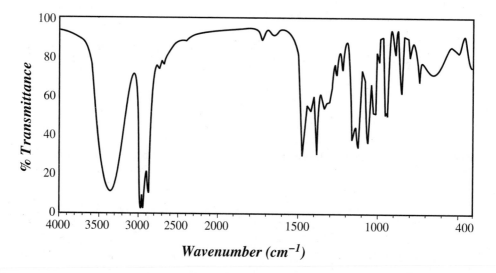

Notice that all signals point down in an IR spectrum. The location of each signal on the spectrum is reported in terms of a frequency-related unit, called wavenumber ($\tilde{\nu}$). The wavenumber is simply the frequency of light (ν) divided by a constant (the speed of light, c):

$$\tilde{\nu} = \frac{\nu}{c}$$

The units of wavenumber are inverse centimeters (cm^{-1}), and the values typically range from 400 cm^{-1} to 4000 cm^{-1}. Don't confuse the terms wavenumber and wavelength. Wavenumber is proportional to frequency, and therefore, a larger wavenumber represents higher energy. Signals that appear on the left side of the spectrum correspond with higher energy radiation, while signals on the right side of the spectrum correspond with lower energy radiation.

Every signal in an IR spectrum has the following three characteristics:

1. the *wavenumber* at which the signal appears

2. the *intensity* of the signal (strong vs. weak)

3. the *shape* of the signal (broad vs. narrow)

We will now explore each of these three characteristics, starting with wavenumber.

2.3   WAVENUMBER

For any bond, the wavenumber of absorption associated with bond stretching is dependent on two factors:

1) *Bond strength* – Stronger bonds will undergo vibrational excitation at higher frequencies, thereby corresponding to a higher wavenumber of absorption. For example, compare the bonds below. The C≡N bond is the strongest of the three bonds and therefore appears at the highest wavenumber:

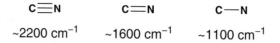

2) *Atomic mass* – Smaller atoms give bonds that undergo vibrational excitation at higher frequencies, thereby corresponding to a higher wavenumber of absorption. For example, compare the bonds below. The C—H bond involves the smallest atom (H) and therefore appears at the highest wavenumber.

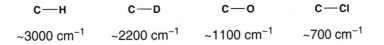

C—H   C—D   C—O   C—Cl
~3000 cm^{-1} ~2200 cm^{-1} ~1100 cm^{-1} ~700 cm^{-1}

Using the two trends shown above, we see that different types of bonds will appear in different regions of an IR spectrum:

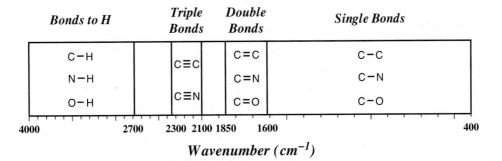

Wavenumber (cm^{-1})

Single bonds appear on the right side of the spectrum, because single bonds are generally the weakest bonds. Double bonds appear at higher wavenumber (1600–1850 cm^{-1}) because they are stronger than single bonds, while triple bonds appear at even higher wavenumber (2100–2300 cm^{-1}) because they are even stronger than double bonds. And finally, the left side of the spectrum contains signals produced by C—H, N—H, or O—H bonds, all of which stretch at a high wavenumber because hydrogen has the smallest mass.

IR spectra can be divided into two main regions, called the diagnostic region and the fingerprint region:

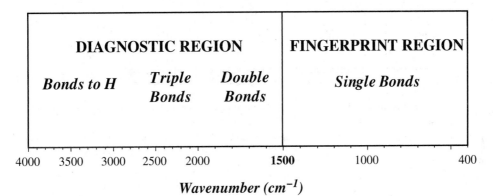

Wavenumber (cm^{-1})

The diagnostic region generally has fewer peaks and provides the most useful information. This region contains all signals that arise from the stretching of double bonds, triple bonds, and bonds to H (C—H, N—H, or O—H). The fingerprint region contains mostly bending vibrations, as well as stretching vibrations of most single bonds. This region generally contains many signals, and is more difficult to analyze. What appears like a C—C stretch might in fact be another bond that is bending. This region is called the fingerprint region because each compound has a unique pattern of signals in this region, much the way each person has a unique fingerprint. For example, IR spectra of ethanol and propanol will look extremely similar in their diagnostic regions, but their fingerprint regions will look different. For the remainder of this chapter, we will focus exclusively on the signals that appear in the diagnostic region, and we will ignore signals in the fingerprint region. You should check your lecture notes and textbook to see if you are responsible for any characteristic signals that appear in the fingerprint region.

PROBLEM 2.1   For the following compound, rank the highlighted bonds in order of decreasing wavenumber.

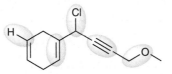

Now let's continue exploring factors that affect the strength of a bond (which therefore affects the wavenumber of absorption). We have seen that bonds to hydrogen (such as C—H bonds) appear on the left side of an IR spectrum (high wavenumber). We will now compare various kinds of C—H bonds. The wavenumber of absorption for a C—H bond is very much dependent on the hybridization state of the carbon atom. Compare the following three C—H bonds:

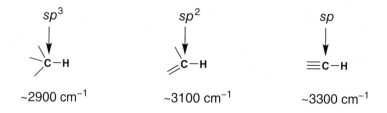

Of the three bonds shown, the C_{sp}—H bond produces the highest energy signal ($\sim 3300 \, cm^{-1}$), while a C_{sp^3}—H bond produces the lowest energy signal ($\sim 2900 \, cm^{-1}$). To understand this trend, we must revisit the shapes of the hybridized atomic orbitals:

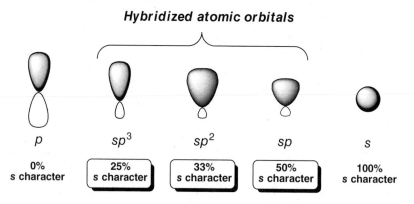

As illustrated, *sp* orbitals have more *s* character than the other hybridized atomic orbitals, and therefore, *sp* orbitals more closely resemble *s* orbitals. Compare the shapes of the hybridized atomic orbitals, and note that the electron density of an *sp* orbital is closest to the nucleus (much like an *s* orbital). As a result, a C_{sp}—H bond will be shorter than other C—H bonds. Since it has the shortest bond length, it will therefore be the strongest bond. In contrast, the C_{sp^3}—H bond has the longest bond length, and is therefore the weakest bond. Compare the spectra of an alkane, an alkene, and an alkyne:

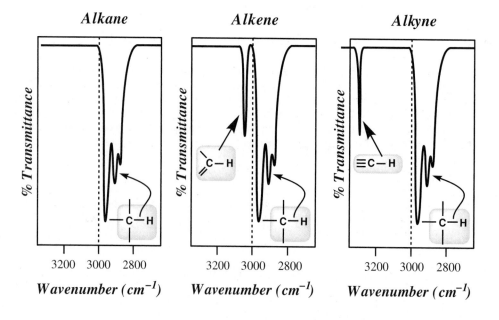

In each case, we draw a line at 3000 cm⁻¹. All three spectra have signals to the right of the line, resulting from C_{sp^3}—H bonds. The key is to look for any signals to the left of the line. An alkane does not produce a signal to the left of 3000 cm⁻¹. An alkene may produce a signal at 3100 cm⁻¹, and an alkyne may produce a signal at 3300 cm⁻¹. But be careful—the absence of a signal to the left of 3000 cm⁻¹ does *not* necessarily indicate the absence of a double bond or triple bond in the compound. Tetrasubstituted double bonds do not possess any C_{sp^2}—H bonds, and internal triple bonds also do not possess any C_{sp}—H bonds.

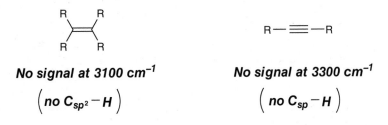

No signal at 3100 cm⁻¹

$\left(no\ C_{sp^2} - H \right)$

No signal at 3300 cm⁻¹

$\left(no\ C_{sp} - H \right)$

PROBLEMS For each of the following compounds, determine whether or not you would expect its IR spectrum to exhibit a signal to the left of 3000 cm⁻¹

2.2 **2.3** **2.4** **2.5**

Now let's explore the effects of resonance on bond strength. As an illustration, compare the carbonyl groups (C=O bonds) in the following two compounds:

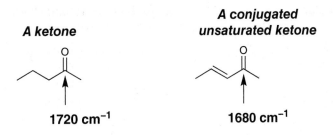

A ketone

1720 cm⁻¹

A conjugated unsaturated ketone

1680 cm⁻¹

The second compound is called a conjugated, unsaturated ketone. It is *unsaturated* because of the presence of a C=C bond, and it is conjugated because the π bonds are separated from each other by exactly one sigma bond. Your textbook will explore conjugated π systems in more detail. For now, we will just analyze the effect of conjugation on the IR absorption of the carbonyl group. As shown, the carbonyl group of an unsaturated, conjugated ketone produces a signal at lower wavenumber (1680 cm⁻¹) than the carbonyl group of a saturated ketone (1720 cm⁻¹). In order to understand why, we must draw resonance structures for each compound. Let's begin with the ketone.

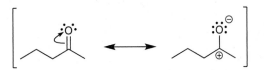

Ketones have two resonance structures. The carbonyl group is drawn as a double bond in the first resonance structure, and it is drawn as a single bond in the second resonance structure. This means that the carbonyl group has some double-bond character and some single-bond character. In order to determine the nature of this bond, we must consider the contribution from each resonance structure. In other words, does the carbonyl group have more double-bond character or more single-bond character? The second resonance structure exhibits charge separation, as well as a carbon atom (C$^+$) that has less than an octet of electrons. Both of these reasons explain why the second resonance structure contributes only slightly to the overall character of the carbonyl group. Therefore, the carbonyl group of a ketone has mostly double-bond character.

Now consider the resonance structures for a conjugated, unsaturated ketone.

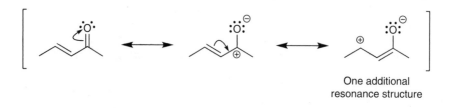

One additional
resonance structure

Conjugated, unsaturated ketones have three resonance structures rather than two. In the third resonance structure, the carbonyl group is drawn as a single bond. Once again, this resonance structure exhibits charge separation as well as a carbon atom (C$^+$) with less than an octet of electrons. As a result, this resonance structure also contributes only slightly to the overall character of the compound. Nevertheless, this third resonance structure does contribute some character, giving this carbonyl group slightly more single-bond character than the carbonyl group of a saturated ketone. With more single-bond character, it is a slightly weaker bond, and therefore produces a signal at a lower wavenumber (1680 cm^{-1} rather than 1720 cm^{-1}).

Esters exhibit a similar trend. An ester typically produces a signal at around 1740 cm^{-1}, but conjugated, unsaturated esters produce signals at a lower wavenumber, usually around 1710 cm^{-1}. Once again, the carbonyl group of a conjugated, unsaturated ester is a weaker bond, due to resonance.

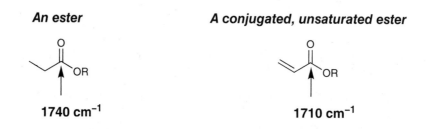

An ester

1740 cm^{-1}

A conjugated, unsaturated ester

1710 cm^{-1}

PROBLEM 2.6   The following compound has three carbonyl groups. Rank them in order of decreasing wavenumber in an IR spectrum:

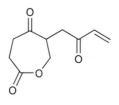

2.4 SIGNAL INTENSITY

In an IR spectrum, some signals will be very strong in comparison with other signals on the same spectrum:

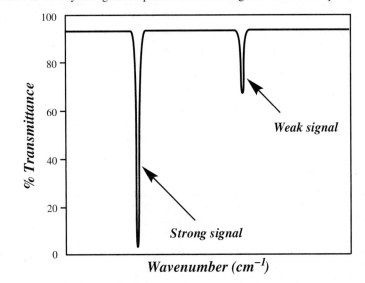

That is, some bonds absorb IR radiation very efficiently, while other bonds are less efficient at absorbing IR radiation. The efficiency of a bond at absorbing IR radiation depends on the magnitude of the dipole moment for that bond. For example, compare the following two highlighted bonds:

Each of these bonds has a measurable dipole moment, but they differ significantly in size. Let's first analyze the carbonyl group (C=O bond). Due to resonance and induction, the carbon atom bears a large partial positive charge, and the oxygen atom bears a large partial negative charge. The carbonyl group therefore has a large dipole moment. Now let's analyze the C=C bond. One vinylic position is connected to electron-donating alkyl groups, while the other vinylic position is connected to hydrogen atoms. As a result, the vinylic position bearing two alkyl groups is slightly more electron-rich than the other vinylic position, producing a small dipole moment.

Since the carbonyl group has a larger dipole moment, the carbonyl group is more efficient at absorbing IR radiation, producing a stronger signal:

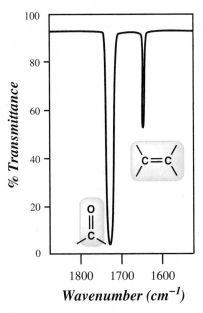

Carbonyl groups often produce the strongest signals in an IR spectrum, while C=C bonds often produce fairly weak signals. In fact, some alkenes do not even produce any signal at all. For example, consider the IR spectrum of 2,3-dimethyl-2-butene:

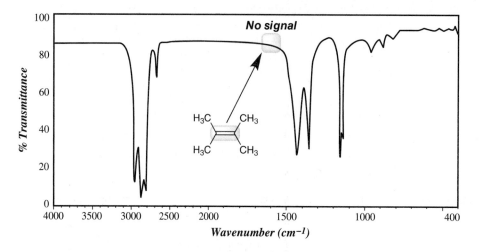

This alkene is symmetrical. That is, both vinylic positions are electronically identical, and the bond has no dipole moment at all. As such, this C=C bond is completely inefficient at absorbing IR radiation, and no signal is observed. The same is true for symmetrical C≡C bonds.

   There is one other factor that can contribute significantly to the intensity of signals in an IR spectrum. Consider the group of signals appearing just below 3000 cm^{-1} in the previous spectrum. These signals are associated with the stretching of the C—H bonds in the compound. The intensity of these signals derives from the number of C—H bonds giving rise to the signals. In fact, the signals just below 3000 cm^{-1} are typically among the strongest signals in an IR spectrum.

PROBLEMS

2.7   Predict which of the following C=C bonds will produce the strongest signal in an IR spectrum:

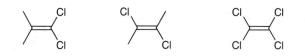

2.8   The C=C bond in the following compound produces an unusually strong signal. Explain using resonance structures:

2.5   SIGNAL SHAPE

In this section, we will explore some of the factors that affect the shape of a signal. Some signals in an IR spectrum might be very broad while other signals can be very narrow:

Broad signal

Narrow signal

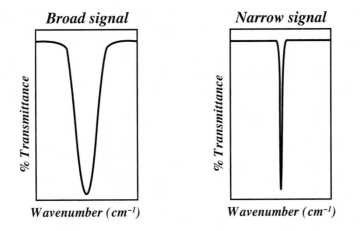

Concentrated alcohols commonly exhibit broad O—H signals, as a result of hydrogen bonding, which weakens the O—H bonds.

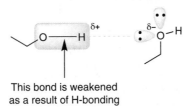

This bond is weakened
as a result of H-bonding

At any given moment in time, the O—H bond in each molecule is weakened to a different extent. As a result, all of the O—H bonds do not have a uniform bond strength, but rather, there is a *distribution* of bond strength. That is, some molecules are barely participating in H-bonding, while others are participating in H-bonding to varying degrees. The result is a broad signal.

The shape of an O—H signal is different when the alcohol is diluted in a solvent that cannot form hydrogen bonds with the alcohol. In such an environment, it is likely that the O—H bonds will not participate in an H-bonding interaction. The result is a narrow O—H signal. When the solution is neither very concentrated nor very dilute, two signals may be observed. The molecules that are not participating in H-bonding will give rise to a narrow signal, while the molecules participating in H-bonding will give rise to a broad signal. As an example, consider the following spectrum of 2-butanol, in which both signals can be observed:

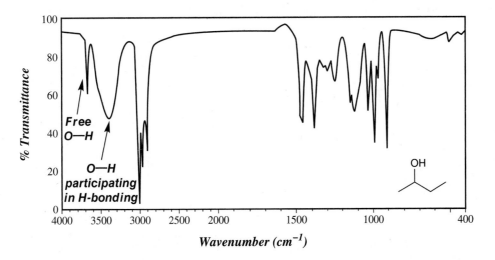

When O—H bonds do not participate in H-bonding, they generally produce a narrow signal at approximately $3600\,\text{cm}^{-1}$. That signal can be seen in the spectrum above. When O—H bonds participate in H bonding, they generally produce a broad signal between $3200\,\text{cm}^{-1}$

and 3600 cm⁻¹. That signal can also be seen in the spectrum. Depending on the conditions, an alcohol will either give a broad signal, or a narrow signal, or both. In most cases that we will encounter, O—H bonds will appear as broad signals.

Carboxylic acids exhibit similar behavior; only more pronounced. For example, consider the following spectrum of a carboxylic acid:

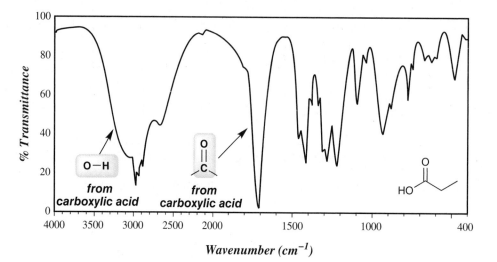

Notice the very broad signal on the left side of the spectrum, extending from 2200 cm⁻¹ to 3600 cm⁻¹. This signal is so broad that it extends over the usual C—H signals that appear around 2900 cm⁻¹. This very broad signal, characteristic of carboxylic acids, is a result of H-bonding. The effect is more pronounced than alcohols, because molecules of the carboxylic acid can form two hydrogen-bonding interactions, resulting in a dimer.

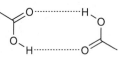

The IR spectrum of a carboxylic acid is easy to recognize, because of the characteristic broad signal that covers nearly one third of the spectrum. This broad signal is also accompanied by a strong C=O signal just above 1700 cm⁻¹.

PROBLEMS   For each IR spectrum below, identify whether it is consistent with the structure of an alcohol, a carboxylic acid, or neither.

2.9

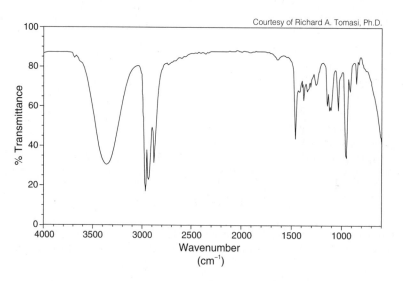

2.10

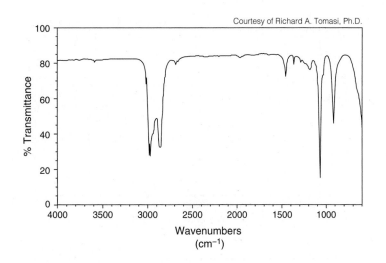

2.11

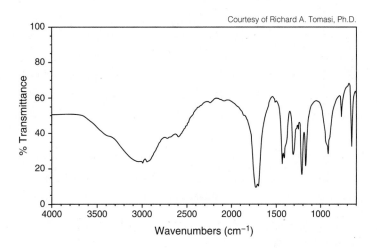

2.12

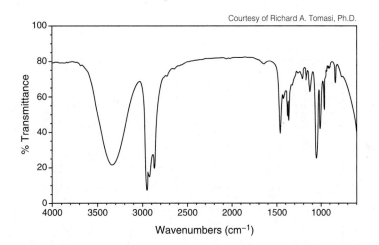

2.13

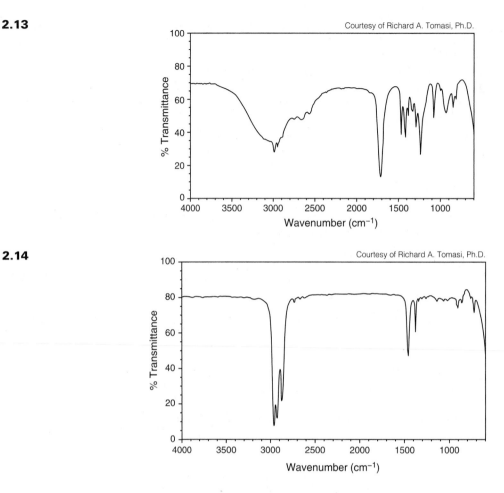

2.14

There is another important factor, in addition to H-bonding, that affects the shape of a signal. Consider the difference in shape of the N—H signals for primary and secondary amines:

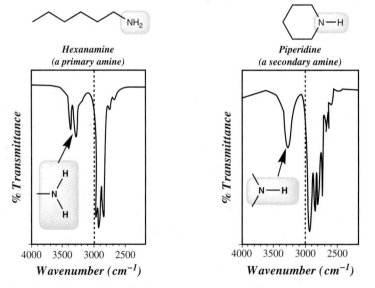

The primary amine exhibits two signals: one at 3350 cm^{-1} and the other at 3450 cm^{-1}. In contrast, the secondary amine exhibits only one signal. It might be tempting to explain this by arguing that each N—H bond gives rise to a signal, and therefore a primary amine gives two signals because it has two N—H bonds. Unfortunately, that simple explanation is not accurate. In fact, both N—H bonds of a single molecule will together produce only one signal. The reason for the appearance of two signals is more accurately explained by considering the two possible ways in which the entire NH_2 group can vibrate. The N—H bonds can be stretching in phase with each other, called *symmetric stretching*, or they can be stretching out of phase with each other, called *asymmetric stretching*:

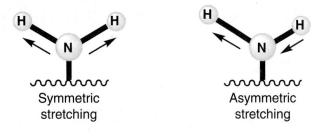

Symmetric stretching   Asymmetric stretching

At any given moment in time, approximately half of the molecules will be vibrating symmetrically, while the other half will be vibrating asymmetrically. The molecules vibrating symmetrically will absorb a particular frequency of IR radiation to promote a vibrational excitation, while the molecules vibrating asymmetrically will absorb a different frequency. In other words, the NH_2 functional group is capable of absorbing two different frequencies. One of the signals is produced by half of the molecules, and the other signal is produced by the other half of the molecules.

For a similar reason, the C—H bonds of a CH_3 group (appearing just below 3000 cm^{-1} in an IR spectrum) generally give rise to a series of signals, rather than just one signal. These signals arise from the various ways in which a CH_3 group can be excited both symmetrically and asymmetrically.

PROBLEMS   For each IR spectrum below, determine whether it is consistent with the structure of a ketone, an alcohol, a carboxylic acid, a primary amine, or a secondary amine.

2.15

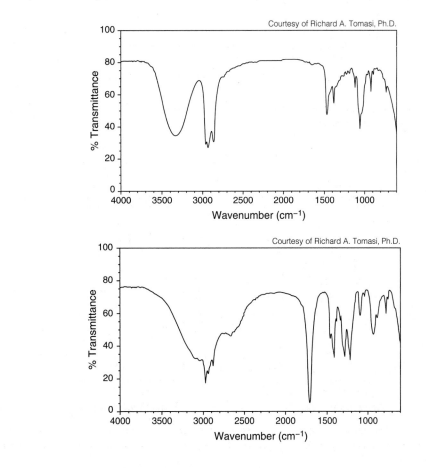

2.17

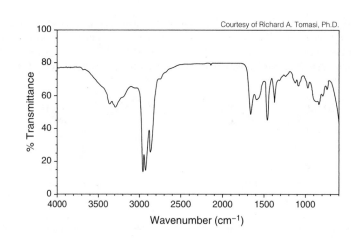

Courtesy of Richard A. Tomasi, Ph.D.

2.18

Courtesy of Richard A. Tomasi, Ph.D.

2.19

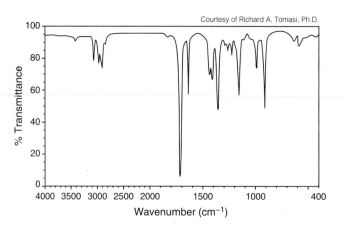

Courtesy of Richard A. Tomasi, Ph.D.

2.20

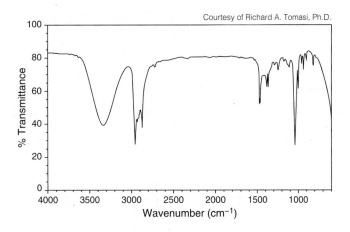

Courtesy of Richard A. Tomasi, Ph.D.

2.6 ANALYZING AN IR SPECTRUM

The following table is a summary of useful signals in the diagnostic region of an IR spectrum:

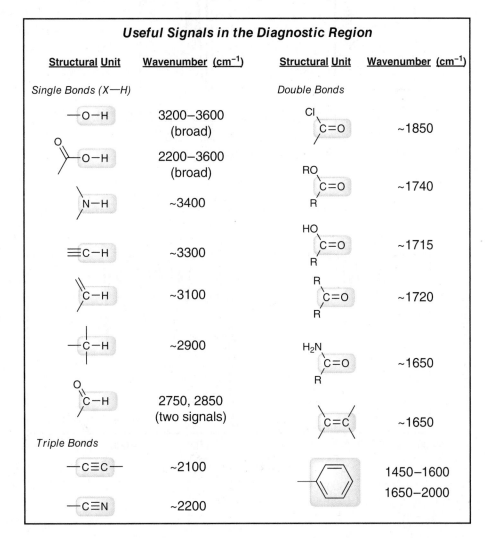

Useful Signals in the Diagnostic Region

Structural Unit	Wavenumber (cm⁻¹)	Structural Unit	Wavenumber (cm⁻¹)
Single Bonds (X—H)		*Double Bonds*	
—O—H	3200–3600 (broad)	Cl—C=O	~1850
O‖—O—H	2200–3600 (broad)	RO—C=O—R	~1740
N—H	~3400	HO—C=O—R	~1715
≡C—H	~3300	R—C=O—R	~1720
C=C—H	~3100	H₂N—C=O—R	~1650
—C—H	~2900		
O‖—C—H	2750, 2850 (two signals)	C=C	~1650
Triple Bonds			
—C≡C—	~2100	⬡	1450–1600 1650–2000
—C≡N	~2200		

When analyzing an IR spectrum, the first step is to draw a line at 1500 cm⁻¹. Focus on any signals to the left of this line (the diagnostic region). In doing so, it will be extremely helpful if you can identify the following regions:

Double bonds: 1600–1850 cm⁻¹

Triple bonds: 2100–2300 cm⁻¹

C—H, N—H, or O—H bonds: 2700–4000 cm⁻¹

Remember that each signal appearing in the diagnostic region will have three characteristics (wavenumber, intensity, and shape). Make sure to analyze all three characteristics.

When looking for bonds to H (C—H bonds, N—H bonds, or O—H bonds), draw a line at 3000 cm⁻¹ and look for signals that appear to the left of the line:

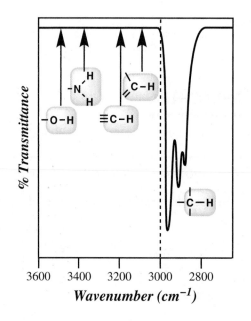

EXERCISE 2.21 A compound with the molecular formula $C_6H_{10}O$ gives the following IR spectrum:

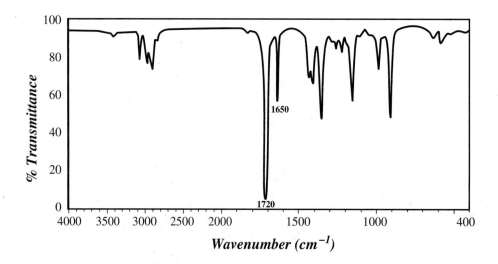

Identify the structure below that is most consistent with the spectrum:

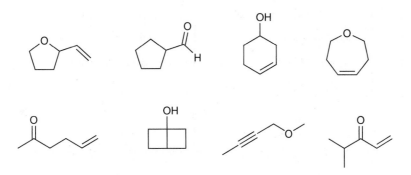

Solution   Draw a line at $1500\,cm^{-1}$, and focus on the diagnostic region (to the left of the line). Start by looking at the double-bond region and the triple-bond region:

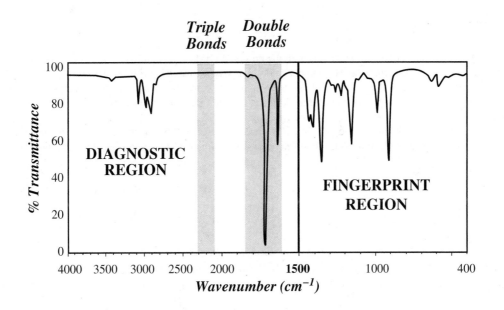

There are no signals in the triple-bond region, but there are two signals in the double-bond region. The signal at $1650\,cm^{-1}$ is narrow and weak, consistent with a C=C bond. The signal at $1720\,cm^{-1}$ is strong, consistent with a C=O bond.

Next, look for C—H bonds, N—H bonds, or O—H bonds that produce signals above $3000\,cm^{-1}$. Draw a line at $3000\,cm^{-1}$, and identify if there are any signals to the left of this line.

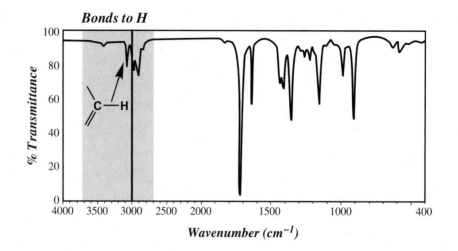

This spectrum exhibits one signal just above 3000 cm^{-1}, indicating a vinylic C—H bond.

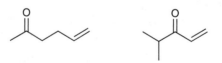

The identification of a vinylic C—H bond is consistent with the observed C=C signal present in the double-bond region (1650 cm^{-1}). There are no other signals above 3000 cm^{-1}, so the compound does not possess any OH or NH bonds.

The little bump between 3400 and 3500 cm^{-1} is not strong enough to be considered a signal. These bumps are often observed in the spectra of compounds containing a C=O bond. The bump occurs at exactly twice the wavenumber of the C=O signal, and is called an overtone of the C=O signal.

The diagnostic region provides the information necessary to solve this problem. Specifically, the compound must have the following bonds: C=C, C=O, and vinylic C—H. Among the possible choices, there are only two compounds that have these features:

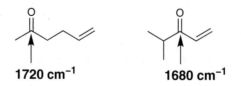

To distinguish between these two possibilities, notice that the second compound is conjugated, while the first compound is *not* conjugated (the π bonds are separated by more than one sigma bond). Recall that ketones produce signals at approximately 1720 cm^{-1}, while conjugated ketones produce signals at approximately 1680 cm^{-1}.

In the spectrum provided, the C=O signal appears at 1720 cm^{-1}, indicating that it is not conjugated. The spectrum is therefore consistent with the following compound:

PROBLEM 2.22 Match each compound with the appropriate spectrum, and identify the signals that justify your choice.

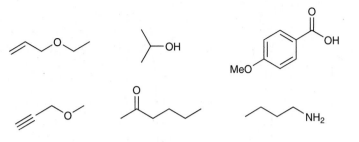

Spectrum A

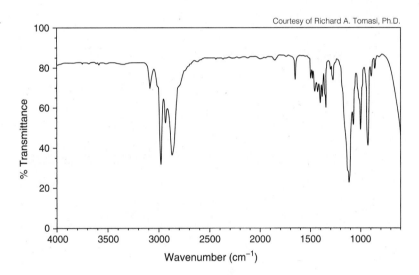

Spectrum B

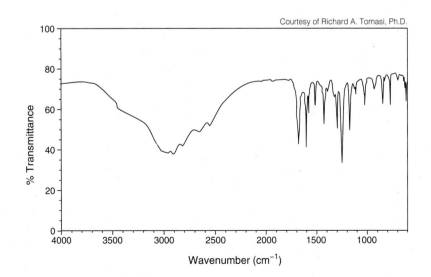

Spectrum C

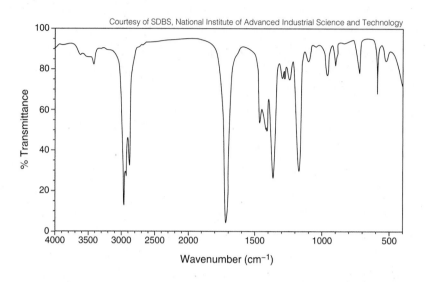

Spectrum D

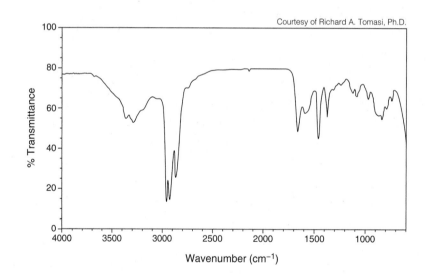

Spectrum E

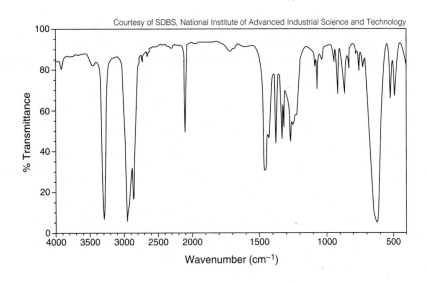

Courtesy of SDBS, National Institute of Advanced Industrial Science and Technology

Spectrum F

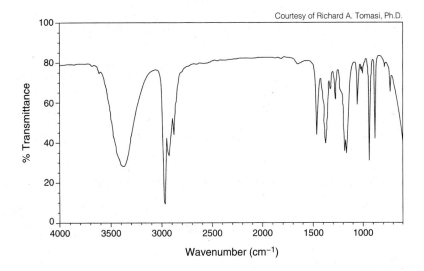

Courtesy of Richard A. Tomasi, Ph.D.

END-OF-CHAPTER PROBLEMS

PRACTICE PROBLEMS *(Problems that involve only one skill)*
Each of the following compounds has a functional group that produces one or more characteristic stretching signals above $1600\,\text{cm}^{-1}$ in IR spectroscopy. In each case, identify the wavenumber of absorption for the stretching signal(s), and place your answer(s) in the box(es) provided.

2.23

2.24

2.25 ☐ + ☐

2.26 ☐ + ☐

2.27 N—H ☐

2.28 H_3C—$C{\equiv}N$ ☐

2.29 H_3C—$C{\equiv}C$—CH_2CH_3 ☐

2.30 H_3C—$C{\equiv}C$—H ☐ + ☐

2.31 H—$C{\equiv}C$—H ☐

2.32 ☐

In each of the following cases, identify how you would use IR spectroscopy to distinguish between the two compounds.

2.33

2.34

2.35

2.36

How would you use IR spectroscopy to monitor the progress of each of the following reactions?

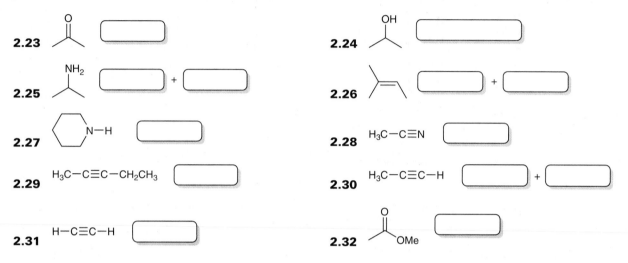

2.37 ⟶ 1) LiAlH₄ 2) H₃O⁺ (workup)

2.38 ⟶ 1) xs LiAlH₄ 2) H₃O⁺ (workup)

2.39 ⟶ 1) NaOH 2) CH₃CH₂I

2.40 All three compounds shown below absorb IR radiation between 1600 and $1850\,\text{cm}^{-1}$. For each compound, identify the bond(s) producing the signal, and then predict which compound will produce the strongest signal.

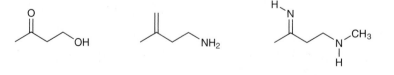

INTEGRATED PROBLEMS *(Problems that involve more than one skill)*

Predict the major product for each of the following reactions, and in each case, identify how you could use IR spectroscopy to monitor the progress of the reaction.

is treated with $Na_2Cr_2O_7$ and aqueous sulfuric acid. Identify the characteristic IR absorption that would confirm the formation of **D**.

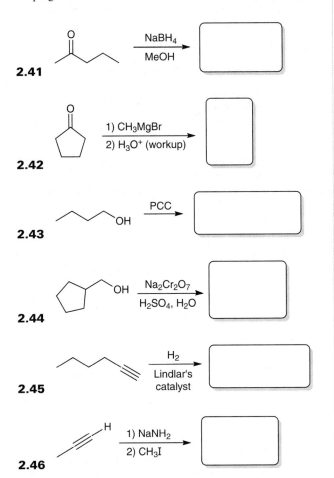

2.41

2.42

2.43

2.44

2.45

2.46

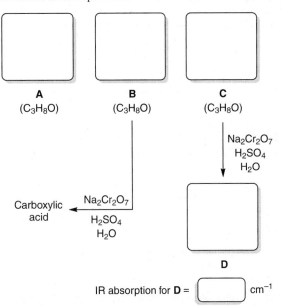

A
(C_3H_8O)

B
(C_3H_8O)

C
(C_3H_8O)

Na$_2$Cr$_2$O$_7$
H$_2$SO$_4$
H$_2$O

Carboxylic
acid

Na$_2$Cr$_2$O$_7$
H$_2$SO$_4$
H$_2$O

D

IR absorption for **D** = ☐ cm^{-1}

2.49 Compounds **A** and **B** are constitutional isomers with the molecular formula C_3H_6. The IR spectrum of compound **A** has signals at approximately $1650\,cm^{-1}$ and $3100\,cm^{-1}$, while the IR spectrum of compound **B** lacks these signals. Draw the structures of compounds **A** and **B**.

A
(C_3H_6)

B
(C_3H_6)

2.47 Draw the major and minor products of the following reaction, and suggest how IR spectroscopy might be used to differentiate between the two products.

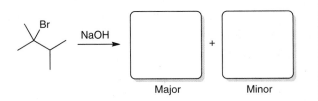

Major   Minor

2.48 Compounds **A**, **B**, and **C** are constitutional isomers with the molecular formula C_3H_8O. The IR spectrum of compound **A** does not have a signal above $3000\,cm^{-1}$. Compound **B** is treated with $Na_2Cr_2O_7$ and aqueous sulfuric acid to give a carboxylic acid. Based on this information, draw the structures of **A**, **B**, and **C**, as well as the structure of compound **D**, which is obtained when **C**

2.50 Below is a synthesis showing the conversion of compound **A** (an alkyl bromide) into compound **D** (an amide). Describe how IR spectroscopy could be used to monitor each step of this synthesis (the conversion of **A** into **B**, the conversion of **B** into **C**, and the conversion of **C** into **D**).

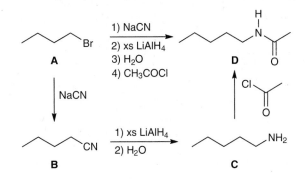

2.51 A compound with the molecular formula C₄H₈O has an IR spectrum with a strong signal at 1720 cm⁻¹. This compound does not react with Na₂Cr₂O₇ and aqueous sulfuric acid. Draw the structure of this compound.

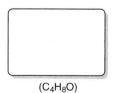

(C₄H₈O)

2.52 A compound with the molecular formula C₄H₆ has an IR spectrum with a signal at 2100 cm⁻¹. Draw the structure of this compound.

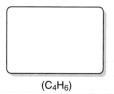

(C₄H₆)

2.53 A compound with a nitro group will typically produce two stretching vibrations in its IR spectrum, at approximately 1370 and 1550 cm⁻¹. Explain why the nitro group produces two signals in this region.

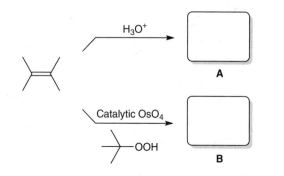

A nitro group

2.54 Compound **X** has the molecular formula C₆H₁₂ and undergoes ozonolysis to give acetone as the only product. Draw the structure of compound **X**, and identify whether you expect there to be any characteristic IR absorptions above 3000 cm⁻¹ in the IR spectrum of compound **X**.

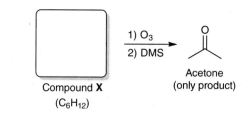

Compound **X**
(C₆H₁₂)

2.55 For the following reaction, draw the major product that is expected, and determine the number of signals that will appear between 1600 and 1850 cm⁻¹ in the IR spectrum of the product.

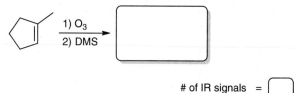

# of IR signals = ☐
1600–1850 cm⁻¹

CHALLENGE PROBLEMS

2.56 The IR spectrum of methanol (CH₃OH) has an O—H stretching signal that is broader (wider) than the O—H stretching signal in the IR spectrum of *tert*-butanol [(CH₃)₃COH]. Explain.

2.57 Draw the structures of compounds **A** and **B** below and suggest how IR spectroscopy might be used to differentiate between compounds **A** and **B**, even though they have similar functional groups.

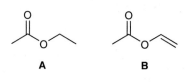

2.58 Will compound **A** or **B** have a higher wavenumber of absorption for the C=O stretching signal in its IR spectrum? Explain.

A   B

CHAPTER 3

NMR SPECTROSCOPY

Nuclear magnetic resonance (NMR) spectroscopy is the most useful technique for structure determination that you will encounter in your textbook. Analysis of an NMR spectrum provides information about how the individual carbon and hydrogen atoms are connected to each other in a molecule. This information enables us to determine the carbon–hydrogen framework of a compound, much the way puzzle pieces can be assembled to form a picture.

Your textbook will provide an explanation of the theoretical underpinnings of NMR spectroscopy (how it works). Here is a brief summary:

The nucleus of a hydrogen atom (which is just a proton) possesses a property called *nuclear spin*. A true understanding of this property is beyond the scope of our course, so we will think of it as a rotating sphere of charge, which generates a magnetic field, called a *magnetic moment*. When the nucleus of a hydrogen atom is subjected to an external magnetic field, the magnetic moment can align either with the field (called the α spin state) or against the field (called the β spin state). There is a difference in energy (ΔE) between these two spin states. If a proton occupying the α spin state is subjected to electromagnetic radiation, an absorption can take place IF the energy of the photon is equivalent to the energy gap between the spin states. The absorption causes the nucleus to *flip* to the β spin state, and the nucleus is said to be in *resonance*; thus the term *nuclear magnetic resonance*. NMR spectrometers employ strong magnetic fields, and the frequency of radiation typically required for nuclear resonance falls in the radio wave region of the electromagnetic spectrum (called rf radiation).

At a particular magnetic field strength, we might expect all nuclei to absorb the same frequency of rf radiation. Luckily, this is not the case, as nuclei are surrounded by electrons. In the presence of an external magnetic field, the electron density circulates, establishing a small, local magnetic field that *shields* the proton. Not all protons occupy identical electronic environments. Some protons are surrounded by more electron density and are more shielded, while other protons are surrounded by less electron density and are less shielded, or *deshielded*. As a result, protons in different electronic environments will absorb different frequencies of rf radiation. This allows us to probe the electronic environment of the hydrogen atoms in a compound.

3.1 CHEMICAL EQUIVALENCE

The spectrum produced by ^{1}H NMR spectroscopy (pronounced "proton" NMR spectroscopy) is called a proton NMR spectrum. Here is an example:

Courtesy of Richard A. Tomasi, Ph.D.

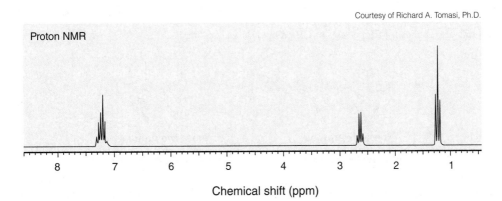

The first valuable piece of information is the number of signals (the spectrum above has three different signals). In addition, each signal has the following important characteristics:

1. The *location* of each signal indicates the electronic environment of the protons giving rise to the signal.
2. The *area* under each signal indicates the number of protons giving rise to the signal.
3. The *shape* of the signal indicates the number of neighboring protons.

We will discuss these characteristics in the upcoming sections. First let's explore the information that is revealed by counting the number of signals in a spectrum.

The number of signals in a proton NMR spectrum indicates the number of different kinds of protons (protons in different electronic environments). Protons that occupy identical electronic environments are called *chemically equivalent*, and they will produce only one signal.

Two protons are chemically equivalent if they can be interchanged by a symmetry operation. Your textbook will likely provide a detailed explanation, with examples. For our purposes, the following simple rules can guide you in most cases.

- The two protons of a CH_2 group will generally be chemically equivalent if the compound lacks chiral centers. But if the compound has a chiral center, then the protons of a CH_2 group will generally not be chemically equivalent:

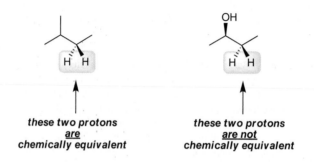

- Two CH_2 groups will be equivalent to each other (giving four equivalent protons) if the CH_2 groups can be interchanged by either rotation or reflection. Example:

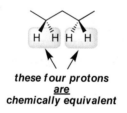

- The previous rule also applies for CH_3 groups or CH groups. Here are examples:

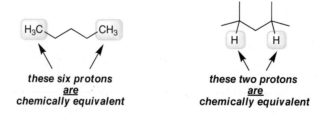

- The three protons of a CH_3 group are always chemically equivalent, even if there are chiral centers in the compound:

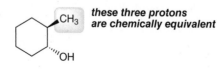

- For aromatic compounds, it will be less confusing if you draw a circle in the ring, rather than drawing alternating π bonds. For example, the following two methyl groups are equivalent, which can be easily seen when drawn in the following way:

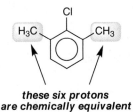

***these six protons
are chemically equivalent***

EXERCISE 3.1   Identify the number of signals expected in the ¹H NMR spectrum of the following compound:

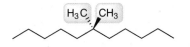

Answer   Let's begin with the two methyl groups at the center of this compound:

These two methyl groups are equivalent to each other, because they can be interchanged by either rotation or reflection (only one type of symmetry is necessary, but in this case we have both, reflection and rotation, so these six protons are certainly chemically equivalent). We therefore expect one signal for all six protons.

Now let's consider the following methylene (CH₂) groups:

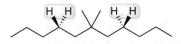

For each of these methylene groups, the two protons are chemically equivalent because there are no chiral centers. In addition, these two methylene groups will be equivalent to each other, since they can be interchanged by rotation or reflection. We therefore expect one signal for all four protons.

The same argument applies for the following two methylene groups:

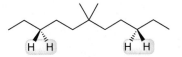

These four protons are chemically equivalent. But they are different from the other methylene groups in the compound, because they cannot be interchanged with any of the other methylene groups via rotation or reflection.

In a similar way, the following four protons are chemically equivalent:

And the following four protons are also chemically equivalent:

And finally, the following six protons are also chemically equivalent:

Note that these six protons are different from the six protons in the center of the compound, because the first set of six protons cannot be interchanged with the other set of six protons via either rotation or reflection.

In total, there are six different types of protons:

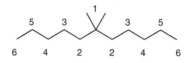

So we would expect the proton NMR spectrum of this compound to have six signals.

PROBLEMS   Identify the number of signals expected in the proton NMR spectrum of each of the following compounds.

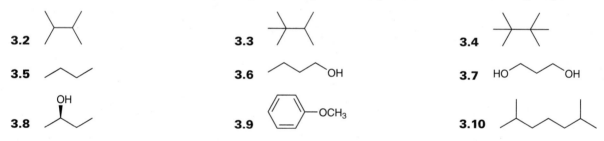

3.2   **3.3**   **3.4**

3.5   **3.6**   **3.7**

3.8   **3.9**   **3.10**

3.11   If you look at your answers to the previous problems, you should find that one of the structures was expected to produce an NMR spectrum with only one signal. In that structure (Problem 3.4), all six methyl groups were chemically equivalent. Using that example as guidance, propose two possible structures for a compound with the molecular formula C_9H_{18} that exhibits an NMR spectrum with only one signal.

3.2   CHEMICAL SHIFT (BENCHMARK VALUES)

We will now begin exploring the three characteristics of every signal in an NMR spectrum. The first characteristic is the location of the signal, called its *chemical shift* (δ), which is defined relative to the frequency of absorption of a reference compound, tetramethylsilane (TMS):

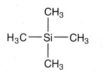

Tetramethylsilane (TMS)

Your textbook will go into greater depth in explaining chemical shift and why it is a unitless number that is reported in parts per million (ppm). For now, we will simply point out that for most organic compounds, the signals produced will fall in a range between 0 and 10 ppm. In rare cases, it is possible to observe a signal occurring at a chemical shift below 0 ppm, which results from a proton that absorbs a lower frequency than the protons in TMS. Most protons in organic compounds absorb a higher frequency than TMS, so all chemical shifts that you encounter in your textbook will be positive numbers.

The left side of an NMR spectrum is described as *downfield*, and the right side of the spectrum is described as *upfield*:

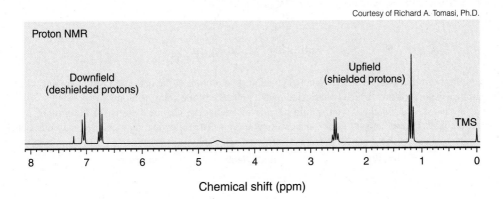

Courtesy of Richard A. Tomasi, Ph.D.

Chemical shift (ppm)

But keep in mind that these terms are used in a relative way. For example, we would say that the signal at 2.5 ppm (in the spectrum above) is downfield from the signal at 1.2 ppm. Similarly, the signal at 6.8 ppm is upfield from the signal at 7.1 ppm.

The protons of alkanes typically produce signals between 1 and 2 ppm. We will now explore some of the effects that can push a signal downfield (relative to the protons of an alkane). Recall that electronegative atoms, such as halogens, withdraw electron density from neighboring atoms:

$$H_3C \longrightarrow X$$

(X = F, Cl, Br, or I)

This inductive effect causes the neighboring protons to be deshielded (surrounded by less electron density), and as a result, the signal produced by these protons is shifted downfield—that is, the signal appears at a higher chemical shift than the protons of an alkane. The strength of this effect depends on the electronegativity of the halogen. Compare the chemical shifts of the protons in the following compounds:

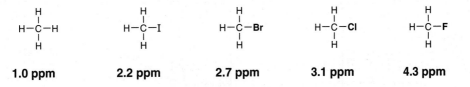

Fluorine is the most electronegative element, and therefore produces the strongest effect. When multiple halogens are present, the effect is generally additive, as can be seen when comparing the following compounds:

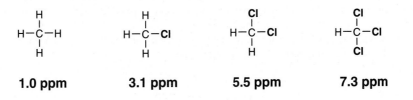

Each chlorine atom adds approximately 2 ppm to the chemical shift of the signal. An important aspect of inductive effects is the fact that they taper off drastically with distance, as can be seen by comparing the chemical shifts of the protons in the following compound:

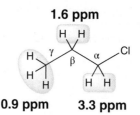

1.6 ppm

0.9 ppm   **3.3 ppm**

The effect is most significant for the protons at the alpha (α) position. The protons at the beta (β) position are only slightly affected, and the protons at the gamma (γ) position are virtually unaffected by the presence of the chlorine atom.

By committing a few numbers to memory, you should be able to predict the chemical shifts for the protons in a wide variety of compounds, including alcohols, ethers, ketones, esters, and carboxylic acids. The following numbers can be used as benchmark values:

methyl   *methylene*   *methine*

~0.9 ppm   ~1.2 ppm   ~1.7 ppm

In the absence of inductive effects, a methyl group (CH_3) will generally produce a signal near 0.9 ppm, a methylene group (CH_2) will produce a signal near 1.2 ppm, and a methine group (CH) will produce a signal near 1.7 ppm. These benchmark values are then modified by the presence of neighboring functional groups, in the following way:

Functional group	Effect on α protons	Example		
Oxygen of an alcohol or ether	+2.5	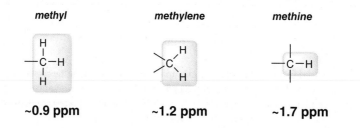	methylene group (CH_2) = 1.2 ppm next to oxygen = +2.5 ppm	**3.7 ppm**
Oxygen of an ester	+3		methylene group (CH_2) = 1.2 ppm next to oxygen = + 3.0 ppm	4.2 ppm
Carbonyl group (C=O) all carbonyl groups, including ketones, aldehydes, esters, etc.	+1		methylene group (CH_2) = 1.2 ppm next to carbonyl = + 1.0 ppm	2.2 ppm

The values above represent the effect of a few functional groups on the chemical shifts of alpha protons. The effect on beta protons is generally about one-fifth of the effect on the alpha protons. For example, in an alcohol, the presence of an oxygen atom adds +2.5 ppm to the chemical shift of the alpha protons, but adds only +0.5 ppm to the beta protons. Similarly, a carbonyl group adds +1 ppm to the chemical shift of the alpha protons, but only +0.2 to the beta protons.

The values above (together with the benchmark values for methyl, methylene, and methine groups), enable us to predict the chemical shifts for the protons in a wide variety of compounds. Let's see an example.

EXERCISE 3.12   Predict the chemical shifts for the signals in the proton NMR spectrum of the following compound:

Solution   First determine the total number of expected signals. In this compound, there are four different kinds of protons, giving rise to four distinct signals. For each type of signal, identify whether it represents methyl (0.9 ppm), methylene (1.2 ppm), or methine (1.7 ppm) groups:

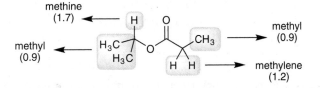

Finally, modify each of these numbers based on proximity to the oxygen and the carbonyl group:

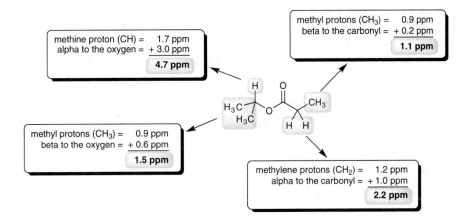

These values are only estimates, and the actual chemical shifts might differ slightly from the predicted values. Note that for the methine proton, we did not count the distant C=O bond and add +0.2, because the ester group is considered as one group, which has the effect of adding +3.0 to the chemical shift of the methine proton. Similarly, note that for the methylene protons, we did not add +0.5 for a distant oxygen. Once again, the ester group is considered as one group, which has the effect of adding +1.0 to the chemical shift of the methylene protons.

PROBLEMS   Predict the chemical shifts for the signals in the proton NMR spectrum of each of the following compounds:

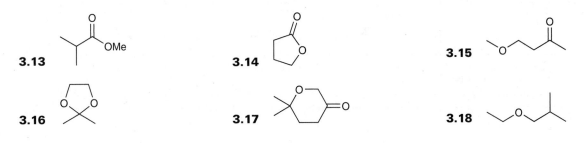

3.13

3.14

3.15

3.16

3.17

3.18

The chemical shift of a proton is also sensitive to the presence of nearby π electrons. This effect is particularly strong for aromatic protons (protons connected directly to an aromatic ring). In your textbook, you will find a diagram showing how (and why) the aromatic protons are affected. Here is a brief summary. The external magnetic field causes the π electrons to circulate, and this flow of electrons causes an induced, local magnetic field. The aromatic protons experience not only the external magnetic field, but they also experience the induced, local magnetic field. As a result, aromatic protons feel a stronger net magnetic field, which causes their signals to be shifted downfield, significantly. In fact, aromatic protons generally produce signals in the neighborhood of 7 ppm (sometimes as high as 8 ppm, sometimes as low as 6.5 ppm) in an NMR spectrum. For example, consider the structure of ethylbenzene:

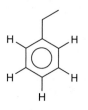

Ethylbenzene has three different kinds of aromatic protons (make sure you can identify them), producing three overlapping signals just above 7 ppm. A complex set of overlapping signals around 7 ppm is characteristic of compounds with aromatic protons.

The methylene group (CH_2) in ethylbenzene produces a signal at 2.6 ppm, rather than the expected benchmark value of 1.2 ppm. These protons have been shifted downfield because of their proximity to the aromatic ring. They are not shifted as much as the aromatic protons themselves, because the methylene protons are farther away from the ring, but there is still a noticeable effect.

All π electrons, whether they belong to an aromatic ring or not, have an effect on neighboring protons. For each type of π bond, the precise location of the nearby protons determines their chemical shift. For example, aldehydic protons produce characteristic signals at approximately 10 ppm. Below are some important chemical shifts. It would be wise to become familiar with these numbers, as they will be required in order to interpret proton NMR spectra:

Type of proton	Chemical shift (δ)	Type of proton	Chemical shift (δ)
methyl R—CH₃	~0.9	alkyl halide R—C(H)(R)—X	2–4
methylene CH₂	~1.2	alcohol R—O—H	2–5
methine —C—H	~1.7	vinylic =C—H	4.5–6.5
allylic C=C—C—H	~2	aryl Ar—H	6.5–8
alkynyl R—≡—H	~2.5	aldehyde R—C(=O)—H	~10
benzylic methyl Ar—CH₃	~2.5	carboxylic acid R—C(=O)—O—H	~12

PROBLEM 3.19   Predict the expected chemical shift for each type of proton in the following compound:

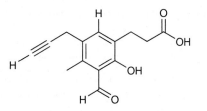

3.3   INTEGRATION

In the previous section, we learned about the first characteristic of every signal, chemical shift. In this section, we will explore the second characteristic, *integration*, which is the area under each signal. This value indicates the number of protons giving rise to the signal. After acquiring a spectrum, a computer calculates the area under each signal, and then displays this area as a numerical value placed under the signal:

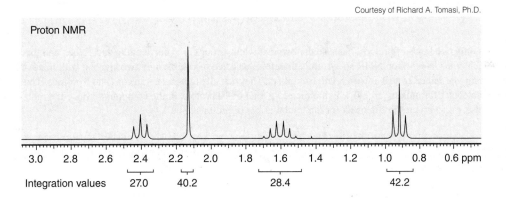

These numbers only have meaning when compared to each other. In order to convert these numbers into useful information, choose the smallest number (27.0 in this case), and then divide all integration values by this number:

$$\frac{27.0}{27.0} = 1 \qquad \frac{40.2}{27.0} = 1.49 \qquad \frac{28.4}{27.0} = 1.05 \qquad \frac{42.2}{27.0} = 1.56$$

These numbers provide the *relative number*, or ratio, of protons giving rise to each signal. This means that a signal with an integration of 1.5 involves one and a half times as many protons as a signal with an integration of 1. In order to arrive at whole numbers (there is no such thing as half a proton), multiply all the numbers above by two, giving the same ratio now expressed in whole numbers, 2 : 3 : 2 : 3. In other words, the signal at 2.4 ppm represents 2 equivalent protons, and the signal at 2.1 ppm represents 3 equivalent protons.

Integration values are often represented by *step-curves*, for example:

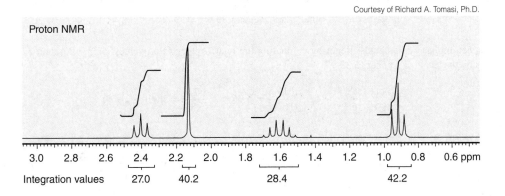

The height of each step curve represents the area under the signal. In this case, a comparison of the heights of the four step curves reveals a ratio of 2 : 3 : 2 : 3.

When interpreting integration values, don't forget that the numbers are only relative. For example, consider the structure of butane:

Butane has two kinds of protons, and will therefore produce two signals in its proton NMR spectrum. The methyl groups give rise to one signal and the methylene groups give another signal. A computer analyzes the area under each signal and provides numbers that allow us to calculate a ratio of 2 : 3. This ratio only indicates the relative number of protons giving rise to each signal, not the exact number of protons. In this case, the exact numbers are 4 (for the methylene groups) and 6 (for the methyl groups). When analyzing the NMR spectrum of an unknown compound, the molecular formula provides extremely useful information because it enables us to determine the exact number of protons giving rise to each signal. If we were analyzing the spectrum of butane, the molecular formula (C_4H_{10}) would indicate that the compound has a total of 10 protons. This information then allows us to determine that the ratio of 2 : 3 must correspond with 4 protons and 6 protons, in order to give a total of 10 protons.

The previous example illustrated the important role that symmetry can play when considering integration values. Here is another example:

This compound has only two kinds of protons, because the two methylene groups are equivalent to each other, and the two methyl groups are equivalent to each other. The proton NMR spectrum is therefore expected to exhibit only two signals, with relative integration values of 2 : 3. But once again, the values 2 and 3 are just relative numbers. They actually represent 4 protons and 6 protons. This can be determined by inspecting the molecular formula ($C_4H_{10}O$) which indicates a total of 10 protons in the compound. Since the ratio of protons is 2 : 3, this ratio must represent 4 and 6 protons, in order for the total number of protons to be 10.

EXERCISE 3.20 A compound with the molecular formula $C_5H_{10}O_2$ has the following proton NMR spectrum. Determine the number of protons giving rise to each signal:

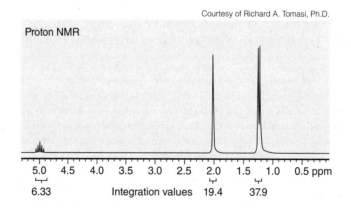

Courtesy of Richard A. Tomasi, Ph.D.

Solution The spectrum has three signals. Begin by comparing the relative integration values: 6.33, 19.4, and 37.9. Divide each of these three numbers by the smallest number (6.33):

$$\frac{6.33}{6.33} = 1 \qquad \frac{19.4}{6.33} = 3.06 \qquad \frac{37.9}{6.33} = 5.99$$

This gives a ratio of 1 : 3 : 6, but these are just relative numbers. To determine the exact number of protons giving rise to each signal, look at the molecular formula, which indicates a total of 10 protons in the compound. Therefore, the numbers 1 : 3 : 6 are not only relative values, but they are also the exact values. Exact integration values are sometimes expressed in the following way:

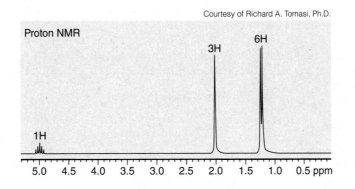

Courtesy of Richard A. Tomasi, Ph.D.

PROBLEMS

3.21 A compound with the molecular formula $C_8H_{10}O$ has the following proton NMR spectrum. Determine the number of protons giving rise to each signal.

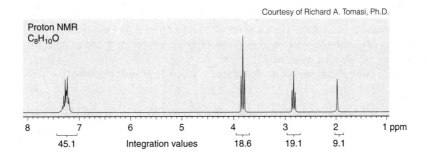

Courtesy of Richard A. Tomasi, Ph.D.

3.22 A compound with the molecular formula $C_7H_{14}O$ has the following proton NMR spectrum. Determine the number of protons giving rise to each signal.

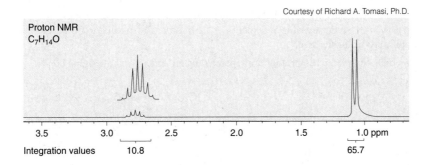

Courtesy of Richard A. Tomasi, Ph.D.

3.23   A compound with the molecular formula $C_4H_6O_2$ has the following proton NMR spectrum. Determine the number of protons giving rise to each signal.

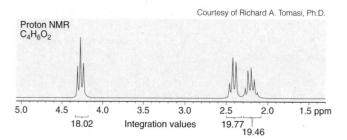

Courtesy of Richard A. Tomasi, Ph.D.

Proton NMR
$C_4H_6O_2$

5.0   4.5   4.0   3.5   3.0   2.5   2.0   1.5 ppm

18.02   Integration values   19.77
19.46

3.4   MULTIPLICITY

The third, and final, characteristic of each signal is its *multiplicity*, which refers to the number of peaks in the signal. A *singlet* has one peak, a *doublet* has two peaks, a *triplet* has three peaks, a *quartet* has four peaks, a *quintet* has five peaks, etc.:

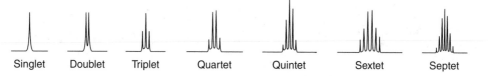

Singlet   Doublet   Triplet   Quartet   Quintet   Sextet   Septet

The multiplicity of a signal is the result of the magnetic effects of neighboring protons, and therefore indicates the number of neighboring protons. Your textbook will explain the cause for this effect in detail. The net effect can be summarized with the $n + 1$ *rule*, which states the following: if n is the number of neighboring protons, then the multiplicity will be $n + 1$. For example, a proton with three neighbors ($n = 3$) will be split into a quarter ($n + 1 = 3 + 1 = 4$ peaks).

It is important to realize that nearby protons do not always split each other. There are two major factors that determine whether or not splitting occurs:

1. Equivalent protons do not split each other. Consider the two methylene groups in the following compound:

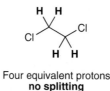

Four equivalent protons
no splitting

All four protons are chemically equivalent, and therefore, they do not split each other. Instead, they produce one signal that has no neighboring protons ($n = 0$), so the signal is a singlet ($n + 1$). In order for splitting to occur, the neighboring protons must be different than the protons producing the signal.

2. Splitting is most commonly observed when protons are separated by either two or three sigma bonds:

          or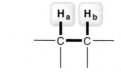

non-equivalent protons   non-equivalent protons
separated by two sigma bonds   separated by three sigma bonds

However, when two protons are separated by more than three sigma bonds, splitting is generally not observed:

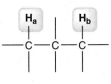

too far apart

Such long-range splitting is only observed in rigid molecules, such as bicyclic compounds, or in molecules that contain rigid structures, such as allylic systems. For purposes of this introductory treatment of NMR spectroscopy, we will avoid examples that exhibit substantial long-range splitting. Make sure to look through your lecture notes to see if any examples of long-range splitting are included.

EXERCISE 3.24 Determine the multiplicity of each signal in the expected proton NMR spectrum of the following compound:

Solution Begin by identifying the different kinds of protons. That is, determine the number of expected signals.

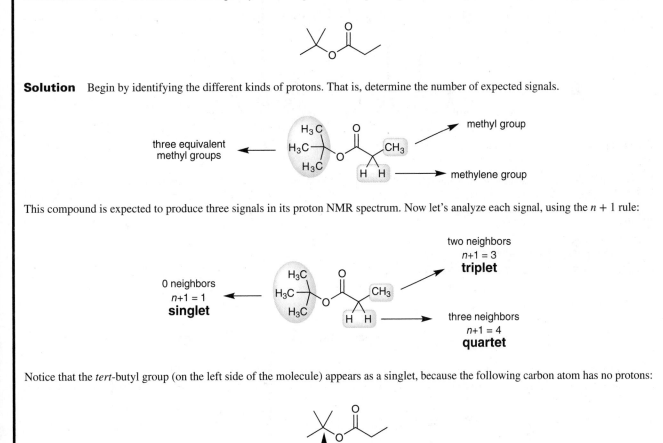

This compound is expected to produce three signals in its proton NMR spectrum. Now let's analyze each signal, using the $n + 1$ rule:

Notice that the *tert*-butyl group (on the left side of the molecule) appears as a singlet, because the following carbon atom has no protons:

no protons

This quaternary carbon atom is directly connected to each of the three neighboring methyl groups, and as a result, each of the three methyl groups has no neighboring protons. This is characteristic of *tert*-butyl groups.

PROBLEMS   Predict the multiplicity of each signal in the expected proton NMR spectrum of each of the following compounds:

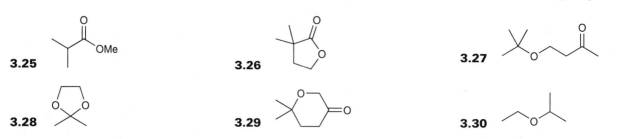

3.25   **3.26**   **3.27**

3.28   **3.29**   **3.30**

When signal splitting occurs, the distance between the individual peaks of a signal is called the *coupling constant*, or *J value*, and is measured in Hz. Neighboring protons always split each other with equal *J values*. For example, consider the two kinds of protons in an ethyl group:

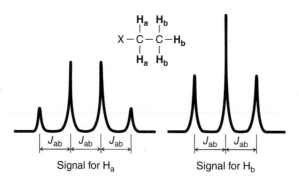

The H_a signal is split into a quartet under the influence of its three neighbors, while the H_b signal is split into a triplet under the influence of its two neighbors. H_a and H_b are said to be coupled to each other. The coupling constant, J_{ab}, is the same for each proton, so the spacing between the peaks is the same in both signals.

3.5   PATTERN RECOGNITION

There are a few splitting patterns that are commonly seen in proton NMR spectra, and you will save yourself time on an exam if you can recognize these patterns:

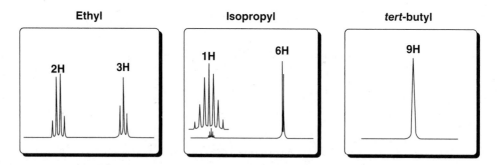

An ethyl group is characterized by a triplet with an integration of 3 and a quartet with an integration of 2. An isopropyl group is characterized by a doublet with an integration of 6 and a septet with an integration of 1. A *tert*-butyl group is characterized by a singlet with an integration of 9.

Let's get some practice recognizing these patterns.

PROBLEMS   Below are proton NMR spectra of several compounds. Identify whether these compounds are likely to contain ethyl, isopropyl, and/or *tert*-butyl groups:

3.31

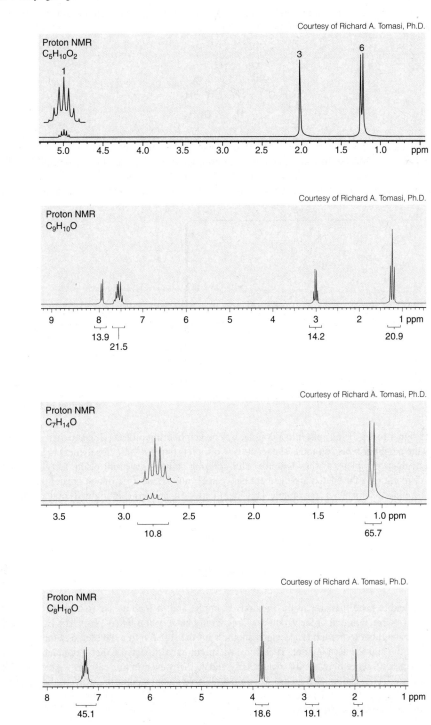

Courtesy of Richard A. Tomasi, Ph.D.

Proton NMR
$C_5H_{10}O_2$

3.32

Courtesy of Richard A. Tomasi, Ph.D.

Proton NMR
$C_9H_{10}O$

3.33

Courtesy of Richard A. Tomasi, Ph.D.

Proton NMR
$C_7H_{14}O$

3.34

Courtesy of Richard A. Tomasi, Ph.D.

Proton NMR
$C_8H_{10}O$

3.6   COMPLEX SPLITTING

Complex splitting occurs when a proton has two different kinds of neighboring protons. For example, consider the splitting pattern that you might expect for the highlighted proton (H_b) in the following compound

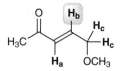

In this case, H_b has two different kinds of neighbors: there is one H_a proton to the left, and there are two H_c protons to the right. As a result, the signal for H_b is being split into a doublet because of the nearby H_a proton, and it is also being split into a triplet because of the nearby H_c protons. The signal is therefore expected to have six peaks (2×3). The shape (multiplicity) of this signal depends greatly on the J values. In this case, J_{ab} is much greater than J_{bc}, so the signal will appear as a doublet of triplets, as shown in the following splitting tree:

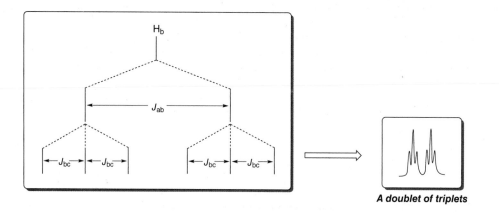

A doublet of triplets

This splitting tree shows the signal for H_b being split into a doublet by one neighboring proton (H_a), and then each peak of this doublet is further split into a triplet by the neighboring H_c protons. The result is two triplets (spaced very closely next to each other in the spectrum), and this pattern is called a doublet of triplets, as shown above. This splitting pattern is a result of the fact that the spatial arrangement between H_b and H_a is very different than the spatial arrangement between H_b and H_c, so J_{ab} is much greater than J_{bc}. Let's now consider a case in which the J values are more similar. As an example, consider the two H_b protons (highlighted) in the following compound:

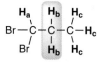

This structure lacks the rigid double bond that was in the previous example, and in such a case (where the compound has only single bonds that are all experiencing free rotation), all of the J values will generally be very similar ($\sim$7 Hz). The H_b protons have two different types of neighbors: there is one neighbor to the left (H_a), which should split the signal into a doublet; and there are three neighbors (H_c) to the right, which should split the signal into a quartet. Therefore, we might have expected either a doublet of quartets, or a quartet of doublets, if there had been a large difference between the J values, J_{ab} and J_{bc}. However, in this case, as we have pointed out, the J values are extremely similar ($J_{ab} \approx J_{ac} \approx 7$ Hz), and in such a case, we do not observe complex splitting. Rather, the H_b protons are considered to have four neighbors, and according to the $n + 1$ rule, the signal for the H_b protons will be a quintet ($n + 1 = 4 + 1 = 5 =$ quintet). This outcome can be justified with the following splitting tree:

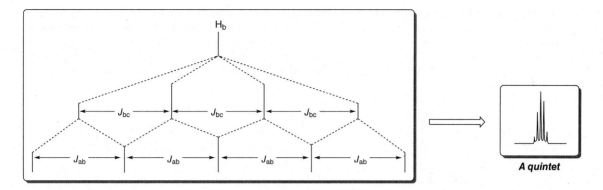

A quintet

This splitting tree shows the signal for the H_b protons first being split into a quartet by the neighboring H_c protons, and then each peak of this quartet is further split into a doublet by the neighboring H_a proton. Since $J_{ab} \approx J_{ac}$, the peaks of the signal end up overlapping perfectly to produce a signal with exactly five peaks (a quintet). Note that we get the same result (a quintet) if we draw a splitting tree where we first split the signal into a doublet, and then split each of the two peaks of the doublet into a quartet. The signal is a quintet because $J_{ab} \approx J_{ac}$.

In each of the previous examples, we have seen how the multiplicity of a signal (its shape) depends greatly on the J values. We saw a case where the J values were very different from each other, giving rise to complex splitting; and we also saw a case where the J values were extremely similar, in which case complex splitting was not observed. Finally, let's consider what would be the case if the J values are similar but not identical. In such a case, the peaks do not overlap nicely, and the result can be a signal that is difficult to analyze. This type of signal is called a multiplet. An example of a multiplet is shown below.

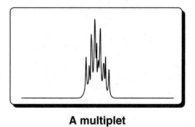

A multiplet

3.7 NO SPLITTING

In the previous section, we saw examples of complex splitting. Now, in this section, we will explore cases where there is no splitting at all, despite the presence of neighboring protons. Consider the proton NMR spectrum of ethanol:

Courtesy of Richard A. Tomasi, Ph.D.

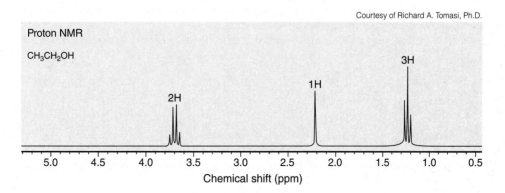

As expected, the spectrum exhibits the characteristic signals of an ethyl group (a quartet with an integration of 2 and a triplet with an integration of 3). In addition, there is another signal at 2.2 ppm, representing the hydroxyl proton (OH). Hydroxyl protons typically produce a signal between 2 and 5 ppm, and it is often difficult to predict exactly where that signal will appear. In the proton NMR spectrum above, notice that the hydroxyl proton is not split into a triplet from the neighboring methylene group. Generally, no splitting is observed across the oxygen of an alcohol, because proton exchange is a very rapid process that is catalyzed by trace amounts of acid or base:

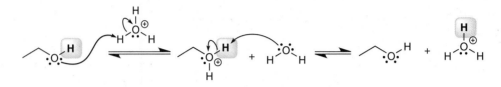

Hydroxyl protons are said to be **labile**, because of the rapid rate at which they are exchanged. This proton transfer process occurs at a faster rate than the timescale of an NMR spectrometer, producing a blurring effect that averages out any possible splitting effect. It is possible to slow down the rate of proton transfer by scrupulously removing the trace amounts of acid and base dissolved in ethanol. Such purified ethanol does in fact exhibit splitting across the oxygen atom, so the signal at 2.2 ppm is observed to be a triplet, and the signal at 3.7 ppm is observed to be a quintet ($n = 4$).

There is one other common example of neighboring protons that often do not produce observable splitting. Aldehydic protons, which generally produce signals near 10 ppm, will often couple only weakly with their neighbors (*i.e.*, a very small J value):

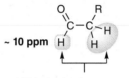

This J value is very small

Because the J value is so small, the aldehyde signal near 10 ppm typically appears to be a singlet, despite the presence of neighboring protons.

3.8 HYDROGEN DEFICIENCY INDEX (DEGREES OF UNSATURATION)

In the previous sections of this chapter, we have learned all of the individual tools that you need for analyzing a proton NMR spectrum (considering the number of signals, analyzing chemical shifts, assessing integration values, interpreting the multiplicity of each signal, pattern recognition, etc.). Now, we are just about ready to put all of these tools together. But there is one more important tool that you will need, and we will cover that tool in this section.

Imagine that you have an unknown compound with a molecular formula of $C_6H_{12}O$. The molecular formula by itself does not provide enough information to draw the structure of the compound. There are many constitutional isomers with the molecular formula $C_6H_{12}O$. Nevertheless, a careful analysis of the molecular formula can often provide helpful clues about the structure of the compound. To see how this works, let's begin by analyzing the molecular formula of several alkanes.

Compare the structures of the following alkanes, paying special attention to the number of hydrogen atoms attached to each carbon atom.

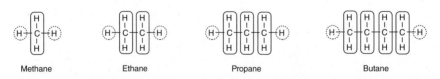

| Methane | Ethane | Propane | Butane |

In each case, there are two hydrogen atoms on the ends of the structures (circled), and there are two hydrogen atoms on every carbon atom. This can be summarized like this:

$$H—(CH_2)_n—H \tag{3.1}$$

where n is the number of carbon atoms in the compound. Accordingly, the number of hydrogen atoms will be $2n + 2$. In other words, all of the compounds above have a molecular formula of C_nH_{2n+2}. This is true even for compounds that are branched rather than having a straight chain.

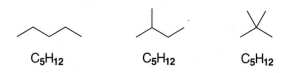

C_5H_{12} C_5H_{12} C_5H_{12}

The compounds above are said to be *saturated*—that is, they possess the maximum number of hydrogen atoms possible relative to the number of carbon atoms present.

A compound with a π bond (a double or triple bond) will have fewer than the maximum number of hydrogen atoms. Such compounds are said to be *unsaturated*.

C_5H_{10} C_5H_8

A compound containing a ring will also have fewer than the maximum number of hydrogen atoms, just like a compound with a double bond. For example, compare the structures of 1-hexene and cyclohexane:

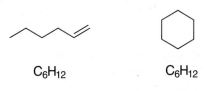

C_6H_{12} C_6H_{12}

Both compounds have the molecular formula C_6H_{12} because both are "missing" two hydrogen atoms [6 carbon atoms should be able to accommodate $(2 \times 6) + 2 = 14$ hydrogen atoms]. Each of these compounds is said to have one *degree of unsaturation*. The *hydrogen deficiency index (HDI)* is a measure of the number of degrees of unsaturation. A compound is said to have one degree of unsaturation for every two hydrogen atoms that are missing. For example, a compound with the molecular formula C_4H_6 is missing four hydrogen atoms (if saturated, it would be C_4H_{10}), so it has two degrees of unsaturation (HDI = 2). There are several ways for a compound to possess two degrees of unsaturation: two double bonds, or two rings, or one double bond and one ring, or one triple bond. Let's explore all of these possibilities for C_4H_6:

Two double bonds	One triple bond	Two rings	One ring and one double bond
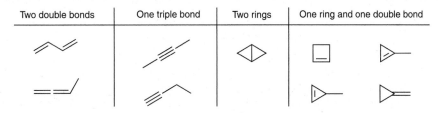			

These are all of the possible constitutional isomers for C_4H_6. With this in mind, let's expand our skills set. Let's explore how to calculate the HDI when other elements are present in the molecular formula.

Halogens: Compare the following two compounds:

ethane chloroethane

Notice that chlorine takes the place of a hydrogen atom. Therefore, for purposes of calculating the HDI, treat a halogen as if it were a hydrogen atom. For example, C_4H_9Cl should have the same HDI as C_4H_{10}.

Oxygen: Compare the following two compounds:

Notice that the presence of the oxygen atom does not affect the expected number of hydrogen atoms. Therefore, whenever an oxygen atom appears in the molecular formula, it should be ignored for purposes of calculating the HDI. For example, C_4H_8O should have the same HDI as C_4H_8.

Nitrogen: Compare the following two compounds:

Notice that the presence of a nitrogen atom changes the number of expected hydrogen atoms. It gives one more hydrogen atom than would be expected (highlighted above). Therefore, whenever a nitrogen atom appears in the molecular formula, one hydrogen atom must be subtracted from the molecular formula. For example, C_4H_9N should have the same HDI as C_4H_8.

In summary:

- Halogens: *Add* one H for each halogen
- Oxygen: *Ignore*
- Nitrogen: *Subtract* one H for each N

These rules will enable you to determine the HDI for most simple compounds. Alternatively, the following formula can be used:

$$HDI = (2 \times C + 2 + N - H - X)/2$$

C is the number of carbon atoms, N is the number of nitrogen atoms, H is the number of hydrogen atoms, and X is the number of halogens. This formula will work for all compounds containing C, H, N, O, and/or halogens.

   Calculating the HDI is particularly helpful, because it provides clues about the structural features of the compound. For example, an HDI of zero indicates that the compound cannot have any rings or π bonds. That is extremely useful information when trying to determine the structure of a compound, and it is information that is easily obtained by simply analyzing the molecular formula. Similarly, an HDI of one indicates that the compound must have either one double bond *or* one ring (but not both). If the HDI is two, then there are several possibilities: two rings, or two double bonds, or one ring and one double bond, or one triple bond. Analysis of the HDI for an unknown compound can often be a useful tool, but only when the molecular formula is known with certainty.

   We will use this technique in the next section. The following exercises are designed to develop the skill of calculating and interpreting the HDI of an unknown compound whose molecular formula is known.

EXERCISE 3.35   Calculate the HDI for a compound with the molecular formula $C_5H_8Br_2O_2$, and identify the structural information provided by the HDI.

Answer   Use the following calculation:

# of H's:	8
Add 1 for each Br:	+2
Ignore each O:	0
Subtract 1 for each N:	0
Total =	10

This compound will have the same HDI as a compound with the molecular formula C_5H_{10}. To be fully saturated, 5 carbon atoms would require $(5 \times 2) + 2 = 12$ H atoms. According to our calculation, two hydrogen atoms are missing, and, therefore, this compound has one degree of unsaturation. HDI = 1.

Alternatively, the following formula can be used:

$$HDI = (2 \times C + 2 + N - H - X)/2 = (10 + 2 + 0 - 8 - 2)/2 = 2/2 = 1$$

With one degree of unsaturation, the compound must contain either one ring or one double bond, but not both. The compound cannot have a triple bond, as that would require two degrees of unsaturation.

PROBLEMS   Calculate the degree of unsaturation for each of the following molecular formulas:

3.36 $C_6H_{10}O_4$ **3.37** $C_7H_{11}N$ **3.38** $C_8H_{14}O_2$

3.39 $C_5H_{12}O_2$ **3.40** $C_6H_{15}N$ **3.41** $C_8H_{10}O$

3.9   ANALYZING A PROTON NMR SPECTRUM

In this section, we will practice analyzing and interpreting NMR spectra, a process that involves four discrete steps:

1. Always begin by inspecting the molecular formula (if it is given), as it provides useful information. Specifically, calculating the hydrogen deficiency index (HDI) can provide important clues about the structure of the compound. An HDI of zero indicates that the compound does not possess any rings or π bonds. An HDI of 1 indicates that the compound has either one ring or one π bond. An HDI of four (or more) indicates the likely presence of an aromatic ring:

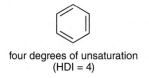

four degrees of unsaturation
(HDI = 4)

2. Consider the number of signals and integration of each signal (this gives clues about the symmetry of the compound).
3. Analyze each signal (chemical shift, integration, and multiplicity), and then draw fragments consistent with each signal. These fragments become our puzzle pieces that must be assembled to produce a molecular structure.
4. Assemble the fragments into a molecular structure.

The following exercise illustrates how this is done.

EXERCISE 3.42 Identify the structure of a compound with the molecular formula $C_9H_{10}O$ that exhibits the following proton NMR spectrum:

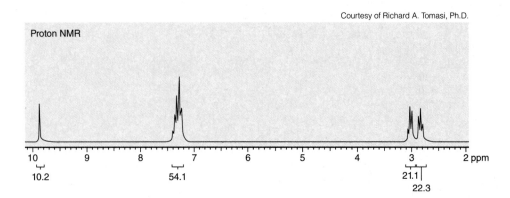

Courtesy of Richard A. Tomasi, Ph.D.

Answer The first step is to calculate the HDI. The molecular formula indicates 9 carbon atoms, which would require 20 hydrogen atoms in order to be fully saturated. There are only 10 hydrogen atoms, which means that 10 hydrogen atoms are missing, and therefore, the HDI is 5. This is a large number, and it would not be efficient to think about all the possible ways of having five degrees of unsaturation. However, anytime we encounter an HDI of 4 or more, we should be on the lookout for an aromatic ring. We must keep this in mind when analyzing the spectrum. We should expect an aromatic ring (HDI = 4) plus one other degree of unsaturation (either a ring or a double bond).

The second step is to consider the number of signals and the integration value for each signal. In this spectrum, we see four signals. In order to analyze the integration of each signal, we must first divide by the lowest number (10.2):

$$\frac{10.2}{10.2} = 1 \qquad \frac{54.1}{10.2} = 5.30 \qquad \frac{21.1}{10.2} = 2.07 \qquad \frac{22.3}{10.2} = 2.19$$

The ratio is 1 : 5 : 2 : 2. Now look at the molecular formula. There are 10 protons in the compound, so the relative integration values represent the actual number of protons giving rise to each signal:

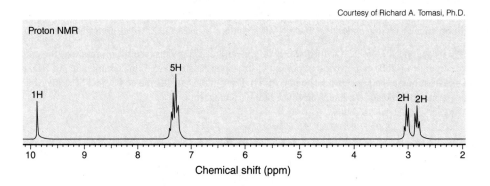

Courtesy of Richard A. Tomasi, Ph.D.

The next step is to analyze each signal. Starting upfield, there are two triplets, each with an integration of 2. This suggests that there are two adjacent methylene groups:

These signals do not appear at 1.2 where methylene groups are expected, so one or more factors is shifting these signals downfield. Our proposed structure must take that into account.

Moving downfield through the spectrum, the next signal appears just above 7 ppm, characteristic of aromatic protons (just as we suspected after analyzing the HDI). The multiplicity of aromatic protons only rarely gives useful information. More often, a messy multiplet of overlapping signals is observed. But the integration value gives important information. Specifically, there are five aromatic protons, which means that the aromatic ring is monosubstituted.

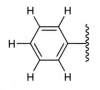

Five aromatic protons

Moving on to the last signal, we see a singlet at 10 ppm with an integration of 1. This is suggestive of an aldehydic proton.

In summary, our analysis has produced the following fragments:

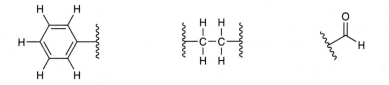

The final step is to assemble these fragments. Fortunately, there is only one way to assemble these three puzzle pieces.

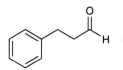

We mentioned before that each methylene group is being shifted downfield by one or more factors. Our proposed structure explains the observed chemical shifts. In particular, one methylene group is shifted significantly by the carbonyl group and slightly by the aromatic ring. The other methylene group is being shifted significantly by the aromatic ring and slightly by the carbonyl group.

PROBLEMS

3.43 Propose a structure for a compound with the molecular formula $C_5H_{10}O$ that is consistent with the following proton NMR spectrum.

Courtesy of Richard A. Tomasi, Ph.D.

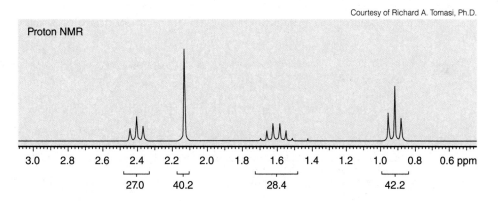

3.44 Propose a structure for a compound with the molecular formula $C_5H_{10}O_2$ that is consistent with the following proton NMR spectrum.

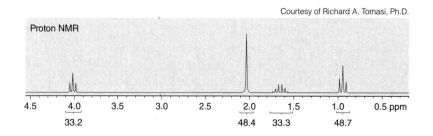

3.45 Propose a structure for a compound with the molecular formula $C_4H_6O_2$ that is consistent with the following proton NMR spectrum.

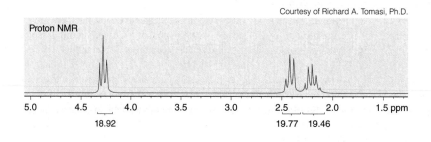

3.46 Propose a structure for a compound with the molecular formula C_9H_{12} that is consistent with the following proton NMR spectrum.

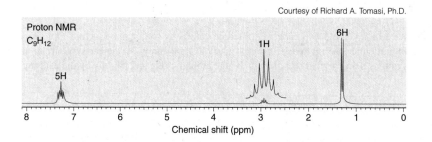

3.47 Propose a structure for a compound with the molecular formula $C_6H_{12}O_2$ that is consistent with the following proton NMR spectrum.

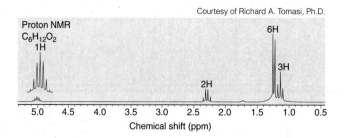

3.48 Propose a structure for a compound with the molecular formula $C_8H_{10}O$ that is consistent with the following proton NMR spectrum.

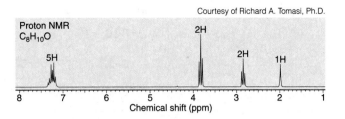

Courtesy of Richard A. Tomasi, Ph.D.

Proton NMR
$C_8H_{10}O$

Your textbook has many more spectroscopy problems, including problems in which you are given both IR and NMR spectra. I recommend that you do ALL of those problems. The skills practiced in this chapter are meant to prepare you for those problems.

3.10 ¹³C NMR SPECTROSCOPY

Many of the principles that apply to ¹H NMR spectroscopy also apply to ¹³C NMR spectroscopy, but there are a few major differences, and we will focus on those. For example, ¹H is the most abundant isotope of hydrogen, but ¹³C is only a minor isotope of carbon, representing about 1.1% of all carbon atoms found in nature. As a result, only one in every hundred carbon atoms will resonate, which demands the use of a sensitive receiver coil for ¹³C NMR.

In ¹H NMR spectroscopy, we saw that each signal has three characteristics (chemical shift, integration, and multiplicity). In ¹³C NMR spectroscopy, only the chemical shift is important. The integration and multiplicity of ¹³C signals are typically not observed, which greatly simplifies the interpretation of ¹³C NMR spectra. Integration values are not routinely calculated in ¹³C NMR spectroscopy because the pulse technique employed by NMR spectrometers has the undesired effect of distorting the integration values, rendering them useless in most cases. Multiplicity is also not a common characteristic of typical ¹³C NMR experiments.

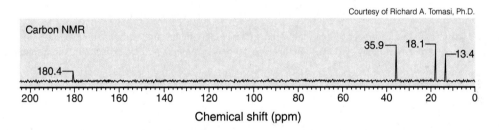

Courtesy of Richard A. Tomasi, Ph.D.

Carbon NMR

Notice that all of the signals are observed as singlets. This is a result of a special technique, called broadband decoupling, that suppresses all ¹³C—¹H splitting. If we did not use this technique, then the signal of each ¹³C atom nucleus would be split not only by the protons directly connected to it (separated by only one sigma bond), but it would also be split by the protons that are two or three sigma bonds removed. This would lead to very complex splitting patterns, and the signals would overlap to produce an unreadable spectrum. The use of broadband decoupling causes all of the ¹³C signals to collapse to singlets, which renders the spectrum more easily interpreted.

In ¹³C NMR spectroscopy, chemical shift values typically range from 0 to 220 ppm. The number of signals in a ¹³C NMR spectrum represents the number of carbon atoms in different electronic environments (not interchangeable by symmetry). Carbon atoms that are interchangeable by symmetry (either rotation or reflection) will only produce one signal. To illustrate this point, consider the compounds below. Each compound has eight carbon atoms, but does not produce eight signals. The equivalent carbon atoms in each compound are labeled with the same letter.

4 signals   5 signals   3 signals

The location of each signal is dependent on shielding and deshielding effects, just as we saw in proton NMR spectroscopy. Below are chemical shifts of several important types of carbon atoms.

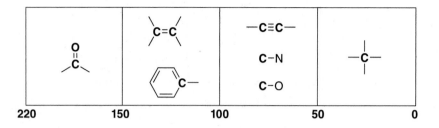

- **0–50 ppm:** This region contains signals from sp^3-hybridized carbon atoms (methyl, methylene, and methine groups).
- **50–100 ppm:** This region contains sp^3-hybridized carbon atoms that are deshielded by electronegative atoms, as well as sp-hybridized carbon atoms.
- **100–150 ppm:** This region contains sp^2-hybridized carbon atoms.
- **150–220 ppm:** This region contains the carbon atoms of carbonyl groups. These carbon atoms are highly deshielded.

Now let's use this information to solve the following exercise.

EXERCISE 3.49 Predict the number of signals and the location of each signal in the expected ^{13}C NMR spectrum of the following compound:

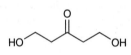

Answer Begin by determining the number of expected signals. The compound has five carbon atoms, but we must look to see if any of these carbon atoms are interchangeable by symmetry. In this case, there is symmetry, and we expect only three signals in the ^{13}C NMR spectrum.

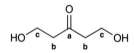

The expected chemical shifts are shown below, categorized according to the region of the spectrum in which each signal is expected to appear:

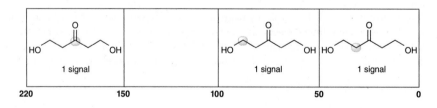

PROBLEMS   For each compound below, predict the number of signals and the location of each signal in the expected ^{13}C NMR spectrum.

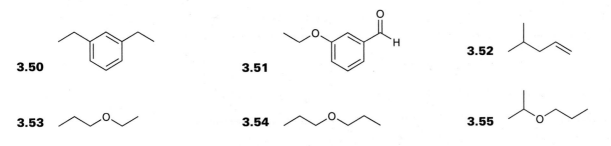

3.50

3.51

3.52

3.53

3.54

3.55

END-OF-CHAPTER PROBLEMS

PRACTICE PROBLEMS *(Problems that involve only one skill)*
How many signals will be present in the ^{1}H NMR and ^{13}C NMR spectra for each of the following compounds?

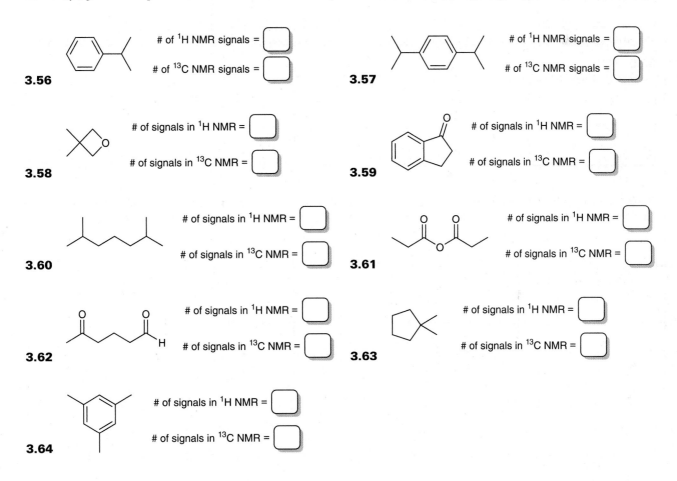

3.56   # of ^{1}H NMR signals = ☐
        # of ^{13}C NMR signals = ☐

3.57   # of ^{1}H NMR signals = ☐
        # of ^{13}C NMR signals = ☐

3.58   # of signals in ^{1}H NMR = ☐
        # of signals in ^{13}C NMR = ☐

3.59   # of signals in ^{1}H NMR = ☐
        # of signals in ^{13}C NMR = ☐

3.60   # of signals in ^{1}H NMR = ☐
        # of signals in ^{13}C NMR = ☐

3.61   # of signals in ^{1}H NMR = ☐
        # of signals in ^{13}C NMR = ☐

3.62   # of signals in ^{1}H NMR = ☐
        # of signals in ^{13}C NMR = ☐

3.63   # of signals in ^{1}H NMR = ☐
        # of signals in ^{13}C NMR = ☐

3.64   # of signals in ^{1}H NMR = ☐
        # of signals in ^{13}C NMR = ☐

For each of the molecular formulas shown below, there is a compound that exhibits a ^1H NMR spectrum with only one signal. In each case, deduce the structure and place your answer in the box provided:

3.65 C_4H_9Cl

3.66 $C_2H_3Br_3$

3.67 $C_2H_4Br_2$

3.68 $C_4H_8O_2$

3.69 C_5H_8

3.70 A compound with the molecular formula C_6H_{12} has a ^1H NMR spectrum with only one signal. Draw two possible structures for this compound, and identify how you could use ^{13}C NMR spectroscopy to differentiate between the two possible structures.

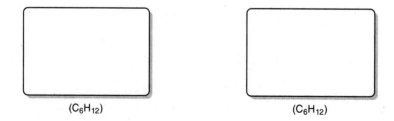

(C_6H_{12})   (C_6H_{12})

3.71   A compound with the molecular formula C_9H_{18} has a ^1H NMR spectrum with only one signal. Draw two possible structures for this compound, and identify how you could use ^{13}C NMR spectroscopy to differentiate between the two possible structures.

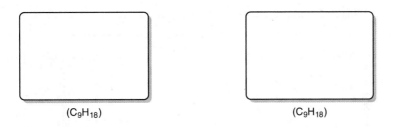

(C_9H_{18}) (C_9H_{18})

3.72   A compound with the molecular formula $C_{11}H_{24}$ has a ^1H NMR spectrum with two signals. Draw the structure of this compound, and identify how many total signals will appear in the ^{13}C NMR spectrum of this compound.

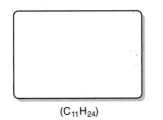

$(C_{11}H_{24})$

INTEGRATED PROBLEMS *(Problems that involve more than one skill)*
In each of the following cases, identify how you would use either ^1H NMR spectroscopy or ^{13}C NMR spectroscopy to distinguish between the two compounds.

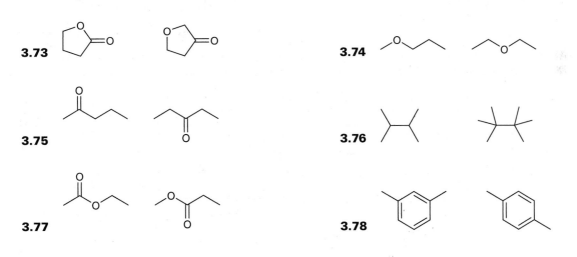

3.79 In the box provided below, draw the structure of a compound with the molecular formula C_3H_8O that has the following IR, ^1H NMR, and ^{13}C NMR spectra:

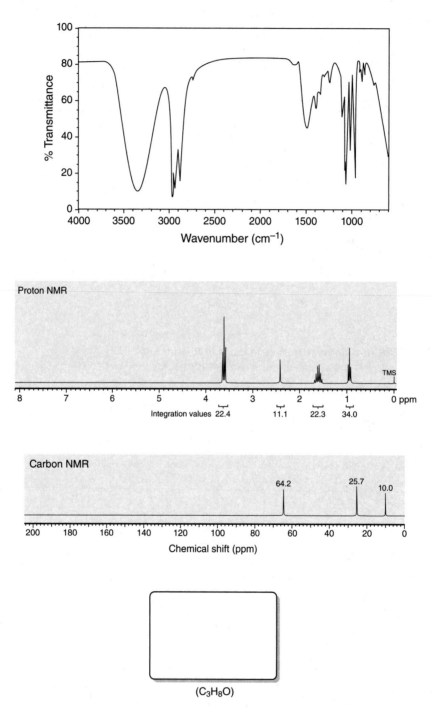

(C₃H₈O)

3.80 In the box provided below, draw the structure of a compound with the molecular formula $C_{13}H_{10}O$ that has the following IR, ^1H NMR, and ^{13}C NMR spectra:

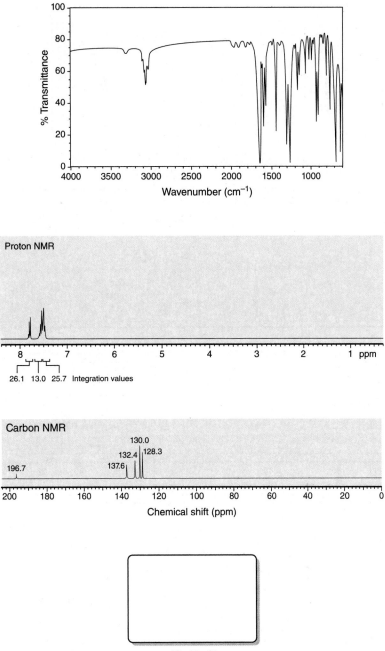

($C_{13}H_{10}O$)

3.81 In the box provided below, draw the structure of a compound with the molecular formula $C_{11}H_{14}O_2$ that has the following IR, ^1H NMR, and ^{13}C NMR spectra:

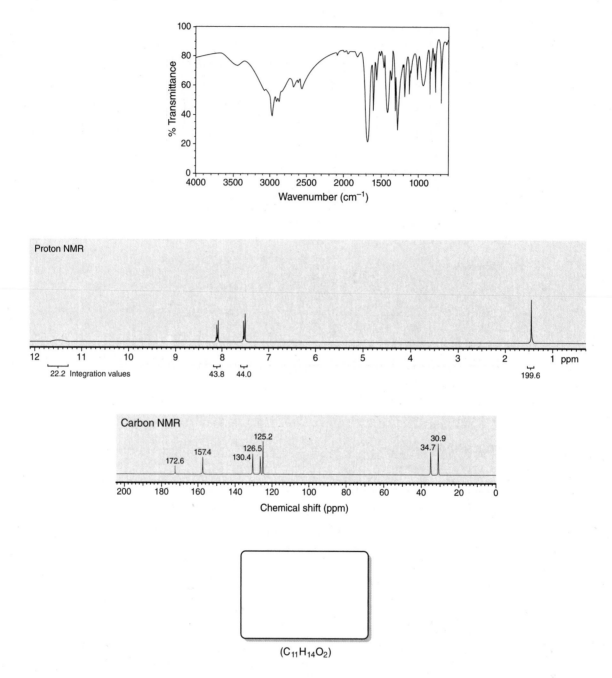

$(C_{11}H_{14}O_2)$

3.82 In the box provided below, draw the structure of a compound with the molecular formula $C_7H_{14}O$ that has the following IR, ^1H NMR, and ^{13}C NMR spectra:

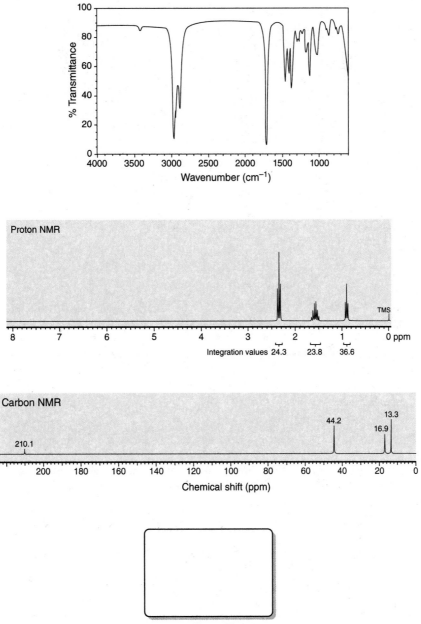

Integration values 24.3   23.8   36.6

$(C_7H_{14}O)$

3.83 In the box provided below, draw the structure of a compound with the molecular formula $C_7H_{14}O_2$ that has the following IR, ^1H NMR, and ^{13}C NMR spectra:

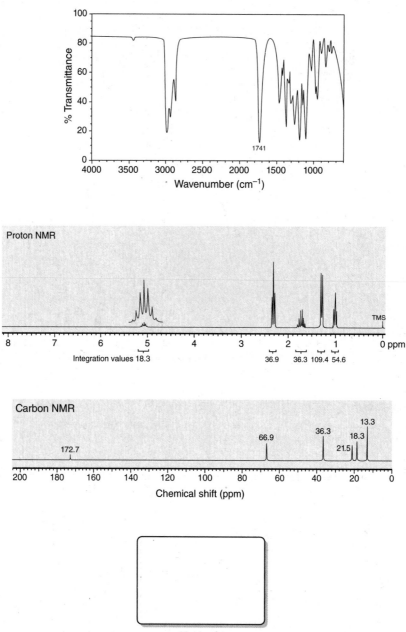

($C_7H_{14}O_2$)

3.84   In the box provided below, draw the structure of a compound with the molecular formula $C_8H_9NO_3$ that has the following ^1H NMR and ^{13}C NMR spectra:

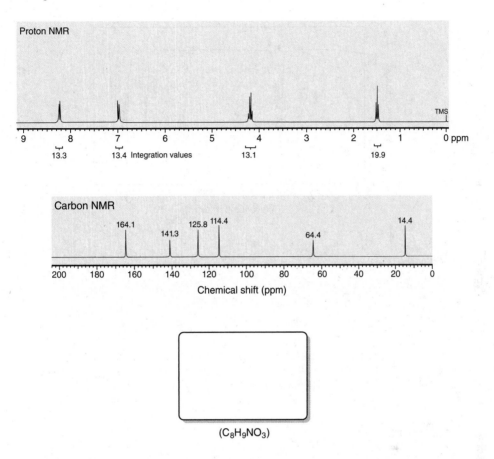

(C_8H_9NO_3)

3.85 In the box provided below, draw the structure of a compound with the molecular formula $C_6H_{14}O$ that has the following IR, ^1H NMR, and ^{13}C NMR spectra:

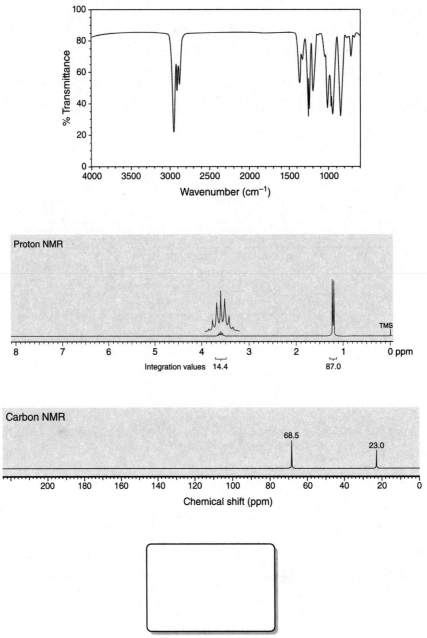

$(C_6H_{14}O)$

3.86 In the box provided below, draw the structure of a compound with the molecular formula $C_9H_{21}N$ that has the following ^1H NMR and ^{13}C NMR spectra:

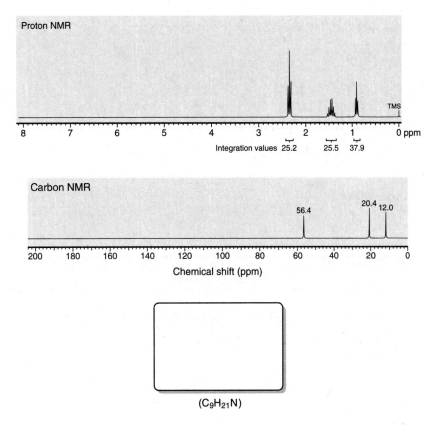

3.87 In the box provided below, draw the structure of a compound with the molecular formula $C_{12}H_{10}$ that has the following ^1H NMR spectrum, as well as a ^{13}C NMR spectrum with only four signals, all between 120 and 150 ppm:

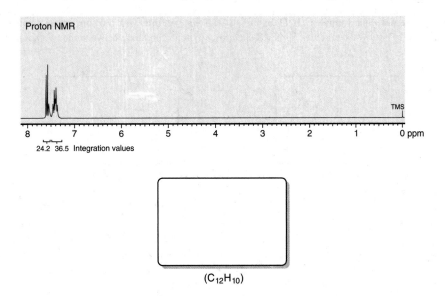

3.88   In the box provided below, draw the structure of a compound with the molecular formula $C_9H_{10}O_2$ that has the following ^1H NMR and ^{13}C NMR spectra:

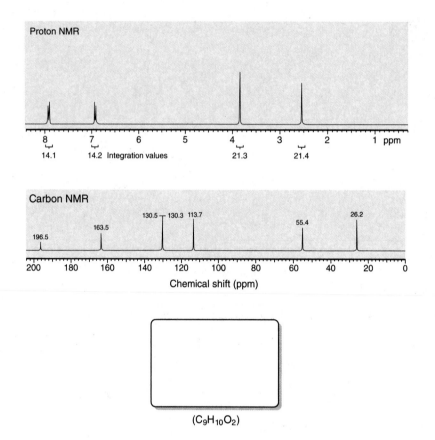

$(C_9H_{10}O_2)$

CHALLENGE PROBLEMS

3.89   A compound with the molecular formula $C_5H_8O_4$ exhibits a ^1H NMR spectrum with only one signal. Draw two possible structures for this compound, and identify how you could use ^{13}C NMR spectroscopy to differentiate between the two possible structures.

$(C_5H_8O_4)$ $(C_5H_8O_4)$

3.90   A compound with the molecular formula $C_{15}H_{24}$ exhibits a ^1H NMR spectrum with the following three signals:

- Singlet, 7.5 ppm, Integration = 1
- Septet, 3.0 ppm, Integration = 1
- Doublet, 1.1 ppm, Integration = 6

Draw the structure of this compound in the box below, and determine how many signals are expected in the ^{13}C NMR spectrum of this compound:

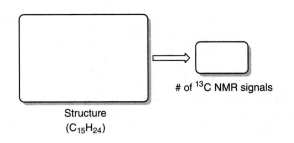

Structure
($C_{15}H_{24}$)

of ^{13}C NMR signals

CHAPTER 4

ELECTROPHILIC AROMATIC SUBSTITUTION

We must begin this chapter with a review of a reaction from the first semester of organic chemistry. Recall the addition of bromine (Br_2) across a double bond:

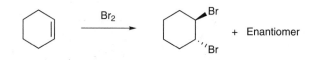

When we learned this reaction in the first semester, we saw that this reaction involves a nucleophile attacking an electrophile. The nucleophile is the double bond, and the electrophile is Br_2. To understand how a double bond can function as a nucleophile, recall that a double bond results from the overlap of two neighboring p orbitals, each with one electron:

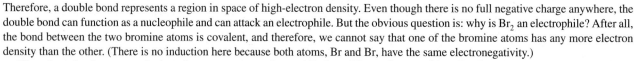

Therefore, a double bond represents a region in space of high-electron density. Even though there is no full negative charge anywhere, the double bond can function as a nucleophile and can attack an electrophile. But the obvious question is: why is Br_2 an electrophile? After all, the bond between the two bromine atoms is covalent, and therefore, we cannot say that one of the bromine atoms has any more electron density than the other. (There is no induction here because both atoms, Br and Br, have the same electronegativity.)

There is a simple reason why bromine can act as an electrophile here. We need to consider what happens when a molecule of Br_2 approaches an alkene. To help us see this, think of Br_2 in terms of the electron cloud surrounding it:

As a molecule of Br_2 gets close to the pi bond of an alkene, the electron density of the pi bond begins to repel the electron cloud around Br_2. This effect gives the Br_2 molecule an induced dipole moment (this is a temporary interaction—it only happens while the bromine molecule is near the alkene):

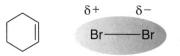

So, we have an electron-rich alkene, which can attack the nearby, electron-poor bromine atom. This gives the reaction that we saw in the first semester of organic chemistry:

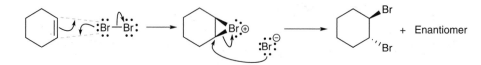

Now let's consider what happens if we try to do this exact same reaction with benzene as our nucleophile. So we are trying to perform this reaction:

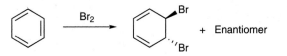

But when benzene is heated in the presence of Br_2, no reaction is observed.

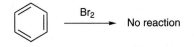

We can understand this because benzene is an aromatic compound. It has a special stability due to its aromaticity. If we add Br_2 across benzene, aromaticity will be lost. And that is why the reaction does not take place—it would be going "uphill" in energy. But is it possible to try to "force" the reaction to happen?

This brings us to a simple and important concept in organic chemistry. One of the driving forces for any reaction between a nucleophile and an electrophile is the difference in the electron density between the two compounds. The nucleophile is electron-rich, and the electrophile is electron-poor. Therefore, they are attracted to each other in space (opposite charges attract). So, if the reaction is not proceeding, we can try to force it along by making the attraction even stronger between the nucleophile and the electrophile. We can accomplish this in one of two ways. We can either make the nucleophile even more electron-rich (more nucleophilic), or we can make the electrophile even more electron-poor (more electrophilic).

In this chapter, we will explore both of these scenarios. For now, let us start by trying to make the electrophile a better electrophile. How do we make Br_2 a better electrophile? Let's remember why Br_2 is an electrophile in the first place. Just a few moments ago, we saw that an induced dipole moment is formed when Br_2 gets close to an alkene. This creates a partial positive charge on one of the bromine atoms. Clearly, if we had Br^+ instead of Br_2, then that would be an even better electrophile. We would not have to wait around for Br_2 to become slightly polarized.

But how do we form Br^+? That is where Lewis acids come into the picture.

4.1 HALOGENATION AND THE ROLE OF LEWIS ACIDS

Consider the compound $AlBr_3$. The central atom in this structure is aluminum. Aluminum is in Column 3A of the periodic table, and therefore, it has three valence electrons. It uses each of these electrons to form a bond, which is why we see three bonds to the aluminum atom in $AlBr_3$:

Br
|
Br — Al — Br

But you should notice that the aluminum atom does not have an octet. If you count the electrons around the aluminum atom, there are only six electrons. That means that aluminum has one empty orbital. That empty orbital is able to accept electrons. In fact, it will exhibit a tendency to accept electrons because that would give aluminum an octet of electrons:

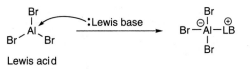

Therefore, we call $AlBr_3$ a *Lewis acid*. To put it simply, Lewis acids are just compounds that can *accept electrons*. Another common Lewis acid is $FeBr_3$:

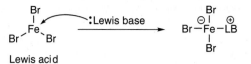

Now let's consider what happens when Br_2 is treated with a Lewis acid, such as $AlBr_3$. The Lewis acid can accept electrons from Br_2:

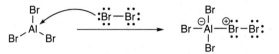

The resulting complex can then serve as a source of Br^+, like this:

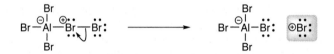

It is probably not accurate to think of this as a free Br^+ that can exist in solution by itself. Rather, the complex can *transfer* Br^+ to an attacking nucleophile:

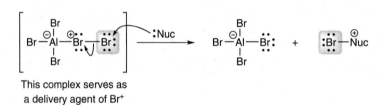

This complex serves as
a delivery agent of Br^+

The important point is that this complex can function as a delivery agent of Br^+, and that is what we needed in order to force a reaction between benzene and bromine. So, now let's try our reaction again. When we treat benzene with bromine in the presence of a Lewis acid, such as $AlBr_3$, a reaction is indeed observed. BUT it is not the reaction that we expected. Look closely at the product:

This is NOT an addition reaction. Rather, it is a *substitution* reaction. One of the aromatic protons was replaced with bromine. Since the ring is being treated with an electrophile (Br^+), we call this reaction an *electrophilic aromatic substitution*.

To see how this reaction occurs, let's take a close look at the accepted mechanism. It is absolutely critical that you fully understand this mechanism, because we will soon see that ALL electrophilic aromatic substitution reactions follow a similar mechanism. The first step shows the ring acting as a nucleophile to attack the complex, thereby transferring Br^+ to the aromatic ring:

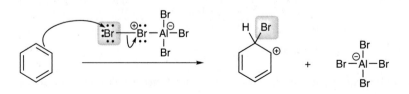

This step generates an intermediate that is not aromatic. It is true that aromaticity has been *temporarily* destroyed, but it will soon be reestablished in the second step (final step) of the mechanism. In this first step of the mechanism (shown above), the ring attacks Br^+ to form an intermediate that has three resonance structures:

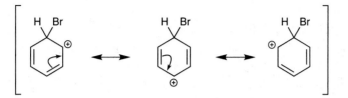

It is important to remember what resonance structures represent. Recall from the first semester that resonance is NOT a molecule flipping back and forth between different states. Rather, resonance is the way we deal with the inadequacy of our drawings. There is no one single drawing that adequately captures the essence of the intermediate, so we draw three drawings, and we meld them all together in our minds to gain a better understanding of the intermediate.

Attempts have been made to draw a single drawing for this intermediate:

You might even see this in your textbook. I usually try to avoid using this drawing because it could easily lead you to think, erroneously, that the positive charge is spread over five atoms in the ring. This is not the case. The majority of the positive charge is actually only spread over three atoms of the ring (which we can clearly see when we look at all three resonance structures above).

This intermediate has some special names. We often call it a sigma complex, or an Arrhenium ion. These are just two different names for the same intermediate. From now on, in this book, we will call it a sigma complex:

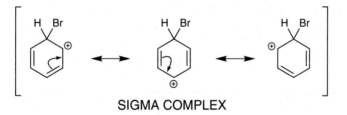

SIGMA COMPLEX

The second step (last step) of the mechanism involves deprotonation to re-form aromaticity:

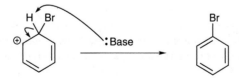

Notice that the proton is removed by a base. Technically, it is not correct to just let a proton fall off into space by itself, like this:

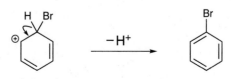

Whenever you are drawing a deprotonation step, you should show the base that is removing the proton. In this particular case, it might be tempting to use Br⁻ to remove the proton. But Br⁻ is not a good base. (In the first semester, we learned the difference between basicity and nucleophilicity, and we saw that Br⁻ is a very good nucleophile but a very poor base.) Instead, aluminum tetrabromide functions as the base that removes the proton. Notice that aluminum tetrabromide functions as a "delivery agent" of Br⁻.

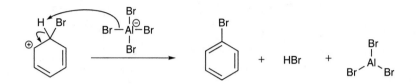

In the end, the Lewis acid (AlBr₃) is regenerated. So the Lewis acid is actually not being consumed by the reaction. It is only there to help the reaction along, which is why we call it a *catalyst* in this case. That is why the presence of even a small quantity of the Lewis acid will suffice.

Now that we have seen both steps of the mechanism, let's take a close look at the mechanism all at once:

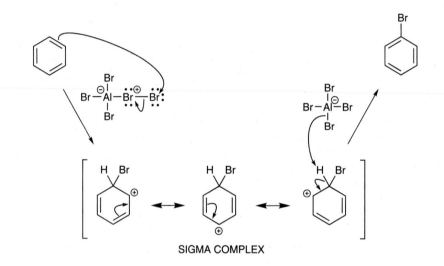

SIGMA COMPLEX

On the surface, it might seem like many steps. However, remember that resonance structures are not actually steps. Those three resonance structures (in the center of the mechanism) are necessary so that we can understand the nature of the one and only intermediate (the sigma complex). Indeed, the mechanism has only two steps. In the first step, benzene acts as a nucleophile attacking Br⁺ to form the sigma complex, and in the second step, a proton is removed from the ring to reestablish aromaticity. In summary, the two steps are: attack, then deprotonate. To put it in other terms: Br⁺ comes on, and then H⁺ comes off. That's all there is to it.

PROBLEM 4.1 Without looking at the mechanism above, try to redraw the entire mechanism on a separate sheet of paper. Don't look above—you can figure it out. Just remember that there are two steps: E⁺ on and then H⁺ off. Don't forget to draw all three resonance structures of the intermediate sigma complex.

PROBLEM 4.2 Consider the following reaction, in which an aromatic ring undergoes chlorination, rather than bromination:

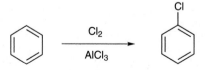

The mechanism is very similar to the mechanism for bromination. First, Cl_2 reacts with $AlCl_3$ to generate a complex that can serve as a source of Cl^+. Draw a mechanism for formation of this complex.

PROBLEM 4.3 Draw a mechanism for the electrophilic aromatic substitution reaction that occurs when benzene is treated with the complex from Problem 4.2. The mechanism is exactly the same as the mechanism for installing a Br on the ring. But PLEASE, do not look back at that mechanism to copy it. Try to do it *without* looking back. Then, when you are finished, compare your answer to the answer in the back of the book (and compare every arrow to make sure that all of your arrows were drawn correctly).

PROBLEM 4.4 Aromatic rings will also undergo iodination when treated with a suitable source of I^+. There are many ways to form I^+; you should look in your textbook and in your lecture notes to see if you are responsible for knowing how to iodinate benzene. If so, be aware that the mechanism is exactly the same as what we have seen. The only difference will be in the mechanism of how I^+ is formed. Draw a mechanism for the reaction between benzene and I^+ to form iodobenzene. In the first step of your mechanism, simply draw I^+ as the electrophile (rather than a complex which delivers I^+), and make sure to draw all resonance structures of the resulting sigma complex. Then, in the last step of your mechanism, use H_2O as the base that removes the proton to restore aromaticity (H_2O will be present for many of the methods that are used to prepare a source of I^+).

4.2 NITRATION

In the previous section, we saw the mechanism of an electrophilic aromatic substitution reaction. We saw that the mechanism is the same, whether you are installing Br^+, Cl^+, or I^+ on the ring. We also said that this same mechanism explains how any electrophile (E^+) can be installed on an aromatic ring. For example, let's say we wanted to convert benzene into nitrobenzene:

In order to form nitrobenzene, we will need NO_2^+ as our electrophile. But how do we make NO_2^+? If we look at how we made Br^+ or Cl^+ in the previous section, we might be tempted to use NO_2Br and $AlBr_3$, to get the following complex:

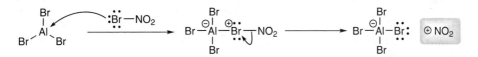

This complex could then serve as a source of NO_2^+. The problem is that NO_2Br is nasty stuff, and you probably would not want to work with it in a lab, especially since there is a much simpler way to make NO_2^+. We can form NO_2^+ by mixing sulfuric acid with nitric acid:

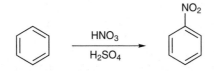

We need to take a close look at how NO_2^+ is formed under these conditions. Let's begin by drawing the structures of sulfuric acid and nitric acid:

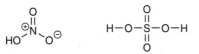

Nitric acid   Sulfuric acid

Notice that nitric acid exhibits charge separation. It might be tempting to remove the charges and draw it like this:

DON'T DRAW THIS

But you cannot do that because it would give five bonds to the central nitrogen atom. Nitrogen cannot EVER have five bonds because it only has four orbitals that it can use to form bonds. So nitric acid must be drawn with charge separation.

Now that we have seen the structures of both nitric acid and sulfuric acid, we must remember that the term *acidic* is a relative term. It is true that nitric acid is acidic, and it is also true that sulfuric acid is acidic. But between the two of them, sulfuric acid is a *stronger* acid. In fact, it is so much stronger as an acid that it is able to give a proton to nitric acid:

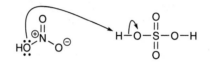

That's right—it might seem weird because nitric acid is essentially functioning *as a base* to remove a proton from sulfuric acid. And it might make us uncomfortable to use nitric acid as a base, but that is exactly what is happening. Why? Because *relative* to sulfuric acid, nitric acid is a base. It's all relative.

OK, so nitric acid removes a proton from sulfuric acid. But the obvious question is: why does the uncharged oxygen remove the proton? Wouldn't it make more sense for the negatively charged oxygen to remove the proton? Like this:

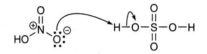

The answer is: yes, this probably would make more sense. And it probably happens like this a lot more often. The oxygen with the negative charge probably removes the proton much more readily than the uncharged oxygen atom does. However, proton transfers are reversible. Protons are being transferred back and forth all of the time. And all of this is happening very quickly. So, it is true that the negatively charged oxygen atom removes the proton more often—but when that happens, the only thing that can happen next is for the proton to be given right back to re-form nitric acid.

Every once in a while, however, the uncharged oxygen atom can remove the proton. And when that happens, something new can happen next: water can leave:

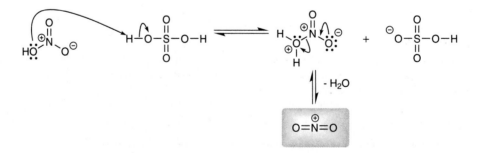

And when this happens, NO_2^+ is generated. So when we mix sulfuric acid and nitric acid, we get a little bit of NO_2^+ in the equilibrium mixture, and that NO_2^+ functions as the electrophile we need in order to install a nitro group on a benzene ring.

Once again, the following mechanism is essentially the same mechanism that we saw in the previous section: NO_2^+ on and then H^+ off.

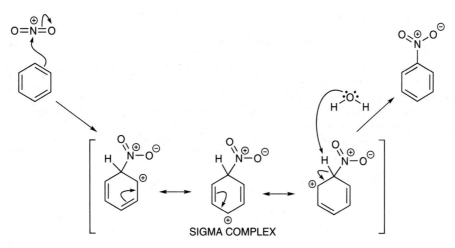

In this case, we are using water to remove the proton (instead of $AlBr_4^-$), which should make sense because we don't have any $AlBr_4^-$ in this reaction. There is plenty of water, because nitric and sulfuric acids are both aqueous solutions. Notice that the mechanism is very similar to what we have already seen in the previous reactions.

So far, we have seen how to install a halogen (Cl, Br, or I) on an aromatic ring, and we have seen how to install a nitro group. Before we move on, let's just make sure that you are familiar with the reagents necessary to perform these reactions. In each of the following cases, identify reagents that you would use in order to achieve the desired transformation.

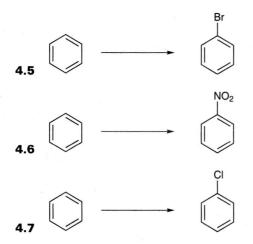

4.5

4.6

4.7

4.8 *Without looking back* at the previous section, try to draw the mechanism for the nitration of benzene. You will need a separate piece of paper to record your answer. Make sure to start by drawing the mechanism for formation of NO_2^+, and then show the reaction of benzene with NO_2^+. When you are finished, look back to see if you drew it correctly.

4.3 FRIEDEL–CRAFTS ALKYLATION AND ACYLATION

In the previous sections, we saw how to install several different groups (Br, Cl, I, or NO_2) on a benzene ring, using an electrophilic aromatic substitution. In each case, the mechanism was the same: E^+ *on* the ring, and then H^+ *off*. In this section, we will learn how to install an alkyl group.

Let's start with the simplest of all alkyl groups: a methyl group. So, the question is: what reagents would we need to achieve the following transformation:

Using the logic that we have developed in this chapter, we would want to use CH_3^+ as our electrophile. But you should probably cringe when you see CH_3^+. After all, you probably remember the trend in the stability of carbocations—that tertiary carbocations are more stable than secondary carbocations, and so on. Certainly, a methyl carbocation would not be very stable at all. In fact, we deliberately avoid using methyl or primary carbocations when drawing mechanisms. But here we are, trying to make a methyl carbocation. Is it even possible? The answer is: yes. In fact, we will make it using the same method we have used in the previous sections.

If we take methyl chloride and we mix it with a pinch of $AlCl_3$, we will have a source of CH_3^+:

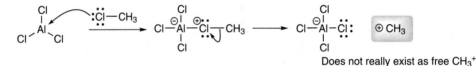

Does not really exist as free CH_3^+

The truth is that we are NOT really forming a free methyl carbocation that can float off into solution. A free CH_3^+ would be too unstable to form. So, instead, we must view this as a complex that can serve as a "source" of CH_3^+.

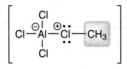

Source of CH_3^+

This provides us with a method for methylating a benzene ring:

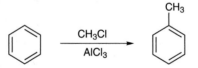

And the mechanism is, once again, the same mechanism that we have seen over and over again. It is an electrophilic aromatic substitution: CH_3^+ *on* the ring and then H^+ *off*:

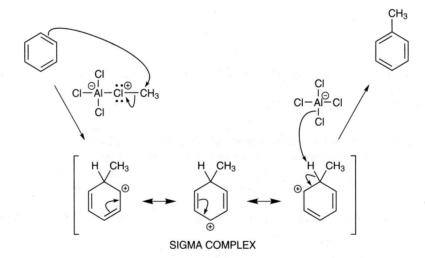

SIGMA COMPLEX

We can use the exact same process to install an ethyl group on an aromatic ring:

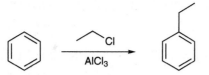

This process (installing an alkyl group onto an aromatic ring) is called a Friedel–Crafts alkylation. It works very well for installing a methyl group or an ethyl group on the ring. BUT we run into problems when we try to install a propyl group on the ring, because a mixture of products is obtained:

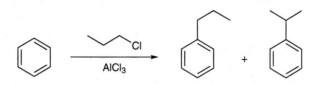

The reason for this is simple. Since we are forming a complex with carbocationic character, it is possible for a carbocation rearrangement to occur. It is not possible for a methyl carbocation to rearrange. Similarly, an ethyl carbocation cannot rearrange to become any more stable. But a propyl carbocation CAN rearrange (via a hydride shift):

And since we are forming a propyl carbocation, we can expect that sometimes it will rearrange before reacting with benzene (while other times, it will not get a chance to rearrange before it reacts with benzene). And that is why we observe a mixture of products. So you need to be careful when using a Friedel–Crafts alkylation to look out for unwanted carbocation rearrangements.

Now, if we wanted to make isopropyl benzene, we could avoid this whole issue by just using isopropyl chloride as our reagent:

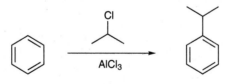

But what if we want to make *n*-propyl benzene?

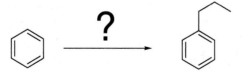

How would we do that? If we use *n*-propyl chloride, we have already seen that we will get some rearrangement, and we will not get a good yield of the desired product. In fact, we can generalize the question like this: how do we install *any* alkyl group and avoid a potential carbocation rearrangement. For example, how could we do the following transformation, ***without*** a carbocation rearrangement?

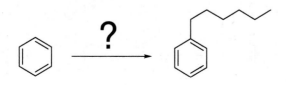

If we just use 1-chlorohexane (and AlCl₃), we will likely get a mixture of products.

Clearly, we need a trick. And there is a trick. To see how it works, we need to take a close look at a similar reaction that also bears the name Friedel–Crafts. But this reaction is not an **alkyl**ation. Rather, it is called an **acyl**ation. To see the difference, let's quickly compare an alkyl group with an acyl group.

We can install an **acyl** group on a benzene ring in exactly the same way that we installed an alkyl group on the ring. We simply use the following reagents:

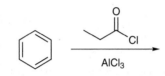

The first reagent is called an acyl chloride (or acid chloride), and we are already familiar with the role of AlCl₃ (the Lewis acid). The Lewis acid interacts with the Cl atom of the acyl chloride, generating a resonance-stabilized acylium ion (highlighted below):

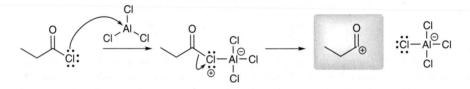

The term **acylium** ion should make sense—"*acyl*" because we can see that this electrophile has an acyl group; and "*ium*" because there is a positive charge. This electrophile actually has an important resonance structure:

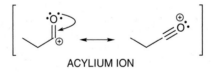

ACYLIUM ION

These resonance structures are important because they indicate that an acylium ion is *stabilized* by resonance, and therefore, it will NOT undergo a carbocation rearrangement. (If it did, it would lose this resonance stabilization.) Compare the following two cases:

CAN REARRANGE   CANNOT REARRANGE

Using a Friedel–Crafts **acyl**ation, we can cleanly install an acyl group on a benzene ring (without any rearrangements):

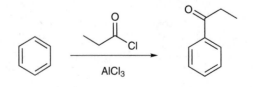

We don't observe any side products that would result from a carbocation rearrangement. Once again, compare a Friedel–Crafts *alkyl*ation with a Friedel–Crafts *acyl*ation:

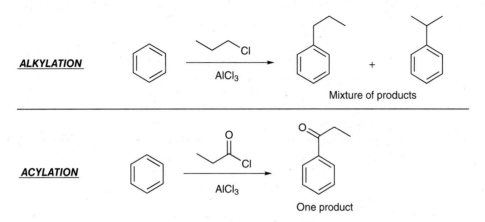

ALKYLATION

Mixture of products

ACYLATION

One product

Now take a close look at the acylation above, and let's point out a very important feature. Notice that we have installed a three-carbon chain on the ring, with the chain being attached by the *first* carbon of the chain, as opposed to the middle carbon of the chain:

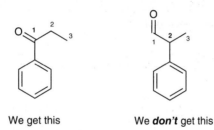

We get this   We *don't* get this

Once again, it is attached by the first carbon because a rearrangement does not occur (the acylium ion is resonance stabilized, and does not rearrange). Now, all we need is a way to reduce the ketone, and then we will have a two-step synthesis for installing an *n*-propyl group on a benzene ring:

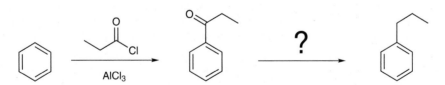

And luckily, there is a simple way to reduce a ketone; in fact, we will see *three* common ways to reduce a ketone. We will just focus on one method right now (which uses acidic conditions), but keep in mind that we will see two other methods of doing this in the upcoming chapters (one method uses basic conditions and the other method uses neutral conditions). When we reduce a ketone under acidic conditions, we call the reaction a Clemmensen reduction:

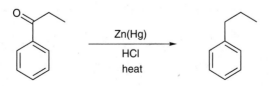

In the presence of HCl and amalgamated zinc (zinc that has been treated so that its surface is an alloy, or mixture, of zinc and mercury), the C═O bond is completely reduced and replaced with two C—H bonds. In this way, a Clemmensen reduction can be used as the second step of a two-step synthesis that installs an alkyl group on an aromatic ring *without* any rearrangement taking place:

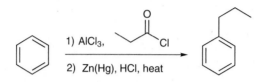

To summarize, we have seen that a Friedel–Crafts acylation can be followed up by a Clemmensen reduction, as a clever way of installing an alkyl group on an aromatic ring (without rearrangements). But there are actually times when you will want to install an acyl group on the ring, and you won't want to do a Clemmensen reduction afterward. For example:

To achieve this transformation, you would just use a Friedel–Crafts acylation, and that's it. There is no need for a Clemmensen reduction, because in this case, we don't want to reduce the C═O bond.

EXERCISE 4.9   Show the reagents you would use to achieve the following synthesis:

Answer   In this problem, we need to install an alkyl group on a benzene ring. So, we first look to see if we could do this in just one step, using a Friedel–Crafts alkylation. In this case, we cannot do it in one step because we have to worry about a carbocation rearrangement. If we think about the electrophile that we would need to make, we will see that it could rearrange:

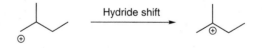

And this would give us a mixture of products:

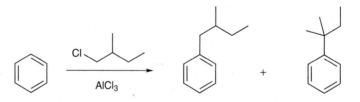

So, instead we will have to use a Friedel–Crafts acylation followed by a Clemmensen reduction:

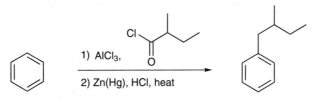

For each of the following problems, show what reagents you would use to accomplish the transformation. In some situations, you will want to use a Friedel–Crafts alkylation, while in other situations, you will want to use a Friedel–Crafts acylation.

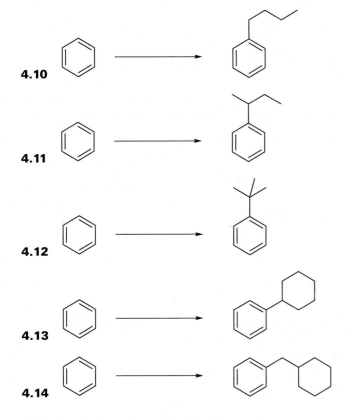

4.10

4.11

4.12

4.13

4.14

4.15   Predict the products of the following reaction.

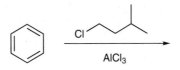

(*Hint*: There should be a mixture of *multiple* products in this case. Be sure to consider all of the possible rearrangements that can take place. If you are rusty on carbocation rearrangements, then you should go back and review them now.)

4.16   On a separate piece of paper, draw a mechanism of formation for each one of the three products from the previous problem.

4.17   On a separate piece of paper, draw a mechanism for the following transformation. Make sure to show the mechanism of formation of the acylium ion that reacts with the ring:

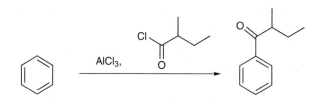

Friedel–Crafts reactions have a few limitations. You should take a moment to read about them in your textbook. The two most important limitations are as follows:

1. When performing a Friedel–Crafts *alkylation*, it is often difficult to install just one alkyl group. Each alkyl group makes the ring *more* reactive toward a subsequent attack on the same ring.

2. When performing a Friedel–Crafts *acylation*, it is generally not possible to install more than one acyl group. The presence of one acyl group makes the ring *less* reactive toward a second acylation.

We need to understand WHY an alkyl group makes the ring more reactive, and WHY an acyl group makes the ring less reactive. We will explain this in greater detail during the upcoming sections. But first, we have one more electrophile to discuss.

4.4   SULFONATION

The reaction we will now explore will be used extensively in synthesis problems later in this chapter. If you do not keep this reaction in mind while you are solving synthesis problems, then you will be at a severe loss. We will explain why this reaction is so important in the upcoming sections. For now, just take my word for it, and let's just master the reaction.

The electrophile is SO_3. Let's take a close look at the structure:

Notice that there are three S=O double bonds here. But these double bonds are not such great double bonds. Recall that a double bond is formed from the overlap of two *p* orbitals:

When we are talking about a carbon–carbon double bond, the overlap of the *p* orbitals is efficient because the *p* orbitals are the same size. But what happens when you try to overlap the *p* orbital of an oxygen atom with the *p* orbital of a sulfur atom? The *p* orbitals are different sizes (oxygen is in the second row of the periodic table, which means that it is using a *p* orbital from the second energy level; but sulfur is in the third row of the periodic table, so it is using a *p* orbital from the third energy level):

Therefore, the overlap is not so efficient, and it is misleading to think of S=O as being a double bond. It is probably much closer in nature to being like this:

When we do this analysis for each of the three double bonds in SO_3, we begin to see that the sulfur atom is VERY electron-poor:

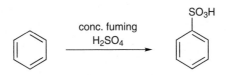

In fact, the sulfur atom is so electron-poor that it is an *excellent* electrophile, even though the compound is overall neutral (no net charge). Now we are going to see a reaction that uses SO_3 as an electrophile.

Sulfuric acid is constantly in equilibrium with SO_3 and water:

$$H_2SO_4 \rightleftharpoons SO_3 + H_2O$$

That means that any bottle of sulfuric acid will have some SO_3 in it. Fuming sulfuric acid is a mixture of sulfuric acid that has a lot of SO_3, and is called "fuming" because of the acrid fumes that it presents when exposed to moisture in the air (SO_3 reacts with moisture to produce the fumes, which are essentially an aerosol of concentrated sulfuric acid). So, from now on, whenever you see concentrated, fuming sulfuric acid, you should realize that we are talking about SO_3 as the reagent.

And here is the reaction:

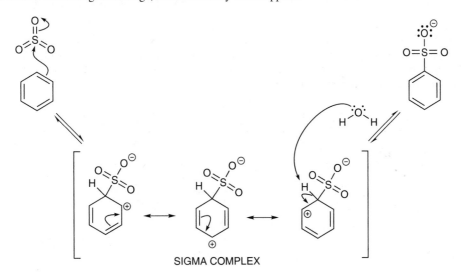

Notice that this process, called sulfonation, installs an SO_3H group on the ring. The obvious question is: why is an H attached to the SO_3 in the end? To see why, let's take a closer look at the mechanism. Remember our two steps for any electrophilic aromatic substitution: E$^+$ goes on the ring, and then H$^+$ comes off. But wait a second .... In this case, we are not using an electrophile with a net positive charge. The electrophile in this case has no net charge. In all of the reactions we have seen so far, we put something positively charged onto the ring, and then we took something positively charged off of the ring. So in the end, our ring never gained or lost any charges. But in this case, we are putting something neutral (SO_3) onto the ring, and then we are removing something positively charged (H$^+$). That should leave our product with an overall negative charge, which is exactly what happens:

So our mechanism will have one additional step. In the presence of sulfuric acid, the negatively charged oxygen atom is protonated. The likely source of the proton is H_3O^+, rather than H_2SO_4, because aqueous sulfuric acid (even if its concentrated) is a mixture comprised mostly of H_3O^+ and HSO_4^-.

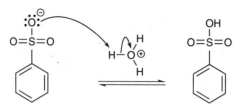

Although this reaction has this one extra step at the end of the mechanism, keep in mind that this extra step is just a proton transfer. The core reaction is still the same as we have seen in all of the previous reactions: the electrophile comes on the ring, and then H^+ comes off of the ring.

An important feature of this reaction (and this is the feature that will make this reaction so important for synthesis problems) is how easily the reaction can be reversed. The amount of product is equilibrium controlled, and it is very sensitive to the conditions. So, if you use dilute sulfuric acid instead, the equilibrium leans the other way:

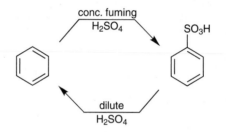

We can use this to our advantage because this provides a way to remove the SO_3H group whenever we want. We would just use dilute sulfuric acid to remove it:

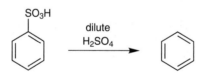

So we now have the ability to install the SO_3H group on the ring whenever we want, AND we can remove it whenever we want as well. You might wonder why you would want to install a group in order just to remove it later. On the surface, that would seem like a waste of time. But in the upcoming sections, we will see that this will become very important in synthesis problems.

For now, let's make sure that we are comfortable with the reagents.

EXERCISE 4.18 Identify the reagents that you would use to achieve the following transformation:

Answer   We know that concentrated fuming sulfuric acid will install an SO_3H group on an aromatic ring, and dilute sulfuric acid will remove the group. In this case, we are removing the group, so we need to use dilute sulfuric acid:

Identify the reagents you would use to achieve each of the following transformations:

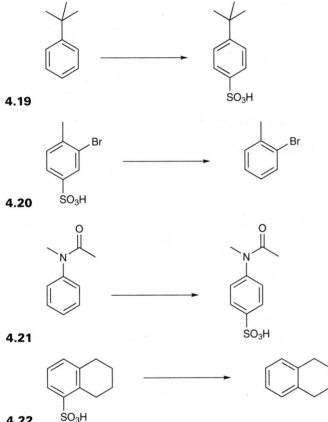

4.19

4.20

4.21

4.22

And to make sure that you are not getting rusty on the other reactions we have learned in this chapter so far, fill in the reagents you would use for the following transformations:

4.23

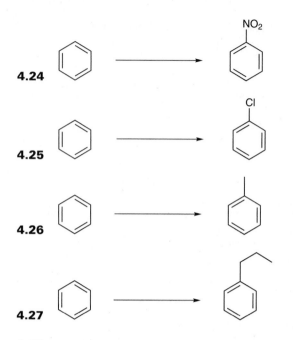

4.24

4.25

4.26

4.27

4.28 Now, let's just make sure that you can draw a mechanism for the sulfonation of benzene (on a separate piece of paper). Remember that there will be three steps: (1) electrophile comes on the ring, (2) H⁺ comes off, and then (3) proton transfer. Don't forget to draw the resonance structures of the intermediate sigma complex. Try to draw the mechanism without looking back to where it is shown earlier in this section.

4.29 And now, for a challenging problem, try to draw the mechanism of a desulfonation reaction (a reaction where we take the SO₃H group off of the ring). The process will be exactly the reverse of what you just drew in the previous problem. There will be three steps: (1) remove the proton from the SO₃H group, (2) H⁺ comes on the ring, and then (3) SO₃ comes off of the ring. The truth is that there are only two core steps here: H⁺ comes on the ring, and then SO₃ comes off of the ring. You can actually pull the proton off of the SO₃H group at the same time that SO₃ comes off of the ring. Try to do it yourself, and if you get stuck, you can look at the answer in the back of the book. Make sure that your mechanism involves an intermediate sigma complex (with a resonance-stabilized positive charge).

4.5 ACTIVATION AND DEACTIVATION

On the first page of this chapter, we saw that benzene is unreactive toward bromine:

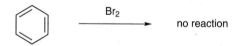

$$\xrightarrow{\text{Br}_2} \text{no reaction}$$

In order to force a reaction to occur, we introduced a Lewis acid into the reaction mixture, which generated a better electrophile (Br⁺ is a better electrophile than Br₂). In fact, all of the reactions we have explored thus far have been examples of benzene reacting with powerful electrophiles (Cl⁺, NO₂⁺, alkyl⁺, acyl⁺, and SO₃). Now, we will turn our attention to the nucleophile—how can we modify the reactivity of the aromatic ring?

To answer this question, we will explore substituted benzene rings, and we will consider the effect that a substituent will have on the reactivity of the ring. Benzene itself (C₆H₆) has no substituents. But consider the structure of phenol:

In this compound, the aromatic ring has one substituent: an OH group. What effect does this substituent have on the nucleophilicity of the aromatic ring? Is this compound a stronger nucleophile or a weaker nucleophile than benzene?

To answer this question, we must explore the effect of an OH group on the electron density of the aromatic ring. There are two factors to consider: *induction* and *resonance*. Let's begin with induction. Recall from the first semester that inductive effects can be evaluated by comparing the relative electronegativity of the atoms. In our example, we are looking specifically at the C—O bond connecting the OH group to the ring. Oxygen is more electronegative than carbon, so there is an inductive effect, shown by the following arrow:

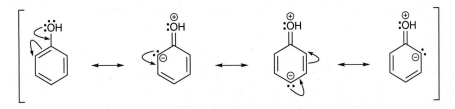

The oxygen atom is withdrawing electron density from the ring. Remember that the aromatic ring is only a nucleophile in the first place because it is electron-rich (from all of those π electrons), so withdrawing electron density from the ring (via induction) should render the ring *less* nucleophilic. But we're not done yet. We need to consider one other factor: resonance.

Below are resonance structures of phenol:

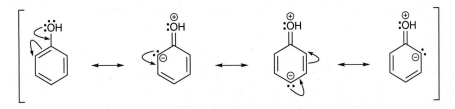

Notice that there is a negative charge spread throughout the ring. When we meld all of these resonance structures together in our minds, we obtain the following information:

The $\delta-$ symbols indicate that there is a partial negative charge spread throughout the ring. Therefore, the effect of resonance is to *donate* electron density to the ring. So, now we have a competition. By induction, the OH group is *electron-withdrawing*, which makes the ring *less* nucleophilic. But by resonance, the OH group is *electron-donating*, which makes the ring *more* nucleophilic. Which effect is stronger? Resonance or induction? This is a common scenario in organic chemistry (where induction and resonance are in competition), and the general rule is: **resonance is usually a stronger factor than induction**. There are some important exceptions. In fact, we will soon see one of these exceptions, but, in general, resonance wins.

Now let's apply this general rule to our case of phenol. If we say that resonance is stronger than induction, then the net effect of the OH group is to *donate* electron-density to the ring. And therefore, the net effect of the OH group is to make the ring *more nucleophilic* (as compared with benzene).

Indeed, experiments reveal that phenol is significantly more nucleophilic than benzene. The effect of the OH group on the nucleophilicity of the aromatic ring is called "activation." The OH group is said to be activating the ring (making it more nucleophilic). So, the OH group is called an *activator*. Alkyl groups (such as methyl or ethyl) are also activators (recall from the first semester that alkyl groups are electron donating because of an effect called hyperconjugation). There are some groups, though, that actually *withdraw* electron density from the ring, and we call those groups *deactivators*, because they deactivate the ring (make the ring *less* nucleophilic). An excellent example is the nitro group. Consider the structure of nitrobenzene:

Once again, the effect of the nitro group can be evaluated by exploring two factors: induction and resonance. Induction is simple. The nitrogen atom is more electronegative than the carbon atom to which it is connected (especially since the nitrogen atom bears a positive charge). So the nitro group withdraws electron density from the ring via induction, which should render the ring less nucleophilic. But we're not done yet. We still have to consider resonance. Below are resonance structures of nitrobenzene:

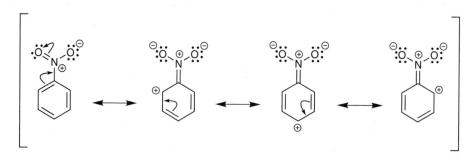

Notice that there is now a positive charge spread throughout the ring (rather than a negative charge, as we saw in the case of phenol). When we meld all of these resonance structures together in our minds, we obtain the following information:

The δ+ symbols indicate that there is a partial positive charge spread throughout the ring. Therefore, the effect of resonance is to *withdraw* electron density from the ring.

In summary, the nitro group is electron-withdrawing via resonance *and* via induction. In other words, there is no competition between resonance and induction. Both factors dictate that a nitro group should *deactivate* the ring. And that is indeed what we observe in the laboratory.

4.6 DIRECTING EFFECTS

Now let's consider electrophilic aromatic substitution reactions with substituted benzene rings. In order to explore this topic, we must first review important terminology that we will use frequently throughout the remainder of this chapter. The various positions on a monosubstituted benzene ring are referred to in the following way:

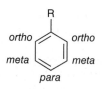

The two positions that are closest to the substituent (R) are called *ortho* positions. Then we have the *meta* positions. And finally, the farthest position from the substituent is called the *para* position. With this terminology in mind, let's consider the products obtained when toluene or nitrobenzene undergo bromination. Both reactions are shown below:

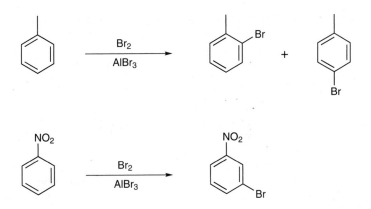

The first reaction (with toluene) is certainly faster, because the methyl group activates the ring toward electrophilic aromatic substitution, while the nitro group deactivates the aromatic ring. But consider the difference in regiochemical outcome. Specifically, notice that the methyl group directs the reaction to occur at the *ortho* and *para* positions, while the nitro group directs the reaction to occur at the *meta* position. Your textbook will give an explanation for this observation, and you should read that explanation, but here is the bottom line:

- Activators are *ortho-para* directors
- Deactivators are *meta* directors

We will not encounter any exceptions to the first rule (all activators that we encounter will be *ortho-para* directors). But there is one important exception to the second rule above. Halogens (F, Br, Cl, or I) are deactivators, so we might expect them to be *meta* directors. But instead, they are actually *ortho-para* directors. Let's try to understand why halogens are the exception.

In the previous section, we analyzed the effect of an OH group on an aromatic ring, and we saw that there were two competing effects: induction and resonance. We saw that induction *withdraws* electron density from the ring, but resonance *donates* electron density to the ring. In order to know which factor dominates, we gave a general rule: **resonance is usually a stronger factor than induction**. We also said that there was an important exception to this general rule that we would see later. Well, it is now later. Halogens are the exception. Let's take a closer look. As an example, consider the structure of chlorobenzene:

The substituent in this case (Cl) is an *ortho-para* director for the same reason that an OH group is an *ortho-para* director. But, unlike OH (which is an activator), Cl is a deactivator. To explain why, we must explore the effect of a halogen (such as Cl) on the electron density of the aromatic ring. As we have seen in the previous section, our analysis must focus on two factors: induction and resonance. Let's start with induction. Just like an OH group, a halogen is also electron-withdrawing by induction:

However, we also need to consider resonance effects. So, we draw the resonance structures:

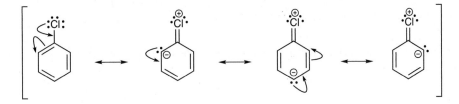

And once again, we see that a halogen is very similar to an OH group. It is donating electron density by resonance:

Now let's consider the net effect of a halogen. Just as we saw with the OH group, there is a competition between resonance and induction. And just as we saw with the OH group, a halogen will *withdraw* electron density by induction, and it will *donate* electron density by resonance. But, in the case of the OH group, we used the argument that resonance beats induction (we said that was a general rule that holds true most of the time). Therefore, the net effect of the OH group was to donate electron density to the ring (thus, the OH group is an activator). But with a halogen, resonance does NOT beat induction. This is one of the rare cases where induction actually beats resonance. Why is resonance not the predominant factor in this case? To answer this question, notice that the resonance structures of chlorobenzene exhibit a positive charge on Cl. Halogens do not easily bear positive charges. So, these resonance structures do not contribute very much character to the overall structure of the compound. Resonance is a weak effect in this case, so induction actually beats resonance. Therefore, the net effect of a halogen substituent is to *withdraw* electron density from the aromatic ring, rendering the ring less nucleophilic (deactivation).

Now we are ready to modify the rules we gave earlier when we said that all activators are *ortho-para* directors and all deactivators are *meta* directors. Here is our new-and-improved formula:

- All activators are *ortho–para* directors.
- All deactivators are *meta* directors, **except for halogens (which are deactivators, but nevertheless, they are <u>ortho–para</u> directors)**.

With that in mind, let's try to predict some directing effects.

EXERCISE 4.30 Look closely at the following monosubstituted benzene ring.

If this compound were to undergo an electrophilic aromatic substitution reaction, predict where the incoming substituent would be installed.

Answer Br is a halogen (remember that the halogens are F, Cl, Br, and I). We have seen that halogens are the one exception to the general rules. That is, they are deactivators but they are not *meta* directors, like most other deactivators. Rather, they are *ortho–para* directors. Therefore, if we use this compound in an electrophilic aromatic substitution, we expect substitution to take place at the *ortho* and *para* positions:

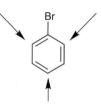

Identify the expected directing effects that would be observed if each of the following compounds were to undergo an electrophilic aromatic substitution reaction.

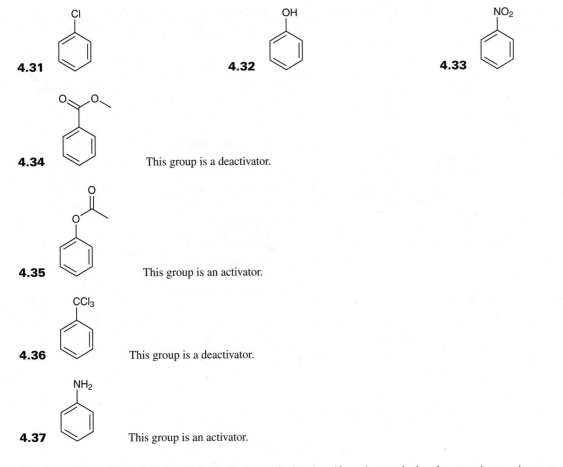

4.31

4.32

4.33

4.34   This group is a deactivator.

4.35   This group is an activator.

4.36   This group is a deactivator.

4.37   This group is an activator.

Clearly, you can only predict where the substitution will take place if you know whether the group is an activator or a deactivator. In the next section, we will learn how to predict whether a group is an activator or a deactivator. But for now, let's get some practice predicting products.

EXERCISE 4.38   Predict the major product(s) of the following reaction:

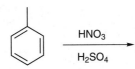

Answer   We begin by looking at the reagents, so that we can determine what kind of reaction is taking place. The reagents are nitric acid and sulfuric acid. We have seen that these reagents generate NO_2^+ as an electrophile, which can react with an aromatic ring in an electrophilic aromatic substitution reaction. The end result is to install a nitro group on an aromatic ring. So, now the question is: where is the nitro group installed?

To answer this question, we must predict the directing effects of the group already present on the ring (before the reaction takes place). There is a methyl group on the ring, and we have seen that methyl groups are activators. Therefore, we predict that the reaction will take place at the *ortho* and *para* positions, relative to the methyl group:

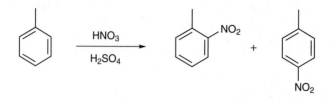

Notice that substitution at either *ortho* position gives the same product:

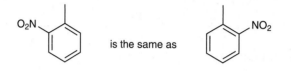

is the same as

Predict the major product(s) for each of the following reactions:

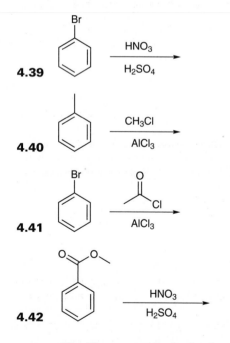

4.39

$$HNO_3$$
$$H_2SO_4$$

4.40

$$CH_3Cl$$
$$AlCl_3$$

4.41

$$AlCl_3$$

4.42

$$HNO_3$$
$$H_2SO_4$$

Hint: The group on the ring is a deactivator.

4.43

$$Br_2$$
$$AlBr_3$$

Hint: The group on the ring is a deactivator.

4.44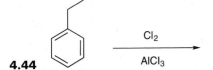

> *Hint*: The group on the ring is an activator.

So far, we have focused on the directing effects when you have *only one group* on a ring. And we have seen that activators direct toward the *ortho* and *para* positions, whereas deactivators direct toward the *meta* positions:

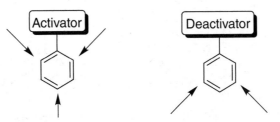

and we saw only one exception (the halogens).

But how do you predict the directing effects when you have *more than one group* on a ring? For example, consider the following compound:

What if we used this compound in an electrophilic aromatic substitution reaction? For example, let's say we try to brominate this compound. Where will the bromine atom be installed?

Let's first consider the effect of the methyl group. We mentioned before that a methyl group is an activator, so we predict that it will direct toward the *ortho* and *para* positions:

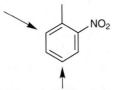

Notice that we do ***not*** point to the *ortho* position that already bears the nitro group (we only look at positions that are unoccupied—remember that in an electrophilic aromatic substitution, E$^+$ comes on the ring and **_H_**$^+$ comes off). So, the methyl group is directing toward *two* spots, as shown above.

Now let's consider the effect of the nitro group. We mentioned before that the nitro group is a powerful deactivator. Therefore, we predict that it should direct to the positions that are *meta **to the nitro group:***

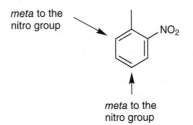

We see that the nitro group and the methyl group are directing toward the same two spots. So, in this case there is no conflict between the directing effects of the nitro group and the methyl group.

But consider this case:

The methyl group and the nitro group are now directing toward different positions:

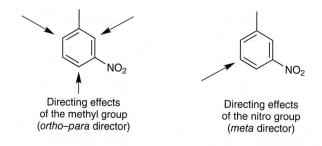

Directing effects
of the methyl group
(*ortho–para* director)

Directing effects
of the nitro group
(*meta* director)

So, the big question is: which group wins? It turns out that the directing effects of the methyl group trump the directing effects of the nitro group. So, if we brominate this ring, we will get the following products (where the Br is installed *ortho* or *para* to the methyl group):

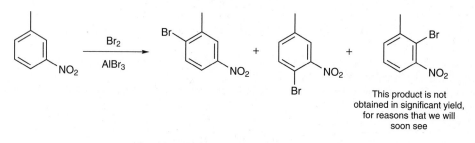

This product is not
obtained in significant yield,
for reasons that we will
soon see

It is common to see a situation where the directing effects of two groups are competing with each other (like the methyl group and nitro group in the above example). So we clearly need rules for determining which group wins. It turns out that you need to know just two simple rules in order to determine which group will dominate the directing effects:

1. *Ortho–para directors always beat meta directors.* The example we just saw is a perfect illustration of this rule. The methyl group is an activator (an *ortho–para* director), and the nitro group is a deactivator (a *meta* director), so the methyl group wins.

2. *Strong activators always beat weak activators.* For example, consider the following case:

The OH group is a *strong* activator, and the methyl group is a *weak* activator. (We will learn in the next section how to predict which groups are strong and which are weak—for now, just take my word for it.) So, the OH group will win, and the directing effects are as follows:

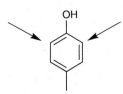

So, we have seen two rules:

- *ortho–para directors always beat meta directors.*
- **Strong** *activators always beat* **weak** *activators.*

Keep in mind that the first rule always trumps the second rule. So if you have a weak activator against a strong deactivator, the weak activator wins. Even though the activator is weak, it still beats a strong deactivator because activators (*ortho–para* directors) always beat deactivators (*meta* directors). This rule was already seen in one of our previous examples:

The methyl group is a weak activator, and the nitro group is a strong deactivator. So, in this case, the methyl group wins (and the directing effects are *ortho* and *para* to the methyl group; and **not** *meta* to the nitro group):

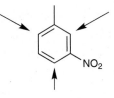

EXERCISE 4.45   Predict the directing effects for an electrophilic aromatic substitution reaction involving the following type of disubstituted aromatic ring.

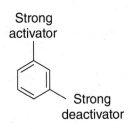

For this problem, you should assume that the deactivator is *not* a halogen.

Answer   We have two groups. The activator will direct toward the positions that are *ortho* or *para* to itself, and the deactivator will direct toward the positions that are *meta* to itself:

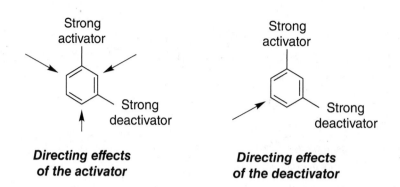

**Directing effects
of the activator**

**Directing effects
of the deactivator**

So, there is a competition in the directing effects. Between the two groups, the strong activator beats the strong deactivator because the strong activator is an *ortho–para* director. So the directing effects are:

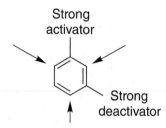

If we performed an electrophilic aromatic substitution on a compound of this type, we might expect three products (because the directing effects are toward three positions, shown above). Here is a specific example of a reaction like this.

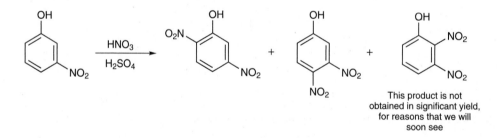

The starting aromatic ring has two substituents. The OH group is a strong activator, and the nitro group is a strong deactivator.

Predict the directing effects for an electrophilic aromatic substitution reaction involving each of the following types of disubstituted aromatic rings. Unless otherwise indicated, assume that anything labeled as a deactivator is not a halogen (unless it is specifically indicated as a halogen).

4.46

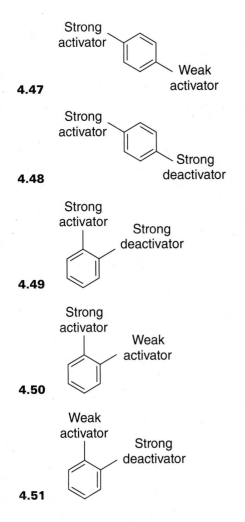

4.47

4.48

4.49

4.50

4.51

4.7 IDENTIFYING ACTIVATORS AND DEACTIVATORS

In the previous section, we learned how to predict the directing effects in a situation where you have more than one group on the ring. But in all of the cases in the previous section, I had to tell you whether each group was an activator or a deactivator and whether it was strong or weak. In this section, we will learn how to predict this, so that you won't have to memorize the characteristics of every possible group. In fact, very little memorization is actually involved here. We will see a few concepts that should make sense. And with those concepts, you should be able to identify the nature of any group, even if you have never seen it before.

We will go through this methodically, starting with strong activators.

Strong activators are groups that have a lone pair next to the aromatic ring. We have already seen an example of this. When an OH group is connected to the ring, there is a lone pair next to the ring, which gives rise to the following resonance structures:

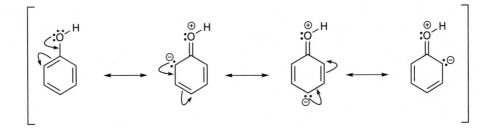

We concluded in the previous section that this resonance effect is very strong and that the OH group is therefore donating a lot of electron density to the ring:

Since the ring is electron-rich, it is a stronger nucleophile than benzene and is more reactive (activated) toward reaction with an electrophile. This effect is observed not only for the OH group, but also for other groups that have a lone pair next to the ring. The same kind of resonance structures can be drawn for an amino group connected to a ring:

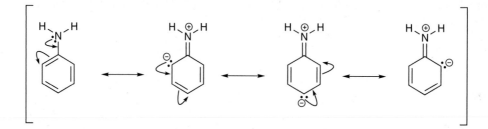

Here are several examples of strong activators. Make sure that you can easily see the common feature (the lone pair next to the ring):

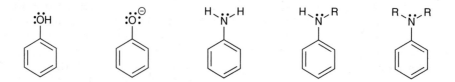

Next, we move on to the *moderate* activators. Moderate activators are groups that have a lone pair next to the ring, BUT that lone pair is already partially tied up in resonance. For example, consider the following group:

This compound has all of the resonance structures that place electron density into the ring (just like an OH group does):

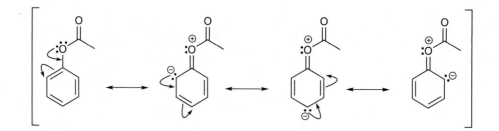

BUT there is an additional resonance structure, which has the electron density *outside* of the ring:

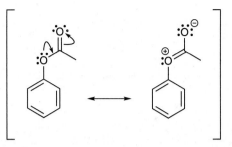

Therefore, the electron density is more spread out (with some in the ring and some out of the ring). This group is therefore not a *strong* activator. Rather, we call it a *moderate* activator. (Some textbooks do not point out this subtle distinction between strong activators and moderate activators.) Here are several examples of moderate activators:

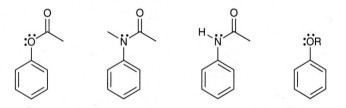

Look closely at the examples above. They all have a lone pair that is tied up in resonance outside of the ring. BUT WAIT A SECOND. What about the last group on this list (the OR group)? This group has a lone pair that is NOT tied up in resonance outside of the ring. We should predict that this group should belong in the first category (*strong* activators), but for some reason, it isn't in that category. It is actually just a *moderate* activator. This is one of the rare examples that departs from the logical explanations that we have given so far. I have spent quite a bit of time trying to figure out why the OR group is a moderate activator (rather than a strong activator). I have come up with several answers over the years, but I am not going to spend several pages dedicated to an esoteric topic that you will certainly not need for your exams (perhaps you might think of it as a brain teaser—something to think about …). For now, you will just have to remember that the OR group doesn't follow the trends we have seen. It is a moderate activator.

Now let's turn our attention to *weak* activators. Alkyl groups (such as methyl, ethyl, propyl, etc.) are weak activators because they donate electron density to the ring via hyperconjugation (a weak effect). We previously encountered hyperconjugation to explain why tertiary carbocations are more stable than secondary carbocations. Similarly here, alkyl groups also exhibit an electron-donating effect (albeit weak), and they are therefore weak activators.

Now we have seen all of the different categories of activators (strong, moderate, and weak). To review, this is what we saw:

Strong activators	Lone pair next to ring
Moderate activators	Lone pair next to ring, but tied up in resonance outside of ring as well
Weak activators	Alkyl groups

Now, we will turn our attention to the different categories of ***deactivators***. This time, we will begin with the *weak* deactivators and work our way toward *strong* deactivators (rather than starting with strong). There is a reason for using this order, and that reason will soon become clear.

Weak deactivators are the halogens. We have already seen that halogens are the one case where induction beats resonance (and therefore, we argued that the net effect of a halogen is to *withdraw* electron density from the ring). So, we saw that halogens are deactivators. But you should know that the competition between induction and resonance (in the case of the halogens) is a close competition, so halogens are only weakly deactivating.

We will summarize all of this information in one complete chart, but for now let's move on to moderate deactivators.

Groups that withdraw electron density from the ring via resonance are moderate deactivators. For example, consider the following group:

This group does *not* have a lone pair next to the ring (so it is *not* an activator). But it does have a pi bond next to the ring, giving rise to the following resonance structures:

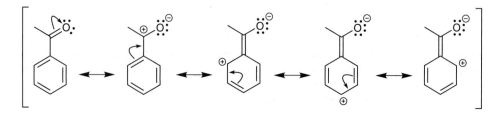

When we look closely at these resonance structures, we can see that the substituent is *withdrawing* electron density from the ring:

Therefore, this group is a *moderate deactivator*. Numerous other similar groups can also withdraw electron density from the ring. Here is a list of many examples:

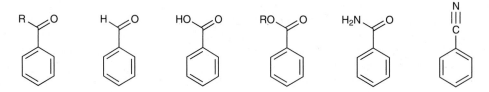

All of these substituents are withdrawing electron density from the ring via resonance. They all have one feature in common: a pi bond to an electronegative atom. Take a close look at the last example. A cyano group is a pi bond (a triple bond) to an electronegative atom (nitrogen). So we see that a triple bond can also be included in this category.

And now for the last category: *strong* deactivators. There are a few common functional groups that fall into this category:

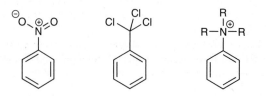

We have already explained why the nitro group is so powerfully electron-withdrawing. The nitro group is electron-withdrawing by resonance *and* induction.

To understand why the second group shown (the trichloromethyl group) is a strong deactivator, we need to focus on the collective inductive effects of all of the chlorine atoms:

The inductive effects of each chlorine atom add together to give one very powerful deactivating group. Be careful not to confuse this group with a halogen on a ring:

When a halogen is connected directly to the ring (above right), then there are resonance effects to consider. (We spent a lot of time talking about the competition between resonance and induction in the case of halogens.) The group we are talking about now (above left) does not have any resonance effects to consider because the lone pairs are *not* directly next to the ring. So, there is only an inductive effect to consider, and this inductive effect is very significant in this case (because there are three inductive effects adding together).

When we consider our final example of a *strong deactivator*, we see a nitrogen atom with a positive charge next to the ring:

The nitrogen atom is so poor in electron density that it is practically sucking electron density out of the ring like a vacuum cleaner:

Now we are ready to summarize everything we have seen into one chart:

Common Feature	**Some Examples**

ACTIVATORS

Strong Lone pair next to ring

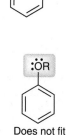

Moderate Lone pair next to ring, but tied up in resonance outside of ring as well

Does not fit the pattern

Weak Alkyl groups

Me   Et   *n*-Pr

DEACTIVATORS

Weak Halogens

Cl   Br   I

Moderate Pi bond to an electronegative atom (next to ring)

Strong Very powerfully electron-withdrawing

Take a close look at this chart and make sure that every category makes sense to you. As you look over the chart, you should be able to remember the arguments that we gave for each category. If you have trouble with this, you might want to review the last few pages of explanation.

And now we can understand why we looked at weak deactivators first (before strong deactivators). When we organize it like this, we can clearly organize the directing effects in our minds:

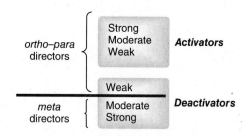

This chart shows that all activators are *ortho–para* directors and all deactivators are *meta* directors, with the exception of the weak deactivators (halogens).

EXERCISE 4.52   Look closely at the following substituent:

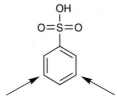

Try to predict what kind of group it is (a strong activator, a moderate activator, a weak activator, a weak deactivator, a moderate deactivator, or a strong deactivator).

· Use this information to predict the directing effects if this compound were to undergo an electrophilic aromatic substitution reaction.

Answer   This group does not have a lone pair next to the ring, and it is not an alkyl group. Therefore, it is not an activator. This group has a pi bond to an oxygen atom (next to the ring), and therefore, it is a moderate deactivator.

Because all deactivators are *meta* directors (except for weak deactivators—halogens), we predict the following directing effects (Note: substitution at either position leads to the same product):

For each of the following substituents, determine what kind of group it is (a strong activator, a moderate activator, a weak activator, a weak deactivator, a moderate deactivator, or a strong deactivator). Place your answer on the space provided. Try to do this **without** looking at the chart that we constructed. You won't have access to this chart on an exam. Try to remember and apply the explanations that we used.

Then, use that information to predict the directing effects. Indicate the directing effects using arrows for pointing to the positions where you would expect an electrophilic aromatic substitution to occur:

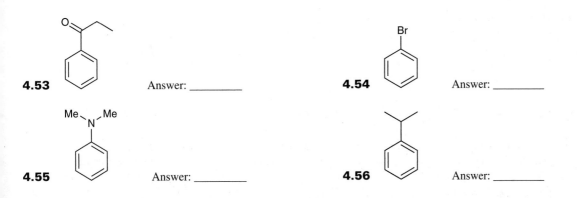

4.53   Answer: _____

4.54   Answer: _____

4.55   Answer: _____

4.56   Answer: _____

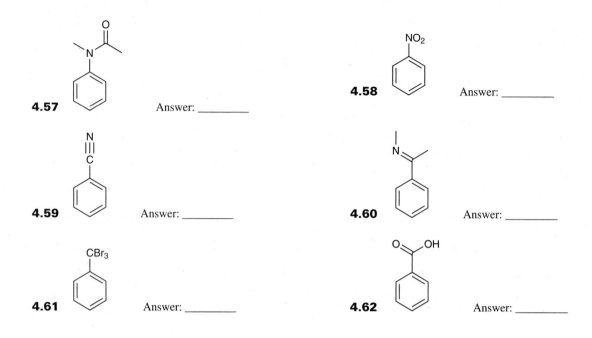

4.57   Answer: _____

4.58   Answer: _____

4.59   Answer: _____

4.60   Answer: _____

4.61   Answer: _____

4.62   Answer: _____

4.63   Can you explain why the following group is a strong activator:

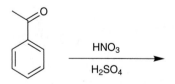

(*Hint*: Think about what strong activators have in common.)

Now we can use the skills that we developed in this section to predict the products of a reaction. Let's see an example:

EXERCISE 4.64   Predict the major product(s) of the following reaction:

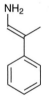

$$\xrightarrow[\text{H}_2\text{SO}_4]{\text{HNO}_3}$$

Answer   We look at the reagents to see what kind of reaction we expect. The reagents are nitric acid and sulfuric acid. These reagents generate NO_2^+, which is an excellent electrophile. So, we know that the reaction will install a nitro group on the ring. But the question is: where?

To answer this question, we must predict the directing effects of the group that is currently on the ring. We recognize that this group is a moderate deactivator, which means that it must be a *meta* director. So, we predict the following product:

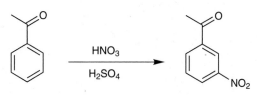

Predict the major product(s) of the following reactions:

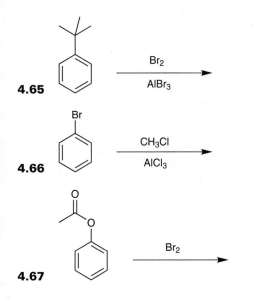

4.65

4.66

4.67

Notice that in this reaction, we do not need a Lewis acid catalyst. Can you explain why not?

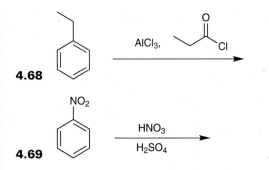

4.68

4.69

Now let's combine what we did in the previous section with the material in this section. Recall from the previous section that we learned how to predict the directing effects when you have more than one group on the ring. When the two groups are competing with each other, we saw that you can determine the directing effects by using the following two rules:

- *ortho–para directors always beat meta directors.*
- ***Strong*** *activators always beat **weak** activators.*

Now that we have learned how to categorize the various kinds of groups, let's get practice using our skills to predict products:

EXERCISE 4.70 Predict the major product(s) of the following reaction:

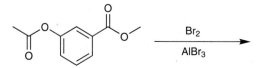

Answer We look at the reagents to see what kind of reaction we expect. The reagents are bromine and aluminum tribromide. These reagents generate Br^+, which is an excellent electrophile. So, we know that we will be installing a Br atom on the ring. But the question is: where?

To answer this question, we must predict the directing effects of the two groups that are currently on the ring. The group on the left is a moderate activator (make sure that you know why), and therefore, it directs *ortho* and *para* to itself:

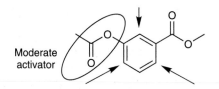

The group on the right is a moderate deactivator (make sure you know why), so it directs *meta* to itself:

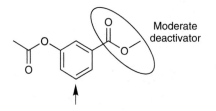

There is a competition between these two groups. Remember our first rule for determining which group wins: *ortho–para* directors beat *meta* directors. So, we expect the following products:

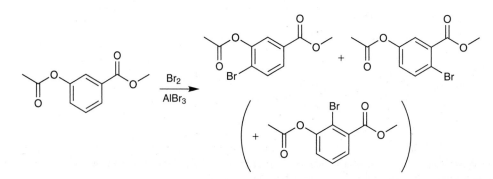

Notice that I placed the last product in parentheses. This product is actually a very minor product. We will see why in the next section. For now, we will just write that we expect three products. Then, we will fine-tune this prediction in the next section.

PROBLEMS Predict the products of the following reactions:

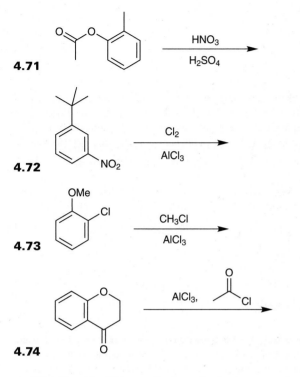

4.71

4.72

4.73

4.74

(*Hint*: Consider this aromatic ring as having two separate substituents, and analyze each separately.)

4.8 PREDICTING AND EXPLOITING STERIC EFFECTS

In the previous sections, we learned the skills that we need in order to predict the products of an electrophilic aromatic substitution. We saw many cases where there is *more* than one product. For example, if the ring is activated, then we expect *ortho and para* products. In this section, we will see that it is possible to predict which product will be the major product and which will be the minor product (*ortho* vs. *para*). It is even possible *to control* the ratio of products (*ortho* vs. *para*). This is VERY important for synthesis problems, which will be the next (and final) section of this chapter.

Consider an electrophilic aromatic substitution with *n*-propyl benzene. The *n*-propyl group is a weak activator, and therefore, we expect the directing effects to be *ortho–para*:

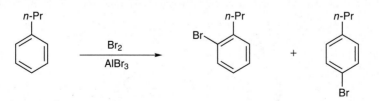

There are two products here. But let's try to figure out which one of these products is the major product? *ortho* or *para*? At first, we might be tempted to say that the *ortho* product should be major. Let's see why. The *n*-propyl group is an *ortho-para* director, so there should be a total of three positions that can get attacked (two *ortho* positions and one *para* position):

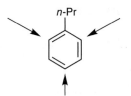

Therefore, the chances of attacking an *ortho* position should be twice as likely as the chances of attacking the *para* position. From a purely statistical point of view, we should therefore expect our product distribution to be 67% *ortho* and 33% *para*. But the product ratio is different from what we might expect, because of steric considerations. Specifically, the *n*-propyl group is fairly large, and it partially "blocks" the *ortho* positions. We do still observe *ortho* products, but much less than 67%. In fact, the *para* product is the major product in this case:

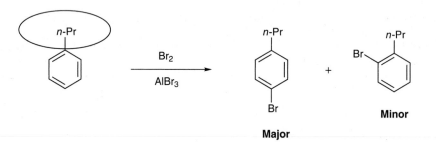

This is usually the case (that *para* is the major product). A notable exception is toluene (methylbenzene), for which the ratio of *ortho* and *para* products is sensitive to the conditions employed, such as the choice of solvent. In some cases, the *para* product is favored; in other cases, the *ortho* product is favored. Therefore, it is generally not wise to use the directing effects of a methyl group to favor a reaction at the *para* position over the *ortho* position.

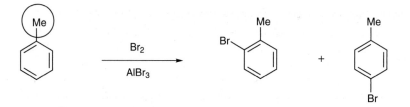

But this is only the case with a methyl group. With just about any other group, we should expect that the *para* product will be the major product. Keep that in mind because it is very important—*para* is usually the major product.

With that in mind, imagine that I asked you to propose an efficient synthesis for the following transformation:

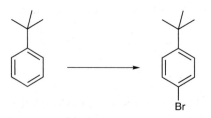

This is simple to do. The *tert*-butyl group is so large that we expect the *para* product to be the major product. So we just use Br_2 and $AlBr_3$, and we should obtain the desired product.

But suppose we wanted substitution to occur at the *ortho* position:

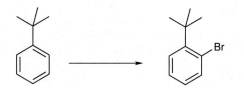

How would we do this? When confronted with this problem, students often suggest using the same reaction as before, with the understanding that the *ortho* product will be a minor product (so *some ortho* product will be formed). But you can't do that. Whenever you have a synthesis problem, you must choose reagents that give you the desired compound as the MAJOR product. If you propose a synthesis that would produce the desired product as a MINOR product, then your synthesis is not efficient. So we have a problem here. How do we run the reaction so that the *ortho* product will be the major product?

The answer is: we cannot do it in one step. There is no way to "turn off" steric effects. However, there is a way to exploit them. At the start of this chapter, we learned about sulfonation (using fuming sulfuric acid to install an SO_3H group on the ring). We saw that this group can be installed on the ring, **and** it can be removed from the ring very easily. We said that this feature (reversibility) would be VERY important in synthesis problems. Now we are ready to see why.

If we perform a sulfonation reaction first, we will expect the SO_3H group to go predominantly in the *para* position (the major product will be from *para* substitution):

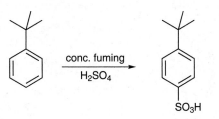

Now think about what we have done. We have "blocked" the *para* position. Now if we brominate, the incoming Br will be installed in the *ortho* position (because the *para* position is already taken). So, the reaction installs Br in the desired location:

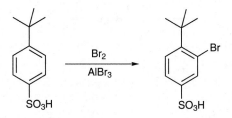

Finally, we can perform a desulfonation to remove the SO_3H group. To do so, remember that we need to use dilute sulfuric acid:

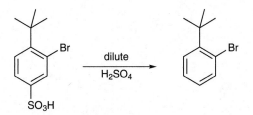

And this is the desired product. In summary, here is our entire synthesis:

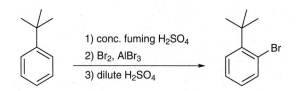

1) conc. fuming H$_2$SO$_4$
2) Br$_2$, AlBr$_3$
3) dilute H$_2$SO$_4$

Notice that it took three steps (where the first step was used to block the *para* position and the third step was used to unblock the *para* position). Three steps might seem inefficient, BUT we did not need to rely on isolating minor products. At each step of the way, we were using the major product to move on to the next step.

If you think about what we have done, you should realize that this trick is really very clever. We recognized that we cannot just "turn off" the steric effects. So, instead, we developed a strategy that *uses* the steric effects. Notice that the SO$_3$H group is not in our final product at all. It was just used temporarily, as a "blocking group." This type of concept is very important in organic chemistry. As you move through the course, you will see a few other examples of blocking groups (in reactions that have nothing to do with electrophilic aromatic substitution). The basic strategy is applicable elsewhere. By temporarily blocking the position where the reaction would primarily occur (and then unblocking after you perform the desired reaction), it is possible to form a product that would otherwise be the minor product.

Now let's get practice using this technique:

EXERCISE 4.75   Propose an efficient synthesis for the following transformation:

Answer   We see that we need to install an acyl group in the *ortho* position. If we just perform a Friedel–Crafts acylation, we would expect the *para* product to be the major product (because of steric effects). So, we must perform a sulfonation reaction to block the *para* position. Our answer is:

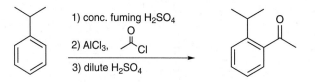

1) conc. fuming H$_2$SO$_4$
2) AlCl$_3$, (acetyl chloride)
3) dilute H$_2$SO$_4$

Propose an efficient synthesis for each of the following transformations. Use sulfonation only when necessary (I am purposefully giving you at least one problem that does not require sulfonation—to make sure that you understand *when* to use this blocking technique):

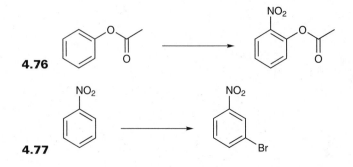

4.76

4.77

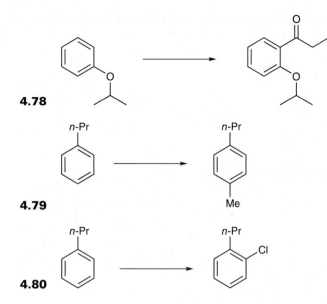

4.78

4.79

4.80

Before we move on to the final section of this chapter, you should be familiar with a few other steric effects. So far, we have seen the steric effects of ONE group on a ring. But what happens when we have two groups on a ring. For example, consider the directing effects of *meta*-xylene:

This compound has *two* methyl groups on the ring. Both methyl groups are directing to the same three positions:

Two of these positions are equivalent because of symmetry:

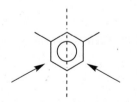

Attacking either of these two positions
would yield the same product

So, if we brominate this compound, we will expect to get only two products (rather than three):

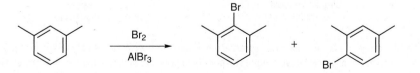

Minor   **Major**

Notice that we have indicated that one of the products is major. To understand why, we must consider steric effects. The position in between the two methyl groups is more sterically hindered than the other positions. Therefore, we primarily get just one product.

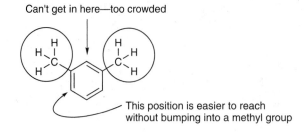

This type of argument can be used in a variety of similar situations. For example, you might remember that we saw the following reaction earlier in this chapter:

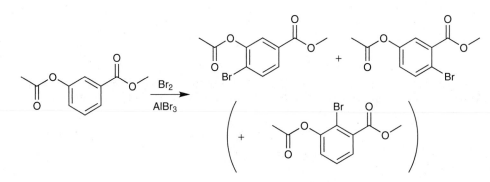

At the time, we said that one of the three products would only be a minor product (the one shown above in parentheses). Now we can understand that it is a minor product, because of steric considerations. The starting compound has two groups that are *meta* to each other, so the spot in between the two groups is sterically hindered.

But suppose you have a disubstituted benzene ring where the two groups are *para* to each other. For example, consider the directing effects for the following compound:

In this case, we have two groups that are *para* to each other. Here is a summary of the directing effects of each group:

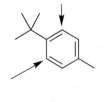
Directing effects
of the *t*-butyl group

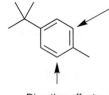
Directing effects
of the methyl group

So, these two groups are directing to all four potential spots. Both groups are weak activators (alkyl groups). So, when we consider electronic factors, we don't really see any preference among the possible spots. However, when we consider steric factors, we notice that the *tert*-butyl group is very large compared to the methyl group. As a result, we observe the following products:

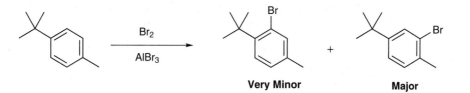

Very Minor   **Major**

In fact, the *tert*-butyl group is so large that you will find some textbooks that do not even show the minor product above at all. It is so minor that it is almost not worth mentioning.

In this section, we have seen many examples where steric effects play a significant role in determining the product distribution. Now let's get some practice using these principles.

EXERCISE 4.81   Predict the major product of the following reaction:

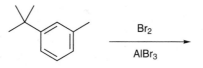

Answer   This example has two groups on the ring: a *tert*-butyl group and a methyl group. Both are weak activators (*ortho–para* directors), and both groups are directing to the same positions:

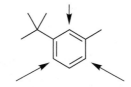

Of these three positions, the one in between the two groups is the most sterically hindered (reaction at that position would be extremely slow). Also, the position next to the *tert*-butyl group is fairly hindered, so we won't expect the reaction to take place there either. Thus, we expect the reaction to take place most often at the least hindered position, next to the methyl group:

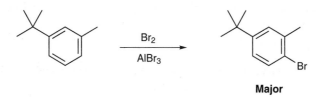

Major

Predict the MAJOR product of each of the following reactions (you do NOT need to show any minor products in these problems):

4.82

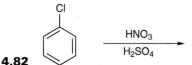

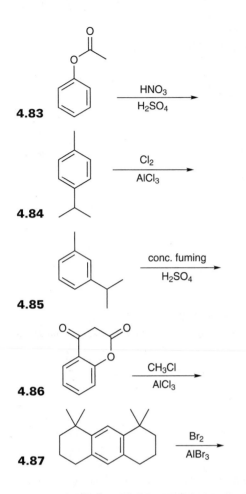

4.9 SYNTHESIS STRATEGIES

In this section, we will discuss some strategies for synthesis problems. Let's begin with a quick review of the reactions we have seen earlier in this chapter. We have seen how to install many different groups on a benzene ring:

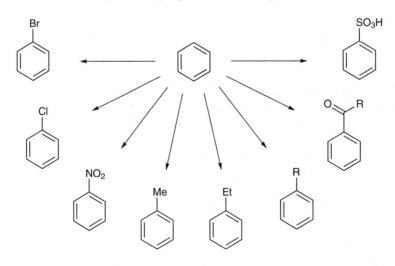

Carefully look at the chart above and make sure that you know the reagents that you would use to achieve each of these transformations. If you are not able to recall the reagents, then you will be totally unable to do synthesis problems.

It would be nice if all synthesis problems were just one-step problems, like this one:

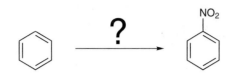

Usually, however, synthesis problems require a few steps, where you must install two or more groups on a ring, like this:

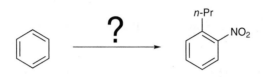

When dealing with such problems, there are many considerations to keep in mind:

- Take a close look at the groups on the ring and make sure you know how to install each group individually.
- Consider the order of events. In other words, which group do you install first? After you install the first group, the directing effects of that group will determine where the next group will be installed. This is an important consideration because it will affect the relative position of the two groups in the product. In the example above, the two groups are *ortho* to each other. So we must choose a strategy that installs the two groups *ortho* to each other.
- Take steric effects into account (and determine when you need to use sulfonation as a blocking technique).

There are certainly other considerations, but these will help you begin to master synthesis problems. The first consideration above is just a simple knowledge of the reagents necessary to install any group on a ring. The last two considerations can be summarized like this: electronics and sterics (hopefully, this will make it easy for you to remember these considerations). Whenever you are solving any problem, you must always consider electronic effects and steric effects. As you move through this course, you will find the same theme in every chapter. You will find that you must always consider electronic effects and steric effects.

Let's try to use these considerations to solve the problem we just saw:

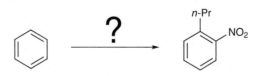

Let's begin by making sure we know how to install both of these groups individually. There are two groups that we need to install on the ring: an *n*-propyl group and a nitro group. The nitro group is easy—we just perform a nitration (using sulfuric acid and nitric acid). The *n*-propyl group is a bit trickier because we *cannot* use a Friedel–Crafts alkylation (remember carbocation rearrangements). Instead, we must use a Friedel–Crafts acylation, followed by a reduction to reduce the C=O double bond. So far, we need at least three steps: one step to install the nitro group and two steps to install the *n*-propyl group.

Now let's focus on electronic considerations. In this case, we can begin to appreciate the importance of "order of events." Imagine that we install the nitro group first. The nitro group is a *meta* director, so the next group will end up being installed *meta* to the nitro group. That doesn't work for us because we want the groups to be *ortho* to each other in the final product. So, we have decided that we cannot install the nitro group first. Instead, let's try to install the *n*-propyl group first. That should work because the *n*-propyl group is an *ortho–para* director. So the *n*-propyl group will direct the incoming nitro group into the correct position (*ortho*). BUT the *n*-propyl group will also direct to the *para* position. And this is where we must consider steric effects.

When we look at the steric effects, we encounter a difficulty. The steric effects are not in our favor here. We should expect the following results:

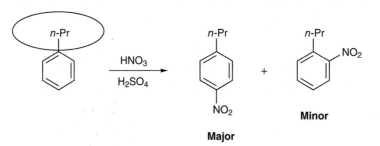

Notice that the compound we want to make is the *minor* product. But we need a way to obtain the *ortho* product as our major product. And we have seen exactly how to do that. We just use sulfonation to block the *para* position. So, our overall synthesis goes like this:

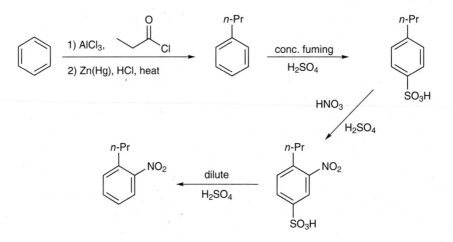

The answer that we just developed can be summarized like this:

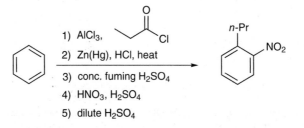

Now let's consider another example in which the order of events is especially relevant. Consider how you might achieve the following transformation:

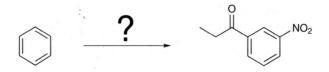

In this case, we must install an acyl group and a nitro group, and the two groups must be *meta* to each other. Both groups are *meta* directing, so it might seem like we can install them in either order (first acylation and then nitration, or vice versa). However, there is a serious limitation to Friedel–Crafts reactions that we have not mentioned until now. And that limitation will dictate the order of events that we must follow in this case. It turns out that a Friedel–Crafts reaction cannot be performed on a ring that is either moderately deactivated or strongly deactivated. You *can* perform a Friedel–Crafts on a weakly deactivated ring (and certainly on an activated ring). But not on a significantly deactivated ring—the reaction just doesn't work (you can perform other reactions with deactivated rings, such as bromination, but not Friedel–Crafts reactions). With that in mind, the nitro group cannot be installed first.

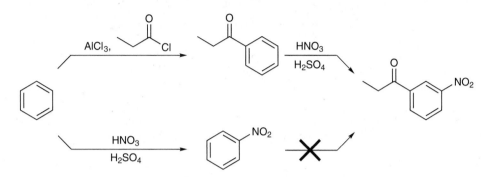

This synthesis teaches us the importance of "order of events." Whenever you are trying to solve a synthesis problem, you must always consider the order of events.

EXERCISE 4.88   Propose an efficient synthesis for the following transformation:

Answer   We must install two groups on the ring: an ethyl group and bromine. Let's first make sure that we know what reagents we would use to install each group individually. To install bromine, we would use Br$_2$ and a Lewis acid. To install the ethyl group, we would use a Friedel–Crafts alkylation (or acylation, followed by a reduction). Whenever we install an ethyl group on a ring, we don't need to worry about carbocation rearrangements, so we can use a simple alkylation (rather than an acylation followed by reduction).

But we immediately see a serious issue when we consider the directing effects of each substituent. The bromine is *ortho–para* directing, so we can't install the bromine on the ring first (if we did, we would not get the groups to be *meta* to each other). And the ethyl group is also *ortho–para* directing. So, whichever group we put on first, there would seem to be no way to get these two groups to be *meta* to each other.

UNLESS we use an acylation (rather than an alkylation). If we do that, we will install an acyl group on the ring first. *And acyl groups are meta directing.* That would allow us to install the bromine in the correct spot. So our strategy would go like this:

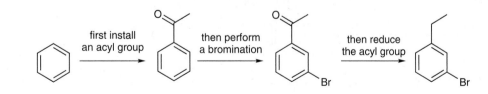

The reagents for our proposed synthesis are as follows:

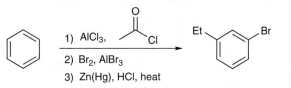

Propose an efficient synthesis for each of the following. You might want to use a separate piece of paper to help you work through each of these problems.

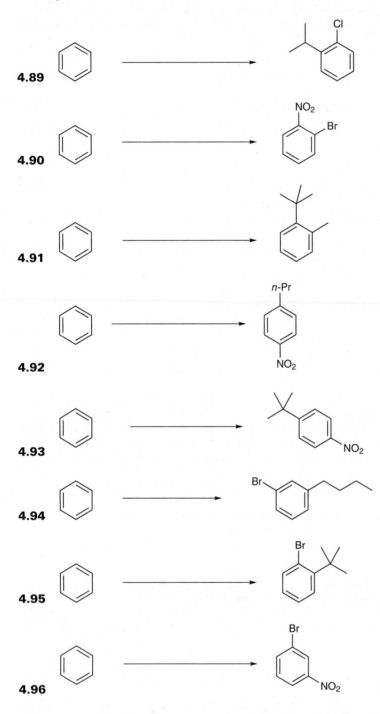

4.97

Before we end this chapter, it is important that you realize what we have covered here and, more importantly, what we have ***not*** covered. We did not cover everything in your textbook chapter on electrophilic aromatic substitution. As you go through your lecture notes and your textbook chapter, you will find a few reactions that we did not cover here. You will need to go through your textbook and your notes carefully to make sure that you learn those reactions. You should find that we covered 80%, or even 90%, of what you read in your textbook.

The purpose of this chapter was not to cover everything but rather to serve as a foundation for your mastery of electrophilic aromatic substitution. If you went through this chapter, then you should feel comfortable with the steps involved in proposing mechanisms, predicting products, and proposing a synthesis. You should know how directing effects work and how to use them when proposing syntheses. You should also know about steric effects and how to use them when proposing syntheses.

With all of that as a foundation, you should now be ready to go through your textbook and lecture notes, and polish off the rest of the material that you must know for your exam. Through the foundation that we have developed in this chapter, you should find (hopefully) that the content in your textbook will seem easy.

Do the problems in your textbook. Do all of them. Good luck.

END-OF-CHAPTER PROBLEMS

PRACTICE PROBLEMS *(Problems that involve only one skill)*

4.98 Arrange the following compounds in terms of reactivity (from least reactive to most reactive) toward electrophilic aromatic substitution.

4.99 Draw the two expected products of the following reaction.

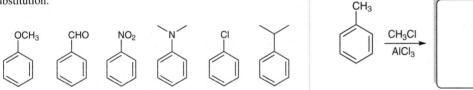

Problems 4.100–4.109 Predict the major product for each of the following reactions.

4.100

4.102

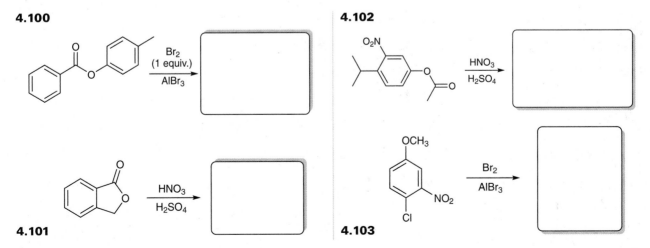

4.101

4.103

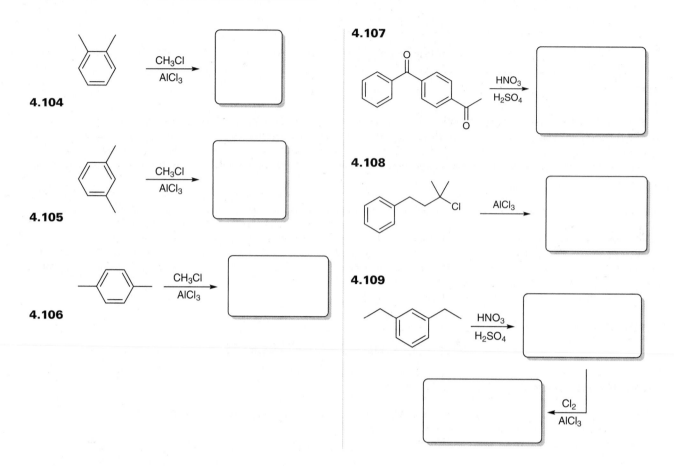

4.104

4.105

4.106

4.107

4.108

4.109

Problems 4.110–4.114 Propose an efficient synthesis for each of the following transformations.

4.110

4.111

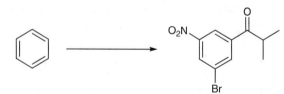

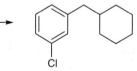

4.112

4.113

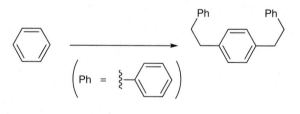

INTEGRATED PROBLEMS *(Problems that involve more than one skill)*

4.114 On a separate piece of paper, draw a plausible mechanism for the following reaction.

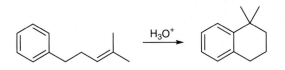

4.115 On a separate piece of paper, draw a plausible mechanism for the following reaction (you will need an entire sheet of paper!).

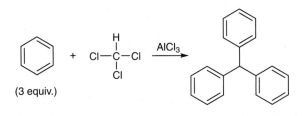

(3 equiv.)

4.116 On a separate piece of paper, draw a plausible mechanism for the following reaction.

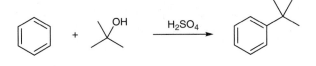

4.117 Predict the major product of the following reaction.

4.118 Predict the major product of the following reaction.

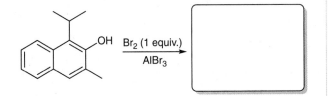

4.119 Propose an efficient synthesis for the following transformation.

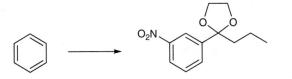

4.120 Propose an efficient synthesis for the following transformation.

4.121 Compounds **A** and **B** will each undergo an electrophilic aromatic substitution reaction upon treatment with molecular bromine in the presence of aluminum tribromide, as shown below. Compound **A** undergoes substitution primarily at the *para* position, while compound **B** undergoes substitution primarily at the *meta* position. Explain this difference in regiochemistry.

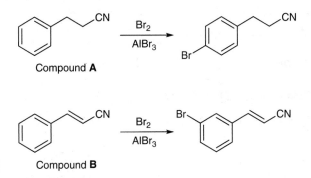

Compound **A**

Compound **B**

4.122 Trichloroisocyanuric acid (TCCA), shown below, can serve as a delivery agent of electrophilic chlorine, and can therefore be used as a chlorinating agent in electrophilic aromatic substitution reactions:

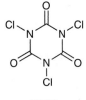

TCCA

(a) Draw the two products that are expected when anisole is treated with TCCA.

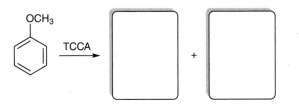

(b) Consider the ^{13}C NMR spectra of the products in part (a). Which spectrum has fewer signals? Explain.

4.123   Draw the structures for compounds **A–D** in the following reaction scheme.

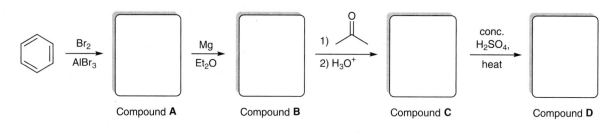

Compound **A**   Compound **B**   Compound **C**   Compound **D**

4.124   Draw the structures for compounds **A–D** in the following reaction scheme.

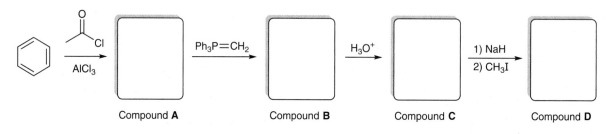

Compound **A**   Compound **B**   Compound **C**   Compound **D**

CHALLENGE PROBLEMS

4.125   Predict the major product of the following reaction, and explain the observed regiochemistry.

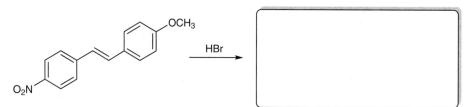

4.126   Compound **A** has the molecular formula C_9H_{12}, and compound **B** has the molecular formula $C_{12}H_{18}$. The ^1H NMR spectra of compounds **A** and **B** are shown here:

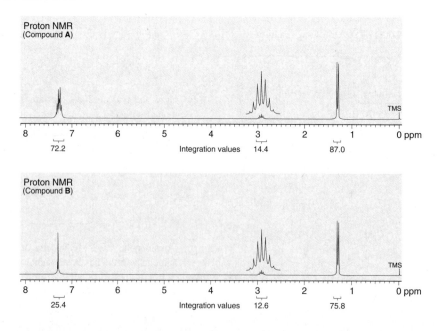

Draw the structures of compounds **A** and **B**, and identify a reagent that can be used (together with AlCl$_3$) to convert compound **A** into compound **B**.

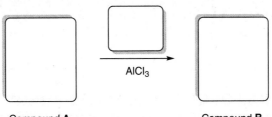

Compound **A**   Compound **B**

4.127 Draw a plausible mechanism for the following transformation:

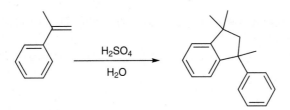

CHAPTER 5

NUCLEOPHILIC AROMATIC SUBSTITUTION

5.1   CRITERIA FOR NUCLEOPHILIC AROMATIC SUBSTITUTION

In the previous chapter, we learned all about *electrophilic* aromatic substitution reactions.

In this short chapter, we will look at the flipside: is it possible for an aromatic ring to function as an electrophile and react with a nucleophile? In other words, is it possible for the aromatic ring to be so electron-poor that it is subject to attack by a nucleophile? The answer is: yes.

But in order to observe this kind of reaction, called nucleophilic aromatic substitution, we will need to meet three very specific criteria. Let's look closely at each one of these criteria:

1. The ring must have a very powerful electron-withdrawing group. The most common example is the nitro group:

We saw in the previous chapter that the nitro group is a strong deactivator toward electrophilic aromatic substitution because the nitro group very powerfully withdraws electron density from the ring (by resonance). This causes the electron density in the ring to be very poor:

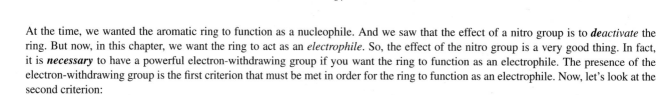

At the time, we wanted the aromatic ring to function as a nucleophile. And we saw that the effect of a nitro group is to ***deactivate*** the ring. But now, in this chapter, we want the ring to act as an *electrophile*. So, the effect of the nitro group is a very good thing. In fact, it is ***necessary*** to have a powerful electron-withdrawing group if you want the ring to function as an electrophile. The presence of the electron-withdrawing group is the first criterion that must be met in order for the ring to function as an electrophile. Now, let's look at the second criterion:

2. There must be a leaving group that can leave.

Reactive toward
nucleophilic aromatic substitution

NOT reactive toward
nucleophilic aromatic substitution

To understand this, let's think back to what happened in the previous chapter, when the ring always functioned as a nucleophile. We saw that all the reactions from the previous chapter could be summarized like this: E^+ comes on the ring, and then H^+ comes off (or, in other words: attack, then deprotonate). But now, in this chapter, we want the ring to function as an electrophile. So we are trying to see if we can get a nucleophile (Nuc^-) to attack the ring. If we can make it happen, and a nucleophile (with a negative charge) actually does attack the ring, then something with a negative charge is going to have to come off of the ring. We can summarize it like this: Nuc^- comes on the ring, and X^- comes off.

There is one main difference between the mechanism here and the one we saw in Chapter 4. The difference is in the kind of charges we are dealing with. In the previous chapter, we dealt with something positively charged coming onto the ring to form a positively charged sigma complex, and then H^+ came off the ring to restore aromaticity. In those mechanisms, everything was positively charged. But now, we are dealing with negative charges. A nucleophile with a negative charge will attack the ring to form some kind of negatively charged intermediate. That intermediate must then expel something negatively charged. And that explains the second criterion for this reaction to occur: we need the ring to have some leaving group that can leave with a negative charge.

If there is no leaving group that can leave with a negative charge, then the ring will have no way of reforming aromaticity. And we cannot just kick off H^- because H^- is a terrible leaving group. NEVER kick off H^-. Good leaving groups include halides (Cl^-, Br^-, I^-) and sulfonates (such as TsO^-). Although we have seen that fluoride (F^-) is generally not a good leaving group for S_N2 and S_N1 reactions, nevertheless, fluoride CAN function as a leaving group in nucleophilic aromatic substitution reactions. Indeed, aryl fluorides are even more reactive than aryl chlorides, aryl bromides, or aryl iodides in nucleophilic aromatic substitution reactions.

3. The final criterion is: the leaving group must be *ortho* or *para* to the electron-withdrawing group:

LG—⟨ring⟩—NO$_2$

Reactive toward
nucleophilic aromatic substitution

LG—⟨ring⟩—NO$_2$

NOT reactive toward
nucleophilic aromatic substitution

To understand why, we will need to take a closer look at the accepted mechanism. In the upcoming section, we will explore this mechanism so that we can understand this last criterion. For now, let's just make sure that we can identify when all three criteria have been met. Once again, the three criteria are:

1. There must be an electron-withdrawing group on the ring.

2. There must be a leaving group on the ring.

3. The leaving group must be *ortho* or *para* to the electron-withdrawing group.

Now let's get some practice looking for all three criteria:

EXERCISE 5.1   Predict whether the following compound can function as a suitable electrophile in a nucleophilic aromatic substitution reaction.

Answer   In order to have a nucleophilic aromatic substitution, all three criteria must be met.

We look at the ring, and we see that it does have a nitro group. Therefore, the first criterion has been met.

We then look for a leaving group. There is NO leaving group here. A methyl group is NOT a leaving group. Why not? Because a carbon atom with a negative charge is a *terrible* leaving group. So criterion 2 has not been met.

Therefore, we conclude that this compound will not function as an electrophile in a nucleophilic aromatic substitution reaction.

Determine whether each of the following compounds can function as a suitable electrophile in a nucleophilic aromatic substitution reaction. If you determine that the three criteria have not been met, then simply write "no reaction."

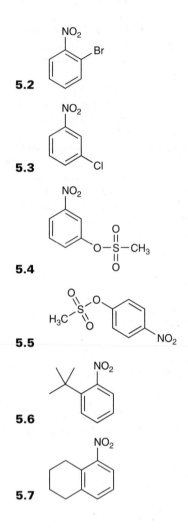

5.2

5.3

5.4

5.5

5.6

5.7

5.2 S_NAr MECHANISM

In the previous section, we saw the three criteria that are necessary in order for an aromatic ring to undergo a nucleophilic aromatic substitution reaction. The following transformation is an example:

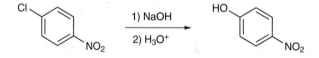

Let's explore some possible mechanisms for this process. It cannot be an S_N2 process because S_N2 processes do not readily occur at an sp^2-hybridized center:

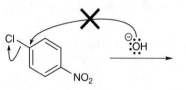

S$_N$2 processes are only effective with *sp*³-hybridized centers. So our reaction cannot be an S$_N$2 mechanism. What about S$_N$1? That would require the loss of the leaving group *first* to form a carbocation:

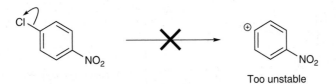

Too unstable

This kind of carbocation is not stabilized by resonance. It is a very high-energy intermediate, and we don't expect the leaving group to leave if it means creating an unstable intermediate. Therefore, we don't expect the mechanism to be an S$_N$1 mechanism either.

So, if it's not S$_N$2 and its not S$_N$1, then what is it? And the answer is: it's a new mechanism, called S$_N$Ar. In many textbooks, it is called an *addition–elimination* mechanism. In the first step of the mechanism, the ring is attacked by a nucleophile, generating a resonance-stabilized intermediate, called a Meisenheimer complex:

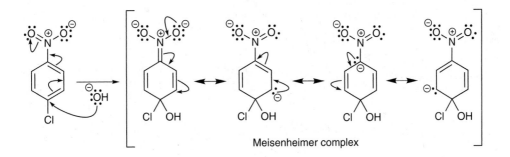

Meisenheimer complex

This intermediate should remind us of the intermediate in an electrophilic aromatic substitution reaction (the sigma complex), but the main difference is that a Meisenheimer complex is *negatively* charged (a sigma complex is positively charged). Let's take a close look at the Meisenheimer complex, and let's focus our attention on one particular resonance structure:

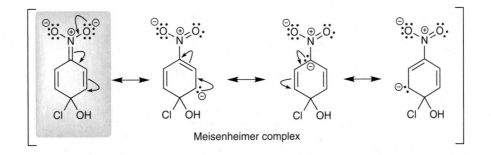

Meisenheimer complex

The highlighted resonance structure is special because it places the negative charge on an *oxygen* atom. Since the negative charge is spread out over three carbon atoms *and an oxygen atom*, the negative charge is fairly stabilized by resonance. You should think of the reaction like this: A nucleophile attacks the ring, kicking the negative charge up into a reservoir:

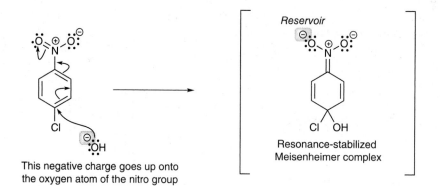

This negative charge goes up onto
the oxygen atom of the nitro group

Resonance-stabilized
Meisenheimer complex

Then, in the second step of the mechanism, the reservoir releases its load by pushing the electron density back down onto a leaving group, thereby restoring aromaticity to the ring:

Negative charge leaves reservoir
to be expelled with the leaving group

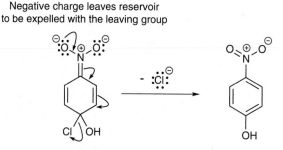

And now we are ready to understand the reason for the third criterion (that the leaving group must be *ortho* or *para* to the electron-withdrawing group). Now we can understand that the reservoir is available only if the nucleophile attacks at the *ortho* or *para* positions. If the nucleophile attacks at the *meta* position, there is no way to place the negative charge up onto the reservoir:

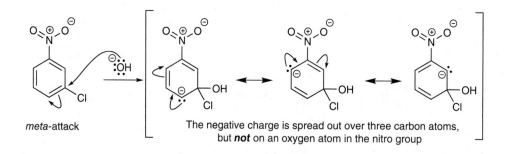

meta-attack

The negative charge is spread out over three carbon atoms,
but **not** on an oxygen atom in the nitro group

Therefore, the intermediate is not stabilized. So the reaction doesn't happen. And that is why the leaving group must be *ortho* or *para* to the electron-withdrawing group.

Before you get practice drawing the complete mechanism of an S$_N$Ar process, there is one subtle point that deserves attention. Let's first summarize what we have seen so far:

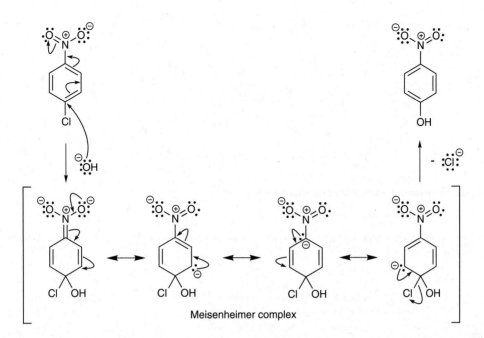

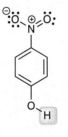

There are two steps: (1) a nucleophile attacks the ring to generate a Meisenheimer complex, followed by (2) loss of a leaving group to restore aromaticity. But inspect the product very carefully. It contains a phenolic proton, highlighted below:

This proton is mildly acidic and cannot survive the strongly basic conditions being employed (hydroxide is a strong base, and hydroxide is present in the reaction flask). So, under these reaction conditions, the product is deprotonated (whether we like it or not), to give the following:

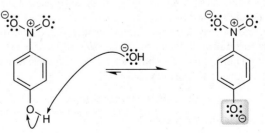

Therefore, in order to regenerate the desired product, a proton source must be introduced into the reaction flask (after the reaction is complete), which is shown like this:

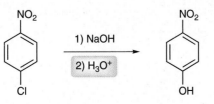

The acid (H_3O^+) is used to achieve the following proton transfer:

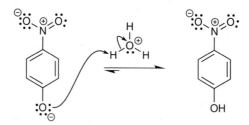

If you draw a complete mechanism for this process, it would look like this:

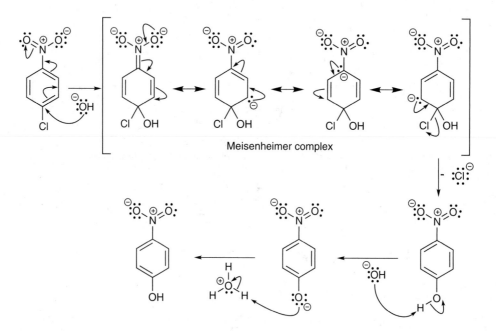

Meisenheimer complex

When you stare at this mechanism, you might be surprised by the last two steps (deprotonation, followed by protonation). Specifically, you might wonder why the mechanism needs to show these last two steps. After all, if we delete these last two steps from the mechanism, isn't the product still correct? Yes, that is true, but these last two steps are necessary. Why? Because under the reaction conditions employed (strongly basic conditions), the phenolic proton does not survive. Deprotonation occurs, whether we like it or not. The mechanism, as drawn, indicates that we understand that subtle point. By drawing the last two steps of the mechanism, you are demonstrating that you understand why a proton source (like H_3O^+) must be added to the reaction flask after the reaction is complete.

Now we are ready to get some practice drawing an S_NAr mechanism.

EXERCISE 5.8 Draw a mechanism for the following reaction.

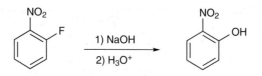

Answer The reagent is a strong nucleophile (hydroxide), and the starting compound meets all three criteria for an S$_N$Ar mechanism: an electron withdrawing group (NO$_2$) and a leaving group (fluoride) that are *ortho* to each other.

In a nucleophilic aromatic substitution reaction, the nucleophile (hydroxide) attacks at the position bearing the leaving group, and electrons are pushed up into the reservoir:

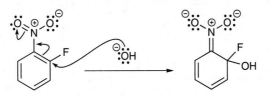

The resulting intermediate is a Meisenheimer complex, and it has the following resonance structures:

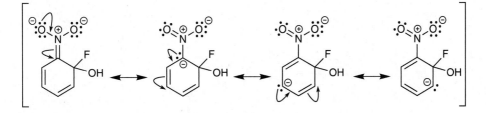

Then, in the second step of the mechanism, the leaving group is expelled to give the product. The first step and the second step of the mechanism are shown here:

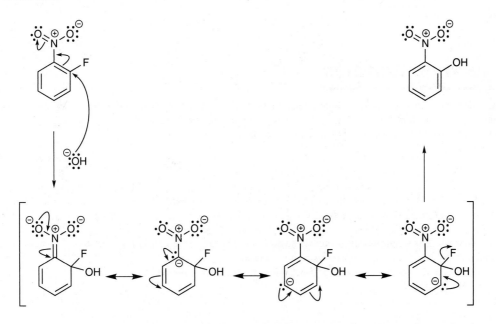

This might appear to be a complete mechanism. But remember, that under basic conditions, the product is deprotonated:

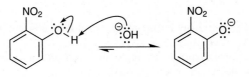

And that is why an acid source is necessary after the reaction is complete:

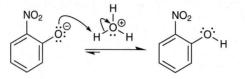

Propose a mechanism for each of the following transformations:

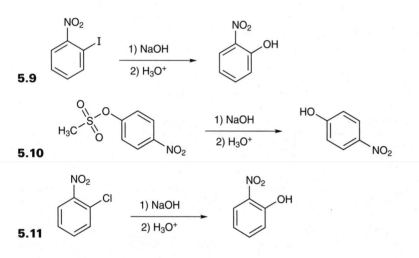

5.9

5.10

5.11

5.3 ELIMINATION–ADDITION

In the previous section, we discussed the three criteria that you need in order to get an S$_N$Ar mechanism. The obvious question is: can it occur without all three criteria? For example, what if there is no electron-withdrawing group?

If we treat chlorobenzene with hydroxide, no reaction is observed:

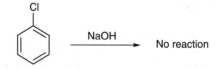

The hydroxide ion does not displace the leaving group via an S$_N$Ar mechanism because there is no "reservoir" to hold the electron density for a moment. In fact, if we try to apply heat, there is still no reaction.

However, at much higher temperatures, such as 350 °C, a reaction is in fact observed:

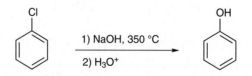

This reaction, called the Dow process, is commercially important because it is an efficient way of making phenol. Let's consider the mechanism for the Dow process. Notice that the starting material (chlorobenzene) lacks a nitro group, so there is no "reservoir" that can temporarily hold electron density (if a nucleophile attacks the ring), so the reaction cannot proceed via an S$_N$Ar mechanism. That leaves us with an important question: What is the mechanism for this process?

To understand the mechanism, chemists have used an important technique called isotopic labeling. All elements have isotopes (for example, deuterium is an isotope of hydrogen because deuterium has a neutron in the nucleus while hydrogen does not). Carbon also has some important isotopes. ^{13}C is an important isotope because we can easily determine the position of a ^{13}C atom in a compound using NMR spectroscopy. So if we enrich a specific spot with ^{13}C, then we can follow where that carbon atom goes during the reaction. For example, let's say we take chlorobenzene, and we enrich one particular site with ^{13}C:

The site with the asterisk is the location where we placed the ^{13}C. When we say that we enriched that spot with ^{13}C, we mean that most of the molecules in the flask have a ^{13}C atom in that position.

Now let's see what happens to that isotopic label as the reaction proceeds. After running the reaction, here are the results that are observed:

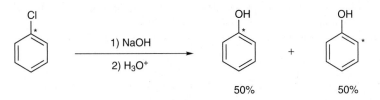

This seems quite strange: how does the isotopic label "move" its position? We will not be able to explain this with a simple nucleophilic aromatic substitution. Even if we could somehow ignore the issue of not having a reservoir for the electron density during the reaction, we would still not be able to explain the isotopic labeling results.

So here is a proposal that explains the isotopic labeling experiments. Imagine that in the first step, the hydroxide ion acts as a base (rather than a nucleophile), giving an elimination reaction:

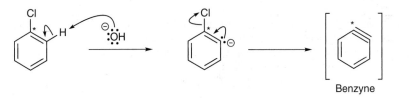

This generates a very strange-looking (and very reactive) intermediate, which is called benzyne. Then, another hydroxide ion is involved, this time acting as a nucleophile to attack benzyne. But there are two places it can attack:

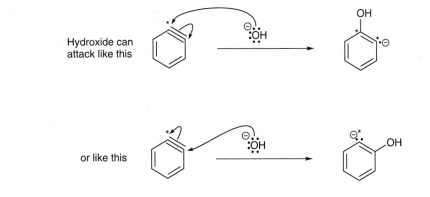

And there is no reason to prefer one site over the other, so we must assume that these two pathways occur with equal probability. That would give a 50–50 mixture of the two anions above, which would undergo proton transfers to generate phenol, with the isotopic labels in the appropriate locations:

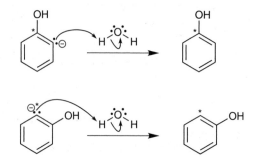

Based on the results of this isotopic labeling experiment, we have proposed a mechanism that is called *elimination–addition*, because it involves an elimination process (to form a benzyne intermediate) followed by an addition process that gives the product. Notice that the product (phenol) is formed under basic conditions (in the presence of hydroxide). Under these conditions, phenol is deprotonated to give a phenolate ion:

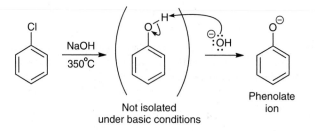

For this reason, we must use aqueous acid as a workup step when the reaction is complete. The function of the workup step is to protonate the phenolate ion and regenerate the product (phenol):

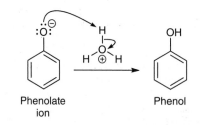

Note that we must use H_3O^+ for this workup step, rather than H_2O, because H_2O is not sufficiently acidic to protonate a phenolate ion.

Also note that the workup step is shown as a separate step, like this:

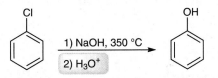

This overall process (elimination–addition) can occur with a variety of nucleophiles, not just hydroxide. For example, a similar reaction can be performed with an alkoxide ion as the nucleophile:

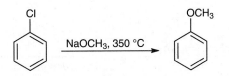

When an alkoxide ion is used as the nucleophile, the product is a phenyl ether. Unlike phenol, an ether lacks acidic protons, so the product is not deprotonated under the conditions of its formation (as we saw when making phenol). Therefore, we do not need a separate workup step to isolate the product.

Another example of an elimination–addition process is the use of an amide ion (H_2N^-) as the nucleophile. Indeed, this reaction can occur below room temperature (heat is not required):

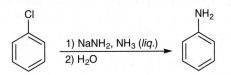

In this case, a workup step is required, just as we saw with the formation of phenol. Once again, this is because the product (aniline) is deprotonated under the conditions of its formation (H_2N^- will deprotonate aniline to give a resonance-stabilized anion). But unlike the Dow process, we cannot use H_3O^+ for the workup step, because H_3O^+ would protonate the amino group of the product to give an ammonium ion:

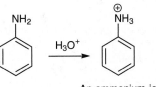

An ammonium ion

Instead, we use water for the workup step, rather than H_3O^+:

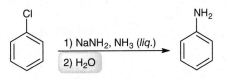

In summary, when we made phenol via the Dow process, we saw that water was not sufficiently acidic to regenerate the product (phenol), so we had to use H_3O^+. In contrast, when $NaNH_2$ is used as the nucleophile, and the product is aniline, we must use water for the workup step, rather than H_3O^+:

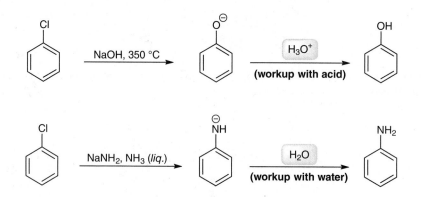

This proposed mechanism is essentially an elimination followed by an addition. So, it makes sense that we call this process an ***elimination–addition*** reaction (as compared with the S_NAr mechanism, which was called *addition–elimination*). This mechanism certainly seems a bit off the wall when you think about. Benzyne? It looks like a terrible intermediate. But chemists have been able to show (with other experiments) that benzyne is in fact the intermediate of this reaction. Your textbook or instructor will most likely provide some evidence for the short-lived existence of benzyne (we use a trapping technique involving a Diels–Alder reaction). If you are curious about the evidence, you can look in your textbook. For now, let's make sure that you can predict the products of elimination–addition reactions.

EXERCISE 5.12   Predict the products for the following reaction:

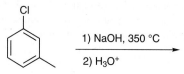

Answer   The ring does *not* have an electron-withdrawing group, so we are not dealing with an S_NAr mechanism. Rather, we must be dealing with an elimination–addition mechanism.

Elimination can occur on either side of the chlorine atom:

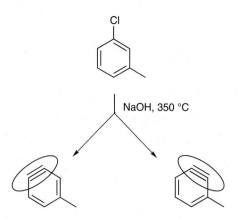

Then, we can add across either triple bond above, giving us the following products:

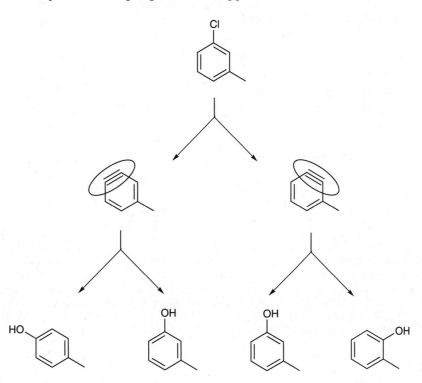

If you look closely at these products, you will see that the middle two are the same. So, we expect the following three products from this reaction:

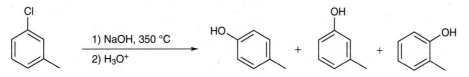

Predict the products for each of the following reactions. Just to keep you on your toes, I will throw in some problems that go through an addition–elimination mechanism (S_NAr), rather than an elimination–addition. In each case, you will have to decide which mechanism is responsible for the reaction (based on whether or not you have all three criteria for an S_NAr mechanism). Your products will be based on that decision.

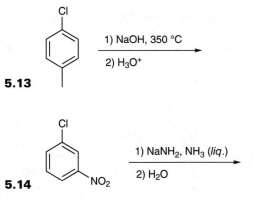

5.13

5.14

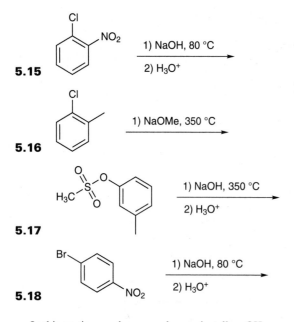

5.15 1) NaOH, 80 °C
2) H₃O⁺ →

5.16 1) NaOMe, 350 °C →

5.17 1) NaOH, 350 °C
2) H₃O⁺ →

5.18 1) NaOH, 80 °C
2) H₃O⁺ →

In this section, we have seen how to install an OH group or an NH₂ group on an aromatic ring. This is important because we did not see how to achieve either of these two transformations in the previous chapter. Here is a summary of how to install an OH group or NH₂ group on an aromatic ring. In each case, it is a two-step process that starts with installing a Cl on the ring:

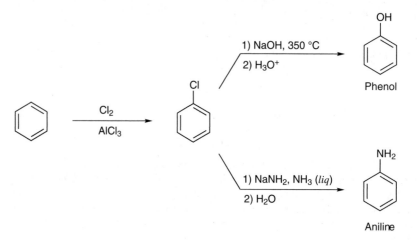

We begin with chlorination of benzene (which is just an electrophilic aromatic substitution reaction), followed by an elimination–addition reaction. When we perform the elimination–addition process, we must carefully choose the reagents. If we use NaOH followed by H₃O⁺, the product will be phenol. If we use NaNH₂ followed by H₂O, the product will be aniline (shown in the scheme above).

5.4 MECHANISM STRATEGIES

So far, we have seen three different mechanisms involving aromatic rings:

1. Electrophilic aromatic substitution
2. S_NAr (also called addition–elimination)
3. Elimination–addition

When you are given a problem, you must be able to look at all of the information and determine which of the three mechanisms is operating. This is not difficult to do. Here is a simple chart that shows the thought processes involved:

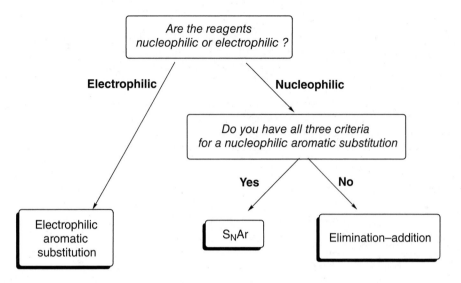

You first look at the reagents that are reacting with the aromatic ring. If the reagents are electrophilic (like all of the reagents we saw in the previous chapter), then expect an electrophilic aromatic substitution. But if the reagents are nucleophilic, then you have to decide between an $S_N Ar$ reaction and an elimination–addition reaction. To do that, look for the three criteria necessary for an $S_N Ar$ reaction.

Let's see an example:

EXERCISE 5.19   Propose a mechanism for the following reaction:

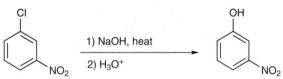

Answer   We begin by looking at the reagents. Hydroxide is a nucleophile, so we do NOT expect an electrophilic aromatic substitution. We must decide between an $S_N Ar$ mechanism and an elimination–addition mechanism. We look for the three criteria that we need for an $S_N Ar$ reaction. (1) We do have an electron-withdrawing group, and (2) we do have a leaving group, BUT (3) the electron-withdrawing group and the leaving group are NOT *ortho* or *para* to each other. That means that an $S_N Ar$ mechanism is unlikely to occur. (When the electron-withdrawing group and the leaving group are *meta* to each other, we don't have the "reservoir" to use.) Therefore, the mechanism must be an elimination–addition:

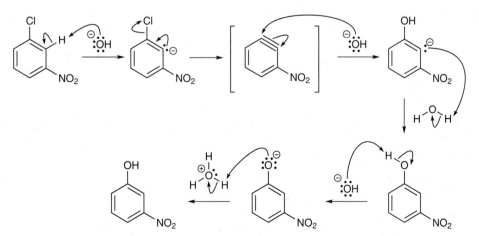

In this particular example, we would expect a total of three products from an elimination–addition mechanism:

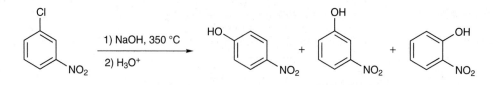

But keep in mind that mechanism problems do not always show you all of the products. The problem will typically show you just one product, and you will need to show the mechanism for forming that product (and only that product). In some cases, it might even be a minor product. But the problem is not making any claims that the product is major or minor. A mechanism problem is simply asking you to justify "how" the product was formed, regardless of how much of it was actually obtained from the reaction.

Propose a mechanism for each of the following reactions:

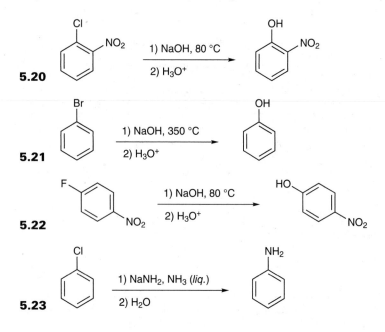

END-OF-CHAPTER PROBLEMS

PRACTICE PROBLEMS *(Problems that involve only one skill)*

Predict the major product for each of the following reactions.

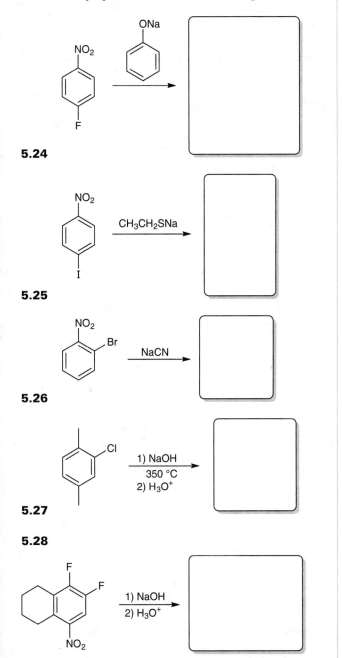

5.24

5.25

5.26

5.27

5.28

5.29 Predict the two major products of the following reaction.

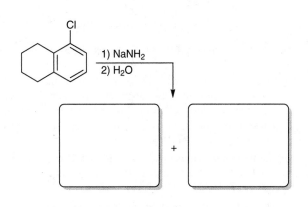

5.30 Draw two aromatic compounds that will react with each other in a nucleophilic aromatic substitution reaction to give the following product.

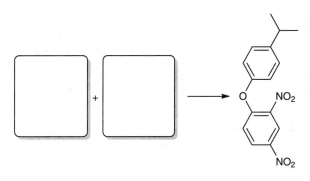

5.31 Which of the following compounds will NOT react with NaNH₂ in liquid ammonia? Explain.

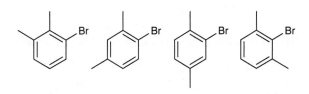

INTEGRATED PROBLEMS *(Problems that involve more than one skill)*

5.32   Consider the structures of compounds **A** and **B** below:

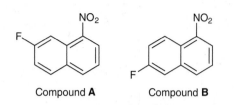

Compound **A**   Compound **B**

(a) One of these compounds undergoes a nucleophilic aromatic substitution reaction upon treatment with NaOH at room temperature, while the other compound is unreactive under the same conditions. Identify the unreactive compound and explain why it is unreactive.

(b) For the compound that will react with NaOH at room temperature, draw the product of this reaction (after workup with aqueous acid).

(c) On a separate piece of paper, draw a complete mechanism for this process.

5.33   On a separate piece of paper, draw a plausible mechanism for the following reaction:

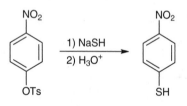

5.34   Propose an efficient synthesis for the following transformation:

5.35   Propose an efficient synthesis for the following transformation:

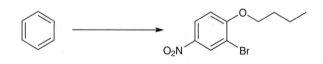

5.36   Propose an efficient synthesis for the following transformation:

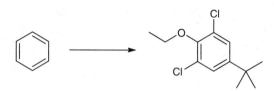

5.37   (a) Draw the two products that are expected when the compound below is heated with sodium hydroxide, followed by aqueous acidic workup:

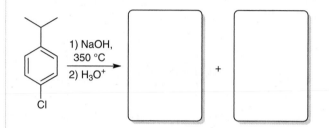

(b) How would you differentiate the two products using ^{13}C NMR spectroscopy?

(c) How would you differentiate the two products using ^{1}H NMR spectroscopy?

5.38   On a separate piece of paper, draw a plausible mechanism for the following reaction:

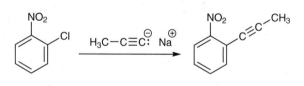

5.39   On a separate piece of paper, draw a plausible mechanism for the following reaction:

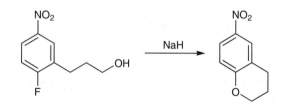

CHALLENGE PROBLEMS

5.40   On a separate piece of paper, draw a plausible mechanism for the following reaction.

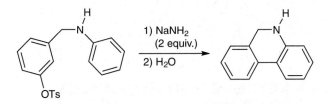

5.41   On a separate piece of paper, draw a plausible mechanism for the following reaction.

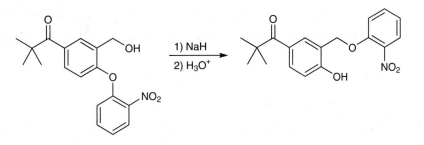

CHAPTER 6

KETONES AND ALDEHYDES

6.1 PREPARATION OF KETONES AND ALDEHYDES

Before we can explore the reactions of ketones and aldehydes, we must first make sure that we know how *to make* ketones and aldehydes. That information will be vital for solving synthesis problems.

Ketones and aldehydes can be made in many ways, as you will see in your textbook. In this book, we will only see a few of these methods. These few reactions should be sufficient to help you solve many synthesis problems in which a ketone or aldehyde must be prepared.

The most useful type of transformation is forming a C=O bond from an alcohol. *Primary* alcohols can be oxidized to form aldehydes:

And *secondary* alcohols can be oxidized to form ketones:

Tertiary alcohols cannot be oxidized, because carbon cannot form five bonds:

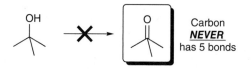

So, we need to be familiar with the reagents that will oxidize primary and secondary alcohols (to form aldehydes or ketones, respectively). Let's start with secondary alcohols.

A secondary alcohol can be converted into a ketone upon treatment with sodium dichromate and sulfuric acid:

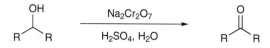

Alternatively, the Jones reagent can be used, which is formed from CrO_3 in aqueous acetone:

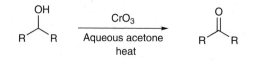

Whether you use sodium dichromate or the Jones reagent, you are essentially performing an oxidation that involves a chromic acid oxidizing agent (the alcohol is being oxidized and the chromium reagent is being reduced). You should look through your lecture notes and textbook to see if you are responsible for the mechanisms of these oxidation reactions. Whatever the case, you should definitely have these reagents at your fingertips, because you will encounter many synthesis problems that require the conversion of an alcohol into a ketone or aldehyde.

Chromic acid oxidations work well for secondary alcohols, but we run into a problem when we try it on a primary alcohol. The initial product is indeed an aldehyde:

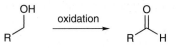

But under these strong oxidizing conditions, the aldehyde does not survive. The aldehyde is further oxidized to give a carboxylic acid:

So clearly, we need a way to oxidize a primary alcohol into an aldehyde, under conditions that will **not** further oxidize the aldehyde. This can be accomplished with a reagent called pyridinium chlorochromate (or PCC):

Pyridinium chlorochromate
(PCC)

This reagent provides milder oxidizing conditions, and therefore, the reaction stops at the aldehyde. That is, PCC will oxidize a primary alcohol to give an aldehyde:

PCC can also be used to convert a secondary alcohol into a ketone.

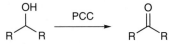

There is another common way to form a C=O bond (other than oxidation of an alcohol). You might remember the following process from last semester:

This process is called ozonolysis. It essentially takes every C=C bond in the compound, and breaks it apart into two C=O bonds:

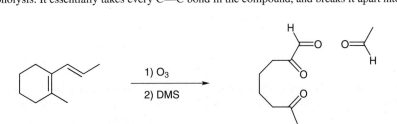

There are many reagents that can be used for the second step of this process (other than DMS). You should look in your lecture notes to see what reagents your instructor (or textbook) used for step 2 of an ozonolysis.

So far, this section has covered only a few ways to make a C=O bond. We saw that an aldehyde can be made by treating a primary alcohol with PCC, and we saw that a ketone can be made by treating a secondary alcohol with a strong oxidizing agent such as sodium dichromate, the Jones reagent, or even PCC. We also saw that ketones and aldehydes can be made via ozonolysis. Let's get some practice with these reactions.

EXERCISE 6.1   Predict the major product(s) of the following reaction:

Answer   The oxidizing agent in this case is PCC, and we have seen that PCC will convert a primary alcohol into an aldehyde:

PROBLEMS   Predict the major product for each of the following reactions:

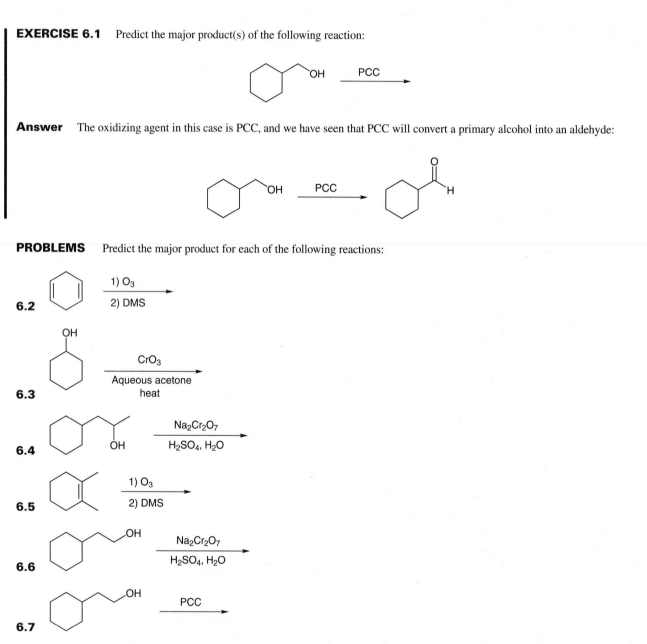

It is not enough to simply "recognize" the reagents when you see them (like we did in the previous problems). But you actually need to know the reagents well enough to write them down when they are not in front of you. Let's get some practice:

EXERCISE 6.8   Identify the reagents you would use to achieve the following transformation:

Answer   In this case, a secondary alcohol must be converted into a ketone. This can be accomplished with a number of reagents, such as sodium dichromate, or the Jones reagent, or even with PCC:

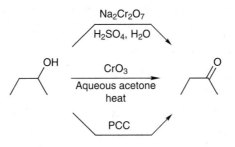

PROBLEMS   Identify reagents that you could use to achieve each of the following transformations. Try not to look back at the previous problems while you are working on these problems.

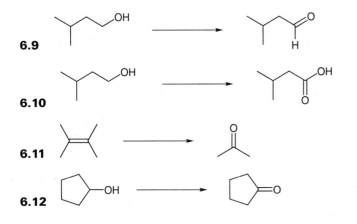

6.9

6.10

6.11

6.12

6.2   STABILITY AND REACTIVITY OF C=O BONDS

Ketones and aldehydes are very similar to each other in structure:

Therefore, they are also very similar to each other in terms of reactivity. Most of the reactions that we see in this chapter will work for both ketones and aldehydes. So, it makes sense to learn about ketones and aldehydes in the same breath.

But before we can get started, we need to know some basics about C=O bonds. Let's start with a bit of terminology that we will use throughout the entire chapter. Instead of constantly using the expression "C=O double bond," we will call it a *carbonyl group*. This term is NOT used for nomenclature. You will never see the term "carbonyl" appearing in the IUPAC name of a compound. Rather, it is just a term that we use when we are talking about mechanisms, so that we can quickly refer to the C=O bond without having to say "C=O double bond" all of the time.

Don't confuse the term "carbonyl" with the term "acyl." The term "acyl" is used to refer to a carbonyl group *together with* one alkyl group:

We will use the term "acyl" in the next chapter. But in this chapter, we will focus on the carbonyl group.

If we want to know how a carbonyl group will react, we must first consider electronic effects (the locations of δ+ and δ−). There are always two factors to explore: induction and resonance. If we start with induction, we notice that oxygen is more electronegative than carbon, and therefore, the oxygen atom will withdraw electron density:

As a result, the carbon atom of the carbonyl group is δ+ and the oxygen atom is δ−.

Next, we look at resonance:

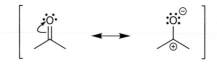

And we see, once again, that the carbon atom is δ+ and the oxygen atom is δ−, this time because of resonance. So, both induction and resonance paint the same picture:

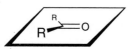

This means that the carbon atom is very electrophilic, and the oxygen atom is electron-rich. We will focus all of our attention in this chapter on the carbon atom of a carbonyl group. We will see *how* the carbon atom functions as an electrophile, *when* it functions as an electrophile, and *what happens after* it functions as an electrophile.

The geometry of a carbonyl group facilitates the carbon atom functioning as an electrophile. We saw in the first semester of organic chemistry that sp^2-hybridized carbon atoms have trigonal planar geometry:

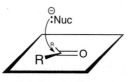

This makes it easy for a nucleophile to attack the carbonyl group, because there is little steric hindrance that would block the incoming nucleophile:

In this chapter, we will see many different kinds of nucleophiles that can attack a carbonyl group. In fact, this entire chapter will be organized based on the kinds of nucleophiles that can attack. We will start with hydrogen nucleophiles and continue with oxygen nucleophiles, sulfur nucleophiles, nitrogen nucleophiles, and, finally, carbon nucleophiles. This approach (dividing the chapter based on the kinds of nucleophiles) might be somewhat different than your textbook. But hopefully, the order that we use here will help you appreciate the similarity between the reactions.

There is one more feature of carbonyl groups that must be mentioned before we can get started. Carbonyl groups are thermodynamically very stable. In other words, forming a carbonyl group is generally a process that is downhill in energy. As a result, the formation of a carbonyl group is often the driving force for a reaction. We will use that argument many times in this chapter, so make sure you are prepared for it. The mechanisms in this chapter will be explained in terms of the stability of carbonyl groups.

Now let's just quickly review the important characteristics that we have seen so far. The carbon atom (of a carbonyl group) is electrophilic, and it is readily attacked by a nucleophile (and there are MANY different kinds of nucleophiles that can attack it). We have also seen that a carbonyl group is very stable. So, the formation of a carbonyl group can serve as a driving force.

These principles will guide us throughout the rest of the chapter, and they can be summarized like this:

- A carbonyl group can be attacked by a nucleophile, and

- After a carbonyl group is attacked, it will try to re-form, if possible.

6.3 H-NUCLEOPHILES

We will now explore the various nucleophiles that can attack ketones and aldehydes. We will divide all nucleophiles into categories, and in this section, we will focus on hydrogen nucleophiles. I call them "hydrogen" nucleophiles, because they are a source of a negatively charged hydrogen atom (which we call a "hydride" ion) that can attack a ketone or aldehyde. The simplest way to get a hydride ion is from sodium hydride (NaH). This compound is ionic, so it is composed of Na^+ and H^- ions (very much the way NaCl is composed of Na^+ and Cl^- ions). So, NaH is certainly a good source of hydride ions.

However, you will not see any reactions where we use NaH as a source of hydride *nucleophiles*. As it turns out, NaH is a very strong base, but it is not a strong nucleophile. This is an excellent example of how basicity and nucleophilicity do NOT completely parallel each other. The reason for this goes back to something from the first semester of organic chemistry. Try to remember back to the difference between basicity and nucleophilicity. Let's review it quickly.

The strength of a base is determined by the *stability* of the negative charge. An unstable negative charge corresponds with a strong base, while a stabilized negative charge corresponds with a weak base. But nucleophilicity is NOT based on stability. Nucleophilicity is based on *polarizability*. Polarizability describes the ability of an atom or molecule to distribute its electron density unevenly in response to external influences. Larger atoms are more polarizable, and are therefore strong nucleophiles; while smaller atoms are less polarizable, and are therefore weak nucleophiles.

With that in mind, we can understand why H^- is a strong base, but not a strong nucleophile. It is a strong base, because hydrogen does not stabilize the charge well. But when we consider the nucleophilicity of H^-, we realize that hydrogen is the smallest atom, and therefore, the least polarizable. As a result, H^- is generally not observed to function as a nucleophile. So, how do we form a hydrogen nucleophile?

Although H^- itself cannot be used as a nucleophile, there are many reagents that can serve as a "delivery agent" of H^-. For example, consider the structure of sodium borohydride ($NaBH_4$):

If we look at the periodic table, we see that boron is in Column 3A, and therefore, it has three valence electrons. Accordingly, it can form three bonds. But in sodium borohydride (above), the central boron atom has *four* bonds. So it must be using one extra electron, and therefore, it has a negative formal charge (for our purposes, we can ignore the sodium ion, Na^+, and treat it as if it was just a counter ion). This reagent can serve as a delivery agent of H^-, as seen in the following example:

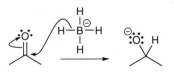

Notice that H^- never really exists by itself in this reaction. Rather, H^- is "delivered" from one place to another. That is a good thing, because H^- by itself would not serve as a nucleophile (as we saw earlier). But sodium borohydride can serve as a source of a hydrogen nucleophile, because the central boron atom is somewhat polarizable. The polarizability of the boron atom allows the entire compound to serve as a nucleophile, and *deliver* a hydride ion to attack the ketone. Now, it is true that boron is not so large, and therefore, it is not very polarizable. As a result, $NaBH_4$ is a somewhat tame nucleophile. In fact, we will soon see that $NaBH_4$ is selective in its reactivity. It will not react with all carbonyl groups (for example, it will not react with an ester). But it will react with ketones **and** with aldehydes (and that is our focus in this chapter).

There is another common reagent that is very similar to sodium borohydride, but it is much more reactive. This reagent is called lithium aluminum hydride ($LiAlH_4$):

This reagent is very similar to $NaBH_4$ because aluminum is also in Column 3A of the periodic table (directly beneath boron). So, it also has three valence electrons. In the structure above, the aluminum atom has four bonds, which is why it has a negative charge. Just as we saw with $NaBH_4$, $LiAlH_4$ is also a source of nucleophilic H^-. But compare these two reagents to each other—aluminum is larger than boron. That means that it is more polarizable, and therefore, $LiAlH_4$ is a much better nucleophile than $NaBH_4$. $LiAlH_4$ will react with almost any carbonyl group (not just ketones and aldehydes).

It will soon become very important that $LiAlH_4$ is more reactive than $NaBH_4$. But for now, we are talking about nucleophilic attack of ketones and aldehydes; and both $NaBH_4$ and $LiAlH_4$ will react with ketones and aldehydes.

In addition to $NaBH_4$ and $LiAlH_4$, there are other sources of hydrogen nucleophiles as well, but these two are the most common reagents. You should look through your textbook and lecture notes to see if you are responsible for being familiar with any other hydrogen nucleophiles.

Now let's take a close look at what can happen after a hydrogen nucleophile attacks a carbonyl group. As we have seen, the reagent (either $NaBH_4$ or $LiAlH_4$) can deliver a hydride ion to the carbonyl group, like this:

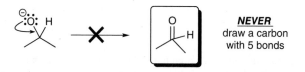

In the beginning of this chapter, we covered two important rules that govern the behavior of a carbonyl group:

- it is easily attacked by nucleophiles (as we just saw in the step above), and
- after a carbonyl group is attacked, it will try to re-form, if possible. Now we need to understand what we mean when we say: "if possible."

In trying to re-form the carbonyl group, we realize that the central carbon atom cannot form a fifth bond:

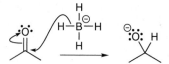

That would be impossible, because carbon only has four orbitals to use. So, in order for the carbonyl group to re-form, a leaving group must be expelled, like this:

When considering which groups can function as leaving groups, avoid expelling H^- or C^- (there are a few exceptions to this rule, which we will see later, but unless you recognize that you are dealing with one of the rare exceptions, do NOT expel H^- or C^-). For example, never do this:

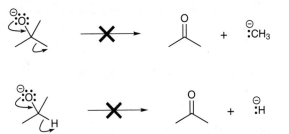

And never do this:

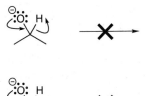

We have just learned a simple general rule. Now let's try to apply this rule to determine the outcome that is expected when a ketone or aldehyde is treated with a hydrogen nucleophile. Once again, the first step was for the hydrogen nucleophile to attack the carbonyl group:

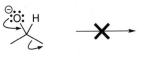

Now let's consider what can possibly happen next. In order for the carbonyl group to re-form, a leaving group must be expelled. But there are no leaving groups in this case. The carbonyl cannot re-form by expelling C^-:

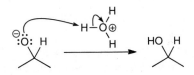

And it cannot re-form by expelling H^-:

And it cannot re-form by expelling C^-:

So we are stuck. Once a hydrogen nucleophile delivers H^- to the carbonyl group, then it will not be possible for the carbonyl group to re-form. So the reaction is complete, and it just waits for us to introduce a source of protons to "work-up" the reaction (to protonate the alkoxide ion). To achieve this protonation, we can introduce either H_2O or H_3O^+ as the source of protons:

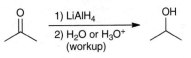

Regardless of the identity of the proton source that we add to the reaction flask after the reaction is complete, the product of this reaction will be an alcohol.

Whenever you are using this transformation in a synthesis, you must clearly show that the proton source is added AFTER the reaction has occurred:

In other words, it is important to show that LiAlH$_4$ and water are **two separate steps**. Do **not** show it like this:

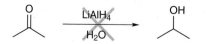

This would mean that LiAlH$_4$ and H$_2$O are present at the same time, and that is not possible. LiAlH$_4$ would react violently with water to form H$_2$ gas (because H$^+$ and H$^-$ would react with each other).

As it turns out, NaBH$_4$ is a milder source of hydride, and therefore, NaBH$_4$ can actually be present at the same time as the proton source:

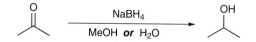

Common proton sources include MeOH and water (sometimes you might see EtOH). Notice that we didn't show it as two separate steps. When you are dealing with LiAlH$_4$, you must show two steps (one step for LiAlH$_4$ and another step for the proton source); but when you are dealing with NaBH$_4$, you should show the proton source in the same step as NaBH$_4$.

LiAlH$_4$ and NaBH$_4$ are very useful reagents. They allow us to *reduce* a ketone or aldehyde, which is important when you realize that we already learned the reverse process:

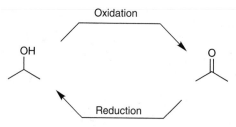

These two transformations will be *tremendously* helpful when you are trying to solve synthesis problems later on. You would be surprised just how many synthesis problems involve the conversion between alcohols and ketones. You need to have these two transformations at your fingertips.

EXERCISE 6.13   Predict the major product of the following reaction:

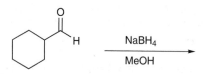

Answer   The starting compound is an aldehyde, and it is being treated with sodium borohydride. This hydrogen nucleophile will *deliver* H$^-$ to the aldehyde, and the carbonyl group will not be able to re-form, because there is no leaving group. In this case, methanol serves as the proton source, and an alcohol will be obtained:

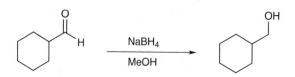

PROBLEMS   Predict the major product for each of the following reactions:

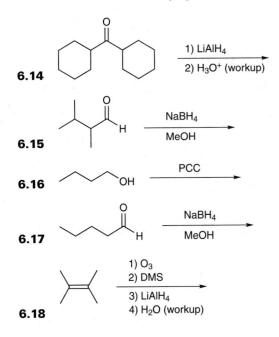

6.14

6.15

6.16

6.17

6.18

EXERCISE 6.19   Draw a mechanism for the following transformation:

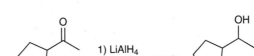

Answer   First, LiAlH₄ delivers a hydride ion to the ketone. Then, the carbonyl group is not able to re-form, so the intermediate waits for a proton from water, during the workup step:

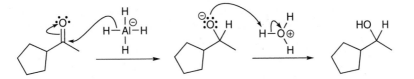

PROBLEMS   Propose a mechanism for each of the following transformations. The following problems will probably seem too easy—but just do them anyway. These basic arrows need to become *routine* for you, because we will step up the complexity in the next section, and you will want to have these basic skills down cold:

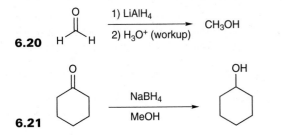

6.20

6.21

6.22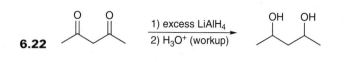

6.4 O-NUCLEOPHILES

In this section, we will focus our attention on oxygen nucleophiles. Let's begin by exploring what happens when an alcohol functions as a nucleophile and attacks a ketone or aldehyde.

Be warned: the mechanism we are about to see is one of the longer mechanisms that you will encounter in this course. But it is incredibly important because it lays the foundation for so many other mechanisms. If you can master this mechanism, then you will be in really good shape to move on. And to be honest, there is no other option; you MUST master this mechanism. So, be prepared to read through the next several pages slowly, and then be prepared to practice the material on these pages as many times as necessary until you know this mechanism extremely well.

Alcohols are nucleophilic because the oxygen atom has lone pairs that can attack an electrophile:

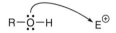

When an alcohol attacks a carbonyl group, an intermediate is generated that should remind us of the intermediate that was formed in the previous section:

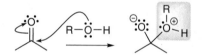

Notice how similar this is to the hydride attack we explored in the previous section:

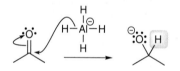

But there is one major difference here. When we saw the attack of a hydrogen nucleophile in the previous section, we argued that the carbonyl group could not re-form after the attack because there was no leaving group. But here, in this section (with an alcohol functioning as a nucleophile), there is a leaving group. So, it *is* possible for the carbonyl group to re-form:

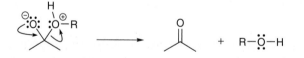

The attacking nucleophile (ROH) can function as the leaving group. But, of course, that gets us right back to where we started. As soon as a molecule of alcohol attacks the carbonyl group, it just gets expelled immediately, and there is no net reaction.

So, let's explore other possible avenues, to see if there is a reaction that can occur. First of all, we should realize that the attack of an alcohol is much slower than the attack of a hydrogen nucleophile, because alcohols do not have a negative charge and are not strong nucleophiles. So, if we want to speed up this reaction, we would want to make the nucleophile more nucleophilic (for example, using RO⁻ instead of ROH):

Theoretically, this would speed up the reaction, but under these conditions, we would have the same problem that we just had a moment ago. We cannot prevent the carbonyl group from re-forming. The initial intermediate will just eject the nucleophile, and we would get right back to where we started:

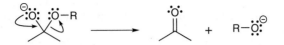

So, we will take a slightly different approach. Rather than making the nucleophile more nucleophilic, we will focus on making the electrophile more electrophilic. So, let's focus on the electrophile of our reaction:

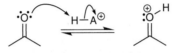

How do we make a carbonyl group even more electrophilic? By introducing a small quantity of catalytic acid into the reaction flask:

The resulting protonated ketone is significantly more electrophilic (this entity bears a full positive charge, rendering the carbonyl group more electron-poor). This is VERY IMPORTANT, because we will see this many times throughout this chapter. Many acids can be used for this purpose, including H_2SO_4. When drawing a mechanism for the protonation of a ketone in the presence of an acid catalyst, we should recognize that the identity of the acid ($H—A^+$) is most likely a protonated alcohol, which received its extra proton from H_2SO_4.

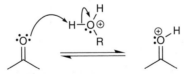

So protonation of the ketone most likely occurs in the following way:

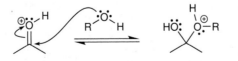

As we continue to discuss this mechanism, we will just show $H—A^+$ as the proton source, and it is expected that you will understand that the identity of $H—A^+$ is likely a protonated alcohol.

Now that the ketone has been protonated, rendering it more electrophilic, let's consider what happens if an alcohol molecule functions as a nucleophile and attacks the protonated ketone:

This gives an intermediate that has a tetrahedral geometry (the starting ketone was sp^2 hybridized, and therefore trigonal planar; but this intermediate is now sp^3 hybridized, and therefore tetrahedral). So, we will refer to this intermediate as a "tetrahedral intermediate."

Doesn't this tetrahedral intermediate give us the same problem? Doesn't it simply expel a leaving group to re-form the protonated ketone?

Yes, this *can* happen. In fact, it ***does*** happen—most of the time. That is in fact why we are using equilibrium arrows, highlighted below:

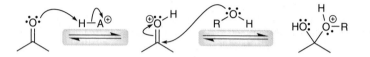

So it is true that there is an equilibrium between the forward and reverse processes. But every now and then, there is something else that can happen. There is a different way in which the carbonyl group can be re-formed:

In other words, we are exploring whether HO$^-$ can be expelled as a leaving group, which should theoretically work because we said before that anything can be expelled except for H$^-$ and C$^-$. Nonetheless, *we cannot expel HO$^-$ in acidic conditions*. Rather, it will have to be protonated first, which converts it into a better leaving group (this is a BIG DEAL—make sure that this rule becomes part of the way you think—NEVER expel HO$^-$ into acidic conditions—always protonate it first). So we draw the following proton transfer steps:

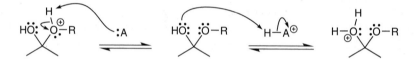

Notice that we first deprotonated to form an intermediate with no charge, and only then protonated. We specifically chose this order (first deprotonate, then protonate) to avoid having an intermediate with two positive charges. This is another important rule that you should make part of the way you think from now on. Avoid intermediates with two similar charges. Now, there are always some clever students who try to combine the two steps above into one step, by transferring a proton intramolecularly, like this:

While this might make sense on paper, it actually doesn't occur that way because the oxygen atom and the proton are simply too far apart in space to interact. An intramolecular proton transfer would require a transition state that resembles a 4-membered ring, which is high in energy and unlikely to occur. So, when drawing a mechanism, you must first remove a proton, and only then, do you protonate (and it is probably not going to be the same exact proton that was removed).

The result of our two separate proton transfer steps is the following intermediate:

And now we are ready to expel the leaving group (which is now H$_2$O, rather than HO$^-$) to re-form a carbonyl group, like this:

This new intermediate now ***does*** have a carbonyl group, ***but*** there is no easy way to remove the charge. You can't just lose R$^+$ the way you can lose a proton:

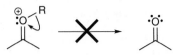

But there is another way for the charge to be removed. This intermediate can be attacked by *another* molecule of alcohol, just like the protonated ketone was attacked at the beginning of the mechanism:

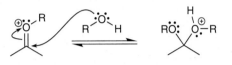

And finally, removal of a proton gives our product:

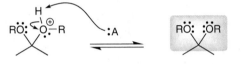

The overall transformation can be summarized as follows:

To make sure that we understand some of the key features of this mechanism, let's take a close look at the whole thing all at once. There are seven steps:

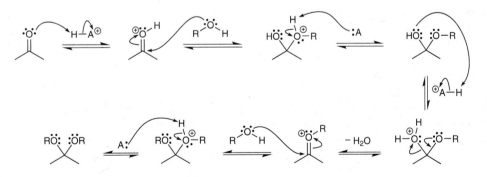

First let's focus our attention on all of the proton transfers in the entire mechanism. Four of the steps above are proton transfer steps. Two of them involve protonation and two involve deprotonation. So, in the end, the acid is not consumed by the reaction. It is a *catalyst* here. From now on, we will place brackets around the acid to indicate that its function is catalytic:

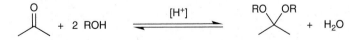

It is interesting to realize that *most* of the steps in the mechanism above are just proton transfer steps. There are only three steps other than proton transfers, and they are: nucleophilic attack (with ROH as the nucleophile), loss of a leaving group (H$_2$O), and another nucleophilic attack (again, with ROH as the nucleophile). All of the proton transfers are simply used to facilitate these three steps (proton transfers make the carbonyl group more electrophilic, to produce water as a leaving group instead of hydroxide, and to avoid multiple charges). It is important that you see the reaction in this way. It will greatly simplify the whole mechanism in your mind.

The drawing below is NOT a mechanism—the arrows in this drawing are only being used to help you review all three critical steps at once:

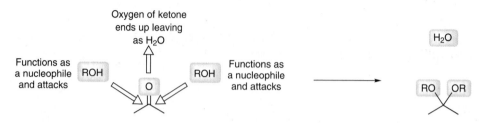

The product of this reaction is called an *acetal*. When we form an acetal from a ketone, there is one intermediate that gets a special name, because it is the only intermediate that does not have a charge. It is called a *hemiacetal*, and you can think of it as "half-way" toward making an acetal:

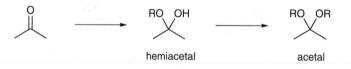

We give it a special name because it is theoretically possible to isolate it and store it in a bottle (although in most cases, this is very difficult to actually do), and because this type of intermediate will be important if/when you learn biochemistry.

Notice that an acetal does not have a carbonyl group. This means that the equilibrium will lean toward the starting materials, rather than the products:

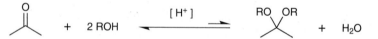

In other words, if we try to perform this reaction in a lab, we will obtain very little (if any) product. So, the question is: how can we force the reaction to form the acetal? There is a clever trick for doing this, and it involves removing water from the reaction as the reaction proceeds. If we remove water as it is formed, we will essentially stop the reverse path at a particular step (highlighted in the following mechanism). It is like putting up a brick wall that prevents the reverse reaction from occurring:

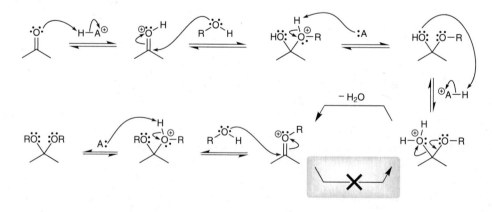

By removing water as it is being formed, we force the reaction to a certain point. Now let's focus on all of the steps (highlighted below) that come after the water-removal step:

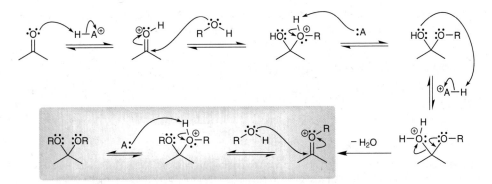

In the highlighted area, we see three structures in equilibrium with each other. Two of them are positively charged, and one of them (the product) is uncharged. This equilibrium now favors formation of the uncharged product.

In summary, formation of the acetal can be favored by removing the water as it is formed. Re-forming the carbonyl group (the reverse reaction) would require water, but there is no water present because it has been removed. This very clever trick allows us to force the equilibrium to favor the products even though they are less stable than the reactants.

In your textbook and in your lectures, you will probably explore the way that chemists remove water from the reaction as it proceeds. It is called azeotropic distillation, and there is a special piece of glassware that is used (called a Dean–Stark trap). I will not go into the details of azeotropic distillation here, but I wanted to just briefly mention it, because you should know how to indicate the removal of water. There are at least two common ways to show it:

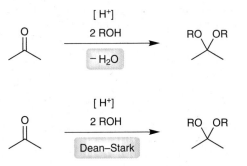

or like this:

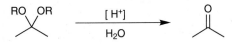

By just writing the words "Dean–Stark," you are indicating that you understand that it is necessary to remove water in order to form the acetal.

Now we can also appreciate how you would reverse this reaction. Suppose you have an acetal, and you want to convert it back into a ketone. You would just add water with a catalytic amount of acid, and the acetal would be converted back into a ketone. This reaction is called acetal hydrolysis:

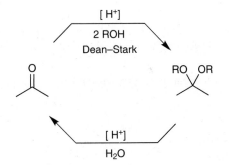

Under these conditions, the equilibrium will favor formation of the ketone. So, now we know how to convert a ketone into an acetal, and we know how to convert the acetal back into a ketone:

It is very important that we are able to control the conditions to push the reaction in either direction. We will soon see why this is so important. But first, let's make sure we are comfortable with the mechanism of acetal formation:

EXERCISE 6.23 Draw a mechanism for the following reaction.

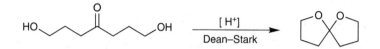

Answer Notice that we are starting with a ketone, and we are ending up with an acetal. It is a bit tricky to see, because it is all happening in an intramolecular fashion. In other words, the two alcoholic OH groups are *tethered* to the ketone:

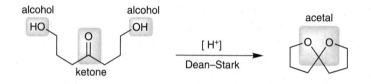

So, the mechanism should follow the same order of steps as the mechanism we have already seen. Namely, there are three critical steps (nucleophilic attack, loss of water, and another nucleophilic attack) surrounded by many proton transfer steps. The proton transfer steps are just there to facilitate these three steps. We use a proton transfer in the very first step to render the carbonyl group more electrophilic. Then, we use proton transfers to form water (so that it can leave). And finally, we use a proton transfer to remove the charge and generate the product.

Perhaps you should try to draw the mechanism for this reaction on a separate piece of paper. Then, when you are done, you can compare your work to the following answer:

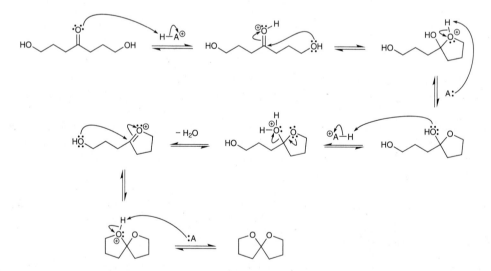

We said before that this type of mechanism is so incredibly important because there will be so many more reactions that build upon the concepts that we developed in this mechanism. To get practice, you should work through the following problems slowly and methodically.

PROBLEMS Draw a mechanism for each of the following transformations. You will need a separate piece of paper for each mechanism.

6.24

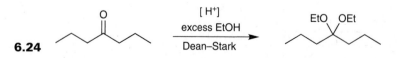

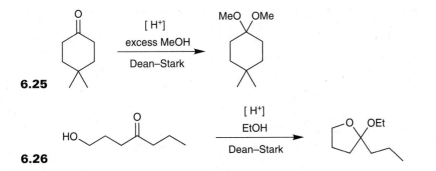

6.25

6.26

6.27   There is one sure way to know whether or not you have mastered a mechanism forward and backwards—you should try to actually draw the mechanism backwards. That's right, backwards. For example, draw a mechanism for the following reaction. Make sure to first read the advice below before attempting to draw a mechanism.

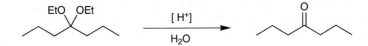

My advice for this mechanism is to start at the end of the mechanism (with the ketone), and then draw the intermediate you would get if you were converting the ketone into an acetal, like this:

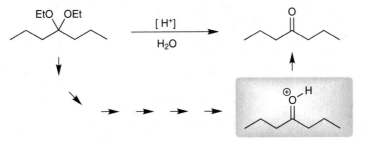

Keep drawing only the intermediates, working your way backwards, until you arrive at the acetal. But ***don't*** draw any curved arrows yet. Draw only the intermediates, working backwards from the ketone to the acetal. Then, once you have all of the intermediates drawn, then come back and try to fill in arrows, starting at the beginning, with the acetal. Use a separate sheet of paper to draw your mechanism. When you are finished, you can compare your answer to the answer in the back of the book.

    In this section, we have seen acetal formation, a reaction that takes place between a ketone and ***two*** molecules of ROH, in the presence of an acid catalyst and under Dean–Stark conditions:

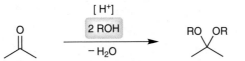

We saw a mechanism, in which the ketone is attacked twice. This same reaction can occur when both nucleophilic OH groups are in the same molecule. This produces a *cyclic* acetal:

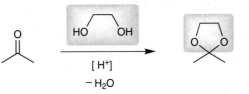

This type of reaction might appear several times throughout your lectures and textbook, so it would be wise to be familiar with this process. The diol in the reaction above is called ethylene glycol, and the transformation can be extremely useful. Let's see why.

We have seen before that we can manipulate the conditions of this reaction to control whether the ketone is favored or whether the acetal is favored. The same is true when we use ethylene glycol to form a cyclic acetal:

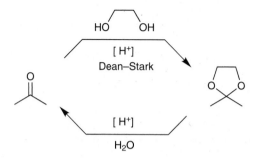

This is important because it allows us to *protect* a ketone from an undesired reaction. Let's see a specific example of this (it will take us a couple of pages to develop this concrete example, so please be patient as you read through this).

Consider the following compound:

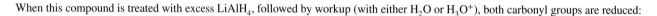

When this compound is treated with excess LiAlH$_4$, followed by workup (with either H$_2$O or H$_3$O$^+$), both carbonyl groups are reduced:

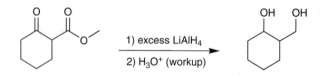

LiAlH$_4$ attacks the ketone *and* the ester. It may be difficult to see why the ester is converted into an alcohol—we will focus on that in the next chapter. But if you are curious to test your abilities, you have actually learned everything you need in order to figure out how an ester is converted into an alcohol in the presence of excess LiAlH$_4$ (remember that you should always re-form a carbonyl group if you can, but never expel H$^-$ or C$^-$).

So, we see that LiAlH$_4$ will reduce both carbonyl groups in the compound above. If instead, we treat the starting compound with excess NaBH$_4$, we observe that only the ketone is reduced:

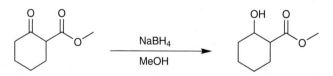

The ester is *not* reduced, because NaBH$_4$ is a milder source of hydride (as we have explained earlier). We will see in the next chapter that NaBH$_4$ will not react with esters (only with ketones and aldehydes) because the carbonyl group of an ester is less reactive than the carbonyl group of a ketone.

Now suppose you want to achieve the following transformation:

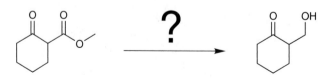

Essentially, you want to reduce the ester, **_but not_** the ketone. That would seem impossible, because esters are less reactive than ketones. Any reagent that reduces an ester should also reduce a ketone.

But there is a way to achieve the desired goal. Suppose we "protect" the ketone by converting it into an acetal:

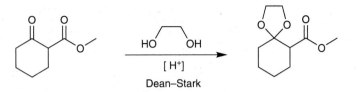

Only the ketone is converted into an acetal. The carbonyl group of the ester is **_not_** converted into an acetal because only aldehydes and ketones can be converted to acetals. Thus, we are able to selectively "protect" the ketone. Now, we can treat this compound with excess LiAlH$_4$, followed by aqueous workup (H$_2$O), and the acetal will not be affected (acetals do not react with bases or nucleophiles under basic conditions):

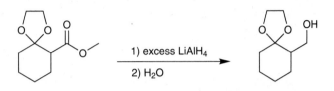

Notice that the acetal survives the water workup above, because the conditions remain strongly basic during workup with water (removal of a proton from H$_2$O gives a hydroxide ion, which is a strong base). To remove the acetal in this case, we would perform the workup with aqueous acid, rather than H$_2$O, after the reduction:

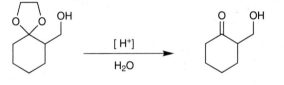

In the end, we have a 3-step process for reducing the ester group **_without affecting_** the ketone:

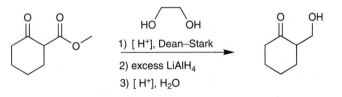

We will talk more about this strategy in the next chapter. For now, let's just focus on knowing the reactions well enough to predict products.

EXERCISE 6.28   Predict the major product of the following reaction:

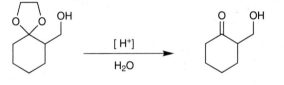

Answer   This reaction utilizes ethylene glycol, so we expect a cyclic acetal. Our starting compound has two carbonyl groups. One is a ketone, and the other is an ester group. We have seen that only ketones (not esters) are converted into acetals. So, the major product should be as follows:

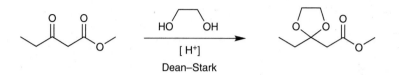

PROBLEMS   Predict the major product of each of the following reactions:

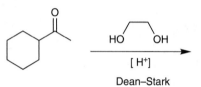

6.29

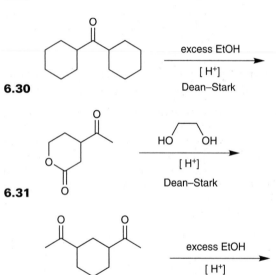

6.30

6.31

6.32

6.5   S-NUCLEOPHILES

Sulfur is directly below oxygen on the periodic table (in Column 6A). Therefore, the chemistry of sulfur-containing compounds is very similar to the chemistry of oxygen-containing compounds. In the previous section, we saw a method for converting a ketone into an acetal:

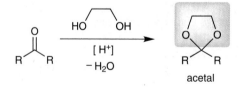

In much the same way, a ketone can also be converted into a ***thio***acetal ("thio" means sulfur instead of oxygen):

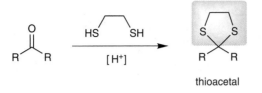

thioacetal

But thioacetals will undergo a transformation not observed for acetals. Specifically, thioacetals are reduced when treated with Raney nickel:

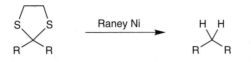

Raney nickel is finely divided nickel that has hydrogen atoms adsorbed to it. The mechanism for this reduction process is beyond the scope of this course. But, this is a useful synthetic transformation. So it is worth remembering, even if you don't know the mechanism. It provides a way to completely reduce a ketone down to an alkane:

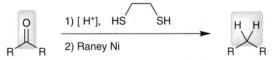

We have actually already seen one way to achieve this kind of transformation. It was called the Clemmensen reduction, which we explored in Chapter 4 (electrophilic aromatic substitution). We will also see one more way to achieve this transformation in the upcoming section.

Why do we need three different ways to do the same thing? Because each of these methods involves a different set of conditions. The Clemmensen reduction employs strongly acidic conditions. The method we learned just now (desulfurization with Raney nickel) employs catalytic acid. And the method in the upcoming section will employ ***basic*** conditions. As we move through the course, we will see times when it won't be good to subject an entire compound to acidic conditions, and we will see other times when it won't be good to subject an entire compound to basic conditions.

PROBLEMS   Predict the major product that is expected when each of the following compounds is treated with ethylene thioglycol (HSCH$_2$CH$_2$SH) in the presence of catalytic acid.

6.33   **6.34** 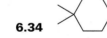   **6.35**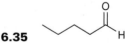

PROBLEMS   Predict the major product that is expected when each of the following compounds is treated with ethylene thioglycol (HSCH$_2$CH$_2$SH) in the presence of catalytic acid, followed by Raney nickel.

6.36   **6.37**   **6.38**

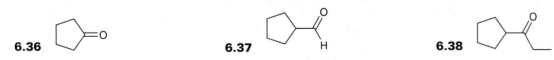

PROBLEMS   Identify reagents that you could use to achieve each of the following transformations:

6.39

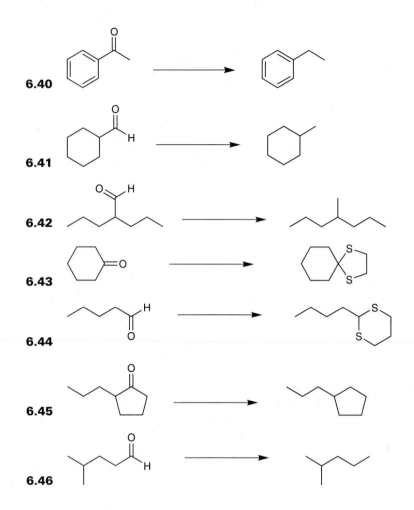

6.6 N-NUCLEOPHILES

In this section, we will focus on reactions between ketones and nitrogen nucleophiles, such as primary and secondary amines:

Primary amine   Secondary amine

A primary amine has one alkyl group connected to the nitrogen atom, while a secondary amine has two alkyl groups. Primary and secondary amines are excellent nucleophiles, as a result of the localized lone pair on the nitrogen atom. Under the right conditions (catalytic acid, and removal of water), both primary amines and secondary amines will react with ketones or aldehydes. The reaction with a primary amine yields a product with a C=N bond, called an imine,

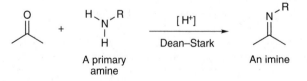

A primary
amine   An imine

while the reaction with a secondary amine yields a product called an enamine:

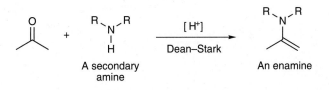

A secondary
amine

An enamine

We will explore each of these reactions now, starting with the reaction of primary amines to give imines. To get started, it will be helpful if we compare imine formation with acetal formation (Section 6.4). These two mechanisms are extremely similar, with only a couple of key differences. By focusing on these differences, it will hopefully be easier to study and remember the mechanisms. Let's begin by comparing the first three steps of imine formation and first three steps of acetal formation, and let's focus our attention specifically on the third step in each case:

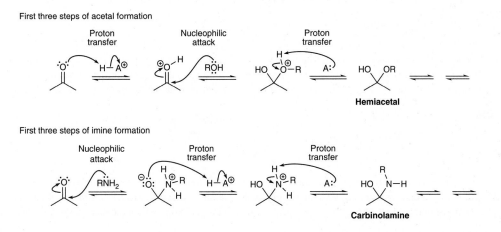

In each case, the third step is a proton transfer step in which the base (represented by the letter A) removes a proton to generate an uncharged intermediate. In the first case (acetal formation), the uncharged intermediate is called a hemiacetal, while in the second case (imine formation), the uncharged intermediate is called a carbinolamine. Notice that these two intermediates are similar in structure. In fact, we can say that these two processes are very much analogous, although one main difference between the two processes is the order of the first two steps. For acetal formation, the ketone is first protonated, and then the protonated ketone is attacked by a nucleophile. But for imine formation, these two steps are reversed. That is, the ketone is first attacked by the nucleophile (before being protonated), and only then it is protonated (after the nucleophilic attack). Why should there be a difference in the order of these two steps? To understand this, we must consider the identity of HA$^+$ in each case. For acetal formation, the identity of HA$^+$ is most likely a protonated alcohol (as we saw in section 6.4). But for imine formation, the identity of HA$^+$ is most likely a protonated amine (called an ammonium ion), which received its extra proton from the acid catalyst:

$$H-A^{\oplus} \quad \equiv \quad H-\overset{\overset{\displaystyle H}{|}}{\underset{\underset{\displaystyle H}{|}}{N}}{}^{\oplus}-R$$

If we compare the pK_a values of an ammonium ion (~ 10) and a protonated ketone (~ -7), we will find that it is extremely unlikely for an ammonium ion to give a proton to a ketone to generate a protonated ketone. Such a process would be forming a much stronger acid from a much weaker one (an unlikely process). It is more likely that the ketone will first be attacked by a molecule of the amine, which is sufficiently nucleophilic to attack the ketone without prior protonation.

Now that we have seen that the first three steps of imine formation are very similar to the first three steps of acetal formation (with the exception of the order of the first two steps), we can move on to the rest of the mechanism. Once again, we will continue to compare the mechanisms for imine formation and acetal formation, because the next two steps are directly analogous in each mechanism:

Next two steps of acetal formation

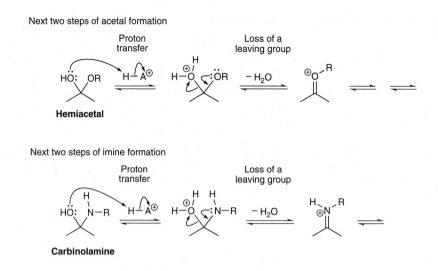

Hemiacetal

Next two steps of imine formation

Carbinolamine

Notice that in each case, the OH group is protonated, followed by loss of a leaving group (water) to give a cationic intermediate (O^+ in the first case, and N^+ in the second case). At this point, our two mechanisms will finally diverge from each other. We can understand why the mechanisms end differently if we focus on the difference between the two intermediates that have been produced thus far (O^+ vs. N^+):

In the first case (during acetal formation), the oxygen atom does not bear a proton, so deprotonation is not an option. This intermediate is extremely electrophilic, so it is readily attacked by another nucleophile (another molecule of ROH), ultimately giving an acetal. However, during imine formation, the intermediate above (N^+) has a proton on the nitrogen atom, which the base removes to yield an imine:

So the mechanism of imine formation is really very similar to the mechanism of acetal formation, with two key differences (the order of the first two steps are reversed, and the ending is different). The following is a summary of the entire mechanism that has been presented for imine formation:

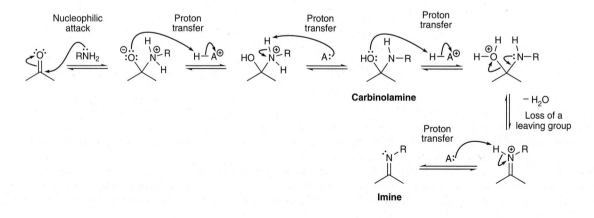

When converting a ketone into an imine, we need to take special notice of whether or not the starting ketone is symmetrical:

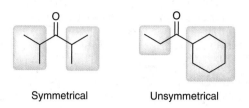

Symmetrical   Unsymmetrical

If the starting ketone is **un**symmetrical, then we should expect two diastereomeric imines, for example:

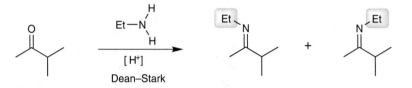

So far, we have seen that *primary* amines will react with ketones (under acid-catalyzed conditions) to give imines. Now, let's focus our attention on the reaction of *secondary* amines with ketones to give enamines. The accepted mechanism for enamine formation is extremely similar to the mechanism that we just saw for imine formation. The first three steps give a carbinolamine (exactly as we saw during imine formation):

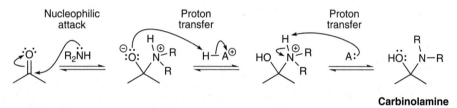

And the last three steps convert the carbinolamine into an enamime:

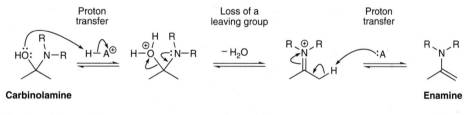

Notice that these three final steps are almost exactly the same as the three final steps of imine formation. But there is one major difference, in the very last step. Let's compare the last step of imine formation with the last step of enamine formation:

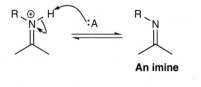

An imine

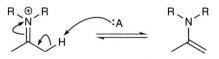

An enamine

In the first case, the nitrogen atom bears a proton, which is removed by a weak base (A) to give the imine. However, in the second case, the nitrogen atom does not bear a proton that can be removed. Instead, the base removes a different proton, as shown, to give an enamine.

The following is a summary of the entire mechanism that has been presented for enamine formation:

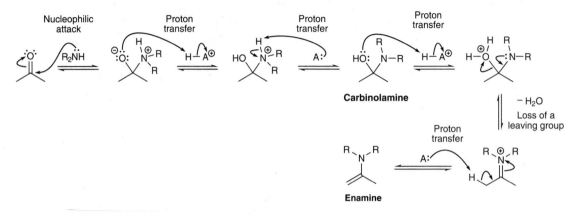

Once again, we need to be careful to check if the ketone is unsymmetrical. If it is, then there will be two ways to form the double bond in the last step of the mechanism. This will give two different enamine products. Here is an example:

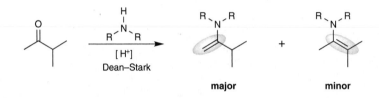

In a situation like this (where we start with an unsymmetrical ketone), the major product will generally be the enamine with the *less-substituted double bond.

So far in this section, we have seen two new reactions (with primary amines, and with secondary amines), and we have seen the similarities in the mechanisms. Now, we will revisit the first reaction (the reaction between a ketone and a *primary* amine):

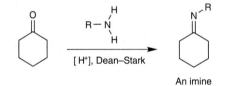

The "R" in RNH$_2$ refers to an alkyl group (that is usually what R means). But imagine that, in place of the R group, there was something *other than an alkyl group*. For example, imagine that R was an OH group, rather than an alkyl group. In other words, imagine that we were starting with the following amine:

This compound is called hydroxylamine, and the product that it forms (when it reacts with a ketone) is not surprising at all:

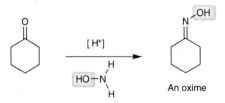

It is the same reaction as if it were a primary amine reacting with the ketone. But, instead of obtaining an imine, the reaction gives a product called an oxime. Remember to always look if the starting ketone is unsymmetrical. If it is, expect two diastereomeric oximes:

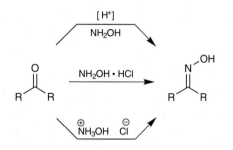

When you see this type of reaction, there are several ways that the presence of hydroxylamine can be indicated:

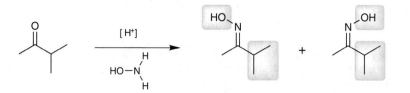

All of these representations are just different ways of showing the same reagent.

Now that we have seen a special N-nucleophile (RNH_2 where R = OH), let's take a close look at another special N-nucleophile. Let's look at a case where R = NH_2. In other words, we are using the following nucleophile:

This compound is called hydrazine, and the product that it forms (when it reacts with a ketone) is not surprising at all:

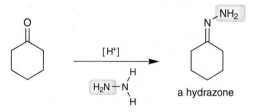

It is the same reaction as if it were a primary amine reacting with a ketone. But, instead of obtaining an imine or an oxime, we obtain a product called a hydrazone.

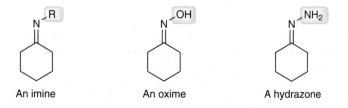

| An imine | An oxime | A hydrazone |

Just like with all of the other reactions we have seen in this section, we need to take special notice of whether or not the starting ketone is symmetrical. If the starting ketone is unsymmetrical, then we should expect to form two diastereomeric hydrazones:

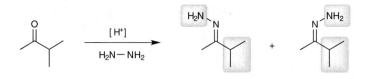

Hydrazones are useful for many reasons. In the past, chemists formed hydrazones as a way of identifying ketones, but nowadays, with the advent of NMR techniques, no one uses hydrazones that way anymore. But the chemistry of hydrazones remains useful in some ways. For example, a hydrazone can be reduced to an alkane under basic conditions:

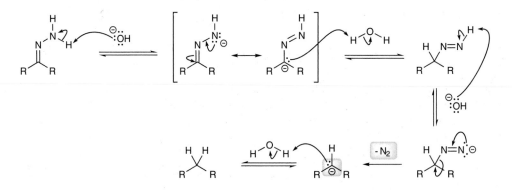

The following is a mechanism for this process:

The formation of a carbanion (highlighted in the second-to-last step) certainly creates an uphill battle (in terms of energy), so we might expect very little product to form. However, notice that the formation of the carbanion is accompanied by loss of N_2 gas (also highlighted in the mechanism above). This explains why the reaction goes to completion. The small amount of nitrogen gas (produced by the equilibrium) will bubble out of the solution and escape into the atmosphere. That forces the equilibrium to produce a little bit more nitrogen gas, which also then escapes into the atmosphere. And the process continues until the reaction reaches completion. Essentially, a reagent is being removed as it is being formed, and that is what pushes the equilibrium over the high energy barrier created by the instability of the carbanion. If you think about it, this concept is not so different from the previous sections where we removed water from a reaction as it was being formed (as a way of pushing the equilibrium towards formation of the acetal).

This now provides a two-step method for reducing a ketone to an alkane:

We have already seen two other ways to do this kind of transformation (the Clemmensen reduction and desulfurization with Raney Nickel). This is now our third way to reduce a ketone to an alkane, and it is called a Wolff-Kishner reduction.

In this section, we have only seen a few reactions involving nitrogen nucleophiles. Here is a short summary. We first saw how a ketone can react with a *primary amine* to form an *imine* (and we saw that the mechanism was very similar to acetal formation, except for the beginning and the end). Then, we saw how a ketone can react with a *secondary amine* to form an *enamine*. We also encountered two special N-nucleophiles (NH_2OH and NH_2NH_2), both of which gave products that we expected. The reaction with NH_2NH_2 was of special interest, because it provided a new method for reducing ketones to alkanes.

EXERCISE 6.47   Predict the major product(s) of the following reaction:

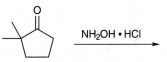

Answer   The starting material is a ketone, and the reagent is hydroxyl amine. As we have seen in this section, the product of this reaction should be an oxime. Since the starting ketone is unsymmetrical, we would expect two diastereomeric oximes:

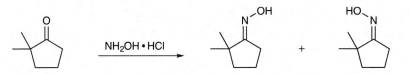

We saw several reactions in this chapter. You must be able to recognize the reagents for these reactions, so that you will be able to predict products. Let's get some practice with the following problems.

PROBLEMS   Predict the major product(s) for each of the following transformations:

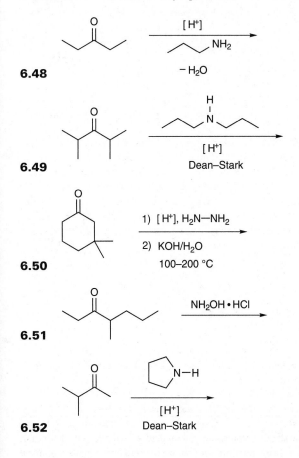

6.48

6.49

6.50

6.51

6.52

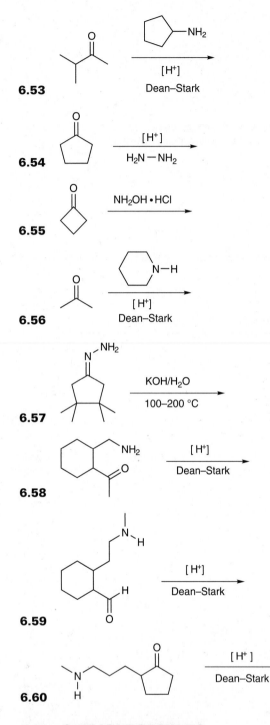

6.53

6.54

6.55

6.56

6.57

6.58

6.59

6.60

6.7 C-NUCLEOPHILES

In this chapter, we have seen many different kinds of nucleophiles that can attack ketones and aldehydes. We started with hydrogen nucleophiles. Then we moved on to oxygen nucleophiles and sulfur nucleophiles. In the previous section, we covered nitrogen nucleophiles. In this section we will discuss carbon nucleophiles. We will see three types of carbon nucleophiles.

Our first carbon nucleophile is the Grignard reagent. You may have encountered this reagent in the first semester. If you didn't, here is a quick overview:

Alkyl (or aryl) halides will react with magnesium in the following way:

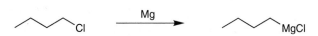

Essentially, an atom of magnesium inserts itself in between the C—Cl bond (this reaction works with other halides as well, such as Br or I). This magnesium atom has a significant electronic effect on the carbon atom to which it is attached. To see the effect, consider the alkyl halide (before Mg entered the picture):

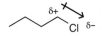

The carbon atom (connected to the halogen) is poor in electron density, or δ+, because of the inductive effects of the halogen. But after magnesium is inserted between C and Cl, the story changes very drastically:

Carbon is much more electronegative than magnesium. Therefore, the inductive effect is now reversed, placing a lot of electron density on the carbon atom, making it very δ−. The C—Mg bond has significant ionic character, so for purposes of simplicity, we will just treat it like an ionic bond:

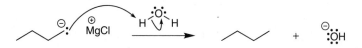

Carbon is not very good at stabilizing a negative charge, so this reagent (called a Grignard reagent) is highly reactive. It is a very strong nucleophile and a very strong base. Now, let's see what happens when a Grignard reagent attacks a ketone or aldehyde.

In the previous section, we always started each mechanism by protonating the ketone (turning it into a better electrophile). That does not occur here, because the Grignard reagent is such a strong nucleophile that it has no problem attacking a carbonyl group directly. In fact, we could **not** use acid catalysis here (even if we wanted to), because protons destroy Grignard reagents. For example, consider what happens when a Grignard reagent is exposed even to a weak acid, such as water:

The Grignard reagent acts as a base and removes a proton from water, to form a more stable hydroxide ion. The negative charge is MUCH more stable on an electronegative atom (oxygen), and as a result, the reaction is irreversible. This means that you can never use a Grignard reagent to attack a compound that has acidic protons. For example, the following reaction would not work:

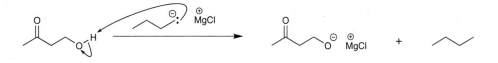

Because this would happen instead:

In general, proton transfers are faster than nucleophilic attack. And when the Grignard reagent removes a proton, it irreversibly destroys the Grignard reagent. Similarly, you could never prepare the following kinds of Grignard reagents:

These reagents could not be formed, because each of these reagents could react with a second molecule of itself to remove the negative charge on the carbon atom, for example:

All of that was a quick review of Grignard reagents. Now let's see how Grignard reagents can attack a ketone or aldehyde. When a ketone or aldehyde is treated with a Grignard reagent, the negatively charged carbon atom of the Grignard reagent can attack the electrophilic carbon atom of the carbonyl group to give an alkoxide intermediate:

This intermediate then will attempt to re-form the carbonyl group, if it can. But let's see if it can. Remember our rules from the beginning of this chapter: re-form the carbonyl if you can, but never expel H⁻ or C⁻. This intermediate is NOT able to re-form the carbonyl group, because there are no leaving groups to expel. This is true whether the Grignard reagent attacks a ketone or an aldehyde:

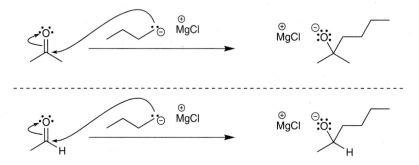

So, in either case, the reaction is complete, and we must now introduce a proton source (either H_2O or H_3O^+) into the reaction flask in order to protonate the alkoxide ion, giving an alcohol:

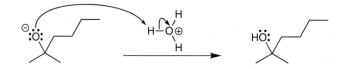

This reaction is not so different from the reactions we saw earlier in this chapter when we explored hydrogen nucleophiles (NaBH₄ and LiAlH₄). We saw a similar scenario there: the nucleophile attacked, and then the carbonyl group was NOT able to re-form because there was no leaving group. Compare one of those reactions to this reaction:

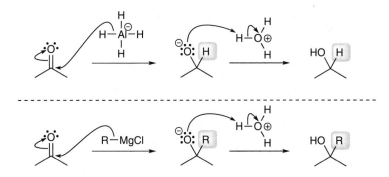

Notice that these mechanisms are similar. And it is worth a minute of time to think about why these reactions are so similar (while the other reactions in this chapter were different from these two reactions). What is special about these two reactions that makes them so similar? Remember our golden rule: never expel H⁻ or C⁻. So, if we attack a ketone (or aldehyde) with either H⁻ or C⁻, then the carbonyl group will be unable to re-form. And that is what these two reactions have in common.

When you write down the reagents of a Grignard reaction (in a synthesis problem), make sure you show the proton source (either H_2O or H_3O^+) *as a separate step*, like this:

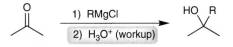

We saw this important subtlety when we learned about $LiAlH_4$, where we also had to show the proton source as a separate step. The same subtlety exists here, because (as we have very recently seen) a Grignard reagent will not survive *in the presence* of a proton source. The proton source must come AFTER the reaction is complete (after the Grignard reagent has been consumed by the reaction).

In order to add this reaction to your toolbox of synthetic transformations, let's compare it one more time to the reaction with $LiAlH_4$. But this time, let's focus on comparing the products, rather than comparing the mechanisms:

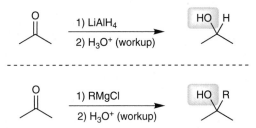

Notice that in both reactions, we are reducing the ketone to an alcohol. But in the case of a Grignard reaction, the reduction is accompanied by the installation of an alkyl group:

This will be helpful as we explore synthesis problems at the end of this chapter.

EXERCISE 6.61   Predict the major product of the following reaction:

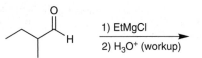

Answer   The starting material is an aldehyde, and it is going to react with a Grignard reagent. First, the Grignard attacks:

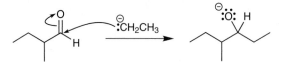

The carbonyl group cannot re-form, because H⁻ or C⁻ cannot be expelled, and there is nothing else that can be expelled in this case. Our product (an alcohol) is obtained when a proton source is introduced, such as H_3O^+:

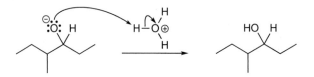

Remember that bond-line drawings don't have to show hydrogen atoms attached to carbon atoms, so we can redraw the product:

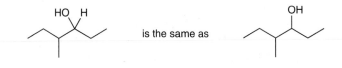

PROBLEMS   Predict the major product(s) for each of the following reactions:

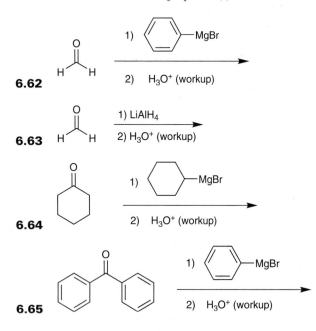

6.62

6.63

6.64

6.65

There are two more carbon nucleophiles that we must explore. Both of them are different from the Grignard reagent. Both reactions involve "ylides" (pronounced "ILL–ids"). Let's take a close look at the first ylide, by exploring how it is prepared:

We start with a compound called triphenylphosphine:

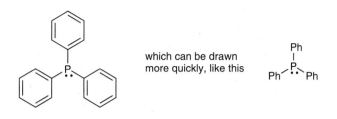

which can be drawn
more quickly, like this

and we treat it with an alkyl halide such as methyl iodide (resulting in an S_N2 reaction):

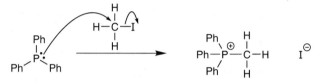

Then, we use *n*-butyllithium (*n*-BuLi), a very strong base, to remove a proton and form the ylide:

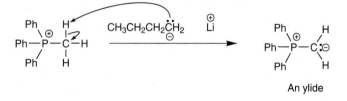

An ylide

An ylide is a compound with two adjacent, oppositely charged atoms (in this case, P^+ and C^-). Notice that this ylide has a region of high electron density on a carbon atom. As a result, this ylide can function as a carbon nucleophile. In a few moments, we will see another type of ylide (one that uses sulfur instead of phosphorus). The ylide above (based on phosphorus) has a special name. It is called a Wittig reagent (pronounced "Vittig"). And when a ketone or aldehyde is treated with a Wittig reagent, the observed reaction is called a Wittig reaction. So, let's take a close look at a mechanism for the Wittig reaction.

The Wittig reaction is believed to occur via the following two-step mechanism:

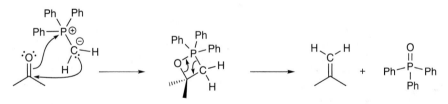

An oxaphosphetane

The first step involves a cycloaddition process, generating an oxaphosphetane (notice that the nucleophilic carbon atom of the Wittig reagent reacts with the electrophilic carbonyl group, as we might expect). Then, in the second step, the oxaphosphetane undergoes fragmentation to give an alkene as the product. This process enables us to achieve the following type of transformation:

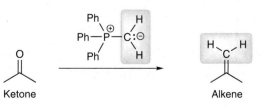

Ketone   Alkene

This reaction is incredibly useful for synthesis. We already saw how to convert an *alkene* into a *ketone*, using an ozonolysis reaction. Now, with a Wittig reaction, we can go either way:

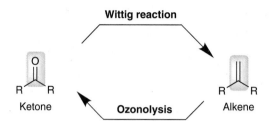

You should always take special notice whenever you learn how to interconvert two functional groups (going in either direction), like above. We have seen several cases like this so far.

EXERCISE 6.66   Predict the product of the following reaction.

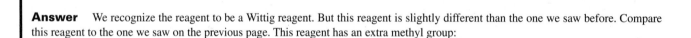

Answer   We recognize the reagent to be a Wittig reagent. But this reagent is slightly different than the one we saw before. Compare this reagent to the one we saw on the previous page. This reagent has an extra methyl group:

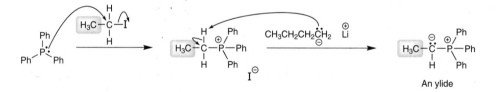

The way to form a reagent like this is to use Et-I instead of Me-I when you are making the Wittig reagent, like this.

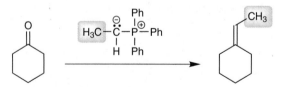

The extra methyl group (highlighted) comes along for the ride, and the final product looks like this:

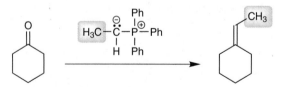

You should try to draw out a mechanism for this reaction to make sure that you can "watch" the extra carbon atom coming along for the ride.

PROBLEMS Predict the major product for each of the following Wittig reactions:

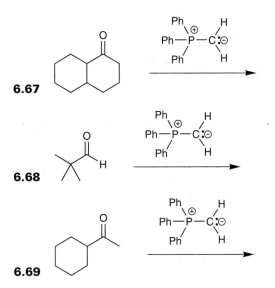

6.67

6.68

6.69

Now we will explore another kind of ylide. This time, it is a *sulfur* ylide, rather than a *phosphorus* ylide. Many instructors (and textbooks) do not cover sulfur ylides, so you might want to look through your lecture notes and textbook to find out if you are responsible for the following reaction.

The mechanism for forming a *sulfur* ylide is very similar to the mechanism for forming a *phosphorus* ylide. Let's compare them:

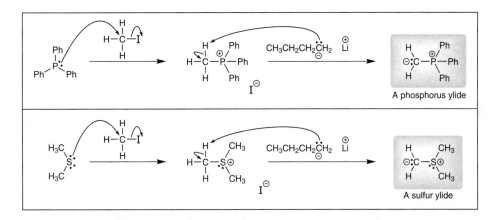

To form a sulfur ylide, we start with dimethyl sulfide (DMS). From that point on, everything is the same: the alkyl halide is attacked, followed by deprotonation with a very strong base to form a sulfur ylide.

But when a sulfur ylide attacks a ketone (or aldehyde), we observe a very different product. We ***don't*** obtain an alkene as our product (like we did in the Wittig reaction). Instead, we obtain an epoxide:

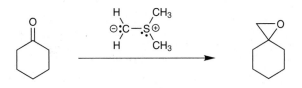

A mechanism for this process is shown here:

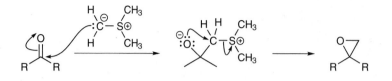

The sulfur ylide attacks the carbonyl group, giving an intermediate that undergoes an intramolecular S_N2-type process to give an epoxide.

This reaction is very useful, because it provides a method for making epoxides from ketones. You should remember from the first semester that you learned how to convert an *alkene* into an epoxide. But now, we see that we can make an epoxide from a ketone as well:

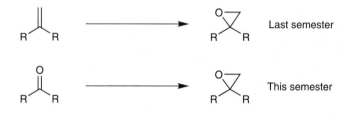

In summary, we have explored three carbon nucleophiles in this section. We started with Grignard reagents, and then we moved on to ylides (phosphorus ylides and sulfur ylides). We have seen that phosphorus ylides and sulfur ylides produce alkenes and epoxides, respectively:

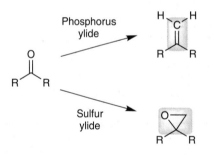

EXERCISE 6.70   Predict the major product of the following reaction:

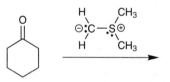

Answer   This reagent is a sulfur ylide, which is used to convert ketones into epoxides. So, our product is an epoxide:

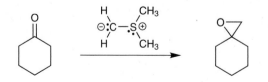

PROBLEMS   Predict the major product for each of the following reactions:

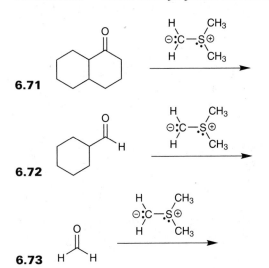

6.71

6.72

6.73

6.8   EXCEPTIONS TO THE RULE

In the beginning of this chapter, we saw a golden rule that helped us understand most of the chemistry that we explored. That rule was: always re-form the carbonyl if you can, but never expel H$^-$ or C$^-$.

The truth is that there are a few, rare exceptions to this rule. In this section, we will look at two of these exceptions.

In the Cannizzaro reaction, it "seems" as though we are expelling H$^-$ to re-form a carbonyl. I am not going to go into great detail on the Cannizzaro reaction, because this reaction has very little synthetic utility. It is unlikely that you will use this reaction more than once, if at all. So, I will just mention it in passing. Look up that reaction in your textbook and in your lecture notes. If you don't need to know that reaction, then you can ignore it. But if you are responsible for knowing that reaction, then you should look carefully at the mechanism in your textbook. If you focus on the step where H$^-$ gets expelled, you will see that H$^-$ is not really expelled by itself. Rather, it is transferred from one place to another. And in that sense, we can understand it a bit better. It is true that H$^-$ is too unstable to ever leave as a leaving group. That is why we never expel H$^-$ into solution. But in the Cannizzaro reaction, it never actually leaves as a leaving group.

We will now explore one other exception to the golden rule. There is a reaction where it seems like we are re-forming a carbonyl group to expel C$^-$. This reaction, called the Baeyer–Villiger reaction, is extremely useful. If you know how to use it properly, you will find that you might use it many times to solve synthesis problems in this course. So, we will spend some time covering that reaction now.

The Baeyer–Villiger reaction uses a peroxy acid (RCO$_3$H) as the reagent. Compare the structure of a peroxy acid with the structure of a carboxylic acid:

A peroxy acid   **A carboxylic acid**

(RCO$_3$H)   (RCO$_2$H)

Notice that a peroxy acid (RCO$_3$H) has one additional oxygen atom, as compared with a carboxylic acid (RCO$_2$H). The R group of the peroxy acid can be a small alkyl group (such as a methyl group), or it can be a much larger group. Below are three commonly used peroxy acids:

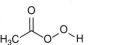

CH$_3$CO$_3$H   **CF$_3$CO$_3$H**

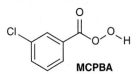

MCPBA

The third example is called *meta*-chloro **per**benzoic **a**cid, or just MCPBA for short. So when you see the letters MCPBA, you should recognize that we are talking about a peroxy acid (RCO_3H). In a Baeyer–Villiger reaction, a peroxy acid is used to insert an oxygen atom next to the carbonyl group of a ketone, producing an ester:

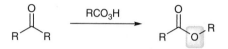

A similar reaction can be used to convert an aldehyde into a carboxylic acid:

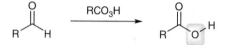

Once again, the outcome of this reaction is to "insert" an oxygen atom next to the carbonyl group. This is very useful, so let's see the accepted mechanism for this process. We will begin with the first three steps of the mechanism, shown here:

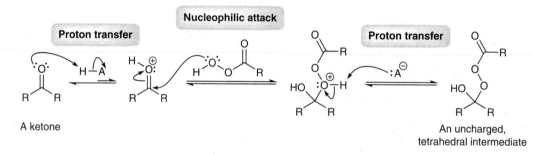

These three steps should seem very familiar, because they are analogous to the first three steps of acetal formation (as seen earlier in this chapter). Under acidic conditions, the ketone is first protonated by an acid (H–A), where H–A represents either the peroxy acid (RCO_3H) or a carboxylic acid (RCO_2H, which accumulates as a byproduct as the reaction proceeds, as we will soon see). The resulting protonated ketone is a powerful electrophile and can be attacked by the peroxy acid (which functions as a nucleophile). The resulting tetrahedral intermediate is then deprotonated by A⁻ (the conjugate base of H–A) to give an uncharged, tetrahedral intermediate.

If we focus on the structure of the uncharged, tetrahedral intermediate, we would conclude that the only way to re-form the carbonyl group would be to expel the group that was recently installed:

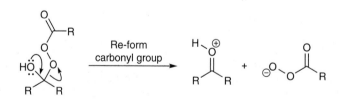

Indeed, this probably happens, but it doesn't lead to the formation of a new product. It effectively regenerates the starting materials. So, we apply our golden rule to see if we can expel any other leaving group. We cannot expel H⁻ or C⁻, and there are no other groups to expel. BUT, this is an apparent exception to the golden rule. Something unique happens, and you will not see this in any other mechanism (so don't worry about trying to apply this next step in any other mechanism). A rearrangement occurs, in which an alkyl group is said to *migrate*:

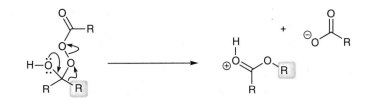

Let's focus on the fate of the migrating alkyl group (highlighted above). Notice that the carbonyl group is re-forming to expel this R group, which migrates to the nearby oxygen atom. In other words, it looks like we are expelling C^-. But the truth is that we are not *really* expelling C^- into solution by itself. After all, C^- is too unstable. Rather, it is just *migrating* from one place to another (it is migrating over to attack the neighboring oxygen atom). It never really becomes C^- for any period of time, and that explains how we can have an exception here.

The final step of the mechanism is a proton transfer step to give the product (an ester), as well as a carboxylic acid by-product (RCO_2H):

An ester
(the product)

A carboxylic acid
(the by-product)

This mechanism is truly unique, and you should not worry if you feel that you would not be able to predict when this could happen in other situations. This is probably the first time you are ever seeing a rearrangement that does not involve a carbocation. So, this is truly different. You will not need to apply this mechanism to any other situations. So for now, don't focus too much on this mechanism. Instead, let's focus on how to use this reaction when you are solving synthesis problems, because it will be a very useful reaction for you to have in your back pocket.

In order to use this reaction properly, you will need to know how to predict where the oxygen atom is installed. For example, consider the following ketone:

This ketone is unsymmetrical, so we must decide where the oxygen atom will be installed during a Baeyer–Villiger reaction. Which of the following two products are expected?

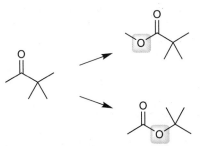

To answer this question, we need to know which alkyl group is more likely to migrate. If you look back at the mechanism above, you will see that the migrating R group is the one that ends up connected to the oxygen atom in the product. So, we just have to decide which alkyl group can migrate faster.

There is an order to how fast alkyl groups can migrate in this reaction, and we call it "migratory aptitude":

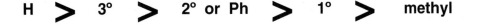

Notice that phenyl groups have similar migratory aptitude as secondary alkyl groups. Also notice that H migrates the fastest. This explains how we can use this reaction to convert aldehydes into carboxylic acids:

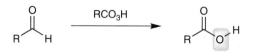

H migrates faster than any other group, so it doesn't even matter what the R group is.

If you are starting with a ketone instead of an aldehyde, then you should look for the most substituted alkyl group. For example:

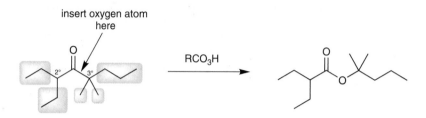

Notice that the oxygen atom is inserted on the side that is more substituted.

In order to use this reaction in synthesis, you must make sure that you can predict where the oxygen atom will be inserted. Let's do some problems to make sure you got it:

EXERCISE 6.74   Predict the major product of the following reaction:

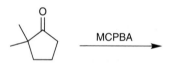

Answer   A ketone is being treated with a peroxy acid, so we expect a Baeyer–Villiger reaction to occur. We look closely at our starting ketone, and we see that it is unsymmetrical. So, we must predict where the oxygen atom will be inserted. We look at both sides, and we see that the left side is more substituted. The more substituted R group will migrate faster, and that is where the oxygen atom will be inserted.

This specific case is an interesting example, because the insertion of an oxygen atom causes a ring expansion:

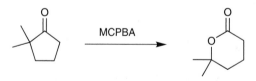

The product is not just an ester, but it is a *cyclic* ester. Cyclic esters are also called *lactones*.

PROBLEMS   Predict the major product for each of the following reactions:

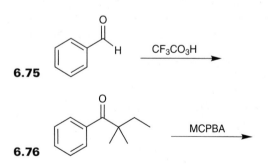

6.75

6.76

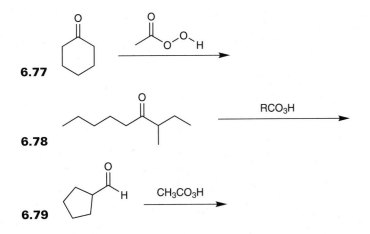

6.77

6.78

6.79

6.9 HOW TO APPROACH SYNTHESIS PROBLEMS

In this chapter, we have seen many reactions. In order to solve synthesis problems, you will need to have all of these reactions at your fingertips. In the beginning of this chapter, we saw a few ways to make aldehydes and ketones. Do you remember those reactions? If you don't, then you are in trouble. This is why organic chemistry can get tough at times. It is not sufficient to be a master of mechanisms. That is an excellent start, and it builds an excellent foundation for understanding the material. But at the end of the day, you have to be able to solve synthesis problems as well. And in order to do that, you must have all of the reactions organized in your mind.

Let's start with a short review of everything we saw:

We started with a few ways of making ketones and aldehydes (two ways to oxidize, and then an ozonolysis). Then, we explored hydrogen nucleophiles (NaBH$_4$ and LiAlH$_4$) and oxygen nucleophiles (making acetals), and we saw that acetals can be used to protect ketones. Then we explored sulfur nucleophiles (to form thioacetals), and we saw how they can be used to *reduce* ketones and aldehydes. Next, we explored nitrogen nucleophiles (primary amines and secondary amines), we examined special primary amines (hydroxylamine and hydrazine), and we saw how hydrazones could be used to reduce ketones to alkanes. Then, we moved on to three kinds of carbon nucleophiles—Grignard reagents, phosphorus ylides, and sulfur ylides. Finally, we examined the Baeyer–Villiger oxidation, focusing on its utility for synthesis. That is everything we saw in this chapter.

In this last section of the chapter, we will bring everything together to solve synthesis problems. The first step is to make sure that you know these reactions well enough to claim that you have them at your fingertips. To ensure that you achieve that goal, try to do the following. Take a separate piece of paper and try to write down all of the reactions listed in the paragraph above (without looking back in the chapter, if possible). Make sure that you can draw all of the reagents, and all of the products. If you cannot do this, then you are simply not ready to even START thinking about synthesis problems. Students often complain that they just don't know how to approach synthesis problems. But the difficulty is usually NOT due to the student's poor abilities, but rather, the difficulty arises from the student's poor study habits. You CAN do synthesis problems. You might even enjoy them, believe it or not. But, you have to walk before you can run. If you try to run before you learn how to walk, you will trip and you will get frustrated. Too many students make this mistake with synthesis problems.

So, take my advice, and focus right now on mastering the individual reactions. Try to fill out a blank sheet of paper with everything that we have done in this chapter. If you find that you have to look back into the chapter to get the exact reagents (or to see what the exact products are), then that is fine. It is part of the studying process—BUT don't trick yourself into thinking that you are ready for synthesis problems once you have filled out the sheet. You are not ready until you can fill out the entire sheet of paper, start-to-finish, without looking back even once into the chapter to get the fine details. Keep filling out a new sheet, again and again, until you can do it all without looking back. Ideally, you should get to a point where you do not even need to look at the short summary that we just gave. You should get to a point where you can reconstruct the summary in your head, and then based on that summary, you should be able to write a list of all the reactions.

It sounds like a lot of work. And it is. It will take you a while. But when you are done, you will be in an excellent position to start tackling synthesis problems. If you get lazy, and you decide to skip this advice, then don't complain later if you are frustrated with synthesis problems. It would be your own fault for trying to run before you have mastered walking.

Once you get to the point where you have all of the reactions at your fingertips, then you can come back to here, and try to prove it, by doing some simple problems. These problems are designed to test you on your ability to list the reagents that you would need in order to do simple one-step transformations. Once you have all of the reactions down cold, then we will be able to move on and conquer some multistep synthesis problems.

EXERCISE 6.80 What reagents would you use to achieve the following transformation:

Answer The starting material is a ketone, and the product is an enamine, so we will need a nitrogen nucleophile. We just need to decide what kind of nitrogen nucleophile. Since our product is an enamine, we will need a secondary amine. When we look at the product, we can determine that we would need to use the following secondary amine:

Finally, we just need to decide if there are any special conditions that should be mentioned. And we did learn that there are special conditions for the reaction between a ketone and a secondary amine. Specifically, we need to have acid-catalysis, and Dean–Stark conditions. So, our answer is:

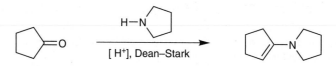

PROBLEMS The following problems are designed to be simple, so that you can prove to yourself that you know these reactions cold. I highly recommend that you photocopy the following problems *before* filling them out. You might find that you get stuck on a few problems, and it might be helpful for you to come back to these problems in the near future to fill them out again. The following problems are not listed in the order in which they appear in this chapter.

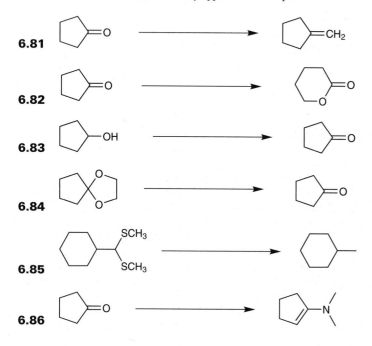

6.81

6.82

6.83

6.84

6.85

6.86

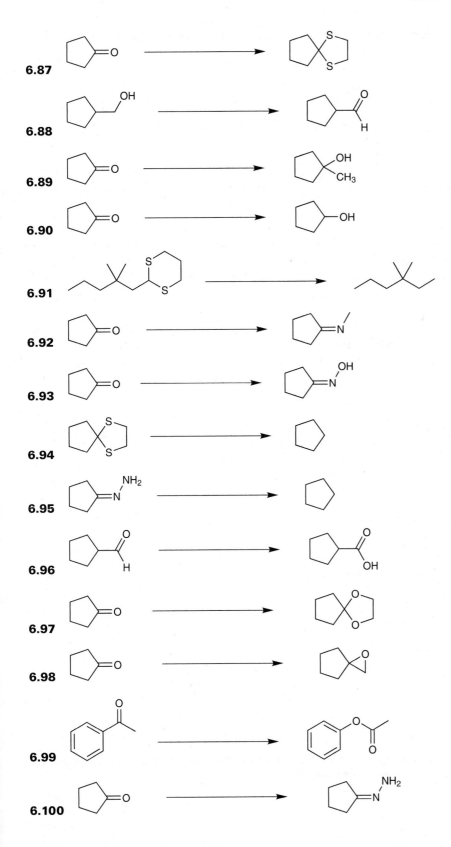

6.87

6.88

6.89

6.90

6.91

6.92

6.93

6.94

6.95

6.96

6.97

6.98

6.99

6.100

If you felt comfortable with those problems, then you should be ready to move on to solving some multistep synthesis problems. Let's see an example:

EXERCISE 6.101 Propose an efficient synthesis for the following transformation:

Answer This transformation is a bit more involved than the previous problems, because it cannot be achieved in one step. We need to install an ethyl group, while maintaining the presence of the carbonyl group. If we use a Grignard reagent, we can install the ethyl group, but we will reduce the ketone to an alcohol in the process:

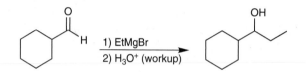

But, this issue can be easily overcome, because we can oxidize the alcohol back up to a ketone:

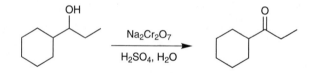

So, we have a two-step synthesis to accomplish this transformation.

Before you try to solve some problems yourself, there is one more important point to make about synthesis problems. Very often, it is helpful to work "backwards." We call this *retrosynthetic analysis*. Let's see an example:

EXERCISE 6.102 Propose an efficient synthesis for the following transformation:

Answer If we focus on the product, we notice that it is an enamine. We have only seen one way to make an enamine—from the reaction between a ketone and a secondary amine. So, we can work our way backwards:

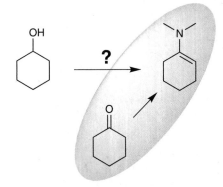

All we need to do is find a way to convert the starting compound into a ketone. And we have seen how to do that. We can convert an alcohol into a ketone using an oxidation reaction. So, our synthesis is:

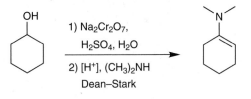

Of course, this last problem was not very difficult, because it had a two-step solution. As you solve problems that require more steps, this approach (retrosynthetic analysis) will become more and more important. But don't worry. You won't have to solve any problems that require ten steps. That is way beyond the scope of this course. You will generally not have to deal with syntheses that require more than three to five steps. So, with a lot of practice, it is definitely realistic to become a master of solving synthesis problems. Once again, it all depends on how well you know all of the reactions.

As you work through the following problems, keep in mind that there is rarely only one answer to a synthesis problem. As we learn more and more reactions, you will find that there are often multiple correct ways to achieve a transformation. Don't get stuck into thinking that you have to find THE answer. You may even come across a perfectly acceptable answer that no one else in the class thought of. Those are the most exciting moments. There actually is room for you to express some creativity when you solve synthesis problems. Now let's get some practice:

PROBLEMS For each of the following problems, suggest an efficient synthesis. Remember that there might be more than one correct answer for each of these problems. If you propose an answer, and it does not match the answer in the back of the book, do not be discouraged. Carefully analyze your answer, because it might also be correct.

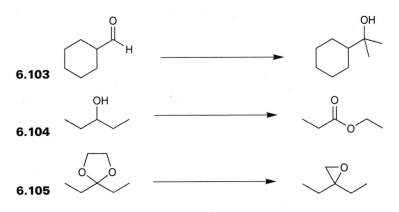

6.103

6.104

6.105

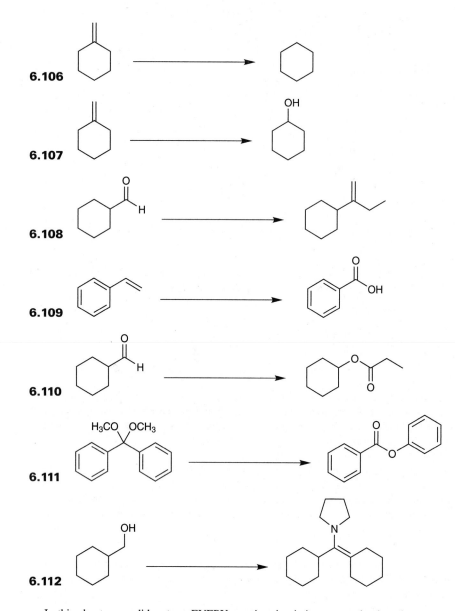

6.106

6.107

6.108

6.109

6.110

6.111

6.112

In this chapter, we did *not* see EVERY reaction that is in your textbook or lecture notes. We covered the core reactions (probably 90% or 95% of the reactions you need to know). The goal of this chapter was NOT to cover every reaction. Rather, our goal was to lay a foundation for you when you are reading your textbook and lecture notes. We saw the similarities between mechanisms, and we saw a simple way of categorizing all nucleophiles (hydrogen nucleophiles, oxygen nucleophiles, sulfur nucleophiles, etc.).

Now you can go back through your textbook and lecture notes, and look for the reactions that we did not cover here in this chapter. With the foundation we have built in this chapter, you should be in good shape to fill in the gaps and study more efficiently.

And make sure to do ALL of the problems in your textbook. You will find more synthesis problems there. The more you practice, the better you will get. Good luck.

END-OF-CHAPTER PROBLEMS

PRACTICE PROBLEMS *(Problems that involve only one skill)*

6.113 Predict the major product of the following reaction.

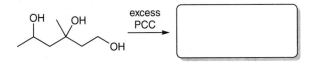

6.114 Predict the major product of the following reaction.

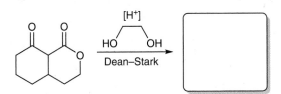

6.115 Predict the major product of the following reaction.

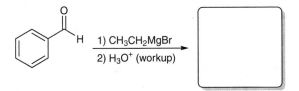

6.116 Identify the missing reagent in the following reaction.

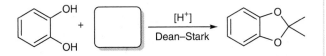

6.117 Identify a starting material that can be used to make the thioacetal shown below.

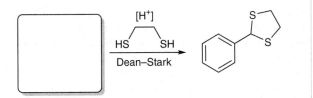

6.118 Consider the reaction shown below:

(a) Identify the missing reactant (in the box provided above).

(b) On a separate piece of paper, draw a mechanism for this transformation.

6.119 Show three different Grignard reactions that can be used to make the tertiary alcohol shown below.

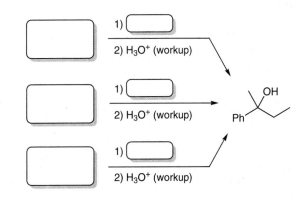

6.120 In the box provided, identify an acyclic starting material that can be used to make the imine shown below:

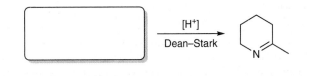

INTEGRATED PROBLEMS *(Problems that involve more than one skill)*

6.121 Show two different ways to achieve the following transformation. Note that one method will require only one step, while the other method will require more than one step.

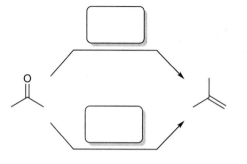

6.122 Show two different ways to achieve the following transformation.

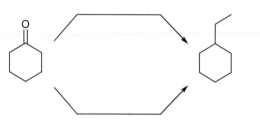

6.123 Propose an efficient synthesis for the following transformation:

6.124 Propose an efficient synthesis for the following transformation:

6.125 Propose an efficient synthesis for the following transformation:

(racemic)

6.126 Propose an efficient synthesis for the following transformation:

6.127 Propose an efficient synthesis for the following transformation:

6.128 Propose an efficient synthesis for the following transformation:

6.129 Propose an efficient synthesis for the following transformation:

6.130 Propose an efficient synthesis for the following transformation:

6.131 Propose an efficient synthesis for the following transformation:

6.132 Propose an efficient synthesis for the following transformation:

6.133 Propose an efficient synthesis for the following transformation:

6.134 Propose an efficient synthesis for the following transformation:

6.135 Propose an efficient synthesis for the following transformation:

6.136 Propose an efficient synthesis for the following transformation:

6.137 Propose an efficient synthesis for the following transformation:

(racemic)

6.138 Propose an efficient synthesis for the following transformation:

6.139 Consider the reaction shown below:

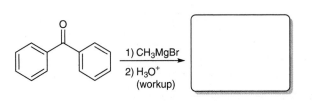

(a) Predict the major product of this reaction.

(b) Identify how you could use ^{1}H NMR spectroscopy to distinguish between the starting material and product.

(c) Identify how you could use ^{13}C NMR spectroscopy to distinguish between the starting material and product.

6.140 Compound **A** has the molecular formula C_8H_8O. The IR spectrum of compound **A** has a strong signal at $1680\,\text{cm}^{-1}$. The ^{1}H NMR spectrum of compound **A** has only two signals: a multiplet at 7.4 ppm with an integration of 5H, and a singlet at 2.5 ppm with an integration of 3H. Compound **A** is converted into compound **B** upon treatment with lithium aluminum hydride, followed by aqueous acidic workup.

(a) Draw the structures of compounds **A** and **B**.

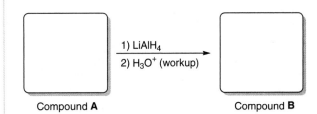

(b) Identify how you could use ^{13}C NMR spectroscopy to distinguish between compounds **A** and **B**.

6.141 Fill in the missing structures and reagents in the following scheme (note that DBU is a non-nucleophilic, bulky base):

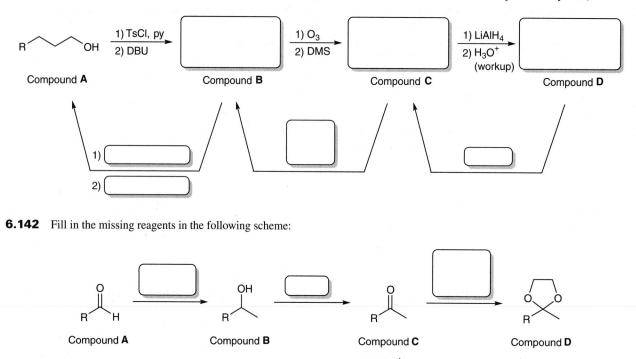

6.142 Fill in the missing reagents in the following scheme:

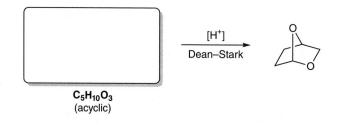

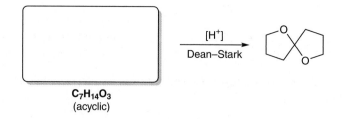

CHALLENGE PROBLEMS

6.143 Identify an acyclic starting material (with the molecular formula $C_5H_{10}O_3$) that can be used to make the acetal shown below.

$C_5H_{10}O_3$
(acyclic)

$\xrightarrow[\text{Dean–Stark}]{[H^+]}$

6.144 The following is a reaction that produces an acetal product.

$C_7H_{14}O_3$
(acyclic)

$\xrightarrow[\text{Dean–Stark}]{[H^+]}$

(a) Identify an acyclic starting material (with the molecular formula $C_7H_{14}O_3$) that can be used to make the acetal shown above.

(b) How many signals would you expect in the ^{13}C NMR spectrum of the acyclic starting material that you drew?

CHAPTER 7

CARBOXYLIC ACID DERIVATIVES

7.1 REACTIVITY OF CARBOXYLIC ACID DERIVATIVES

Carboxylic acid derivatives are similar to carboxylic acids, but the OH group has been replaced with a different group (Z),

Carboxylic acid   Carboxylic acid
                         derivative

where Z is a heteroatom (an atom other than C or H, such as Cl, O, N, etc.). Several types of carboxylic acid derivatives are shown below:

Acid halide   Acid anhydride   Ester   Amide

The chemistry of carboxylic acid derivatives is different from the chemistry of ketones and aldehydes, because carboxylic acid derivatives possess a built-in leaving group, which allows the carbonyl group to re-form after being attacked:

The identity of the leaving group will determine how reactive the compound is. For example, acid chlorides are extremely reactive because the built-in leaving group (chloride) is very stable:

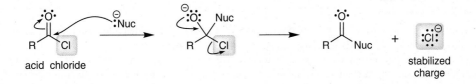

acid chloride   stabilized
                                                            charge

Chloride is a great leaving group because it is a weak base. For this reason, acid halides (also called acyl halides) are the most reactive of the carboxylic acid derivatives. This is very useful, because it means that we can use acid halides to form any of the other carboxylic acid derivatives.

Carboxylic acid derivatives can be viewed as having a "wild card" next to the carbonyl group. I am calling it a "wild card," because it can be easily exchanged for a different group:

In this chapter, we will learn how to exchange the groups so that we can convert one carboxylic acid derivative into another. In order to do this, we will have to know something about the order of reactivity of carboxylic acid derivatives:

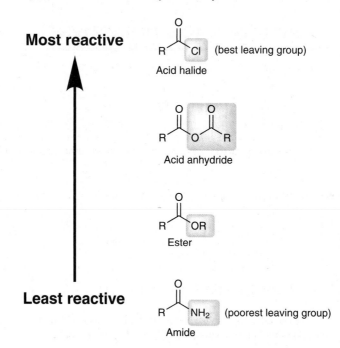

As we learn to exchange the "wild card," we will see that there are just a few simple rules that will determine everything. We will see dozens of reactions, but all of these reactions are completely predictable and understandable if you know how to apply just a few simple rules, covered in the following section.

We will NOT cover every reaction in your textbook or lecture notes. Rather, we will focus on the core skills you need. When you are finished with this chapter, you must make sure to go through your textbook and lecture notes to learn any reactions that we did not cover here in this chapter. This chapter will arm you with the skills you need in order to master the material in your textbook.

7.2 GENERAL RULES

The most important rule was already covered in the previous chapter. We called it our "golden rule" and it went like this: after attacking a carbonyl group, always try to re-form the carbonyl group if you can, but never expel H⁻ or C⁻.

With this rule, we can now appreciate that a carboxylic acid derivative will react differently with H⁻ or C⁻ than it will with any other type of nucleophile. When we use a hydrogen nucleophile or a carbon nucleophile to attack a carboxylic acid derivative, *two equivalents* of the nucleophile are consumed. Let's see why:

Consider, for example, the reaction that occurs when an acid halide is treated with an excess of Grignard reagent. First, the Grignard reagent attacks the carbonyl group, just as we would expect:

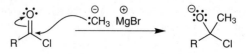

Then, we apply our golden rule: re-form the carbonyl group, if you can, but don't expel H⁻ or C⁻. In this case, Cl⁻ can be expelled, so we expect the carbonyl group to re-form:

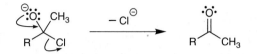

This step generates a ketone, and under these conditions, the ketone can be attacked by a second equivalent of the nucleophile, generating an alkoxide ion:

an alkoxide ion

The resulting alkoxide ion is now unable to re-form the carbonyl group, because there is no leaving group. We have already said numerous times that H⁻ or C⁻ cannot be expelled to re-form the carbonyl group. So, the reaction is complete. In order to protonate the alkoxide ion, a proton source must be introduced into the reaction flask (note that the proton source must be introduced into the reaction flask AFTER the reaction is complete):

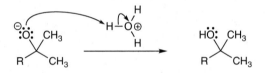

In the end, the product is an alcohol.

A similar result is expected when an acid halide is attacked by a hydrogen nucleophile. Once again, the acid halide can react with two equivalents of the nucleophile to generate an alcohol:

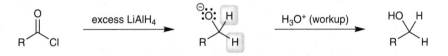

The situation is very different when we use any other nucleophile (not H⁻ or C⁻). Suppose, for example, an acid halide is treated with RO⁻:

As we might expect, the resulting tetrahedral intermediate is capable of re-forming the carbonyl group by expelling a chloride ion, generating an ester:

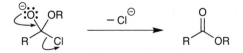

Now, it might be true that the resulting carbonyl group (of the ester) CAN be attacked by another alkoxide ion:

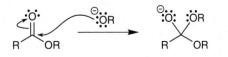

But the resulting tetrahedral intermediate will simply re-form the carbonyl group, by expelling the second RO⁻ that just attacked. This brings us right back to the ester:

The second attack is only irreversible when the nucleophile is H⁻ or C⁻. And this should make sense based on our golden rule; if the second attack is by H⁻ or C⁻, then the carbonyl group will not be able to re-form after the second attack occurs.

So, we have seen that there is a difference in the types of products we obtain when we use H⁻ or C⁻ vs. when we use any other nucleophile. Keep this in mind: H⁻ or C⁻ will attack twice, but all other nucleophiles will only attack once:

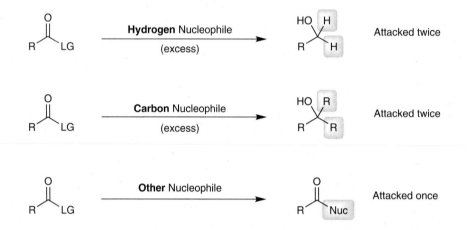

Now let's focus our attention on the last case above (nucleophiles other than H⁻ or C⁻), in which the nucleophile attacks only once. In those reactions, the outcome will be to exchange one type of carboxylic acid derivative for another. For example:

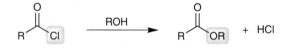

The mechanism for this process (and all others like it) involves two core steps: attack the carbonyl group, and then re-form the carbonyl group. That's it. Just two core steps. But very often, proton transfer steps are necessary when drawing a mechanism. Proton transfers can only occur at three different moments: the beginning, the middle, or the end:

1. A proton transfer step may or may not be required at the *beginning* of the mechanism, before the nucleophile has attacked the carbonyl group.

2. Proton transfer steps may or may not be required in the *middle* of the mechanism, after the carbonyl group has been attacked, but before the carbonyl group has re-formed.

3. A proton transfer step may or may not be required at the *end* of the mechanism, after the carbonyl group has re-formed.

Sometimes, a mechanism won't involve any proton transfers at all. For example, when an acid halide is treated with an alkoxide ion, no proton transfers occur:

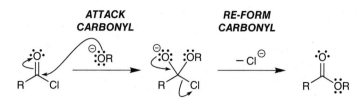

And sometimes, a mechanism will involve only one proton transfer. For example, when an acid halide is treated with water, one proton transfer occurs at the end of the mechanism:

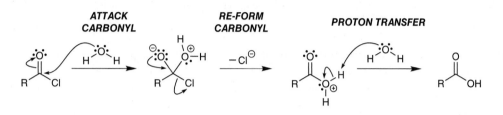

But sometimes, proton transfers occur at all three moments in the mechanism (beginning, middle, and end). For example, consider the following mechanism for the reaction in which an ester is treated with aqueous acid, generating a carboxylic acid:

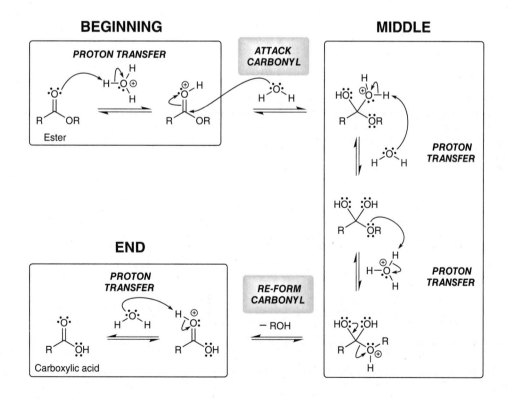

This is a long mechanism, with six steps. The two core steps (attack the carbonyl, and then re-form the carbonyl) are highlighted in gray. Notice that proton transfer steps occur at all three possible moments throughout the mechanism (beginning, middle, and end). Later in this chapter, we will study this reaction in greater detail, and we will explain the rationale behind each and every proton transfer step in this mechanism.

7.3   ACID HALIDES

As we have mentioned, acid halides are the most reactive of the carboxylic acid derivatives, because they produce the most stable leaving groups. Therefore, we can prepare any of the other carboxylic acid derivatives from acid halides. So, it is critical that you know how to make an acid halide. It is very common to encounter a synthesis problem where you will need to make an acid halide at some point in the synthesis.

To make an acid halide, it would be nice if Cl⁻ would attack a carboxylic acid directly, expelling HO⁻ as a leaving group:

But this doesn't work, because HO⁻ is less stable than Cl⁻, so this would be an uphill battle. Cl⁻ is not going to expel HO⁻. So, we must first convert the OH group into a different group that CAN be expelled by Cl⁻. So, here is our strategy:

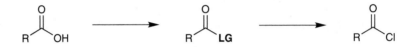

Thionyl chloride, SOCl₂, is a reagent that can be used to execute both steps of this strategy:

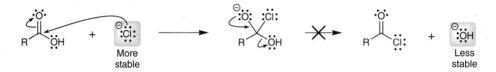

Thionyl chloride
(SOCl₂)

An acid halide can generally be obtained in good yields by treating a carboxylic acid with thionyl chloride. This reagent achieves two objectives in one reaction flask: (1) it converts the OH group into a better leaving group and (2) it serves as a source of chloride ions that will attack the carbonyl group and expel the newly formed leaving group. The first objective (converting OH into a better leaving group) is achieved via the following mechanism:

EXCELLENT
LEAVING GROUP

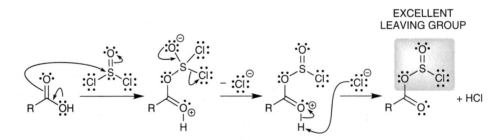

+ HCl

The carboxylic acid functions as a nucleophile and attacks the S═O bond. The S═O bond is then re-formed (by expelling a chloride ion), followed by a proton transfer step. The net result of these three steps is the conversion of an OH group into a better leaving group, thereby achieving the first objective.

The second objective is then achieved when the chloride ion (generated during the steps above), attacks the carbonyl group, expelling the leaving group:

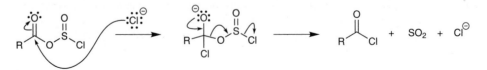

Notice that SO_2 is formed as a by-product. This is important because SO_2 is a gas, which bubbles out of solution, thereby forcing the reaction to completion.

Now that we have covered a method for preparing acid halides, we are ready to explore reactions of acid halides. We will see many reactions. BUT don't try to memorize them. Instead, try to appreciate that they all follow the same general rules. We saw in the previous section that there are two core steps (attack, and re-form the carbonyl). Then, we just need to be careful about proton transfers in the three possible moments when they might be necessary:

With acid halides, it gets much easier, because proton transfers generally only occur at the end of the mechanism (you can skip "1" and "2" on the diagram above). Consider the following example, in which an acid halide is treated with water to generate a carboxylic acid:

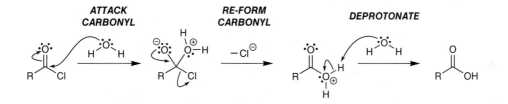

Carefully consider the three steps of this mechanism: attack, re-form, and deprotonate. Say that out loud 10 times real fast (attack, re-form, and deprotonate). You will find that this order of events keeps repeating itself in many of the reactions we are about to see.

Notice that the by-products of the reaction are Cl^- and H_3O^+ which, together, represent an aqueous solution of HCl. The build-up of acid can often produce undesired reactions (depending on what other functional groups might be present in the compound), so pyridine is used to remove the acid as it is produced:

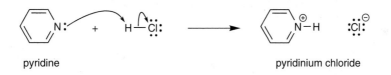

Pyridine is a base that reacts with HCl to form pyridinium chloride. This process effectively traps the HCl so that it is unavailable for any other side reactions. The use of pyridine is often represented with the letters "py," like this:

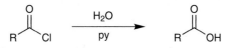

When an acid halide is treated with an alcohol, in the presence of pyridine, the product is an ester:

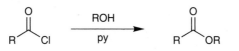

The mechanism for this process has the familiar three steps: attack, re-form, and deprotonate:

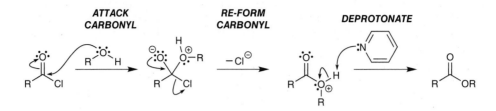

When an acid halide is treated with an amine, the product is an amide:

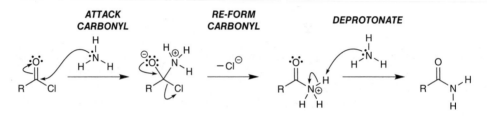

Notice that pyridine is not used in this case. Instead, we use two equivalents of ammonia (twice as much ammonia as acid chloride). One equivalent serves as a nucleophile in the first step of the mechanism, and the other equivalent serves as a base in the last step of the mechanism:

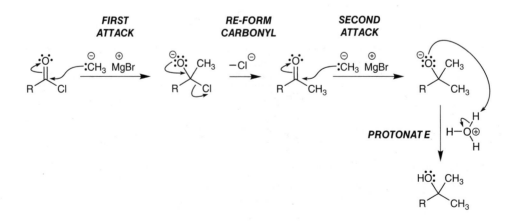

Once again, notice that this process has the familiar three steps: attack, re-form, and deprotonate.

Now let's consider what happens when an acid halide is treated with H⁻ or C⁻. We have already seen that H⁻ and C⁻ are special, because they will attack twice:

This leaves us with an obvious question. What if we wanted to attack with C⁻ just once? In other words, suppose we wanted the following outcome:

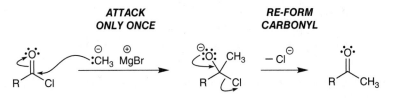

What if the ketone is the desired product? We have a problem here, because the starting acid halide will react with two equivalents of the nucleophile, giving an alcohol. If we just try to use exactly one equivalent of the Grignard reagent, we will observe a mess of products (some molecules of the acid halide will get attacked twice, and others will not get attacked at all). In order to prepare the ketone, we need a carbon nucleophile that will only react with an acid halide, but will **not** react with a ketone. And we are in luck, because there is a class of compounds that will do exactly that. They are called lithium dialkyl cuprates (R_2CuLi).

You might have been introduced to these compounds in the first semester of organic chemistry. Lithium dialkyl cuprates are carbon nucleophiles, but they are less reactive than Grignard reagents. Lithium dialkyl cuprates will react with acid halides, but not with ketones. And therefore, we can use these reagents to convert acid halides into ketones:

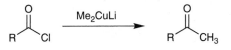

Thus far, we have seen a lot of reactions, so let's just quickly review them:

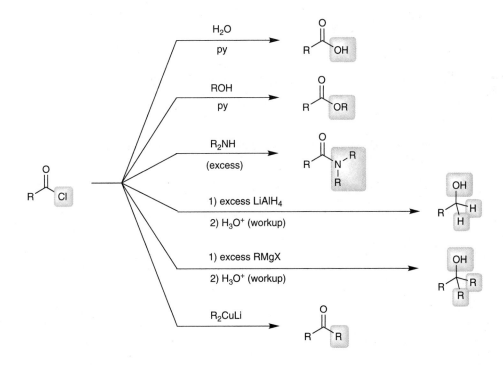

EXERCISE 7.1 Propose a mechanism for the following reaction:

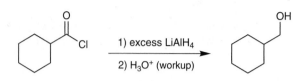

Answer The starting material is an acid chloride, so let's carefully consider the reagents. LiAlH$_4$ is a hydrogen nucleophile (source of H$^-$), and it is expected to attack the carbonyl group twice:

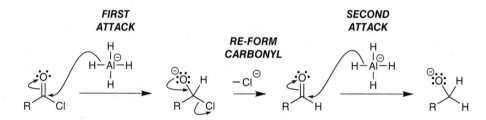

Finally, a proton transfer will be required at the end of the mechanism:

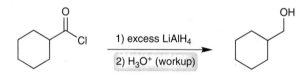

That is, in fact, the reason why H$_3$O$^+$ is shown as a separate step:

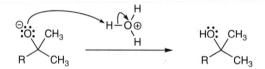

PROBLEMS Propose a plausible mechanism for each of the following reactions. Space has been provided for you to record your answer directly below each reaction:

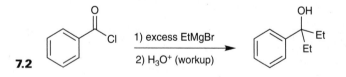

7.2

7.3
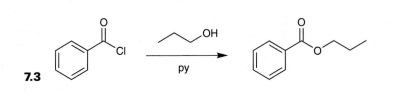

7.4
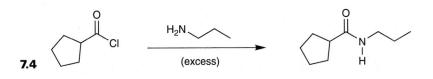

7.5
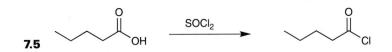

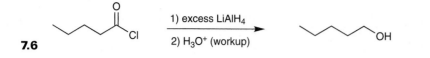

7.6

EXERCISE 7.7 Predict the major product of the following reaction:

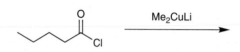

Answer We are starting with an acid halide. So, in order to predict the major product of this reaction, we will have to determine if the nucleophile attacks once or twice. The reagent is a lithium dialkyl cuprate. This reagent is a carbon nucleophile, but it is not like a Grignard reagent—it does not attack twice. Instead, it will only attack once, because it is a very *tame* carbon nucleophile, as we saw earlier in this section. And the product is expected to be a ketone:

PROBLEMS Predict the major product in each of the following cases:

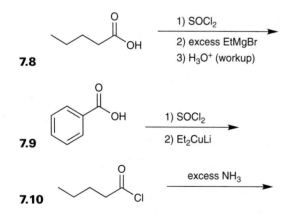

7.8

7.9

7.10

7.11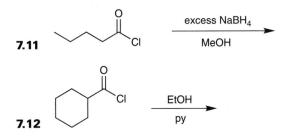
excess NaBH$_4$
MeOH

7.12
EtOH
py

EXERCISE 7.13   Identify the reagents you would use to achieve the following transformation:

Answer   We are starting with a carboxylic acid, and the final product is an alcohol. We also notice that there are two methyl groups in the product:

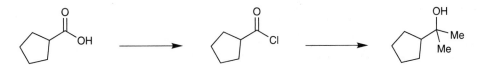

We therefore need to install *two* methyl groups, which means that we need to attack the carbonyl group *twice*. And in the process, the carbonyl group must be reduced to an alcohol. This sounds like a Grignard reaction.

But we have to be careful. We cannot perform a Grignard reaction with a carboxylic acid. Remember that a Grignard reagent is sensitive to its conditions—if there are any acidic protons available, the Grignard reagent will be destroyed. Since our starting compound (a carboxylic acid) has an acidic proton, we must first convert the carboxylic acid into an acid halide. So our strategy goes like this:

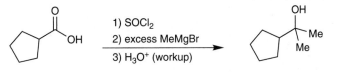

And to achieve this, we could use the following reagents:

1) SOCl$_2$
2) excess MeMgBr
3) H$_3$O$^+$ (workup)

PROBLEMS   Identify the reagents you would use to achieve each of the following transformations:

7.14

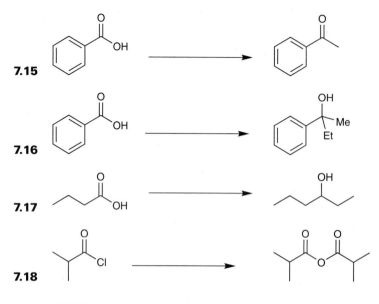

7.15

7.16

7.17

7.18

7.4   ACID ANHYDRIDES

Anhydrides can be prepared by the reaction between a carboxylic acid and an acid halide:

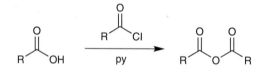

Notice that pyridine is used, once again, to remove the HCl that is formed as a by-product. We can avoid the need for pyridine by using a carboxylate ion (a deprotonated carboxylic acid) instead of a carboxylic acid:

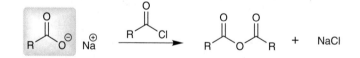

Notice that the by-product is NaCl, rather than HCl, so pyridine is no longer needed.

Acid anhydrides are almost as reactive as acid halides. So, the reactions of acid anhydrides are very similar to the reactions of acid halides. You just have to train your eyes to see the leaving group:

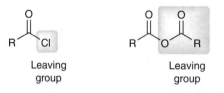

When a nucleophile attacks an acid anhydride, the carbonyl group can re-form to expel a leaving group that is resonance-stabilized:

So, an acid anhydride can be treated with any of the nucleophiles that we saw in the previous section to give the same products that we saw in the previous section:

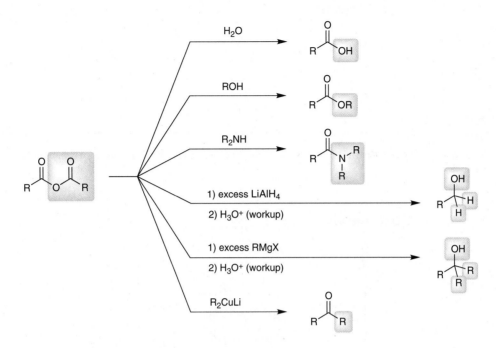

Notice that pyridine is not required in any of these reactions, because HCl is not a by-product.

EXERCISE 7.19   Predict the major product of the following reaction:

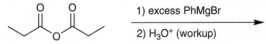

Answer   In order to predict the major product of this reaction, we will have to determine if the nucleophile attacks once or twice. The reagent is a Grignard reagent, which is a strong carbon nucleophile. So we expect that the anhydride will react with two equivalents of the Grignard reagent, which will produce an alcohol:

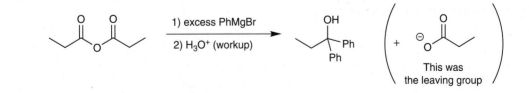

PROBLEMS   Predict the major product for each of the following reactions:

7.20

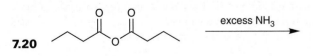

excess NH_3

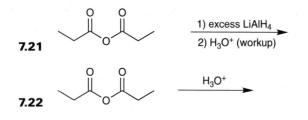

7.21

7.22

7.5 ESTERS

Esters can be made from carboxylic acid derivatives that are more reactive than esters. In other words, we can make an ester from an acid halide, or from an anhydride:

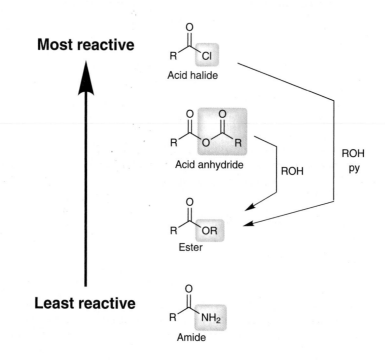

And we have already seen how to make acid halides—we can make them from carboxylic acids (using thionyl chloride). So, this provides a two-step method for making an ester from a carboxylic acid:

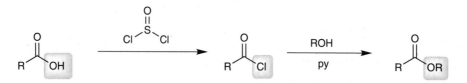

The carboxylic acid is first converted into an acid halide, which is then converted into an ester. But this begs the question: can we achieve the desired transformation *in one reaction*? That is, can we convert a carboxylic acid into an ester in just one reaction, avoiding the need to prepare an acid halide? If we simply try to mix an alcohol and a carboxylic acid, we do *not* observe a reaction:

So, let's see what we can do to force this reaction along. Let's consider what happens if we try to make the nucleophile more nucleophilic. In other words, suppose we try to use RO⁻ instead of ROH:

Recall that carboxylic acids have a mildly acidic proton, and alkoxide ions (RO⁻) are strong bases. So, the alkoxide ion would just function as a base and deprotonate the carboxylic acid:

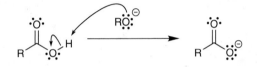

So, once again, an ester would NOT be produced.

But there is one more thing we can try. Rather than making the nucleophile more nucleophilic, we can try to make the electrophile more electrophilic. Do you remember how to do that? We saw in the previous chapter (Section 5.4) that a carbonyl group becomes more electrophilic when it is protonated:

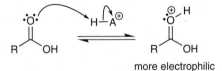

more electrophilic

As we will soon see, the acid is not consumed during the reaction, so its function is catalytic (just as we saw in Section 5.4). And under these conditions (acidic conditions), we *do* observe the desired reaction (a one-step synthesis of an ester from a carboxylic acid):

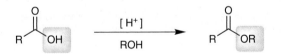

This reaction is incredibly useful and important (and it is considered to be a staple of any organic chemistry course), so let's explore the accepted mechanism, step-by-step.

The mechanism has six steps, but you should notice that there are only two core steps here: attack, and then re-form. All other steps are just proton transfers:

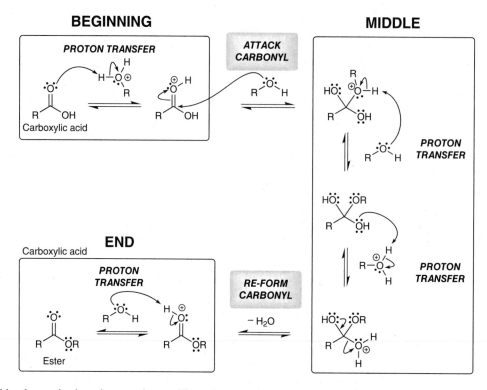

Carefully consider the mechanism above, and you will see there are three moments at which proton transfer steps occur. This exactly follows the pattern that we described in the beginning of this chapter:

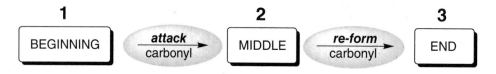

1. *In the beginning of the mechanism,* a proton transfer step is necessary in order to protonate the carbonyl group, rendering it more electrophilic.

2. *In the middle of the mechanism* (after attack of the carbonyl, but before re-forming the carbonyl), two proton transfers are required to protonate the leaving group (HO⁻ cannot be expelled in acidic conditions). Notice that two proton transfer steps are required. It would not be OK to simply protonate the OH group without first deprotonating the OR group, because that would give an intermediate with two positive charges. So, first we deprotonate to remove the positive charge, and then, we protonate the OH group to give a good leaving group.

3. *In the end of the mechanism,* a proton transfer is required to remove the positive charge and generate the product.

This reaction is called a Fischer esterification. The position of equilibrium is very sensitive to the concentrations of starting materials and products. Excess ROH favors formation of the ester, while excess water favors the carboxylic acid:

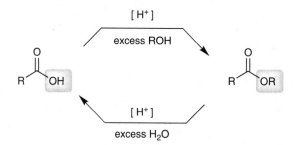

This is very helpful, because it provides a way to convert an acid into an ester, *or* an ester into an acid.

For now, let's focus on converting an acid into an ester:

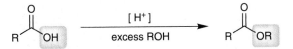

We will spend some time on the reverse process very soon.

We have seen that a Fischer esterification is the reaction between a carboxylic acid and an alcohol (with acid catalysis). If a single compound contains both functional groups (COOH and OH), it is possible to observe an intramolecular Fischer esterification. For example, consider the following compound:

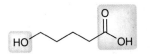

This compound has both a COOH group and an OH group. And, in this case, an intramolecular reaction is observed:

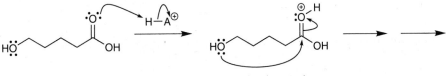

Intramolecular attack

The rest of the mechanism is directly analogous to what we have already seen. The mechanism has two core steps (attack and re-form), with proton transfer steps in the beginning, middle, and end.

PROBLEM 7.23 In the space provided on the next page, draw a mechanism for the following transformation:

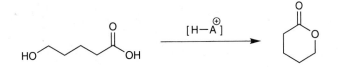

Remember that the general steps are:

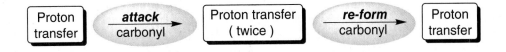

EXERCISE 7.24 Identify the reagents you would use to make the following compound via a Fischer esterification:

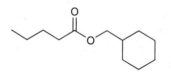

Answer To make an ester using a Fischer esterification, we need to start with a carboxylic acid and an alcohol. The question is: how do we decide which carboxylic acid and which alcohol to use? To do this, we must identify the bond that will be formed during the reaction (highlighted):

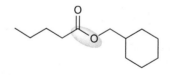

To form the highlighted bond via a Fischer esterification, we will need the following reagents:

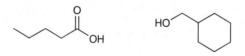

And, don't forget that we need acid catalysis. So our synthesis would look like this:

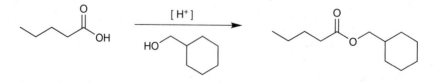

PROBLEMS   Identify the reagents you would use to make each of the following esters:

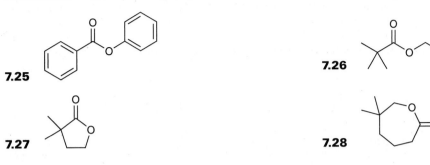

7.25

7.26

7.27

7.28

Now that we have seen how to make esters, let's focus our attention on the reactions of esters. We will center on two reactions in particular. Esters can be hydrolyzed to give carboxylic acids, under two different sets of conditions:

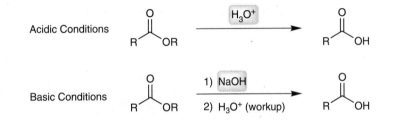

The first set of conditions above (acidic conditions) should seem very familiar to you. This reaction is simply the reverse of a Fischer esterification. Let's explore the accepted mechanism for this process:

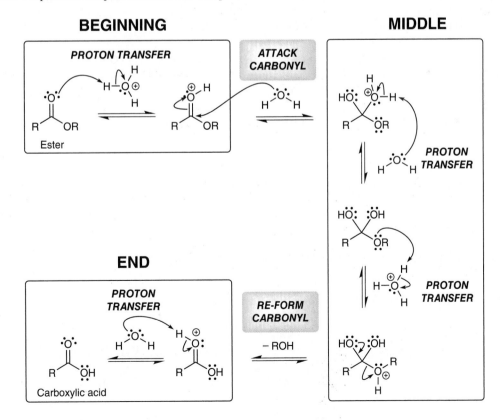

Once again, we see the same pattern again and again. Look closely at this mechanism. There are two core steps: attack, and re-form. All other steps in the mechanism are just proton transfers that facilitate the reaction. One proton transfer is required in the beginning (to protonate the carbonyl group), two proton transfers are required in the middle (so that the leaving group can leave as a neutral species, ROH), and one proton transfer is required at the end (to deprotonate).

The process above occurs under acidic conditions. But it is also possible to hydrolyze an ester under basic conditions as well. The following is a mechanism for that process:

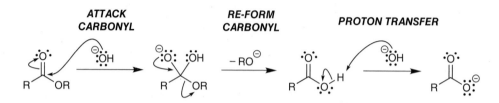

Once again, we have two core steps (attack and re-form), followed by a deprotonation. This proton transfer at the end is unavoidable under these conditions. In basic conditions, a carboxylic acid will be deprotonated. In fact, formation of a more stable anion is the driving force for this reaction:

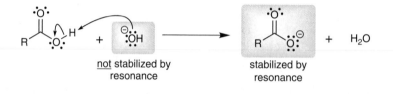

This resonance-stabilized anion is called a carboxylate ion, and its formation serves as a driving force. In order to protonate the carboxylate ion, a proton source must be introduced into the reaction flask (note that the proton source is introduced into the reaction flask AFTER the reaction is complete, and must be indicated as a separate step):

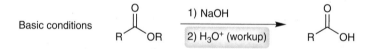

For this workup step, we must use H_3O^+ (we cannot use H_2O), because a carboxylate ion will not remove a proton from water:

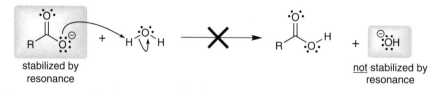

This process (hydrolysis of an ester under basic conditions) has a special name: *saponification*.

EXERCISE 7.29 Propose a mechanism for the following transformation:

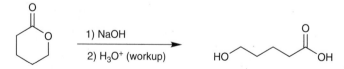

Answer   This reaction utilizes basic conditions to hydrolyze an ester (a process called saponification). We begin by attacking the carbonyl group, and then re-forming it:

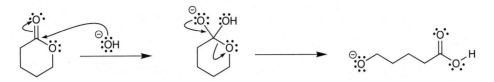

Notice that we just produced a carboxylic acid (a weak acid) under strongly basic conditions. Under these conditions, the carboxylic acid is deprotonated to give a carboxylate ion:

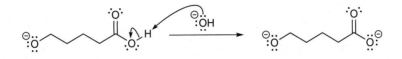

After the reaction is complete, an acid is introduced into the reaction flask to protonate the dianion (notice that the stronger base is protonated first):

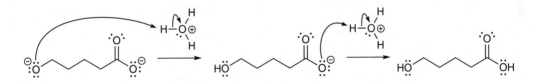

PROBLEMS   In the space provided, draw a mechanism for each of the following transformations:

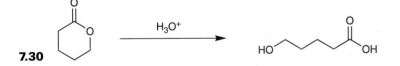

7.30

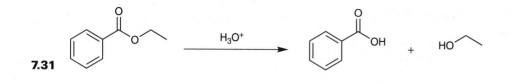

7.31

EXERCISE 7.32 Predict the products of the following reaction:

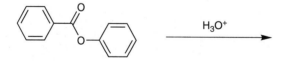

Answer We are starting with an ester, and we are subjecting it to aqueous acidic conditions. This will convert the ester into a carboxylic acid and an alcohol (the reverse of a Fischer esterification). We expect the following products:

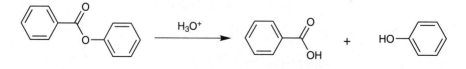

PROBLEMS   Predict the products of each of the following reactions:

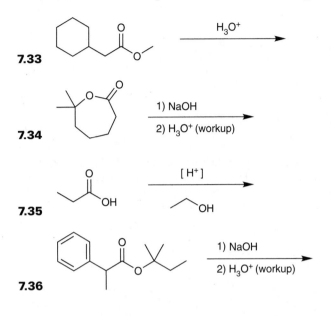

7.33

7.34

7.35

7.36

7.6   AMIDES AND NITRILES

We have said before that a carboxylic acid derivative can be prepared from any other carboxylic acid derivative that is more reactive. Let's go back to our reactivity chart to see what this means practically:

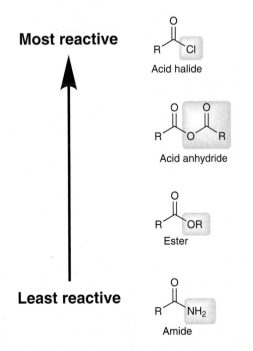

Since amides are the least reactive of the carboxylic acid derivatives (shown on the chart above), we can therefore make amides from any carboxylic acid derivatives that are higher on the chart. In other words, we can make amides from acid halides, from anhydrides, or from esters.

Earlier in this chapter, we saw how to make amides from acid halides or anhydrides:

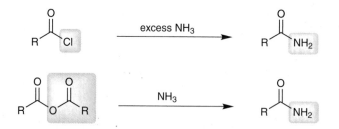

But now the question is: how do we make amides from esters? Esters are less reactive than acid halides or anhydrides. So, we have to use some kind of trick to coax the reaction along. We cannot use acid or base (an acid would just protonate the attacking amine, rendering it useless; and a base would cause other side reactions that we will learn in the next chapter). Instead, we use brute force and patience. We just heat the reaction for a long time, and a reaction is observed, which can occur via the following mechanism:

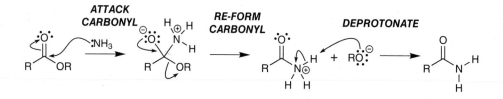

Notice that RO⁻ is expelled to re-form the carbonyl group. That might seem strange, because there is a much better leaving group available to leave:

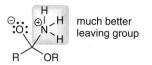

But this gets back to something we have said many times before. Of course it is possible for the amine to leave. In fact, it happens all of the time. The amine attacks, and then it gets expelled. It attacks, and is then expelled again. Every time this happens, there is no change to observe. But every once in a while, something else can happen. RO⁻ can be expelled, which is then immediately protonated, as shown in the mechanism above. We are allowed to expel RO⁻ when re-forming a carbonyl group, because the tetrahedral intermediate is so high in energy (negative charge on an oxygen atom).

The equilibrium for this process favors the products (amide + alcohol) over the reactants (ester + amine):

So, this is another method for making amides, however, the process is extremely slow and is not really a practical synthetic technique. Therefore, you should avoid using this reaction in a synthesis, if possible. A better method to make amides is to start with a more reactive carboxylic acid derivative.

So far, in this section, we have seen that we can make amides from acid halides or from acid anhydrides. Now that we know how to make amides, let's explore some important reactions of amides. Specifically, we will explore hydrolysis of amides (first under acidic conditions, and then under basic conditions). It is worth mentioning that much of biochemistry is dependent on how, when, and why amides will undergo hydrolysis. So, if you plan on taking biochemistry, you should certainly be familiar with the hydrolysis of amides, which can occur under either basic conditions or acidic conditions:

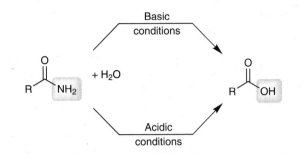

Let's begin with acid-catalyzed conditions. This reaction is really no different than the other acid-catalyzed reactions we have seen. Take a close look at the following mechanism:

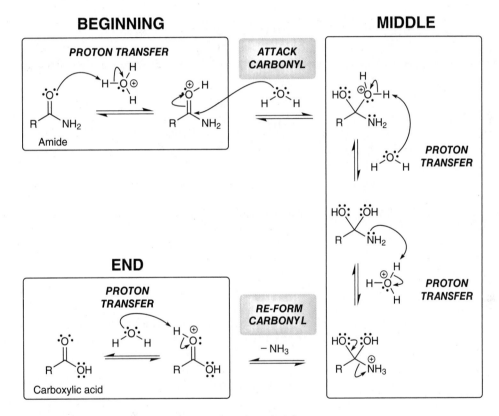

Notice that it follows the exact same pattern that we have seen again and again:

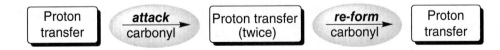

This pattern is common among the acid-catalyzed reactions that we have seen so far in this chapter.

Now let's explore base-catalyzed conditions for the hydrolysis of an amide:

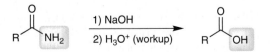

This transformation can occur via the following mechanism:

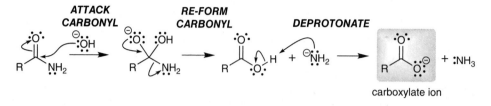

Under basic conditions, the product is a carboxylate ion (highlighted above), and that is why an acid is listed as a reagent (in a separate step):

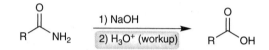

When the reaction is complete and H_3O^+ is then added to the reaction flask, the carboxylate ion is protonated to generate the carboxylic acid:

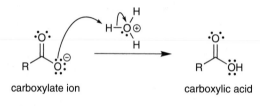

Before we do some problems, let's look at one last carboxylic acid derivative that we have not yet seen. Compounds containing a cyano group are called nitriles:

cyano
group

You might be wondering why nitriles are considered to be carboxylic acid derivatives. After all, a nitrile looks very different from the other carboxylic acid derivatives. To make sense of this, we need to consider oxidation states. Each of the carboxylic acid derivatives has three bonds to electronegative atoms:

The carbon atom of the carbonyl group has two bonds with oxygen, and it also has one more bond with some heteroatom, Z (O, N, Cl, etc.). That gives a total of three bonds to heteroatoms. The carbon atom of a cyano group also has three bonds to a heteroatom. So, nitriles are at the same oxidation level as the other carboxylic acid derivatives.

Nitriles can be prepared using cyanide as a nucleophile to attack an alkyl halide:

This is an S_N2 process, so you can only use this method with primary or secondary alkyl halides (primary halides are much better). Don't use this method with tertiary alkyl halides. There are other ways to make nitriles. You should look through your textbook (and your lecture notes) to see if you are responsible for knowing any other ways to make nitriles.

We can easily see that nitriles really are at the same oxidation level as other carboxylic acid derivatives, because hydration (which is *not* an oxidation–reduction reaction) produces an amide:

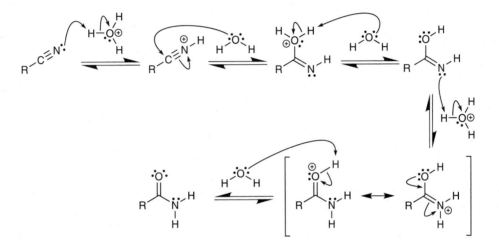

Hydration can occur either under acidic conditions or under basic conditions.

Whether we perform an acid-catalyzed hydration or a base-catalyzed hydration, the core steps are slightly different than the core steps of the mechanisms we have seen in this chapter. So far, all of our reaction mechanism have had at least *two* core steps (attack the carbonyl, and then expel a leaving group to re-form the carbonyl), with all other steps being proton transfers. But now, we will see a mechanism that has just *one* core step (attack the carbonyl). When it comes to the hydration of nitriles, no leaving group is expelled. The carbonyl group can form simply through proton transfers (after the nucleophilic attack):

This is the mechanism for the hydration of a nitrile under *acidic* conditions, and it is analogous to hydration of an alkyne under acidic conditions to give a ketone. Notice that the second step in the mechanism shows a nucleophile (H$_2$O) attacking a protonated cyano group (very much the way a protonated carbonyl group can be easily attacked). All other steps in the mechanism are just proton transfers. When you think of it this way, it greatly simplifies the mechanism. The many proton transfers are necessary to avoid the formation of a strong base, which cannot occur under acidic conditions.

Now let's consider the hydration of nitriles under *basic* conditions. The mechanism is actually VERY similar to the mechanism above. There is also only *one* core step (attacking the cyano group), and all the other steps are just proton transfers. But in basic conditions, the cyano group is *not* first protonated. Rather, it is attacked by hydroxide first:

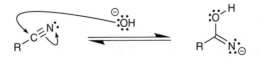

The rest of the mechanism is all just proton transfer steps. In order to draw the proton transfers properly, you must keep one thing in mind: stay consistent with the conditions. In acidic conditions, avoid the formation of a strong base. In basic conditions, avoid the formation of a strong acid.

With this in mind, complete the following exercise (Problem 7.37) to see if you can draw a plausible mechanism for the hydration of a nitrile under basic conditions.

PROBLEM 7.37   Based on everything we have just seen, propose a mechanism for the hydration of a nitrile under basic conditions:

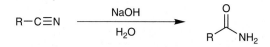

Remember, there is just one core step (attacking the cyano group with hydroxide). After that, all other steps are just proton transfers. Use the space below to record your answer. When you have finished, you can look in the back of the book (or in your textbook) to see if you got it right.

EXERCISE 7.38   Draw a plausible mechanism for the following transformation:

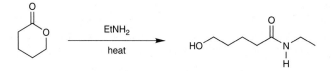

Answer   In this reaction, an ester is being treated with an amine under conditions of heating. We have seen these conditions before. A source of protons (an acid) has not been indicated among the reagents, so the carbonyl group is not protonated. The first step of the mechanism will involve the amine directly attacking the carbonyl group of the ester:

Then the carbonyl group is re-formed via expulsion of an alkoxide ion (RO⁻) as a leaving group:

This intermediate is then converted into the product via two successive proton transfer steps:

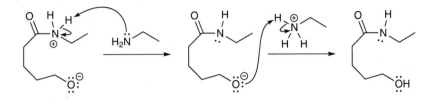

PROBLEMS   Propose a plausible mechanism for each of the following transformations.

7.39

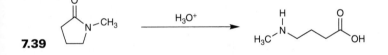

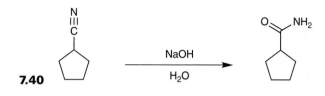

7.40

7.41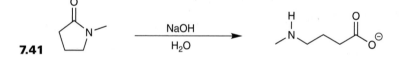

And now let's just do one more challenging mechanism. I say "challenging" not because it is difficult, but because you have not seen this exact mechanism before. Rather, you should be able to work your way through the mechanism, using all of the skills we have developed in this chapter:

PROBLEM 7.42 Propose a mechanism for the following reaction:

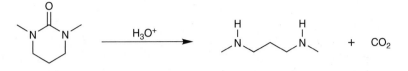

EXERCISE 7.43 Predict the products of the following reaction:

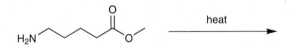

Answer We were not given any reagents here (just conditions of heat), so we look carefully at the starting material to see if we can have an intramolecular reaction. We notice that there are two functional groups in our starting compound (an ester and an amine). And we have seen that an ester can react with an amine under conditions of heating. The products should be an amide and an alcohol:

PROBLEMS Predict the products for each of the following reactions:

7.44

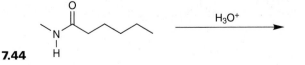

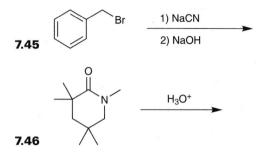

7.45

$$\text{1) NaCN} \quad \text{2) NaOH}$$

7.46

$$\text{H}_3\text{O}^+$$

7.7 SYNTHESIS PROBLEMS

We have seen a lot of reactions in this chapter. Almost all of them involved the conversion of one carboxylic acid derivative into another. We saw that you can make a carboxylic acid derivative from any other derivative that is more reactive. In other words, you can always step your way **down** the following chart:

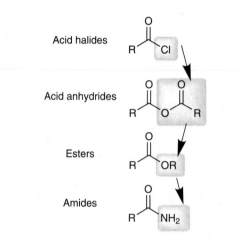

You can even jump down the chart if you want:

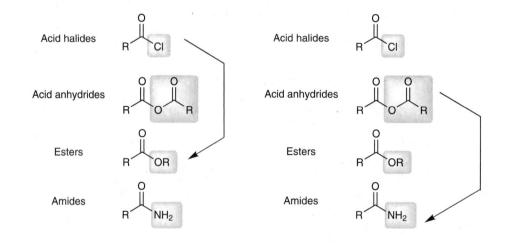

But travelling *up* the chart is more difficult:

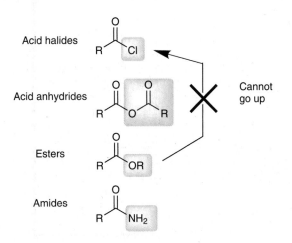

So, how do you travel *up* the chart, if you need to? Here is the way to do it: You can exit this chart by converting into a carboxylic acid, and then come back into the chart, like this:

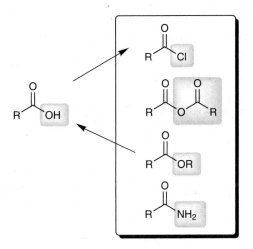

Let's get some practice with this:

EXERCISE 7.47 Propose an efficient synthesis for the following transformation:

Answer We must convert an amide into an ester, but we have not learned a way to do this directly in one step (because that would involve going *up* the chart). Amides are less reactive than esters, so we cannot go directly from an amide to an ester. Instead, we can first convert the amide into a carboxylic acid, and then we can convert the carboxylic acid into an ester:

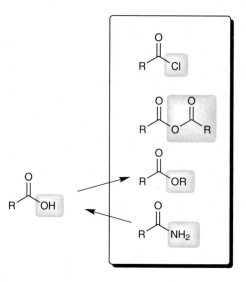

To achieve this transformation, we would use the following reagents:

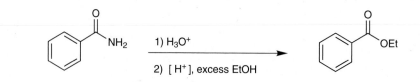

PROBLEMS Propose an efficient synthesis for each of the following transformations:

7.48

7.49

7.50

7.51

7.52

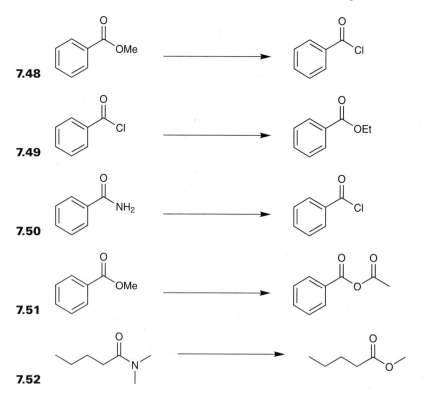

There is one other important strategy to keep in mind when you are proposing a synthesis. In this chapter, we explored the chemistry of carboxylic acid derivatives; and in the previous chapter, we explored the chemistry of ketones/aldehydes. These two chapters represent two different realms:

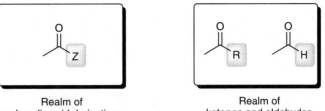

Realm of
carboxylic acid derivatives

Realm of
ketones and aldehydes

But these realms are not completely isolated from one another, because we have seen ways to convert from one realm into another. In this chapter, we saw how to convert an acid halide into a ketone:

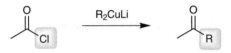

There is also another way to cross over from the realm of carboxylic acids into the realm of ketones and aldehydes. Rather than making a ketone (like above), we can make an aldehyde using the following two steps:

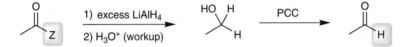

Some textbooks and instructors will teach you a reagent that can achieve this overall transformation in one step (converting from an acid halide into an aldehyde). There are actually many hydride reagents that are sufficiently selective to convert an acid halide into an aldehyde (very much the way lithium dialkyl cuprates are sufficiently selective to convert an acid halide into a ketone, without attacking the carbonyl group a second time). You should look through your textbook and lecture notes to see if you have covered a selective hydride nucleophile. If you haven't, you can always use the two-step method (shown above) for converting an acid halide into an aldehyde.

With the reactions above, we have seen how to "cross over" from the realm of carboxylic acid derivatives into the realm of ketones and aldehydes:

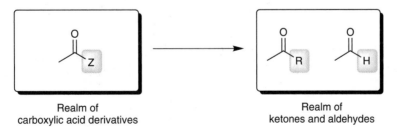

Realm of
carboxylic acid derivatives

Realm of
ketones and aldehydes

But what about the reverse direction? Do we have a way to "cross over" from the realm of ketones and aldehydes into the realm of carboxylic acid derivatives?

Realm of
carboxylic acid derivatives

Realm of
ketones and aldehydes

Yes, we have seen a way to do this also. Recall (from the previous chapter) that a Baeyer–Villiger oxidation will convert a ketone into an ester:

We can also use a Baeyer–Villiger oxidation to convert an aldehyde into a carboxylic acid (remember migratory aptitude?):

So, we now have reactions that allow us to "cross over" from one realm into the other (in either direction). Let's see an example of how to use this:

EXERCISE 7.53   Propose an efficient synthesis for the following transformation:

Answer   The final product is an alcohol, and in the process of converting the carboxylic acid into an alcohol, an ethyl group must be installed. It might be hard to see, at first glance, how this transformation can be achieved. But don't get discouraged. You are not expected to know how to solve problems like this instantly. Synthesis problems require thought and strategizing. Remember that you always want to try to go backwards as much as possible (retrosynthetic analysis). So, let's work our way backwards.

Did we learn any simple ways to make alcohols? In the previous chapter, we learned how to make an alcohol from a ketone, using LiAlH$_4$:

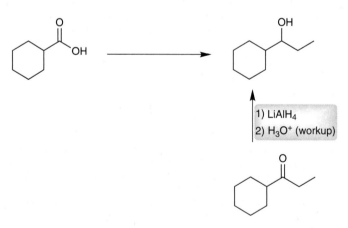

With this one important step, we are now in a position to realize that this problem can be thought of as a "cross-over" problem. The starting material is a carboxylic acid, and we need to turn it into a ketone:

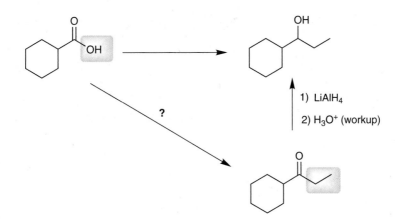

In other words, we need to cross over from the realm of carboxylic acids into the realm of ketones. And we did see one reaction that allows us to do that. We can make a ketone from an acid halide, using a lithium dialkyl cuprate. So, now we have worked backwards again:

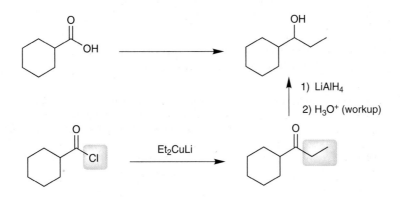

To complete the synthesis, we just need to convert the carboxylic acid into an acid halide, and we can do that in one step with thionyl chloride.

So, our answer is:

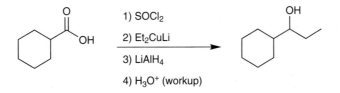

Now let's get some practice with some more "cross-over" problems. In order to do these problems, you will need to review this chapter **and** the previous chapter (ketones and aldehydes)—you will need to have all of the reactions from both chapters at your fingertips.

At first, you might find it difficult to identify the following problems as cross-over problems, but hopefully, you will start to see some trends as you solve these problems.

In solving these problems, make sure that you are familiar with the ways that we have seen for crossing over. We have seen four such reactions so far, all of which are summarized in the following chart:

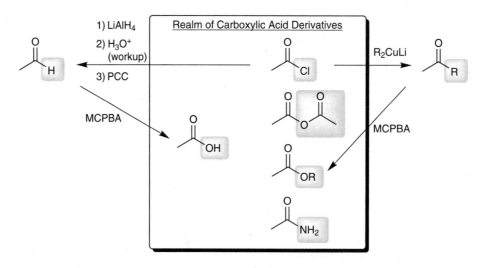

Study this chart carefully. To help you remember them, you should notice one thing that all four reactions have in common. They are all reduction–oxidation reactions. This should make sense because carboxylic acid derivatives are at a different oxidation state than ketones and aldehydes.

You will need some time to do the following problems, so don't sit down to do these problems when you only have 5 minutes to study. That would just frustrate you. Make sure that you have some time to spend when you sit down to work through these problems.

PROBLEMS Propose an efficient synthesis for each of the following transformations. In each case, remember to work backwards (retrosynthetic analysis), and try to determine which cross-over reaction to use. When you compare your answers to the answers in the back of the book, keep in mind that there is often more than one way to solve a synthesis problem. If your answer is different from the answer in the back of the book, you should not necessarily conclude that your answer is wrong.

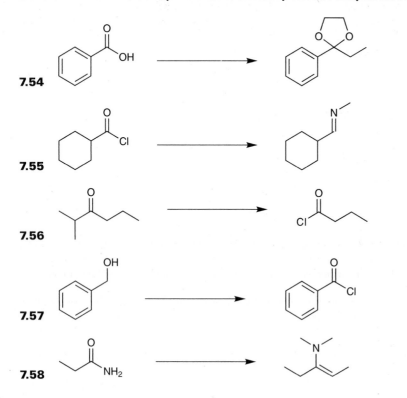

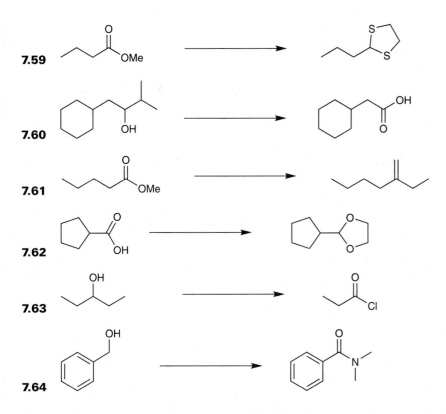

7.59

7.60

7.61

7.62

7.63

7.64

The goal of this chapter was to lay a foundation that will enable you to study your textbook and lecture notes more efficiently. We saw a few simple rules that govern all of the mechanisms in this chapter, and we learned several synthesis strategies.

Now you can go back through your textbook and lecture notes, and look for those reactions that we did not cover here in this chapter. With the foundation we have built in this chapter, you should be in good shape to fill in the gaps and study more efficiently.

And make sure to do ALL of the problems in your textbook. You will find more synthesis problems there. The more you practice, the better you will get. Good luck.

END-OF-CHAPTER PROBLEMS

PRACTICE PROBLEMS *(Problems that involve only one skill)*

7.65 Predict the major product of the following reaction.

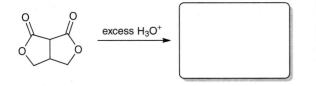

7.66 Predict the major product of the following reaction.

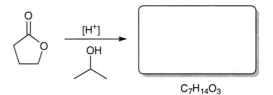

$C_7H_{14}O_3$

7.67 Predict the major product of the following reaction.

7.68 Draw the major products of the following reaction.

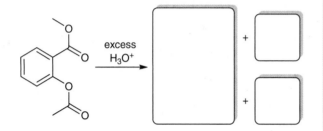

7.69 On a separate piece of paper, draw a mechanism for the following reaction:

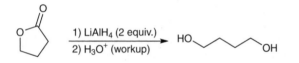

7.70 Predict the major product of the following reaction.

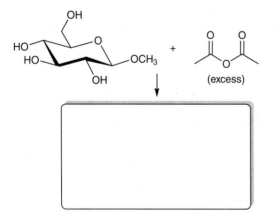

7.71 Draw the missing structures in the following reaction scheme:

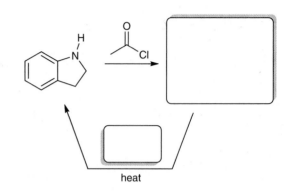

7.72 Draw a starting material that can be used to prepare the product below via an intramolecular Fischer esterification:

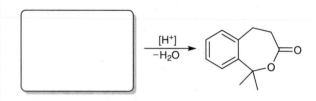

7.73 Draw the missing structure in the following reaction scheme:

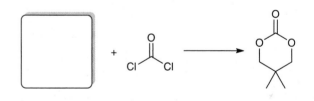

INTEGRATED PROBLEMS *(Problems that involve more than one skill)*

7.74 Propose an efficient synthesis for the following transformation:

7.75 Propose an efficient synthesis for the following transformation:

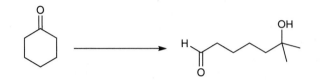

7.76 Propose an efficient synthesis for the following transformation:

7.77 Propose an efficient synthesis for the following transformation:

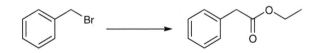

7.78 Propose an efficient synthesis for the following transformation:

7.79 Consider the following reaction:

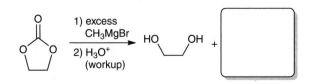

(a) This reaction has two products. One of the products is shown. Draw the other product in the box provided.

(b) On a separate piece of paper, draw a complete mechanism for this reaction.

7.80 Consider the following reaction:

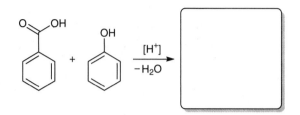

(a) Draw the major product of this reaction.

(b) Identify how you could use IR spectroscopy to determine whether the reaction was successful.

(c) Identify how you could use ^{1}H NMR spectroscopy to determine whether the reaction was successful.

7.81 Compound **A** has the molecular formula $C_6H_{12}O_2$, and the IR spectrum of compound **A** has a signal at $1740\,cm^{-1}$. When treated with aqueous acid, compound **A** is hydrolyzed to give compounds **B** and **C**.

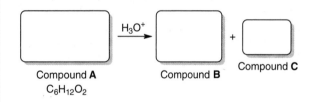

(a) The ^{1}H NMR spectrum of compound **B** has three signals: a singlet at 12 ppm (I = 1H), a septet at 2.7 ppm (I = 1H), and a doublet at 1.3 ppm (I = 6H). Draw the structure of compound **B** in the box provided above.

(b) In the box provided above, draw the structure of compound **A**.

(c) In the box provided above, draw the structure of compound **C**.

(d) How many signals will appear in the ^{1}H NMR spectrum of compound **C**?

(e) How many signals will appear in the ^{13}C NMR spectrum of compound **C**?

(f) How many signals will appear in the ^{13}C NMR spectrum of compound **B**?

7.82 Consider the following reaction, which generates a product with the molecular formula $C_9H_{14}O_2$:

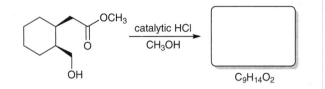

(a) In the box provided above, draw the product of this reaction.

(b) On a separate piece of paper, draw a complete mechanism for this reaction.

(c) Describe how you could use ^{13}C NMR spectroscopy to differentiate the starting material from the product.

7.83 The reaction below is part of a larger process called the bromoform reaction, which will be covered in the next chapter. On a separate piece of paper, draw a reasonable mechanism for the reaction shown. Your mechanism should be consistent with the observation that Br_3C^- is a weaker base than hydroxide but is a stronger base than a carboxylate ion (RCO_2^-).

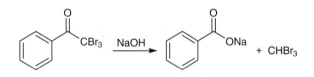

7.84 On a separate piece of paper, draw a reasonable mechanism for the following reaction. (Hint: Your mechanism should have only six steps and should not involve the use of water as a nucleophile, because an intramolecular nucleophilic attack is faster than an intermolecular one).

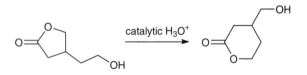

7.85 When the lactone (cyclic ester) below is treated with a catalytic amount of hydroxide, a polymerization process occurs, generating the polymer shown below:

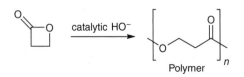

A polymer is a very large molecule (a macromolecule) comprised of repeating units. In the polymer above, the repeating unit is shown in brackets, and the "n" indicates that a large number of these repeating units are connected to each other. Below is another drawing of the same polymer, showing four repeating units.

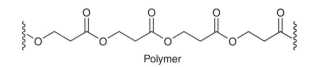

Polymer

On a separate piece of paper, draw a mechanism showing how the lactone is converted into a polymer in the presence of catalytic hydroxide. Your mechanism should include enough steps to show how at least two repeating units are joined together in the growing polymer.

7.86 Acetaminophen, sold under the trade name Tylenol, can be made by treating 4-aminophenol with one equivalent of acetic anhydride. The IR spectrum of acetaminophen has a broad signal between 3200 and 3600 cm^{-1}.

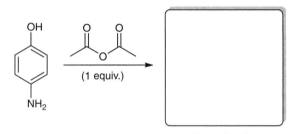

Acetaminophen

(a) Draw the structure of acetaminophen in the box provided above.

(b) On a separate piece of paper, draw a complete mechanism for the reaction between 4-aminophenol and acetic anhydride to give acetaminophen.

7.87 Compare the structures of an amide and an imidazolide:

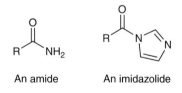

An amide   An imidazolide

Predict whether an amide or an imidazolide will be more reactive toward a nucleophilic acyl substitution reaction, and explain your reasoning. Note that one is significantly more reactive than the other.

7.88 Draw all of the major products of the following reaction:

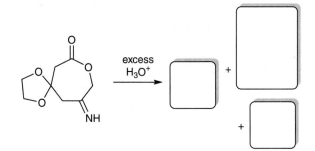

7.89 Draw all of the major products that are expected when cocaine is treated with aqueous acid:

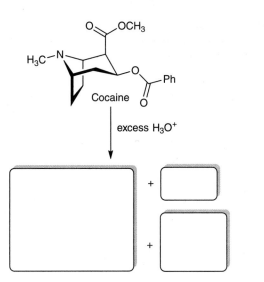

7.90 On a separate piece of paper, draw a plausible mechanism for the following reaction. (Note: This is a very long mechanism, and you will likely need an entire sheet of paper.)

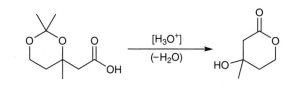

7.91 On a separate piece of paper, draw a plausible mechanism for the following reaction.

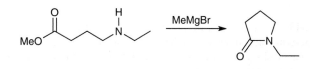

7.92 Fill in the missing structures and reagents in the following synthetic scheme:

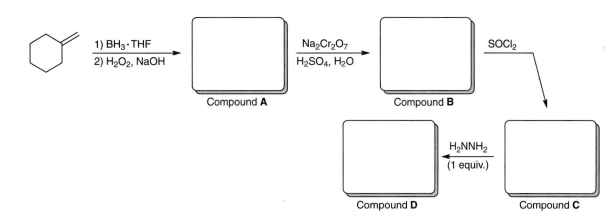

7.93   Fill in the missing structures and reagents in the following scheme:

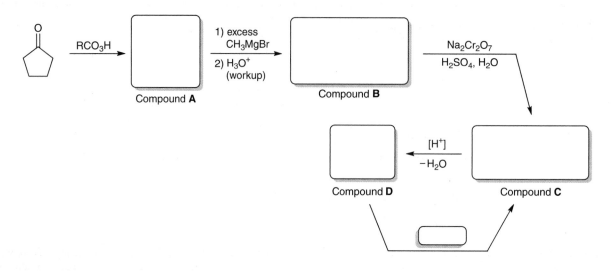

CHALLENGE PROBLEMS

7.94   On a separate piece of paper, draw a plausible mechanism for the following transformation:

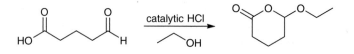

7.95   On a separate piece of paper, draw a plausible mechanism for the following transformation:

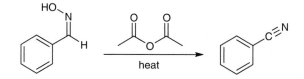

CHAPTER *8*

ENOLS AND ENOLATES

8.1 ALPHA PROTONS

In the previous two chapters, we focused on the reactions that can take place when a nucleophile attacks a carbonyl group:

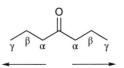

We first learned about nucleophilic attack on ketones and aldehydes (in Chapter 6). Then, in Chapter 7, we explored reactions of carboxylic acid derivatives. Now, we are ready to move away from the carbonyl group, and explore the chemistry that can take place at the alpha (α) carbon:

We call this the alpha carbon, because it is the carbon atom directly connected to the carbonyl group. We use the Greek alphabet to label carbon atoms, moving away from the carbonyl group, in either direction:

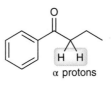

Notice that in this compound, there are two alpha positions. In this chapter, we will focus on the chemistry that can take place at the alpha positions.

Before we get started, we should discuss one more piece of terminology. Any protons connected to an alpha carbon are called alpha protons:

Not all alpha carbon atoms will have alpha protons. For example, consider the following compound:

253

This compound has no alpha protons. If you look just to the right of the carbonyl group, you will see that there is no alpha carbon (it is just an aldehyde). That aldehydic H is NOT an alpha proton because it is not connected to an alpha carbon. And if you look just to the left of the carbonyl group, you will see that there IS an alpha carbon, but this carbon has no protons connected to it.

It is important to recognize the presence or absence of alpha protons. We will see a lot of reactions in this chapter, and most of these reactions will be based on the presence of alpha protons. It turns out that alpha protons are somewhat acidic; and the removal of an alpha proton generates an anion that is fairly reactive. We will see this in greater detail very soon. For now, let's just make sure that we can identify alpha protons when we see them.

EXERCISE 8.1   Identify all alpha protons in the following compound:

Answer   To see if there are any α protons, we must first identify the alpha carbon atoms:

The alpha carbon on the right does *not* have any protons connected to it. The alpha carbon on the left *does* have a proton (highlighted):

So, there is just one alpha proton in this compound.

PROBLEMS   For each of the compounds below, identify all alpha protons (some compounds may not have any alpha protons).

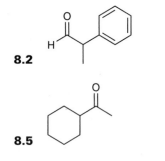

8.2

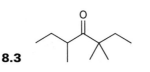

8.3

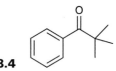

8.4

8.5

8.6

8.7   H   H

8.2 KETO-ENOL TAUTOMERISM

When a ketone has an alpha proton, there is an interesting thing that can happen. In the presence of either acid or base, the ketone exists in equilibrium with another compound:

This other compound is called an **enol**, because it has a C=C bond ("ene") and an OH group ("ol"). The equilibrium shown above is actually very important, because you will see it in many mechanisms. So, let's take a closer look.

If we focus on the connections of atoms, we will find that the two compounds differ from each other in the placement of one proton. The ketone has the proton attached to an alpha carbon, and the enol has the proton connected to oxygen:

It is true that the π bond is also in a different location. But when we just focus on the atoms (which atoms are connected to which other atoms), we find that the difference is in the placement of just one proton. We have a special name to describe the relationship between compounds that differ from each other in the placement of just one proton. We call them *tautomers*. So, the enol above is said to be the *tautomer* of the ketone, and similarly, the ketone is the *tautomer* of the enol. The equilibrium shown above is called *keto-enol tautomerism*.

Keto-enol tautomerism is ***NOT*** resonance. The two compounds shown above are NOT two representations of the same compound. They are, in fact, different compounds. These two compounds are in equilibrium with each other.

In most cases, the equilibrium greatly favors the ketone:

This should make sense, because the last two chapters focused on the formation of C=O bonds as a driving force for reactions. A ketone has a C=O bond, but an enol does not. So we should not be surprised that the equilibrium favors the ketone.

There are some situations where the equilibrium can favor the enol. For example:

In this case, the enol is an aromatic compound, and it is much more stable than the ketone (which is not aromatic). There are many other situations where an enol can be more stable than its tautomer. You will probably find some of these examples in your textbook (such as 1,3-diketones). But in most cases (other than these few exceptional cases), the equilibrium will favor a ketone over an enol.

It is very hard (close to impossible) to prevent the equilibrium from being established. Imagine that you are performing a reaction that generates an enol as the product, and you take great efforts to remove all traces of acid or base. Your hope is that you can prevent the equilibrium from being established, so as to avoid the conversion of the enol into a ketone. But you will find that your efforts will likely be unsuccessful. Even trace amounts of acid or base adsorbed on the glassware (that you cannot remove) will allow the equilibrium to be established.

We will now explore a mechanism for keto-enol tautomerism. We said that compounds will tautomerize in the presence of either acid or base, so we will need to explore two mechanisms: one under *acidic* conditions, and one under *basic* conditions.

We saw that, by definition, tautomers will differ in the position of one proton. So, the conversion of a ketone into an enol requires two steps: 1) introduce a proton, and 2) remove a proton:

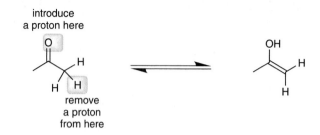

Similarly, the conversion of an enol into a ketone also requires the same two steps: 1) introduce a proton, and 2) remove a proton:

You might wonder why two separate steps are required. Why can't the proton just move over, in one step (in an intramolecular proton-transfer reaction), like this:

This doesn't work, because the oxygen atom is just too far away (in space) from the proton it is trying to remove:

The mechanism requires two separate proton transfer steps, but the order of these steps depends on the conditions. Under acidic conditions, the first step is protonation. But under basic conditions, the first step is deprotonation. Here is the mechanism under basic conditions:

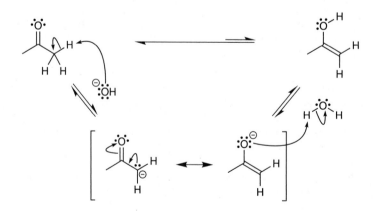

Notice that there is one intermediate (for which we must draw resonance structures), and this intermediate is negatively charged. If you look at the second resonance structure, it looks like an enol that is missing a proton. So, we call this intermediate an **enolate**. Don't be fooled into thinking that the mechanism above has more than two steps. Resonance (of the intermediate) is NOT a step. Our mechanism only has two steps, like this:

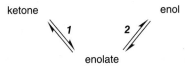

The enolate is very important for the rest of this chapter, because it can function as a nucleophile. We will see many examples in the coming sections. For now, let's finish our discussion of keto-enol tautomerism.

In the mechanism above, the first step was deprotonation, because the conditions were basic. Under acidic conditions, the first step is protonation:

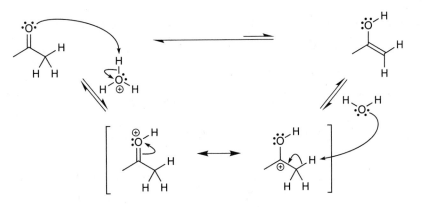

Once again, there are just two steps here. Don't be fooled by the resonance of the intermediate. Resonance is not a step. Resonance is just our way of dealing with the fact that we cannot draw the intermediate with only one drawing. We need two drawings to capture its character. And if you look at these resonance structures, you will see that this intermediate is positively charged.

Notice the difference between these two mechanisms (acidic vs. basic conditions). The first mechanism (basic conditions) has a negatively charged intermediate, and the second mechanism (acidic conditions) has a positively charged intermediate. Other than that, the difference between these two mechanisms is pretty small. Each mechanism has only two steps. And both steps are just proton transfers. The only question is the sequence of events. Is it: deprotonate, then protonate? Or is it: protonate, then deprotonate?

When drawing the mechanism of a keto-enol tautomerization, we must look carefully at the conditions. In acidic conditions, protonation occurs first, giving a positively charged intermediate, which is consistent with acidic conditions. But in basic conditions, deprotonation occurs first, giving a negatively charged intermediate, which is consistent with basic conditions.

EXERCISE 8.8   It is not possible to isolate and purify the following compound, because upon formation, it will rapidly tautomerize to form a ketone. Draw a mechanism for the conversion of this enol into a ketone under acidic conditions.

Answer   This compound is an enol, and its tautomer will be the following ketone:

To convert the enol into a ketone, our mechanism will have two steps: protonate and deprotonate. But we must decide what order to use. Do we first protonate? Or do we first deprotonate? To answer this question, we look at the conditions. Since we are in acidic conditions, we should first protonate (forming a positively charged intermediate), and only then do we remove the other proton.

Now that we have determined the order of events, we must also decide *where* to protonate, and *where* to deprotonate. To figure this out, we look at the overall reaction:

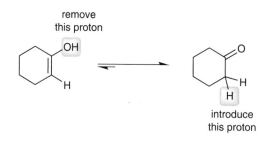

When we analyze the reaction like this, it is easy to see *where* to introduce a proton and *where* to remove a proton. This might seem trivial, but it is extremely important because it showed us that we must protonate the double bond (rather than protonating the oxygen atom), like this:

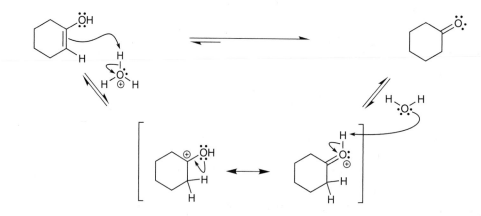

So often, students will start this problem by protonating the OH group. Although that might make sense at first, you will find that this step will NOT lead to formation of the ketone. The first step is to protonate the double bond; *not* the OH group. Think about why this is the case. Protonation of the double bond leads to a resonance-stabilized intermediate, while protonation of the OH group would lead to an intermediate that is not resonance-stabilized (and therefore higher in energy).

When drawing a mechanism, make sure to never use HO⁻ and H₃O⁺ in the same mechanism. When acidic conditions are indicated, use H_3O^+ to protonate and use *H_2O* to deprotonate. Don't use hydroxide to remove a proton, because there are not many hydroxide ions present under acidic conditions.

Similarly, when basic conditions are indicated, use HO⁻ to remove the proton and use *H_2O* to protonate. Don't use H₃O⁺ to protonate, because we are in basic conditions. Here is the take home message: always stay consistent with your conditions.

So, to recap, there are three things to consider in order to correctly draw the mechanism of a keto-enol tautomerization: 1) what *order* to use (first protonate or first deprotonate), 2) *where* to protonate and where to deprotonate, and 3) *what reagents* to show when drawing the proton transfer steps (stay consistent with the conditions).

PROBLEMS   For each of the following transformations, propose a mechanism that is consistent with the conditions indicated (you will need a separate piece of paper to record your answers):

8.9
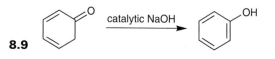

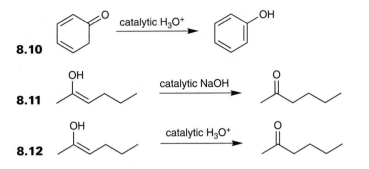

8.10

8.11 catalytic NaOH

8.12 catalytic H₃O⁺

8.13 Propose a plausible mechanism for the following keto-enol tautomerization. Remember to ask three important questions: 1) what *order* to use (first protonate or first deprotonate), and 2) *where* to protonate and deprotonate, and 3) *what reagents* to show for each step of the mechanism. You will need a separate piece of paper to record your answer.

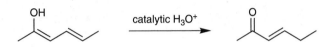

8.3 REACTIONS INVOLVING ENOLS

It is hard to see how the alpha carbon of a ketone can be nucleophilic:

The alpha carbon does *not* have a lone pair or a π bond that can function as a nucleophilic center. However, when we examine the structure of the enol (that is in equilibrium with the ketone), we get a different picture:

The enol has a π bond on the alpha carbon, which renders it nucleophilic. Also, consider the resonance structure of the enol:

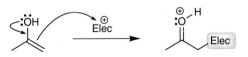

Notice that there is a negative charge on the alpha position, and therefore, the alpha carbon can function as a nucleophile to attack some electrophile:

In order for the attack to occur, we are relying on the ability of a ketone to tautomerize. But, not every ketone will exist in equilibrium with an enol. A ketone that lacks alpha protons will **not** tautomerize to form an enol:

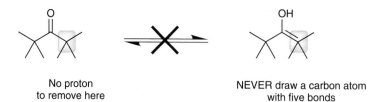

No proton
to remove here

NEVER draw a carbon atom
with <u>five</u> bonds

Most ketones do in fact have alpha protons, and therefore, a typical ketone will exist in equilibrium with an enol. In the previous section, we saw some rare cases where the equilibrium can actually favor the enol, but in general, the equilibrium favors the ketone. Therefore, you will generally only have trace amounts of the enol present in equilibrium with the ketone.

This small amount of enol is able to react as a nucleophile and attack some electrophile. After the enol attacks the electrophile, the keto-enol equilibrium is re-established by producing some more enol (to account for the enol that "disappeared" as a result of the reaction). Slowly but surely, most of the ketone molecules end up converting into enols and reacting with the electrophile. One example is alpha-halogenation, which can occur when a ketone is treated with Br_2 in aqueous acid (H_3O^+) to generate an α-halo ketone:

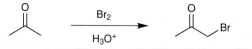

Let's explore how this process occurs. The ketone tautomerizes to generate a small amount of enol. Then comes the critical step: the enol functions as a nucleophile to attack Br_2 (the electrophile):

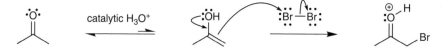

Deprotonation then generates the product:

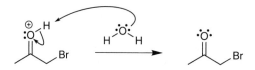

Notice that most of the steps in this mechanism are just proton transfers. Our mechanism represents the following pattern: tautomerize, attack, deprotonate. But "tautomerize" is just a new name for a special combination of two proton transfer steps. There is really only one step where an attack takes place (when the enol attacks the electrophile).

In the end, this provides a method for installing a halogen at the alpha position of a ketone:

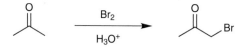

Alternatively, we might see a different reagent other than H_3O^+, for example:

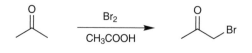

Acetic acid, CH$_3$COOH, can be used as a mild acid to facilitate the tautomerization. We don't have to worry about this acid undergoing halogenation itself (at its alpha position), like this:

THIS REACTION IS TOO SLOW

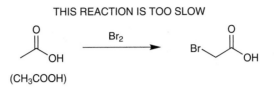

(CH$_3$COOH)

We don't have to worry about this, because carboxylic acids are much slower to react in this kind of reaction.

If we *want* to halogenate the alpha position *of a carboxylic acid*, it is possible, but it will require some extra steps. First, we must convert the carboxylic acid into an acid halide. We do this because the enol of an acid halide will rapidly attack a halogen. Then, in the end, we just convert the acid halide back into a carboxylic acid:

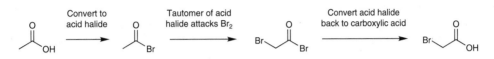

This strategy (for halogenating carboxylic acids) is called the Hell–Volhard–Zelinsky reaction.

Here is the bottom line: in this section, we have seen two reactions that exploit the nucleophilic nature of enols. These reactions can be used to install a halogen at the alpha position of a ketone,

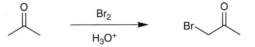

or to install a halogen at the alpha position of a carboxylic acid:

Notice the reagents that we used. For halogenation of a ketone, the reagents were Br$_2$ in acidic conditions. But to halogenate a carboxylic acid, we use a different set of reagents. We use Br$_2$ and PBr$_3$, followed by H$_2$O. The function of Br$_2$ and PBr$_3$ is to make the acid halide, form the enol, and then have the enol attack Br$_2$. Then, water is used in the last step to convert the acid halide back into a carboxylic acid.

EXERCISE 8.14   Predict the product of the following reaction:

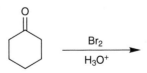

Answer   We are starting with a ketone, and we are subjecting it to Br$_2$ in acidic conditions. The acid promotes tautomerization to the enol, which then attacks the Br$_2$ in an alpha halogenation. So, in the end, our product will have a Br at one of the alpha positions. Either side is the same, so we can just pick a side:

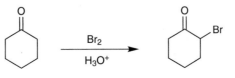

PROBLEMS   Predict the products of each of the following reactions. Remember that you can only halogenate an alpha position that has protons.

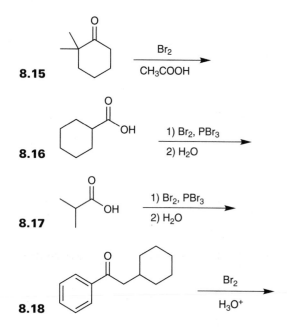

8.15

8.16

8.17

8.18

8.4 MAKING ENOLATES

In the previous section, we saw that enols can be nucleophilic. But enols are only mild nucleophiles. So, the question is: how can we make the alpha position even more nucleophilic (so that we can have a broader range of possible reactions)? There is a way to do this. We just need to give the alpha position a negative charge. To see how this can be achieved, let's quickly review the mechanism we saw for tautomerization under basic conditions, and let's focus on the intermediate (highlighted below):

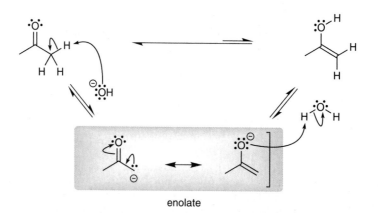

enolate

The intermediate is negatively charged, and we mentioned before that it is called an enolate. In order to capture the essence of the enolate, we must draw resonance structures. Remember what resonance structures represent. We cannot draw this *one* intermediate with any single drawing, so we draw two drawings, and we meld these two images together in our minds in order to get a better picture of this intermediate. And that picture shows the enolate as being electron rich in two locations: the alpha carbon **and** the oxygen atom:

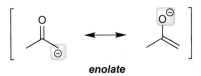

enolate

So, we expect **both** of these locations to be very nucleophilic. Nevertheless, we won't explore any reactions in which the oxygen atom functions as a nucleophile (called O-attack). Most textbooks and instructors do not teach the conditions for O-attack, because reactions at that site are far less common than reactions at the alpha carbon. So, from now on, we will only explore examples of C-attack (where the alpha carbon acts as the nucleophile, attacking some electrophile):

Notice that, in showing the attack, we have drawn only one resonance structure of the enolate. If we had used the other resonance structure, it would have looked like this:

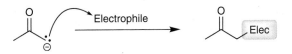

This is just another way of showing the same step. Many textbooks will show it the second way (starting with the resonance form that has the negative charge on oxygen). Perhaps this is more appropriate, because this resonance form is contributing more to the overall character of the enolate. However, in this book, we will use the resonance structure where the negative charge is on carbon:

We will do it this way, because it will make the mechanisms easier to follow. To be absolutely correct, we should actually draw *both* resonance forms, like this:

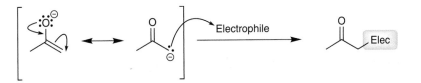

But for simplicity, we will just show one resonance structure for the enolate (in most of the mechanisms that we will see in this chapter).

Now let's think about what kind of base we would need **to make** an enolate. If we use bases such as HO⁻ or RO⁻ (bases with a negative charge on oxygen), we find that these bases are *not* strong enough to completely convert the ketone into an enolate. Rather, an equilibrium is established between the ketone and enolate. This equilibrium only produces very small amounts of the enolate, but that doesn't matter. Once an enolate reacts with an electrophile, the equilibrium produces more enolate to replenish the supply. Over time, all of the ketone can convert into the enolate and then react with some electrophile. This is very similar to the situation we saw with enols. Once again, we are relying on the equilibrium to continuously produce more of the enolate. The major difference here is that enolates are so much more reactive than enols. Therefore, the chemistry of enolates is more robust than the chemistry of enols.

As an example for the richness of enolate chemistry, consider this: some enolates are much more stabilized than other enolates. These "stabilized" enolates are more "tame" nucelophiles (more selective in what they react with). For example, a compound with two carbonyl groups (separated by one carbon) can be deprotonated to form an intermediate described as a "stabilized" enolate:

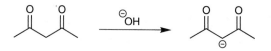

The negative charge in this intermediate is delocalized over both carbonyl groups:

And therefore it is extremely stable. In fact, it is even more stable than HO^- or RO^-. So, when we use bases such as HO^- or RO^-, the equilibrium greatly favors the enolate:

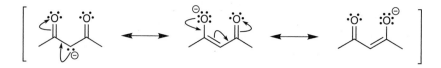

We will soon see that the position of this equilibrium will be a driving force in the Claisen condensation (later in this chapter).

EXERCISE 8.19 Consider the following compound:

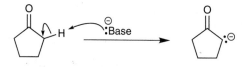

Draw the enolate that is formed when this compound is deprotonated. Make sure to draw all resonance structures.

Answer We just need to identify the alpha proton, and then remove it:

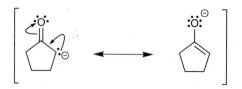

And then we draw the resonance structures:

PROBLEMS   Draw the enolate that would be generated when each of the following compounds is treated with hydroxide. Make sure to draw all significant resonance structures.

8.20

8.21

8.22

8.23

8.5   HALOFORM REACTION

In the previous section, we learned how to make enolates. Now, we will begin to see what an enolate can attack. In this section, we will explore the reaction between an enolate and a halogen (such as Br, Cl, or I). In the following sections, we will explore the reactions between an enolate and other electrophiles.

Consider what might happen under the following conditions:

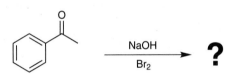

We have a ketone and hydroxide, which means that the equilibrium will involve a small amount of enolate:

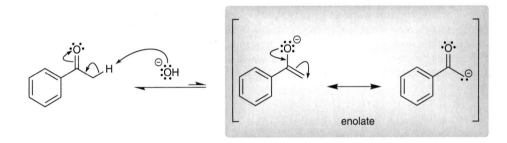

enolate

This enolate is formed in the presence of Br$_2$, which can function as an electrophile. The initial product is not surprising:

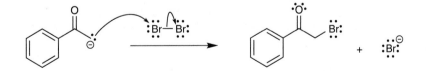

The enolate attacks Br$_2$ and expels Br$^-$ as a leaving group. The result is that we have installed a bromine atom at the alpha position:

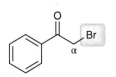

But the reaction doesn't stop there. Remember that the base (hydroxide) is still present in solution. So hydroxide can remove another alpha proton. In fact, it is even easier to remove this proton, because the inductive effect of the bromine atom serves to further stabilize the resulting enolate, which can then attack Br_2 again:

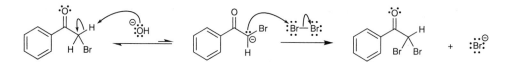

Now we have *two* Br atoms in our compound. And then, it happens again:

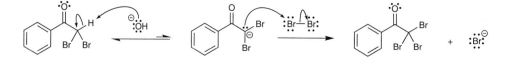

Think about what has happened so far. A methyl group (CH_3) group has been converted into a CBr_3 group. This transformation is very significant, because a CBr_3 group is able to function as a leaving group:

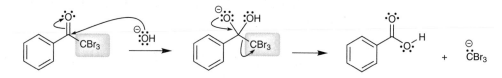

At this point, you should be feeling uncomfortable. You probably remember our golden rule from the previous chapters (don't expel H^- or C^-), and it seems as though we are breaking our golden rule. Aren't we kicking off a C^- here? Yes, we are. This is actually one of the rare exceptions to the golden rule. In general, the golden rule holds true ***most*** of the time, because C^- is generally too unstable to serve as a leaving group. But there are cases where a C^- can be stabilized enough for it to serve as a leaving group, and this is one of those rare situations. Br_3C^- is actually a pretty good leaving group, because of the combined electron-withdrawing effects of all three bromine atoms. In addition, the loss of this leaving group is driven by the formation of the stable carbonyl group. But even though it can leave, it is not the most stable anion on the planet. In fact, it is not even as stable as a carboxylate ion (the conjugate base of a carboxylic acid). So, the following proton transfer occurs:

This forms a carboxylate anion and $CHBr_3$ (called bromoform). And this is the end of our mechanism. If we want to isolate the carboxylic acid, we will have to introduce a source of protons into the reaction flask in order to protonate the carboxylate anion.

When the same reaction is performed with iodine instead of bromine, iodoform is obtained as a by-product, instead of bromoform:

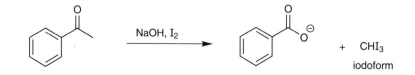

Iodoform is a yellow solid that will precipitate out of solution. Therefore, this reaction can be used to probe the identity of an unknown compound. If the unknown compound is a methyl ketone, then it will produce iodoform under these conditions (NaOH and I_2). This iodoform test is not really used anymore (we now have spectroscopy techniques that give us this information and much, much more). So, this chemical test is really a relic of the past. But for some reason, it is still used in textbook problems. You will usually see it like this: "An unknown compound tests positive for iodoform, and ...." The beginning of this problem is telling you that you have a methyl ketone. If you see this in a problem in your textbook, you should know what it means.

But there is a much more important use for this reaction. You can use it when solving synthesis problems. This reaction provides a way to convert a methyl ketone into a carboxylic acid:

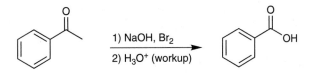

The haloform reaction is most efficient when the other side of the ketone has no α protons, for example.

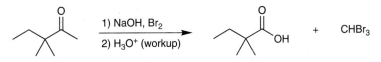

This should stick out in your mind, because it is a new example of a "cross-over" reaction. In the previous chapter, we talked about ways of converting ketones into carboxylic acid derivatives (cross-over reactions). The process that we explored in this section can be used to convert a methyl ketone into a carboxylic acid. You should add this to your synthetic toolbox.

EXERCISE 8.24 Predict the products of the following reaction:

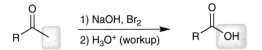

Answer The reactant is a methyl ketone and the reagents will convert a methyl ketone into a carboxylic acid (with a by-product of bromoform):

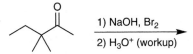

PROBLEMS Predict the products for each of the following reactions:

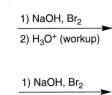

8.25

8.26

8.27 On a separate piece of paper, draw a mechanism for the transformation in the previous Problem (8.26).

8.6 ALKYLATION OF ENOLATES

In this section, we will continue to explore reactions between enolates and electrophiles. Specifically, we will learn how to install an alkyl group at an alpha position:

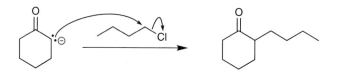

In order to alkylate the alpha position, it makes sense to use an enolate to attack an alkyl halide, for example:

This is just an S_N2 reaction, so it should work with *primary* alkyl halides (if a secondary alkyl halide is used, the enolate will function as a base, and elimination will be favored over substitution).

But we run into a major obstacle when we try to make the enolate by treating a ketone with hydroxide. Remember that when we use hydroxide as the base to form our enolate, we find that the equilibrium lies very far to the side of the ketone:

At equilibrium, there is a very small amount of enolate, but there is a lot of ketone and a lot of hydroxide present. So, if we introduce some alkyl halide into the reaction flask, we run into a major obstacle. The excess hydroxide can react with the alkyl halide (elimination or substitution), which creates competing side reactions that generate a mixture of undesired products.

In order to avoid this problem, we will need to form the enolate under conditions where most of the ketone molecules are converted into enolates. If we are able to do this, we will have very little base left over, and therefore, we won't have to worry about the base reacting with the alkyl halide. It is possible to do this, but we will need to use a base that is much stronger than the bases we have been using so far (HO^- and RO^-). We can use the following base instead:

The name of this compound is lithium diisopropylamide, or LDA for short. LDA is a very strong base, because the negative charge is on a nitrogen atom (which is less stable than a negative charge on an oxygen atom). The two isopropyl groups are sterically bulky, so LDA is *not* a good nucleophile. LDA is primarily used as a strong, sterically hindered base, which is exactly what we need in our situation. By using LDA, we can achieve an efficient conversion of the ketone into the enolate.

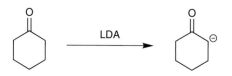

So, we will have mostly enolates in our reaction flask (and very little ketone or base). Now when we introduce some alkyl halide into our reaction flask, the risk of competing side reactions is greatly reduced.

So, to alkylate a ketone, we use the following reagents:

In step 1, we use LDA to deprotonate the ketone, to form an enolate. When you see THF in the reagents above, don't get confused. THF (tetrahydrofuran) is just the solvent that is typically used with LDA. In step 2 above, we use an alkyl halide (RX) to install the alkyl group, where R is some primary alkyl group, and X is a halogen (Cl, Br, or I).

In the situation above, the starting ketone was symmetrical. But what happens when we start with an unsymmetrical ketone? For example, consider the following situation:

Where will the incoming alkyl group be installed? On the left side, or on the right? In order to answer this question, we need to take a close look at the two possible enolates:

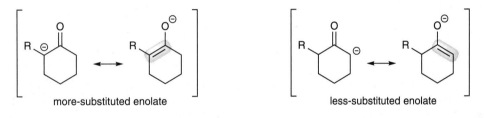

| more-substituted enolate | less-substituted enolate |

The more-substituted enolate (above left) is the more stable enolate because it has a more substituted pi bond. However, the less-substituted enolate (above right) can form faster, because there are twice as many protons available on the less-substituted side:

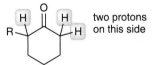

So from a probability point of view, we expect the less-substituted enolate to form more rapidly. Also, we expect the sterically hindered base to have a much easier time removing one of these protons. So, we have two competing arguments:

| More stable | Forms faster |

This is a classic example of thermodynamics vs. kinetics. Thermodynamics is all about stability and energy levels. So, a thermodynamic argument says that we should predominantly form the more stable enolate. However, a kinetic argument tells us to expect the other enolate, simply because it forms faster. Which argument wins? The truth is that a mixture of products is observed. But with LDA at low temperature, there is a clear preference to form the kinetic enolate:

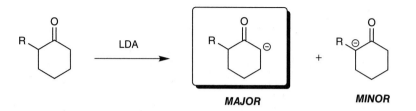

MAJOR   +   **MINOR**

When we introduce the alkyl halide to the reaction flask, alkylation will occur primarily at the less-substituted alpha position:

That works very well if we *want* to install the alkyl group at the less-substituted position. But what if we want to install the alkyl group at the more-substituted position? In other words, what if we want to do this:

There are many different ways to achieve this transformation. Essentially, you need to form the thermodynamic enolate, rather than the kinetic enolate. Some textbooks will teach one or two ways to do this, while other textbooks will skip it altogether. You should look through your textbook and lecture notes to see if you are responsible for knowing how to alkylate the more-substituted side.

EXERCISE 8.28   Predict the major product of the following reaction:

Answer   This is an alkylation reaction. In step 1, we are using LDA to form an enolate. And then in step 2, we are using an alkyl halide to alkylate.

Because the alkyl halide is ethyl chloride in this case, we will be installing an ethyl group on an alpha carbon. The only question is: which alpha carbon? The more-substituted carbon or the less-substituted carbon? The use of LDA as our base predominantly gives the kinetic enolate (the less-substituted enolate). Therefore, our major product will have the ethyl group at the less-substituted alpha position:

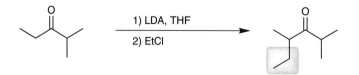

PROBLEMS   Predict the major product for each of the following reactions:

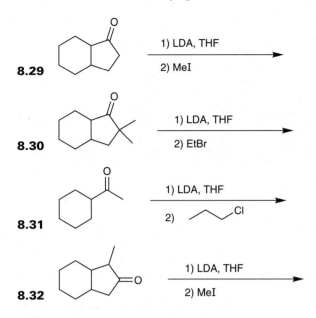

8.29   1) LDA, THF / 2) MeI

8.30   1) LDA, THF / 2) EtBr

8.31   1) LDA, THF / 2) Cl

8.32   1) LDA, THF / 2) MeI

EXERCISE 8.33   What reagents would you use to achieve the following transformation:

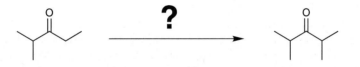

Answer   If we look at the difference between the starting material and the product, we will see that there is an extra methyl group that was introduced at an alpha position. This methyl group was installed at the less substituted position, which can be achieved by treating the ketone with LDA, followed by methyl iodide:

1) LDA, THF
2) MeI

PROBLEMS   Identify reagents that can be used to achieve each of the following transformations:

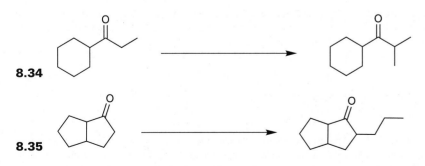

8.34

8.35

8.36

8.7 ALDOL REACTIONS

So far in this chapter, we have learned how to make enolates, and we have used them to attack various electrophiles (including halogens and alkyl halides). In this section, we will explore what happens when an enolate attacks a ketone or aldehyde.

Suppose we start with a simple ketone, and we subject it to basic conditions, using hydroxide as a base. We have already seen that an equilibrium will be established between the ketone and the enolate:

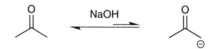

If we do this in the presence of an electrophile, the enolate can attack the electrophile. And then the equilibrium will produce more enolate to replenish the supply. But what if we do not add any other electrophiles to the reaction mixture? What if we just treat the ketone with hydroxide?

It turns out that there actually is an electrophile present. We said that the enolate is in equilibrium with the ketone (and there is a lot of ketone present). Well, ketones are electrophilic, aren't they? We devoted an entire chapter to the reactions that take place when ketones get attacked. So, what happens when an enolate attacks a ketone?

The enolate attacks the ketone, forming an alkoxide intermediate. Now, our golden rule tells us to try and re-form the carbonyl group, but don't expel H⁻ or C⁻. In this case, we have no leaving groups that can be expelled. So, the only way to remove the charge is to protonate. In these basic conditions, the proton source is water (not H_3O^+, because the concentration of H_3O^+ is insignificant in basic conditions):

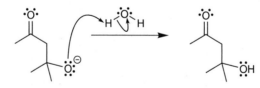

This is the initial product of this reaction. Notice that the OH group is at the β position relative to the surviving carbonyl group:

This will always be the case whenever an enolate attacks a carbonyl group, regardless of the structure of the starting ketone and the structure of the enolate. The alpha carbon of the enolate is directly attacking the carbonyl group of the ketone. That will always place the OH group in the beta position. Always. This product is called a *β-hydroxy ketone*, and the reaction is called an **aldol** addition.

In general, the reaction doesn't stop there (at the β-hydroxy ketone). With heating, the basic conditions favor an elimination to form a double bond:

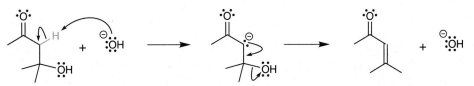

This product has a double bond in conjugation with the carbonyl group. The double bond is located between the α and β positions. So, the product is called an *α,β-unsaturated ketone*.

In the laboratory, we can often control how far the reaction goes. By carefully controlling the conditions of the reaction (temperature, concentrations, etc.), we can usually control whether the reaction stops at a *β-hydroxy ketone*, or whether it continues to form an *α,β-unsaturated ketone*.

But you should be familiar with the proper terminology. When we go all the way to the *α,β-unsaturated ketone*, we call the reaction an aldol *condensation*. By definition, a condensation is any reaction where two molecules come together, and in the process, a small molecule is liberated. The small molecule can be N_2 or CO_2 or H_2O, etc. In this case, we have two molecules of ketone coming together, and in the process, a molecule of water is liberated:

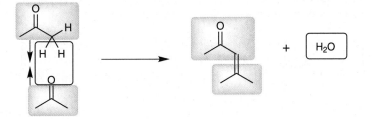

Therefore, we call this reaction an aldol *condensation*. But what if we control the reaction conditions so that we stop at the *β-hydroxy ketone*?

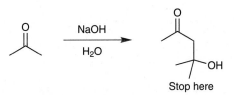

If we stop here, then we cannot call it a condensation reaction anymore, because a water molecule was not lost in the process. So, instead, we call it an aldol addition. The difference between an aldol *condensation* and an aldol *addition* is how far we go in the process:

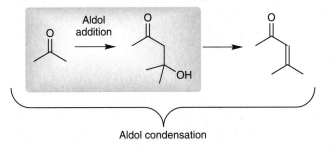

This distinction (between the aldol *addition* and the aldol *condensation*) is often absent in textbooks, and you might find the terms being used interchangeably in your textbook. I am taking the time to point out the distinction, because I believe that it will help you to remember and master the mechanism.

The mechanism of an aldol condensation is fairly straightforward. But sometimes, it can get hard to see what reagents to use when proposing a synthesis. So try to think of it the way we showed it just a few moments ago:

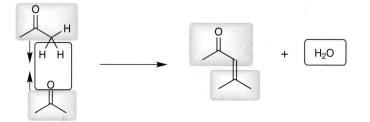

We are removing two alpha protons from one ketone, and we are removing the oxygen atom from the other ketone. Do *not* confuse the drawing above with a mechanism. A mechanism is when you show all of the curved arrows and intermediates. But this way of thinking about the reaction might come in handy when proposing syntheses.

In a case where two stereoisomers are expected, the major product will be the one that exhibits fewer steric interactions, for example:

Let's get some practice:

EXERCISE 8.37   Draw the aldol condensation product that is obtained when the following compound is heated in the presence of hydroxide ions:

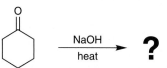

Answer   We can certainly work through the mechanism to get the answer, and we will soon get practice with that. But for now, let's just make sure that we can use our simple method for drawing the expected product.

We start by drawing two molecules of the ketone, so that the oxygen of one ketone is pointing directly at the alpha protons of the other ketone:

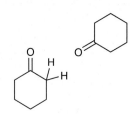

Then, we erase the two alpha protons and the oxygen atom, and we push the fragments together (connecting them with a double bond):

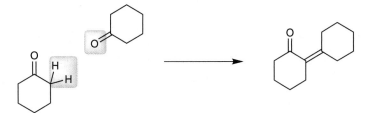

That's all there is to it. It is an efficient way of drawing the product.

PROBLEMS Predict the major product for each of the following reactions. In each case, assume that a condensation takes place, and draw the α,β-unsaturated ketone that is produced.

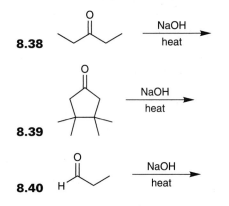

8.38

8.39

8.40

In all of the aldol condensations that we have seen so far, two molecules *of the same ketone* (or aldehyde) were reacting with each other. One molecule of ketone was deprotonated to give an enolate, which then attacked another molecule of the same ketone. But what if we had used two different ketones? For example, what if we try to do this:

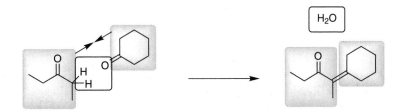

Notice that the ketones are different from each other. We call this a *crossed*-aldol. This can work, but care must be taken to avoid generating many different products. To see why this is the case, we must realize that it is possible for enolates and ketones to exchange protons:

So, you can't really control which ketone will be converted into the enolate. This means that there will be more than one type of enolate and more than one type of ketone present in solution. So there are a number of possible reactions that can take place, and this will give a mixture of undesired products.

So, practically, it is important to try to avoid these types of situations. There is one very easy way to avoid this issue. If one of the ketones has no alpha protons, then it cannot form an enolate. For example, consider the following compound:

This compound, called benzaldehyde, has no alpha protons. Therefore, it cannot be converted into an enolate. It will just wait to be attacked. Here is another example of a compound with no alpha protons:

So, one way to achieve a crossed aldol is to make sure that one of the reagents has no alpha protons. That will minimize the number of potential products. Your textbook may or may not show methods for achieving a crossed aldol where both starting ketones have alpha protons. You should look through your textbook and lecture notes to see if you are responsible for such methods.

EXERCISE 8.41   Identify the starting materials that you would use to make the following compound, using an aldol condensation:

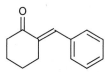

Answer   We can use the same method we used earlier. We just need to do it in reverse. We break the molecule apart into two fragments in order to insert water. We break it apart at the C=C bond:

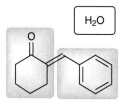

And we just have to decide which fragment gets the oxygen atom and which fragment gets the protons. The fragment on the left already has a carbonyl group, so that fragment must get the two alpha protons. The fragment on the right will get a carbonyl group in place of the C=C bond:

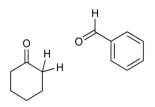

PROBLEMS   Identify the starting materials that you would use to make each of the following compounds (using an aldol condensation):

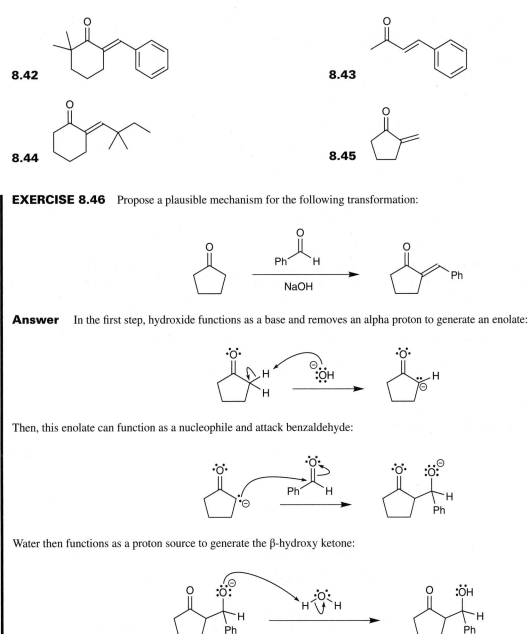

8.42

8.43

8.44

8.45

EXERCISE 8.46   Propose a plausible mechanism for the following transformation:

Answer   In the first step, hydroxide functions as a base and removes an alpha proton to generate an enolate:

Then, this enolate can function as a nucleophile and attack benzaldehyde:

Water then functions as a proton source to generate the β-hydroxy ketone:

Finally, water is eliminated to generate the α,β-unsaturated ketone:

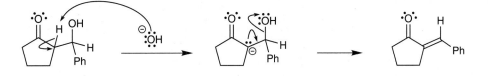

PROBLEMS Now let's get some practice drawing mechanisms for aldol condensations. Draw a mechanism for each of the following transformations. You will need a separate piece of paper to record each of your answers:

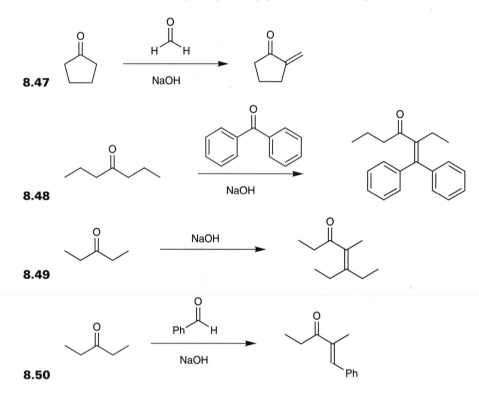

8.47

8.48

8.49

8.50

8.8 CLAISEN CONDENSATION

In the previous section, we saw that an enolate can attack a ketone:

In this section, we will explore what happens when an *ester enolate* attacks an ester:

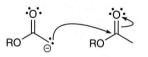

An ester enolate is similar to a regular enolate: an ester enolate is nucleophilic, and it will also attack a carbonyl group. When an ester enolate attacks an ester (shown above), the reaction that takes place is called a Claisen condensation. Here is the overall transformation:

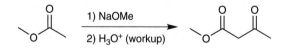

The product is called a β-keto ester:

The ester gets priority over the other carbonyl, so we label the carbon atoms (α, β, γ, etc.) moving away **from the ester**

The "keto" group is located at the *beta* position

At first glance, this product seems very different from the α,β-unsaturated ketones obtained from aldol condensations. But when we explore the mechanism, we will see the parallel between the aldol and Claisen condensations.

Let's start with the first step: preparing the enolate:

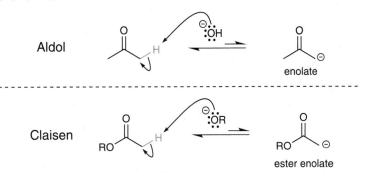

So far, both mechanisms are almost identical. The only difference is the choice of base (the aldol condensation uses hydroxide, and the Claisen condensation uses an alkoxide), and we will discuss the reason for this shortly. For now, let's continue comparing the mechanisms. In the next step, the enolate attacks:

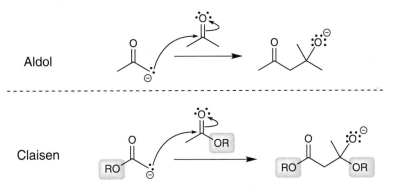

Once again, we see that both mechanisms are essentially identical. In the Claisen condensation, the alkoxy groups seem to just come along for the ride.

But now the two reactions take different routes. And we can use our golden rule to understand why. In the aldol reaction, the carbonyl group cannot re-form, so the oxygen atom must be protonated with a suitable proton source. But in a Claisen condensation, the carbonyl group CAN re-form, because there is a group that can leave:

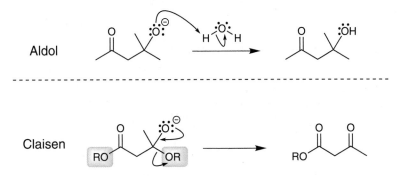

And this is why the product of a Claisen condensation looks very different from the product of an aldol condensation. But when you understand the mechanisms, you can appreciate that these reactions are very similar. The difference between these two reactions stems from the fact that Claisen condensations involve esters; and esters have a "built-in" leaving group:

Built-in
leaving group

Now let's go back and explore the mechanism in more detail. In the first step, we made an ester enolate. To do this, we used a strong base. But we pointed out at the time that we did NOT use hydroxide. Instead, we used an alkoxide ion. Let's try to understand why.

If we had used hydroxide, then we might have observed a competing reaction. Instead of hydroxide acting as a base to remove a proton, it is possible for hydroxide to function as a nucleophile, attacking the carbonyl group of the ester:

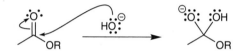

After the initial attack, the carbonyl group could re-form by expelling an alkoxide ion. This unwanted side reaction would hydrolyze the ester and produce a carboxylate ion (we saw this saponification reaction in the previous chapter):

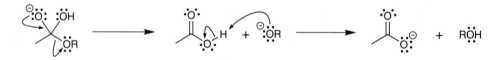

In order to avoid this, we use an alkoxide as our base. It is true that alkoxides can *also* function as nucleophiles, but think about what happens if the alkoxide ion functions as a nucleophile and attacks:

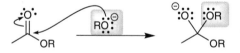

When the carbonyl group re-forms, it doesn't matter which alkoxy group gets expelled. Either way, the original ester will be regenerated:

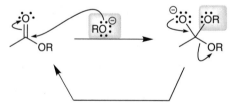

Although the alkoxide *can* attack the carbonyl group, we don't have to worry about it, because it does not actually lead to new products. So, we can avoid unwanted side reactions by using an alkoxide ion as the base for a Claisen condensation.

Be careful, though. We cannot just use *any* alkoxide ion. We must choose the alkoxide ion carefully. If we are dealing with a methyl ester, then we would use methoxide:

The reason for this is simple. Suppose we used ethoxide in this case. This would actually change some of our ester:

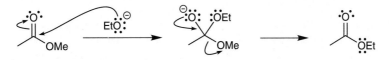

This is called *transesterification*, and we can avoid this by choosing our base to match the alkoxy group of the ester. That way, we avoid unwanted side reactions. If we are dealing with an ethyl ester, then we just use ethoxide as our base:

Now that we know what base to choose for a Claisen condensation, let's talk about another special role that the base plays in a Claisen condensation. We said that the final product is a β-keto ester. But remember that this reaction is performed under basic conditions (in the presence of alkoxide ions). Under these conditions, the β-keto ester will be deprotonated to form an enolate that is especially stabilized:

In the beginning of this chapter, we talked about this kind of "stabilized" enolate. This enolate is much more stable than an alkoxide ion. That is an important point, because it means that the reaction will favor formation of the product. Why? Because the reaction is converting alkoxide ions into enolate ions (which are more stable):

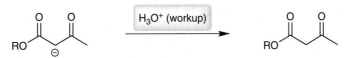

More stabilized negative charge

The formation of this stabilized enolate is a strong driving force pushing this reaction toward the formation of products.

So, when the reaction is finished, a proton source must be introduced into the reaction flask in order to protonate the enolate and obtain the product:

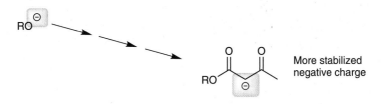

The Claisen condensation is important because it gives us a way to make β-keto esters. And we will soon see that there is a clever synthetic trick that you can perform with β-keto esters. So, let's make sure that we have mastered the Claisen condensation.

Overall, here is what is happening:

We are removing an alpha proton from one ester, and we are removing an alkoxy group from the other ester. The remaining fragments are then joined together. Notice that a small molecule is liberated in the process (ROH). That is why we call this reaction a Claisen *condensation*.

Now let's practice predicting products. We will soon come back and master the mechanism. But for now, let's make sure that you train your eyes to see the products of a Claisen condensation in an instant. You will need that skill for proposing syntheses.

EXERCISE 8.51   Draw the product of the following Claisen condensation:

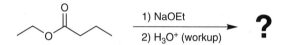

Answer   The following two fragments are adjoined, while EtOH is expelled, like this:

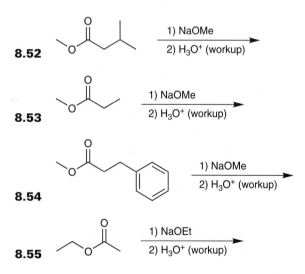

PROBLEMS   Draw the product for each of the following Claisen condensations:

8.52

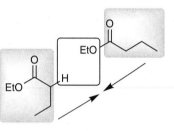

8.53

8.54

8.55

EXERCISE 8.56   Identify the starting materials and reagents you would use to prepare the following compound using a Claisen condensation:

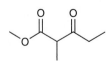

Answer   We break the molecule apart into two fragments in order to insert MeOH. We break it apart between the α and β positions:

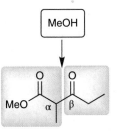

And we just have to decide which fragment gets the methoxy group and which fragment gets the proton. The fragment on the left already has its alkoxy group, so that must be the fragment that gets the proton. The fragment on the right will get the alkoxy group:

These two esters are identical, which is good. That means that we just need one kind of ester. We choose our base to match the alkoxy group (methoxide in this case), so our synthesis would look like this:

It is possible to achieve crossed Claisen condensations (just like we can achieve crossed aldol condensations), but we would have the same concerns as before. We would have to worry about potential side reactions. A crossed Claisen condensation will be more efficient when one of the esters has no alpha protons. You will see that some of the problems below are the products of crossed Claisen condensations. Keep an eye out for them.

PROBLEMS   Identify the starting materials and reagents you would use to prepare each of the following compounds using a Claisen condensation:

8.57

8.58

8.59

8.60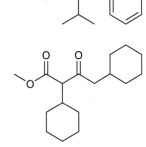

EXERCISE 8.61   Propose a mechanism for the following transformation:

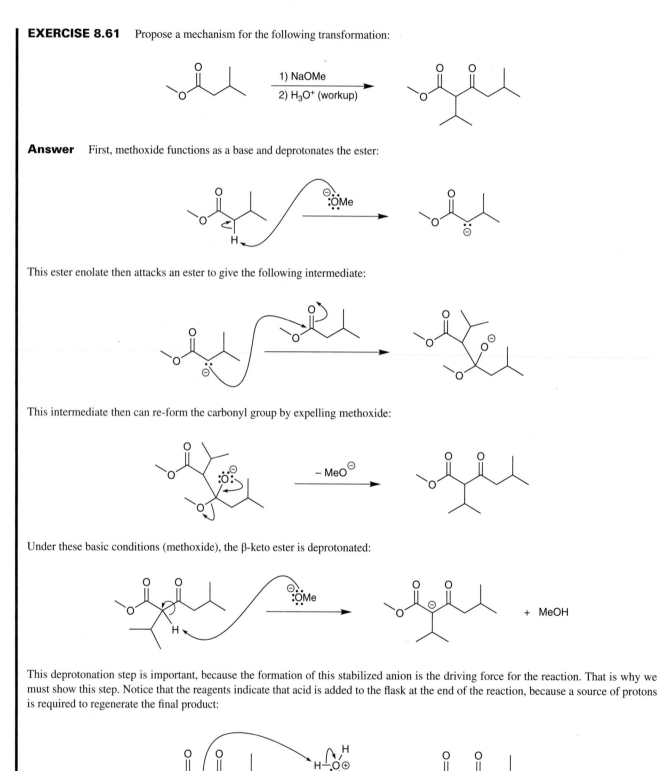

Answer   First, methoxide functions as a base and deprotonates the ester:

This ester enolate then attacks an ester to give the following intermediate:

This intermediate then can re-form the carbonyl group by expelling methoxide:

Under these basic conditions (methoxide), the β-keto ester is deprotonated:

This deprotonation step is important, because the formation of this stabilized anion is the driving force for the reaction. That is why we must show this step. Notice that the reagents indicate that acid is added to the flask at the end of the reaction, because a source of protons is required to regenerate the final product:

PROBLEMS Propose a mechanism for each of the following reactions. You will need a separate piece of paper to record your answers:

8.62

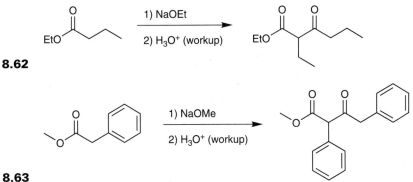

8.63

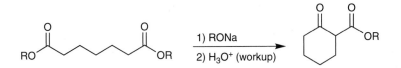

8.64 When a diester is used as a starting material, it is possible to achieve an ***intramolecular*** Claisen condensation:

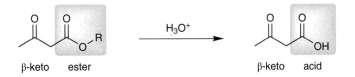

Notice once again that the product is just a β-keto ester. This reaction has its own name (the Dieckmann condensation). But it is really just an intramolecular Claisen condensation. Therefore, the steps of this mechanism are identical to the steps of a regular Claisen condensation. Propose a mechanism for the Dieckmann condensation. Try to do it without looking back at your previous work. You will need a separate piece of paper to record your answer.

8.9 DECARBOXYLATION

In the previous section, we learned how to use a Claisen condensation to prepare a β-keto ester:

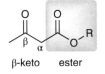

β-keto ester

Now let's see what we can do with β-keto esters. There are some very useful synthetic techniques that start with β-keto esters. In order to see how they work, we will need to remind ourselves of one reaction that we saw in the previous chapter. When we explored the chemistry of carboxylic acid derivatives, we saw that esters can be hydrolyzed to give carboxylic acids. We can use the exact same process to hydrolyze a β-keto ester, like this:

And the product is a β-keto acid, which will undergo a unique reaction when heated. Specifically, the carboxyl group is completely expelled:

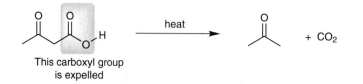

This carboxyl group
is expelled

We call this process a *decarboxylation*. This reaction is the basis for the synthetic techniques we will learn in this section, so let's make sure we understand how a decarboxylation occurs. The process begins with a pericyclic reaction that liberates CO_2 as a gas:

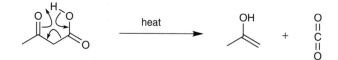

Pericyclic reactions are characterized by a ring of electrons moving around in a circle. There are many kinds of pericyclic reactions (for example, the Diels–Alder reaction is an important pericyclic reaction that we will cover in Chapter 10). Pericyclic reactions truly deserve their own chapter, and unfortunately, many textbooks do not devote an entire chapter to pericyclic reactions (they are just scattered throughout the various chapters). Perhaps your instructor will spend some time on pericyclic reactions. We will not cover them right now, as we must continue with the topic at hand.

In the reaction above, CO_2 gas is liberated (that's how the carboxyl group is expelled), generating an enol. And we know that enols will quickly tautomerize to give ketones:

So, when a β-keto acid is heated, the carboxyl group is expelled, and we end up with a ketone.

Now consider what we have just done. We took a β-keto ester (which is the product of a Claisen condensation), and we hydrolyzed it to produce a β-keto *acid*. Then, we heated this compound, and we expelled the carboxyl group:

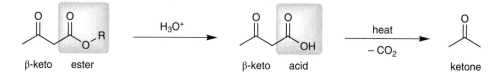

β-keto   ester   β-keto   acid   ketone

In the end, the product is a ketone. To see why this is so useful, we need to add just one more step at the very beginning of the overall process. Imagine that we first alkylate the β-keto ester:

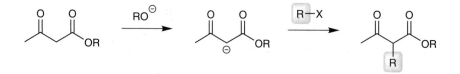

We have already seen this kind of reaction before (Section 8.6). It is just an alkylation. We used an alkoxide ion to produce a stabilized enolate, which then attacks the alkyl halide in an S_N2 reaction. If we then continue with the rest of the strategy (hydrolysis, followed by decarboxylation), the following product is obtained:

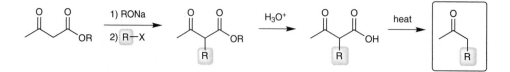

Take a close look at the product. This compound is a substituted derivative of acetone:

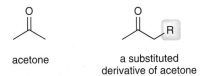

acetone   a substituted
derivative of acetone

This provides a method to make a wide variety of substituted derivatives of acetone.

This is useful, because we would encounter an obstacle if we tried to alkylate acetone directly:

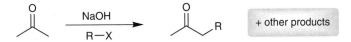

The desired product would be produced together with many other undesired products (from polyalkylation and from elimination reactions). So the strategy we have learned provides a clean way to make substituted derivatives of acetone. But be careful—remember that the alkylation step is an S_N2 process, so primary alkyl halides will be more efficient. In other words, you *could* use this strategy to prepare the following compound:

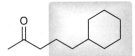

But you could *not* use this synthetic strategy to make this compound:

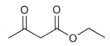

because that would have required an alkylation step involving a tertiary alkyl halide, which cannot function as a substrate in an S_N2 process. In order to use this synthetic strategy, we will always start with the following compound:

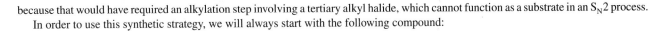

This compound is called ethyl acetoacetate. This compound belongs to a class of compounds called acetoacetic esters. Therefore, we call our strategy the *acetoacetic ester synthesis*.

To summarize what we have seen, the acetoacetic ester synthesis has three main steps: alkylate, hydrolyze, and then decarboxylate. Now let's get some practice using this synthetic strategy:

EXERCISE 8.65 Starting with ethyl acetoacetate, show how you would prepare the following compound:

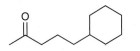

Answer   Remember that the acetoacetic ester synthesis has the following steps: alkylate, hydrolyze, and then decarboxylate. So, in order to solve this problem, we must identify the alkyl group that should be installed:

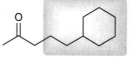

So, we will need the following alkyl halide:

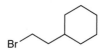

Now that we have determined what alkyl halide to use, we are ready to propose our synthesis:

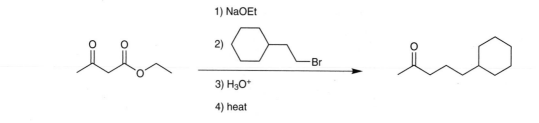

1) NaOEt

2)

3) H₃O⁺

4) heat

PROBLEMS   Show how you would prepare each of the following compounds from ethyl acetoacetate:

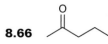

8.66

8.67

8.68

8.69   Explain why the following compound *cannot* be made with an acetoacetic ester synthesis.

In all of the problems we have done so far, we have focused on alkylating *once*. But it is also possible to alkylate twice, which would give a product with two alkyl groups:

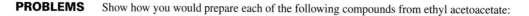

And the R groups don't even have to be the same. In the following sequence, the alkyl groups may be different.

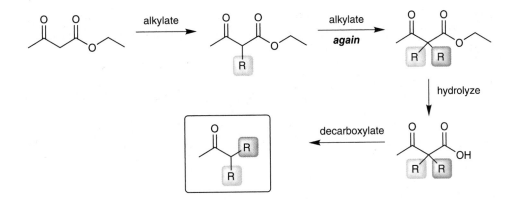

8.70 Show how you would prepare the following compound from ethyl acetoacetate:

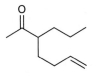

8.71 Show how you would prepare the following compound from ethyl acetoacetate:

8.72 Propose a synthesis for the following transformation. (*Hint:* This reaction is similar to an acetoacetic ester synthesis, but we are just starting with a different β-keto ester):

There is another common synthetic strategy that utilizes the same concepts as the acetoacetic ester synthesis. So, let's now focus on this other strategy. It is called the *malonic ester synthesis*, because the starting material is a malonic ester (called diethyl malonate):

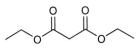

We follow the same three steps that we followed in our previous strategy: alkylate, hydrolyze, and then decarboxylate. The only difference is that we start with a slightly different starting material (malonic ester, instead of acetoacetic ester), and therefore, our product will be slightly different. Compare the structures of ethyl acetoacetate and diethyl malonate:

Ethyl acetoacetate   Diethyl malonate

Notice that diethyl malonate has *two* carboxyl groups (as opposed to ethyl acetoacetate, which has one carboxyl group and one carbonyl group). To see how this extra carboxyl group affects the structure of our end product, let's go through the three steps: alkylate, hydrolyze, and then decarboxylate.

We start with an alkylation:

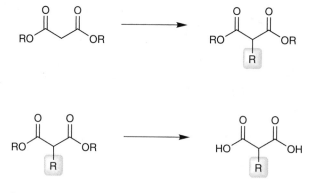

Then, we hydrolyze:

Notice that *both* sides get hydrolyzed.

Then, finally, we decarboxylate:

Only one side undergoes decarboxylation. Why? Remember how a decarboxylation works. It is a pericyclic reaction that can occur when a C=O bond is α,β to a carboxylic acid group. After the first carboxyl group is expelled, there is no longer a C=O bond that is β to the remaining carboxylic acid group. Try to draw a mechanism for the second carboxyl group leaving, and you should find that you can't do it.

Notice that the product is now a substituted carboxylic acid. This is the power of the malonic ester synthesis. It provides a method for making a wide variety of substituted carboxylic acids:

This synthesis can also be used to install *two* alkyl groups (just like we did with the acetoacetic ester synthesis). We would just alkylate twice at the beginning of our procedure:

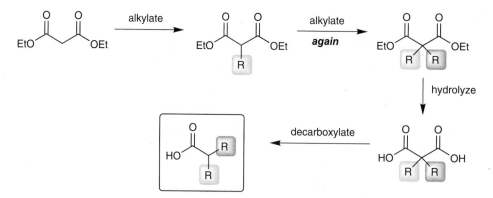

Once again, the process is most efficient for primary R groups, because alkylation is an S$_N$2 process.

This strategy is very useful, because it would be very difficult to alkylate a carboxylic acid directly. If we try to alkylate a carboxylic acid directly, we immediately run into an obstacle, because we cannot form an enolate of a carboxylic acid:

Cannot form this enolate

You cannot form an enolate in the presence of an acidic proton. So, the malonic ester synthesis gives us a way around this obstacle. It provides a method for making substituted carboxylic acids. Let's get some practice with this:

EXERCISE 8.73   Starting with diethyl malonate, show how you would prepare the following compound:

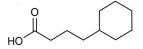

Answer   Remember that the malonic ester synthesis has the following steps: alkylate, hydrolyze, and then decarboxylate. So, in order to solve this problem, we must identify the alkyl group that should be installed.

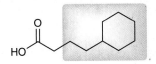

So, we will need the following alkyl halide:

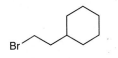

Now that we have determined what alkyl halide to use, we are ready to propose our synthesis:

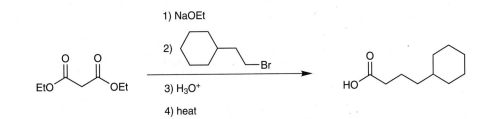

PROBLEMS   Identify what reagents you would use to achieve each of the following transformations:

8.74

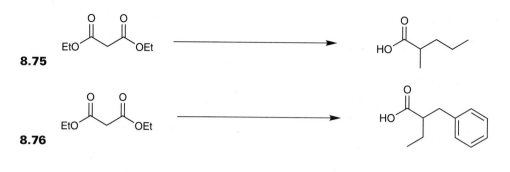

8.75

8.76

8.10 MICHAEL REACTIONS

In this chapter, we have seen that enolates can attack a wide variety of electrophiles. We started the chapter with the reaction between enolates and halogens. Then we looked at the reaction between enolates and alkyl halides. We also saw that enolates can attack ketones or esters. In this section, we will conclude our discussion of enolates by looking at a special kind of electrophile that can be attacked by an enolate. Consider the following compound:

This compound is an α,β-unsaturated ketone, and we have seen that compounds of this type can be made with an aldol condensation. This compound is a special kind of electrophile. To understand why it is special, let's take a close look at the resonance structures:

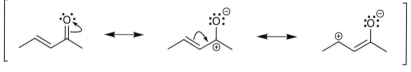

These resonance structures paint the following picture:

We see that there are **two** electrophilic centers. We already knew that the carbonyl group itself is electrophilic. But now, we can appreciate that the β position is also electrophilic. So, an attacking nucleophile has two choices. It can attack at the carbonyl group (as we have seen many times already), **or** it can attack at the β position. Let's look at both possibilities, and we will compare the products.

Consider what happens if the nucleophile attacks the carbonyl group, and the resulting intermediate is then protonated:

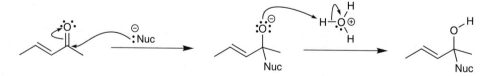

Notice that we had a π system that spanned 4 atoms, and we installed the nucleophile and the H in positions 1 and 2:

Therefore, we call this a **1,2-addition**.

Now consider what happens if the nucleophile attacks the β position, rather than the carbonyl group. The initial intermediate is an enolate:

an enolate

Then, when this enolate is protonated, an enol is formed:

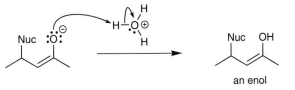

an enol

Once again, we have added the nucleophile and H across the π system. But this time, we have added them across the *ends* of this system:

So, we call this a *1,4-addition*. Chemists have given this reaction other names as well. A 1,4-addition is often called a ***conjugate addition***, or a ***Michael addition***.

We know that the product of a 1,4-addition is not going to stay in the form of an enol, because an enol will tautomerize to form a ketone:

When you look at this ketone, it is hard to see why we call it a 1,4-addition. After all, it looks like the nucleophile and the H have added across the C=C bond:

You need to draw the entire mechanism in order to see why we call it a 1,4-addition.

Now that we know the difference between a 1,2-addition and a 1,4-addition, let's take a look at what happens when our attacking nucleophile is an enolate.

If an α,β-unsaturated ketone is treated with an enolate, a mixture of products is obtained. Not only do we observe both possibilities (the enolate attacking the carbonyl group, or the enolate attacking the β position), but it gets even more complicated. The product of the 1,4-addition is a ketone, which can be attacked again by an enolate. You can get crossed aldol condensations, and all sorts of unwanted products. So, we can't use an enolate to attack an α,β-unsaturated ketone. The enolate is simply too reactive, and we observe a mixture of undesired products.

The way around this problem is to create an enolate that is more stabilized. A more stable enolate will be less reactive, and therefore, it will be more selective in what it reacts with. But how do we make a more stabilized enolate? We have actually already seen such an example in this chapter. Consider the following enolate:

We argued that this enolate is more stable than a regular enolate, because the negative charge is delocalized over *two* carbonyl groups. If we use this enolate to attack an α,β-unsaturated ketone, we find that the predominant reaction is a 1,4-addition:

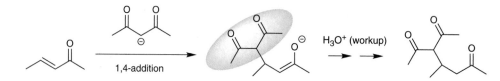

We said earlier that a 1,4-addition is also called a Michael addition. In order to get a Michael addition, you need to have a stabilized nucleophile, like the stabilized enolate shown in the reaction above. This stabilized enolate is called a Michael donor. There are many other examples of **Michael donors**. Look at them carefully, because you will need to recognize them as being Michael donors when you see them:

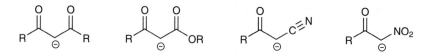

All of these nucleophiles are sufficiently stabilized to function as Michael donors. Organocuprates, R_2CuLi, are also stabilized nucleophiles and will also attack an α,β-unsaturated ketone to give a 1,4-addition reaction.

In any Michael reaction, there is always a Michael donor *and* a Michael acceptor:

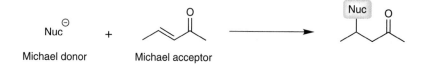

In the reaction above, the Michael acceptor is an α,β-unsaturated ketone. But there are other compounds that can also function as Michael acceptors:

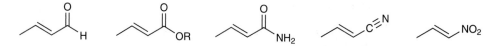

You can use any Michael donor to attack any Michael acceptor. The following reaction would also be considered a conjugate addition reaction:

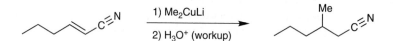

In order to do this next set of problems, you will need to review the lists of Michael donors and Michael acceptors.

EXERCISE 8.77   Draw the expected product of the following conjugate addition reaction:

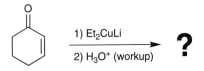

Answer   First identify the nucleophile and the electrophile. The α,β-unsaturated ketone is the electrophile, and the lithium dialkyl cuprate is the nucleophile, and we expect a conjugate addition reaction.

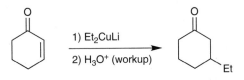

Stare at this transformation for a few moments. It might be helpful to you on an exam.

PROBLEMS   For each of the reactions below, determine whether you expect an efficient conjugate addition reaction. If so, then draw the product you expect.

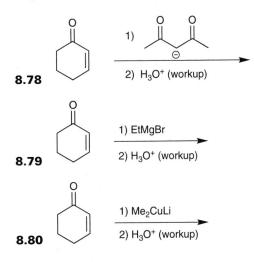

There is one more Michael donor that requires special mention. Enamines are very special Michael donors, because they provide us with a useful synthetic strategy.

When we learned about ketones and aldehydes (Chapter 6), we saw that you can prepare an enamine by treating a ketone with a secondary amine, under the following conditions:

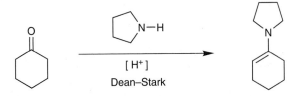

This was our way of converting a ketone into an enamine. To understand how an enamine can function as a Michael donor, let's take a close look at the resonance structures of an enamine:

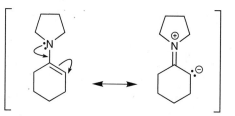

When we meld these two images together in our minds, we see that the carbon atom is nucleophilic (it has some partial negative character). But, it is a fairly weak nucleophile, because the compound does not have a *full* negative charge. Rather, the carbon atom only has *partial* negative character. Therefore, this compound is a *stabilized* nucleophile (in other words, it will be selective in its reactivity). This means we must add it to our list of Michael donors. If we use an enamine as a nucleophile to attack a Michael acceptor, a Michael reaction will occur:

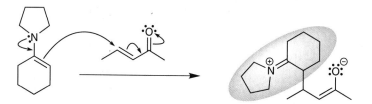

And then, the resulting iminium group can be removed by adding water under acidic conditions (and under these conditions, the enolate is protonated to form an enol, which tautomerizes to form a ketone):

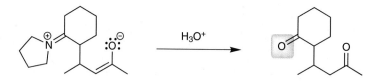

But why is this so important? Why am I singling out this one Michael donor, and why are we learning about enamines again? To understand the usefulness of enamines here, let's imagine that we wanted to achieve the following transformation:

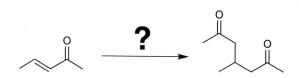

You decide that it should be simple. Your plan is to use a nucleophile that can attack an α,β-unsaturated ketone in a 1,4-addition:

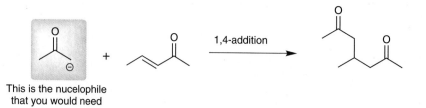

This is the nucelophile
that you would need

But, when you try to perform this reaction, you obtain a mixture of undesired products. Why? Because this enolate is **not** a Michael donor, and therefore, it will not efficiently attack in a 1,4-addition. So, how do you get around this problem? This is where our enamine comes in handy.

Instead of using an enolate as the nucleophile, suppose we convert the ketone into an enamine:

This enamine *is* a Michael donor, and it *will* attack cleanly in a 1,4-addition:

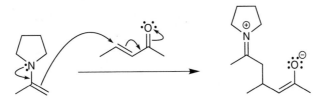

Finally, we use H_3O^+ to remove the iminium group and protonate the enolate (which then turns into a ketone):

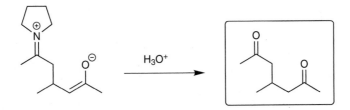

In the end, we prepared the desired product. This synthetic strategy is called the ***Stork enamine synthesis***, and it can come in handy when you are proposing syntheses. Whenever you are trying to propose a synthesis, and you decide that you need an enolate to attack as a nucleophile in a 1,4-addition, you will have a problem. Regular enolates are not stable enough to be Michael donors. But, you can convert it into an enamine, which *is* stable enough to be a Michael donor. Then, you can remove the enamine in the end. The enamine serves as a way of temporarily modifying the reactivity of the enolate so that we can achieve the desired result. It is very clever when you really think about it.

EXERCISE 8.81   Propose a plausible synthesis for the following transformation:

Answer   When we inspect this transformation, we see that we need to install the following fragment:

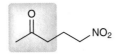

and at the same time, we need to remove the double bond that was in the starting material. We can accomplish both of these at the same time by performing a 1,4-addition with the appropriate nucleophile. When we look carefully to see what nucleophile we would need, we realize that we will need to use the following enolate:

This enolate is *not* stable enough to function as a Michael donor, so we need to use a Stork enamine synthesis:

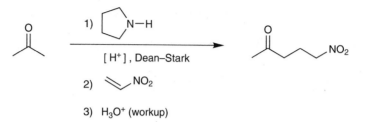

PROBLEMS   Propose a synthesis for each of the following transformations. In some cases, you will need to use a Stork enamine synthesis, but in other cases, it will not be necessary. Analyze each problem carefully to see if a Stork enamine synthesis is necessary.

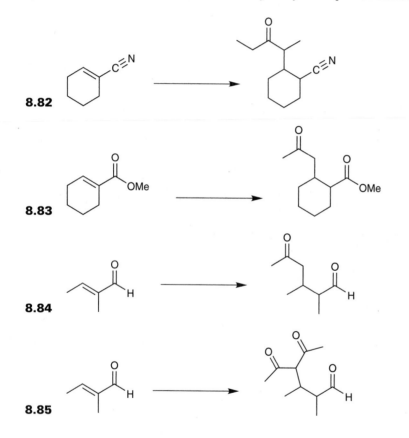

8.82

8.83

8.84

8.85

END-OF-CHAPTER PROBLEMS

PRACTICE PROBLEMS *(Problems that involve only one skill)*

8.86 Draw the missing structures in the following reaction scheme:

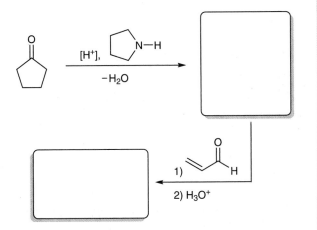

8.87 Draw a starting material that can be used to prepare the product below via an intramolecular aldol condensation:

8.88 Predict the major product of the following reaction.

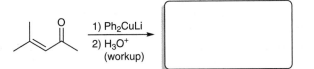

8.89 Draw a starting material that can be used to prepare the product below via an intramolecular Claisen condensation:

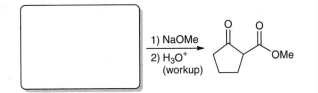

8.90 Consider the structures of the following two compounds:

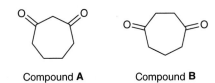

Compound **A**   Compound **B**

(a) Draw all resonance structures for the conjugate base of compound **A**.

(b) Compare the stability of the conjugate base of compound **A** with the conjugate base of compound **B**. Which conjugate base is more stable?

(c) Which is more acidic, compound **A** or compound **B**?

8.91 Which of the following compounds will undergo decarboxylation upon heating? For each compound that reacts, draw the resulting product.

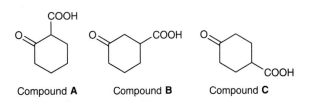

Compound **A**   Compound **B**   Compound **C**

8.92 On a separate piece of paper, draw a mechanism for the following isomerization process, which occurs in the presence of catalytic base:

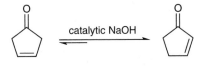

8.93 Consider the following aldol condensation:

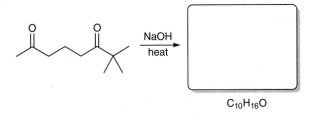

$C_{10}H_{16}O$

(a) Draw the major product of this reaction in the box provided above.

(b) On a separate piece of paper, draw a mechanism for this process.

8.94 Identify reagents that can be used to convert diethyl malonate into the carboxylic acid shown:

8.95 On a separate piece of paper, draw a mechanism for the following process. Note: Make sure to show the role that aqueous acid plays during the workup step.

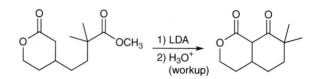

INTEGRATED PROBLEMS (Problems that involve more than one skill)

8.96 Propose an efficient synthesis for the following transformation:

8.97 Propose an efficient synthesis for the following transformation:

8.98 Propose an efficient synthesis for the following transformation:

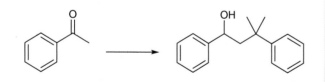

8.99 Propose an efficient synthesis for the following transformation:

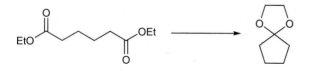

8.100 Propose an efficient synthesis for the following transformation:

8.101 Propose an efficient synthesis for the following transformation:

8.102 Propose an efficient synthesis for the following transformation:

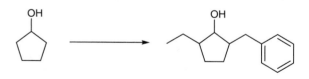

8.103 Propose an efficient synthesis for the following transformation. Note that acetone (2-propanone) is your only allowed source of carbon atoms:

Only source
of carbon

8.104 Propose an efficient synthesis for the following transformation. Note that ethanol is your only allowed source of carbon atoms. HINT: This transformation can be achieved in just a few steps.

Only source
of carbon

8.105 Consider the following aldol condensation.

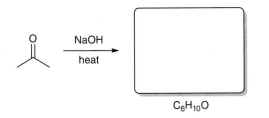

$C_6H_{10}O$

(a) Draw the condensation product in the box provided above.

(b) Identify the number of signals that will appear in the ^1H NMR spectrum of the product.

(c) Identify the number of signals that will appear in the ^{13}C NMR spectrum of the product.

8.106 The malonic ester synthesis was used to prepare a carboxylic acid with the molecular formula $C_4H_8O_2$. The ^1H NMR spectrum of the product has four signals, and the ^{13}C NMR spectrum of the product has four signals.

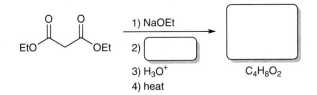

1) NaOEt
2)
3) H_3O^+
4) heat

$C_4H_8O_2$

(a) Draw the product of this process in the box provided.

(b) Draw a reagent that can be used in the second step of the process to give the desired carboxylic acid.

8.107 Consider the isomerization of compound **A** to give compound **B**, shown below:

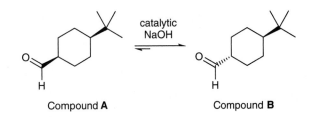

catalytic
NaOH

Compound **A**   Compound **B**

(a) On a separate piece of paper, draw a mechanism for the conversion of compound **A** into compound **B** in the presence of catalytic base.

(b) Provide an explanation for why the equilibrium favors compound **B** over compound **A**.

8.108 On a separate piece of paper, draw a mechanism for the following reaction.

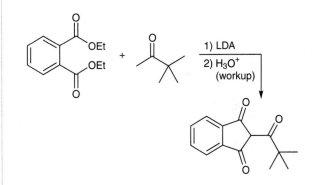

1) LDA
2) H_3O^+
(workup)

8.109 Fill in the missing reagents in the following scheme:

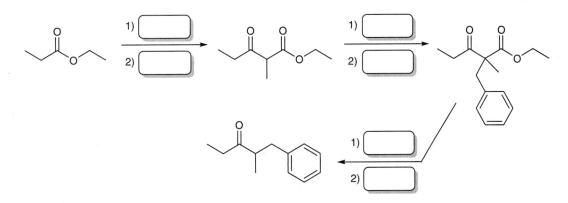

8.110 Fill in the missing structures in the following scheme:

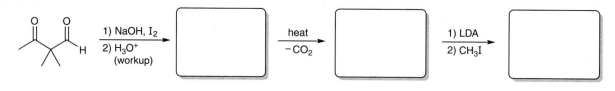

8.111 Fill in the missing structures in the following scheme:

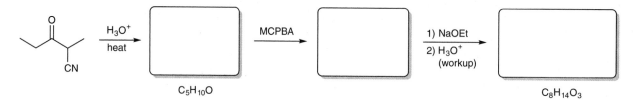

8.112 Consider the following reaction sequence:

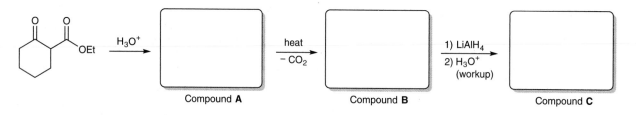

(a) Draw the structures of compounds **A**, **B**, and **C** in the boxes provided above.

(b) Describe how you could use IR spectroscopy to confirm the conversion of **A** to **B**, as well as the conversion of **B** to **C**.

(c) Describe how you could use ^{13}C NMR spectroscopy to confirm the conversion of **A** to **B**.

(d) Describe how you could use ^{13}C NMR spectroscopy to confirm the conversion of **B** to **C**.

CHALLENGE PROBLEMS

8.113 On a separate piece of paper, draw a plausible mechanism for the following transformation. Note: The starting material is a thioester (which is similar to an ester, but the OR group has been replaced with an SR group). Much like an ester, a thioester will react with LDA to give an enolate.

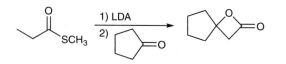

8.114 On a separate piece of paper, draw a plausible mechanism for the following transformation:

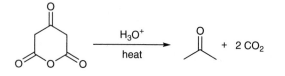

8.115 On a separate piece of paper, draw a plausible mechanism for the following transformation:

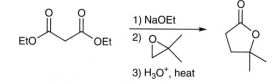

CHAPTER 9

AMINES

9.1 NUCLEOPHILICITY AND BASICITY OF AMINES

Amines are classified based on the number of alkyl groups attached to the nitrogen atom:

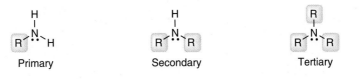

The reactivity of all amines derives from the presence of a lone pair on the nitrogen atom. All amines have this lone pair:

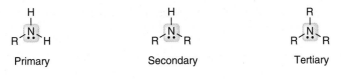

This lone pair can function as a nucleophile (attacking an electrophile):

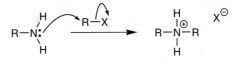

or it can function as a base (receiving a proton):

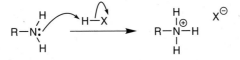

By focusing on this lone pair, we can understand why amines are good nucleophiles *and* good bases. When an amine participates in a reaction, the first step will always be one of these two possibilities: either the lone pair will receive a proton or the lone pair will attack an electrophile.

Of course, if we had a negative charge on the nitrogen atom, it would be an even stronger base. To get a negative charge on the nitrogen atom, we must deprotonate the amine. Tertiary amines don't have a proton that can be removed, but primary and secondary amines can be deprotonated to give the following anions:

These anions are stronger bases than uncharged amines. We call these structures *amides*. Here are two examples of amides that we have seen so far in this course:

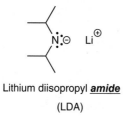

Lithium diisopropyl *amide*
(LDA)

Sodium *amide*

The term *amide* is actually a terrible name, because we have already used this term to describe a type of carboxylic acid derivative:

An *amide*

Amide ion

Perhaps chemists could have assigned different names to these structures (so that it would be less confusing to students). But historically, chemists have used the term *amide* to refer to both of these types of structures. And old habits die hard. Since we are probably not going to convince chemists around the world to change their terminology, we will just have to get used to the terminology that is in use. Don't let this confuse you.

Now let's get back to uncharged amines. Some amines are actually less nucleophilic and less basic than regular amines. For example, compare the following two amines:

An <u>alkyl</u> amine

An <u>aryl</u> amine

The first amine is called an *alkyl* amine, because the nitrogen atom is connected to an alkyl group. The second compound is called an *aryl* amine, because the nitrogen atom is connected to an aromatic ring. Aryl amines are less nucleophilic and less basic, because the lone pair is delocalized into the aromatic ring. We can see this when we draw the resonance structures:

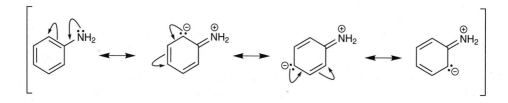

Since the lone pair is delocalized, it will be less available to function as a nucleophile or as a base. That does not mean that an aryl amine can't attack something. In fact, we will soon see a reaction where an aryl amine *is* used as a nucleophile. It *can* function as a nucleophile—but it is just *less* nucleophilic than an alkyl amine.

Now that we have had an introduction to amines, let's focus on a few ways to make amines.

9.2   PREPARATION OF AMINES THROUGH S$_N$2 REACTIONS

Suppose you wanted to make the following primary amine:

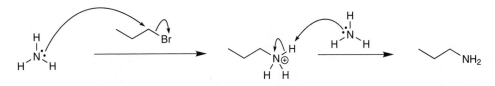

It might be tempting to suggest the following synthesis:

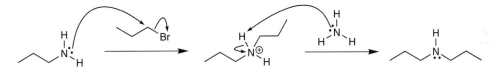

This is just an S$_N$2 reaction, followed by a deprotonation. This approach does work, BUT it is difficult to get the reaction to stop after monoalkylation. The product is also a nucleophile and it competes with NH$_3$ for the alkyl halide. In fact, the product is an even better nucleophile than NH$_3$, because the alkyl group is electron donating. Therefore, it will attack again:

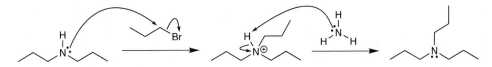

And then it attacks again:

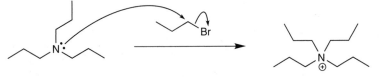

And then one last time:

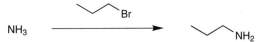

This time,
there is no proton to remove

The final product has four alkyl groups (which is referred to as **quaternary**), and the nitrogen has a positive charge (called an **ammonium ion**). So, we call this a **quaternary ammonium ion**. This structure does not have a lone pair, so it is neither nucleophilic nor basic.

If our intention is to make a quaternary ammonium salt, then the synthesis above is a good approach. But what if we want to make a primary amine? We cannot simply alkylate ammonia:

NH$_3$ →   (with alkyl bromide)   →   amine-NH$_2$

because it is too difficult to stop the reaction at this stage. The alkyl halide will react again with the primary amine. Even if we try to use only one mole of alkyl halide and one mole of ammonia, we will still get a mixture of undesired products. We will get some polyalkylated products, and we will get some ammonia that could not find any alkyl halide to attack. So what do we do?

To get around this problem, we use a clever trick. We use a starting amine that already has two dummy groups on it:

And we choose our dummy groups so that they are easily removable after we alkylate. So, we first alkylate, and then we remove the dummy groups:

This strategy is called the Gabriel synthesis. It is a very good way of making primary amines from primary alkyl halides, so let's explore this strategy in more detail.

We start with the following compound, called phthalimide:

We use a base (KOH) to remove the proton, which gives the following anion:

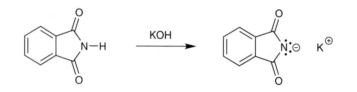

Notice that this negative charge is very stabilized (delocalized) by resonance. This is similar to the Michael donors that we saw in the end of the previous chapter. It is a stabilized nucleophile. We use this nucleophile to attack an alkyl halide:

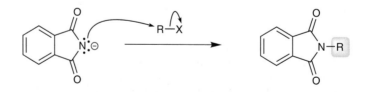

This reaction proceeds via an S_N2 process, so tertiary alkyl halides cannot be used. Acid-catalyzed or base-catalyzed hydrolysis is then performed to release the amine. Acidic conditions are more common than basic conditions.

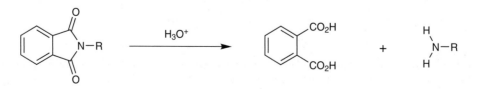

The mechanism of hydrolysis is directly analogous to the hydrolysis of amides, as seen in Section 6.6. The hydrolysis step is slow, and many alternative approaches have been developed. One such alternative employs hydrazine to release the amine:

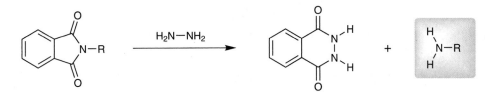

In this step, the dummy groups are removed, generating the desired product.
   The Gabriel synthesis is very useful for making primary amines from primary alkyl halides:

EXERCISE 9.1   Identify how we would use a Gabriel synthesis to prepare the following amine:

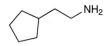

Answer   In order to perform a Gabriel synthesis, we must identify the alkyl halide that is necessary. To do this, we just draw a halogen instead of NH$_2$, like this:

Now that we know what alkyl halide to use, we are ready to propose our synthesis:

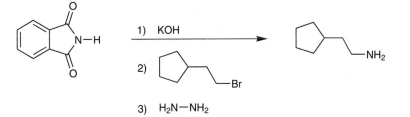

PROBLEMS   Identify how you would use a Gabriel synthesis to prepare each of the following compounds.

9.2

9.3

9.4

9.5

The Gabriel synthesis has its limitations, though. Since it relies on an S$_N$2 reaction, it will work well with ***primary*** alkyl halides. But it is not so great for secondary alkyl halides, and it will not work at all for tertiary alkyl halides.

In addition, it will not work for aryl halides, because you cannot perform an S_N2 process on an aryl halide:

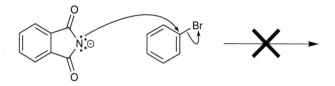

EXERCISE 9.6   Is it possible to prepare the following compound with a Gabriel synthesis?

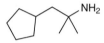

Answer   We draw the alkyl halide that would be necessary,

and we see that it is a tertiary alkyl halide, so we ***cannot*** use a Gabriel synthesis to make the desired product in this case.

PROBLEMS   Identify whether each of the following compounds could be made with a Gabriel synthesis, and provide a Gabriel synthesis for those that are possible.

9.7

9.8

9.9

9.10

9.3   PREPARATION OF AMINES THROUGH REDUCTIVE AMINATION

In the previous section, we learned how to make amines via an S_N2 process. That method works very well for making primary amines. In this section, we will learn a way to make secondary amines, using a two-step synthesis (where the first step is a reaction that we have already seen). When we learned about ketones and aldehydes, we saw how to make *imines*:

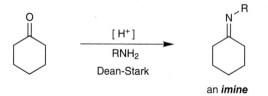

an *imine*

We saw that a ketone can be converted into an imine. Now, we will use this reaction to make amines.

When we form an imine, we are forming the essential C—N connection that you need in order to have an amine:

But, the oxidation state is not correct. In order to get an amine, we need to do the following conversion:

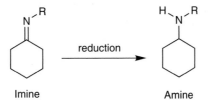

In order to convert an *i*mine into an *a*mine, we will need to perform a reduction. One way to do so is to reduce the imine in much the same way that a ketone is reduced, using LiAlH$_4$:

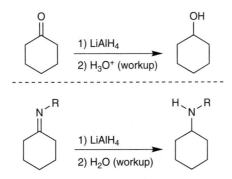

When reducing an imine with LiAlH$_4$, notice that the workup step is H$_2$O, rather than H$_3$O$^+$, because the product is an amine (and the use of H$_3$O$^+$ would cause the amine to be protonated to give an ammonium ion).

Alternatively, the C=N bond can undergo hydrogenation (in the presence of a catalyst):

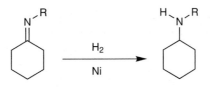

There are many other ways to reduce an imine as well. But the bottom line is that we now have a two-step synthesis for preparing amines:

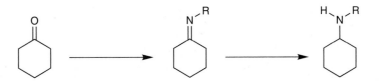

This process is called ***reductive amination***, because we are forming an amine (a process called *amination*) via a *reduction*.

EXERCISE 9.11   Suggest an efficient synthesis for the following transformation:

Answer   This problem involves the conversion of an aldehyde into a secondary amine. This should alert us to the possibility of a reductive amination. If we used a reductive amination, our strategy would go like this:

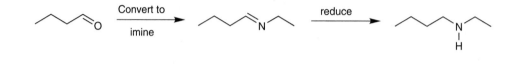

So, our synthesis will look like this:

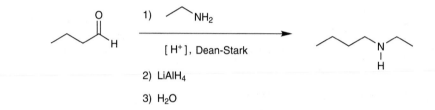

PROBLEMS   Suggest an efficient synthesis for each of the following transformations.

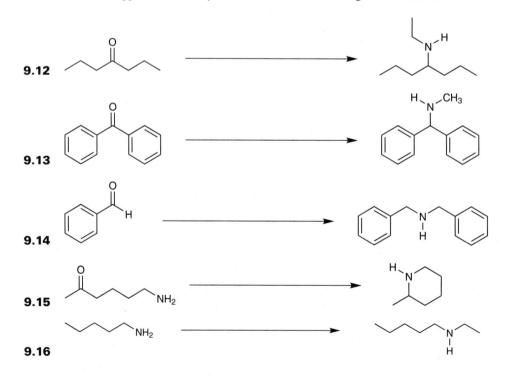

9.12

9.13

9.14

9.15

9.16

Reductive amination is a useful technique, because the starting material is a ketone or aldehyde. And we have seen many ways to make ketones or aldehydes. This gives us a way to make amines from a variety of compounds:

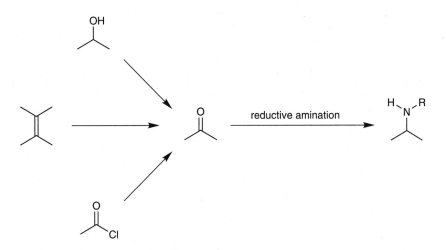

Students often have difficulty combining reactions from different chapters to propose a synthesis, so let's get some practice:

EXERCISE 9.17   Suggest an efficient synthesis for the following transformation:

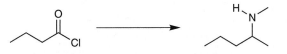

Answer   Our product is a secondary amine, so we will explore if we can achieve this synthesis using a reductive amination. Let's work backwards.

If we did use a reductive amination, our last step would need to be the reduction of the following imine:

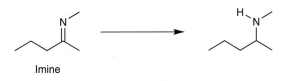

Imine

So, we would need to make the imine above, and we could have done that starting with the following ketone:

So, our goal is to make this ketone. If we can make this ketone, then we can use a reductive amination to form our product:

reductive amination

But how do we make this ketone from the starting material?

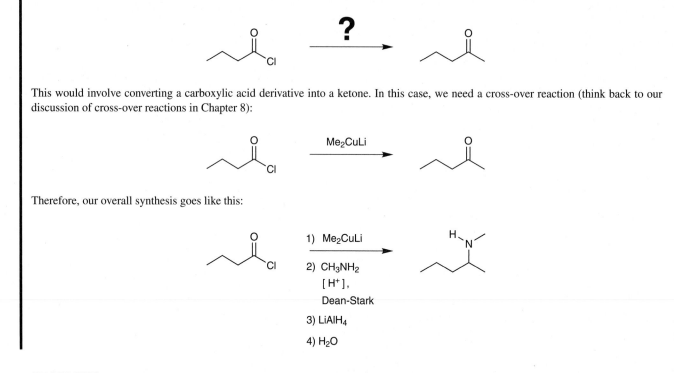

This would involve converting a carboxylic acid derivative into a ketone. In this case, we need a cross-over reaction (think back to our discussion of cross-over reactions in Chapter 8):

Therefore, our overall synthesis goes like this:

PROBLEMS Suggest an efficient synthesis for each of the following transformations. In each case, you should work backwards. Start by asking what ketone or aldehyde you would need in order to make the desired product via a reductive amination. Then, ask yourself how you could make that ketone from the starting material.

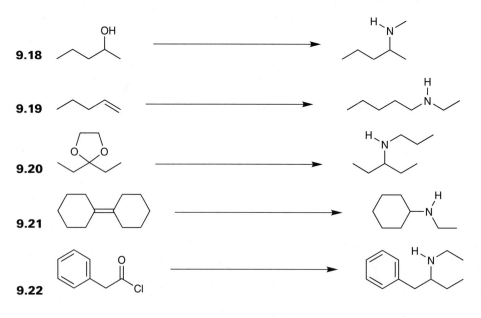

9.18

9.19

9.20

9.21

9.22

9.4 ACYLATION OF AMINES

So far in this chapter, we have focused on ways of *making* amines. For the rest of the chapter, we will shift our focus. We will now explore reactions of amines.

We will begin our survey with a reaction that we have actually already seen in a previous chapter. When we learned about carboxylic acid derivatives (Chapter 7), we saw that you can convert an acid halide into an amide. For example:

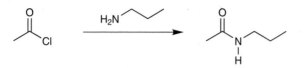

When we draw it this way (with the amine over the reaction arrow), the focus is on what happens to the acid halide (it is converted into an amide). But what if we choose to focus on the amine instead? In other words, let's rewrite the same reaction a bit differently. Let's put the acid halide on top of the reaction arrow, like this:

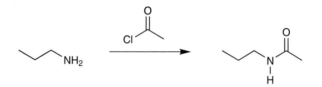

We have not changed the reaction at all. It is still the same reaction (an amine reacting with an acid halide). But when we draw it like this, our attention focuses on converting the *amine* into an *amide*. The acid halide is just the reagent that we use to accomplish this conversion.

Overall, we have introduced an *acyl* group onto the amine:

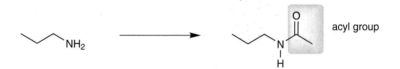

Therefore, we call this an *acylation* reaction.

Most primary and secondary amines can be acylated. Here is another example:

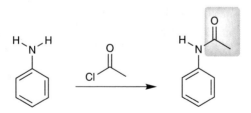

Now that we have seen how to acylate an amine, let's take a look at how to remove the acyl group. This reaction was also covered in the chapter on carboxylic acid derivatives. It is just the hydrolysis of an amide:

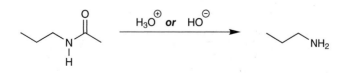

Notice that we have removed the acyl group to regenerate the amine. So, now we know how to install an acyl group, and we know how to remove it:

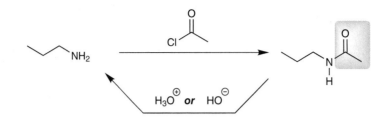

But the obvious question is: why would we want to do that? Why would we ever install a group, just to remove it later? The answer to this question is very important, because it illustrates a common strategy that organic chemists use. Let's try to answer this question through a specific example.

Imagine that we want to achieve the following transformation:

This seems easy to do. Do you remember how to install a nitro group on an aromatic ring (Chapter 4)? We just used a mixture of nitric acid and sulfuric acid. The amino group is an activator, so it will direct to the *ortho* and *para* positions, with a preference for *para* substitution (for steric reasons). So, we propose the following:

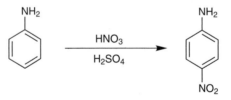

But when we try to perform this reaction, we find that it does not work. It is true that an amino group is a strong activator, but under the strongly acidic reaction conditions, the amino group is protonated to give an ammonium ion, which is a strong deactivator (and a *meta*-director). Under those conditions, any reaction that occurs will be undesired. So, how can we generate the desired product?

The way to do it is actually very clever. We first acylate the amino group:

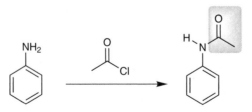

This converts the amino group into an amide group, which is a weaker base, because the lone pair is further delocalized into the carbonyl group of the amide. As such, the nitrogen atom of the amide group is not protonated under acidic conditions. And since the amide group is an activator, the ring will undergo nitration in the desired location:

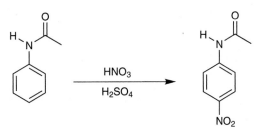

Then, we remove the acyl group to obtain the desired product:

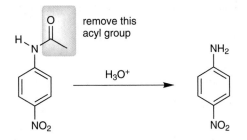

Think about what we just did. We used an acylation process as a way of ***temporarily modifying*** the electronics of the amino group, so that it would not interfere with the desired reaction. This strategy is used all of the time by organic chemists. This idea of temporarily modifying a functional group (and then converting it back later) is an idea that is used in many other situations as well (not just for acylation of amines).

EXERCISE 9.23   Suppose we want to achieve the following transformation:

We try to perform this transformation using Br_2, but we find that aniline is too reactive, and we get a mixture of mono-, di-, and tri-brominated products. What can we do to obtain the desired product and avoid polybromination?

Answer   The problem is the amino group. It is too strongly activating. To circumvent this obstacle, we use the strategy we have developed in this section. We acylate the amino group, and that makes the ring less activated (temporarily):

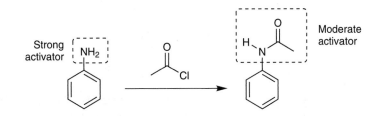

Now we are able to brominate:

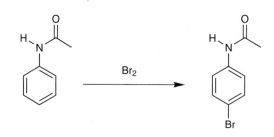

and that installs one bromine atom in the *para* position. Finally, we remove the acyl group to obtain the desired product:

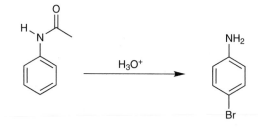

PROBLEMS Propose an efficient synthesis for each of the following transformations:

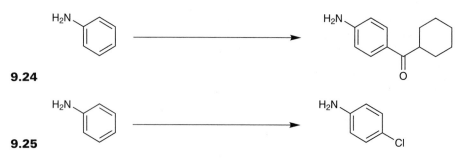

9.24

9.25

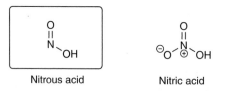

9.5 REACTIONS OF AMINES WITH NITROUS ACID

In this section, we will begin to explore the reactions that take place between amines and nitrous acid. Compare the structures of nit*rous* acid and nit*ric* acid:

Nitrous acid   Nitric acid

When nitrous acid reacts with amines, the products are very useful. We will soon see that these products can be used in a large number of synthetic transformations. So, let's make sure that you are comfortable with the reactions between amines and nitrous acid.

Let's start by looking at the source of nitrous acid. It turns out that nitrous acid is fairly unstable and, therefore, we cannot just store it in a bottle. Rather, we have to make nitrous acid in the reaction flask. To do this, we use sodium nitrite ($NaNO_2$) and HCl:

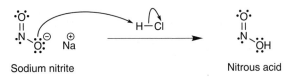

Sodium nitrite   Nitrous acid

Under these (acidic) conditions, nitrous acid is protonated again, to produce a positively charged intermediate:

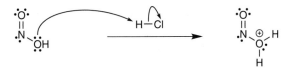

This intermediate can then lose water to give a highly reactive intermediate, called a nitrosonium ion:

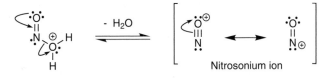

Nitrosonium ion

This intermediate (the nitrosonium ion) is the intermediate we must focus on. Whenever we talk about an amine reacting with nitrous acid, we really mean to say that the amine is reacting with a nitrosonium ion (NO^+). You might notice the similarity between this intermediate and the NO_2^+ intermediate (that we used in nitration reactions). Do not confuse these two intermediates. NO^+ and NO_2^+ are different intermediates. In this section, we are only talking about the reactions of amines with the nitrosonium ion (NO^+).

As we said a few moments ago, nitrosonium ions cannot be stored in a bottle. Instead, we must make them *in the presence of an amine*. That way, as soon as the nitrosonium ion is formed, it will immediately react with the amine before it has a chance to do anything else. This is called an *in situ* preparation.

So, now the question is: what happens when an amine reacts with a nitrosonium ion? Let's begin by exploring secondary amines (and then we will explore primary amines).

A secondary amine can attack a nitrosonium ion like this:

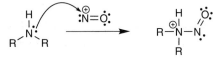

Deprotonation then generates the product:

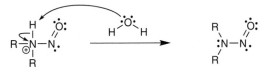

This product is called an *N-nitroso amine*. For short, chemists often call it a *nitrosamine*.

This reaction is not very useful. But when a **primary** amine attacks a nitrosonium ion, the resulting reaction is extremely important. The amine attacks, to initially form a nitrosamine:

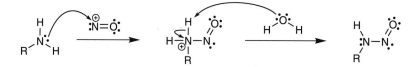

Since we started with a primary amine, we notice that we have a proton in our nitrosamine:

Because of this proton, the nitrosamine continues to react in the following way:

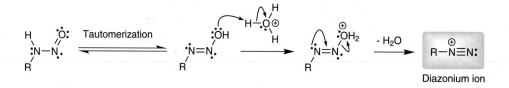

This cation is called a **diazonium** ion. The term **azo** means nitrogen, so **diazo** means two nitrogen atoms. And, of course, **onium** means a positive charge. That is what we have here: two nitrogen atoms connected to each other, and a positive charge; thus, the name **diazonium**.

Primary **alkyl** amines will give **alkyl** diazonium salts, and primary **aryl** amines will give **aryl** diazonium salts:

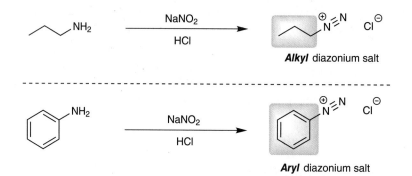

Alkyl diazonium salts are not terribly useful. They are very explosive, and as a result, they are very dangerous to prepare. But **aryl** diazonium salts are much more stable, and they are incredibly useful, as we will see in the upcoming section. For now, let's just make sure that we know how to make diazonium salts:

EXERCISE 9.26   Predict the product of the following reaction:

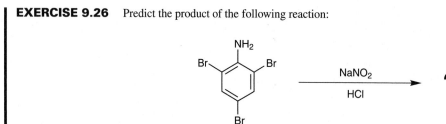

Answer   Our starting material is a primary amine. These reagents (sodium nitrite and HCl) are used to form nitrous acid, which then forms a nitrosonium ion. Primary amines react with a nitrosonium ion to give a diazonium salt. So, the product of this reaction is:

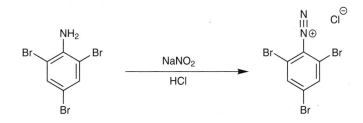

PROBLEMS   Predict the major product for each of the following reactions:

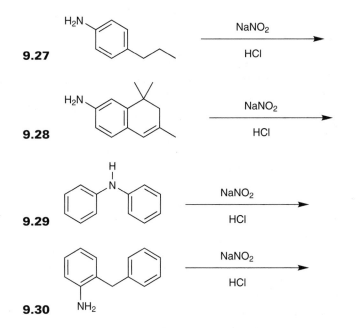

9.27

9.28

9.29

9.30

9.6   AROMATIC DIAZONIUM SALTS

In the previous section, we learned how to make aryl diazonium salts:

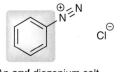

An *aryl* diazonium salt

Now we will learn what we can do with aryl diazonium salts. Here are a just few reactions:

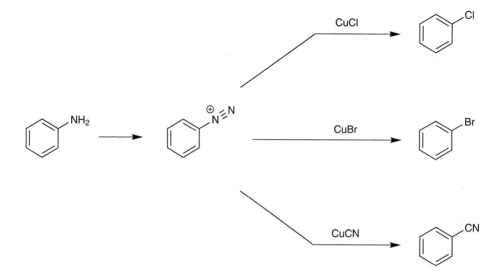

In all of these reactions, we are using copper salts as the reagents. These reactions are called *Sandmeyer reactions*. They are useful, because they allow us to achieve transformations that we could not otherwise achieve with the chemistry that we learned in Chapters 4 and 5 (electrophilic and nucleophilic aromatic substitution). As a case in point, we did not see how to install a cyano group on an aromatic ring. This is our first way to do this.

EXERCISE 9.31 What reagents would you use to achieve the following transformation:

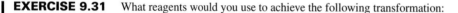

Answer If we brominate aniline, we will find that the amino group is so activated that we will obtain a tribrominated product:

Then, we can convert the amino group into a chloro group. To do this, we make a diazonium salt, followed by a Sandmeyer reaction:

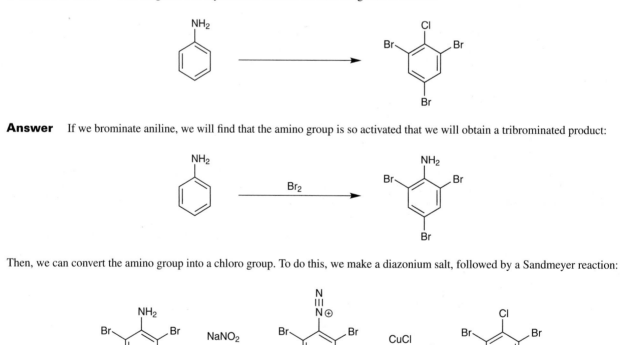

PROBLEMS   What reagents would you use to achieve each of the following transformations:

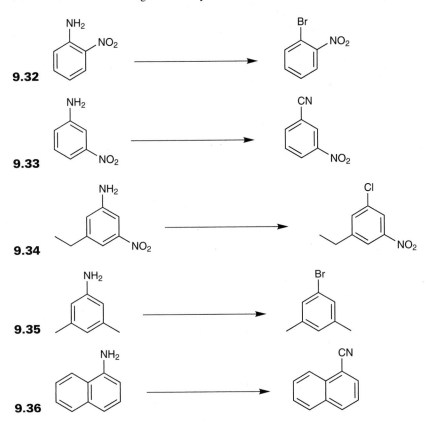

9.32

9.33

9.34

9.35

9.36

So far, we have seen a few reactions that you can perform with aryl diazonium salts. Many instructors will cover additional reactions of aryl diazonium salts. You should look through your lecture notes to see what you are responsible for. Your textbook will certainly show you several more reactions that can be performed with aryl diazonium salts.

These reactions are very useful in synthesis problems. You will find that some problems will combine these reactions with electrophilic aromatic substitution reactions. These problems can range in difficulty, and they can get really tough at times. You will find many such problems in your textbook. As you go through some of the challenging problems in your textbook, I will leave you with a bit of last-minute advice:

There are two important activities that you must do in order to master these types of problems.

1. You must review the reactions and principles from the entire course. Go through your textbook and your lecture notes again and again and again. Make sure that you get to a point where you know all of the reactions cold (you should have a strong command over all of the reactions).

2. You must do as many problems as possible. If you don't get practice, you will find that even a very strong grasp of the reactions will be insufficient. In order to truly master the art of problem solving, you must ***practice, practice, practice***. I recommend that you do as many problems as possible. You might even find that they can be fun, believe it or not.

END-OF-CHAPTER PROBLEMS

PRACTICE PROBLEMS *(Problems that involve only one skill)*

9.37 In the following structure, identify the most basic nitrogen atom, and explain your choice:

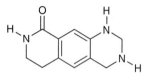

9.38 Predict the major product of the following reaction sequence:

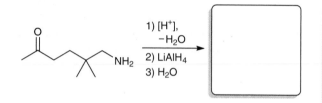

1) [H⁺], —H₂O
2) LiAlH₄
3) H₂O

9.39 Propose an efficient synthesis for the following transformation:

9.40 Draw the missing reagents and products in the following reaction scheme:

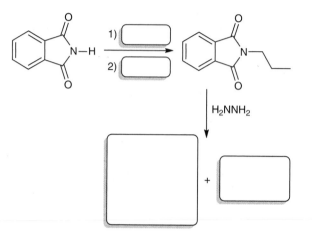

9.41 Starting with phthalimide, draw the reagents that you would use to make the product below via a Gabriel synthesis:

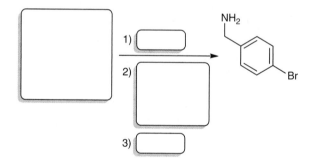

INTEGRATED PROBLEMS *(Problems that involve more than one skill)*

9.42 Identify which of the following compounds is the weaker base, and explain your choice:

Compound **A**   Compound **B**

9.43 Propose an efficient synthesis for the following transformation:

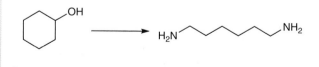

9.44 Propose an efficient synthesis for the following transformation:

9.45 Propose an efficient synthesis for the following transformation:

9.46 Propose an efficient synthesis for the following transformation:

9.47 Propose an efficient synthesis for the following transformation:

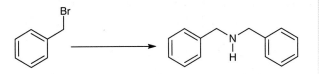

9.48 The process below can be used to convert a nitro group into an amino group:

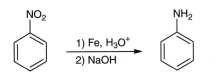

Propose an efficient synthesis for the following transformation, using the reaction above.

9.49 Ammonia (NH_3) can function as a Michael donor and attack a Michael acceptor to give a Michael reaction. For example, ammonia will attack the following α,β-unsaturated ketone at the β position to give a product with the molecular formula C_4H_9NO:

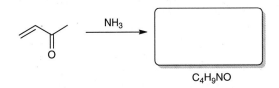

(a) In the box provided, draw the product of this Michael reaction.

(b) How would you use IR spectroscopy to differentiate between the starting material and the product?

(c) How would you use ^1H NMR spectroscopy to differentiate between the starting material and the product?

(d) How would you use ^{13}C NMR spectroscopy to differentiate between the starting material and the product?

(e) On a separate piece of paper, draw a mechanism for this reaction.

9.50 Below are four constitutional isomers (labeled **A–D**) with the molecular formula C_3H_9N.

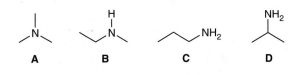

(a) How would you use IR spectroscopy to differentiate between compounds **A**, **B**, and **C**?

(b) How would you use ^1H NMR spectroscopy to differentiate between compounds **C** and **D**?

(c) How would you use ^{13}C NMR spectroscopy to differentiate between compounds **C** and **D**?

(d) Which of the compounds above (**A–D**) will react with $NaNO_2$ and HCl to give a diazonium ion?

(e) When treated with excess methyl iodide, which of the compounds above (**A–D**) will react with only one equivalent of methyl iodide?

9.51 Fill in the missing structures in the following reaction scheme:

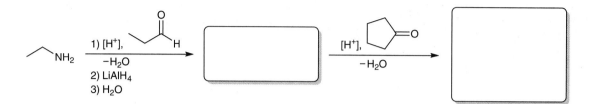

9.52 Fill in the missing structures in the following reaction scheme:

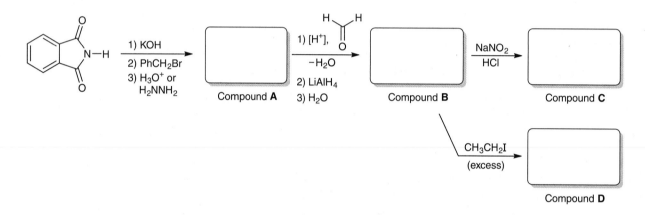

CHALLENGE PROBLEMS

9.53 On a separate piece of paper, draw a plausible mechanism for the following transformation, in which the starting compound is first converted to carbonic acid ($HOCO_2H$), which then decomposes to carbon dioxide and water.

9.54 On a separate piece of paper, draw a plausible mechanism for the following transformation:

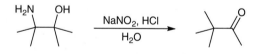

CHAPTER 10

DIELS–ALDER REACTIONS

10.1 INTRODUCTION AND MECHANISM

During your organic chemistry course, you will encounter hundreds of reactions. Most (~95%) of these reactions can be characterized as *ionic reactions*, because they involve the participation of ions as reactants, intermediates, or products. However, you will encounter some reactions that do not involve ions. These reactions can generally be classified into one of two categories: *radical reactions* or *pericyclic reactions*. In this chapter, we will discuss one type of pericyclic reaction, called a *Diels–Alder* reaction. Other pericyclic reactions will be briefly introduced in the last section of this chapter.

The following reaction, between 1,3-butadiene and ethylene, is a Diels–Alder reaction:

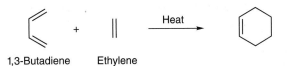

Notice that there are two reactants, but only one product, because the two reactants join together to form a six-membered ring. Also notice that the reactants have a total of three π bonds, while the product has only one π bond:

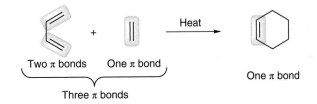

The disappearance of two π bonds is accompanied by the appearance of two new σ bonds, which form the six-membered ring:

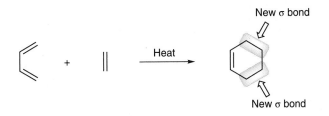

The yield of this reaction is actually quite poor (~20%). However, the yield is dramatically increased (making the reaction useful as a synthetic technique) if we use a substituted ethylene, in which the substituent is an electron-withdrawing group (EWG):

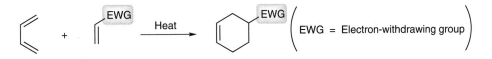

The following highlighted group is an example of an electron-withdrawing group:

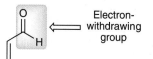

To rationalize why this group is electron withdrawing, we draw the resonance structures, and we focus specifically on the third resonance structure shown below:

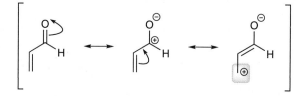

Notice the position of the positive charge (highlighted) in the third resonance structure. The location of this charge indicates that the substituent withdraws electron density from the electron-rich π bond. When this compound reacts with 1,3-butadiene in a Diels–Alder reaction, the yield is observed to be very high (~100%):

Other examples of electron-withdrawing groups are shown below (highlighted). Each of these compounds will react with 1,3-butadiene in a Diels–Alder reaction to give a high yield of product:

To rationalize why each of these groups is electron withdrawing, resonance structures can be drawn (just as we did for the aldehyde group a moment ago).

Diels–Alder reactions proceed via a mechanism that does NOT involve ions. Rather, these reactions occur via a single, concerted step, in which the π electrons move in a ring, as shown by the following curved arrows:

This process does not involve intermediates, because everything happens in one, concerted step. The electrons can be shown moving in a clockwise fashion, as shown above, or in a counterclockwise fashion, as shown here:

Depending on what textbook you are using, you may also see the curved arrows drawn like this (without dotted lines):

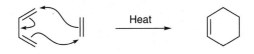

When a monosubstituted ethylene is used in a Diels–Alder reaction, the product will contain a chiral center (highlighted below):

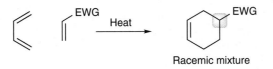

Racemic mixture

Since the reactants are achiral, the product is a racemic mixture. In other words, the product is a pair of enantiomers:

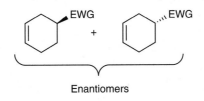

Enantiomers

PROBLEMS   Predict the major product(s) for each of the following reactions:

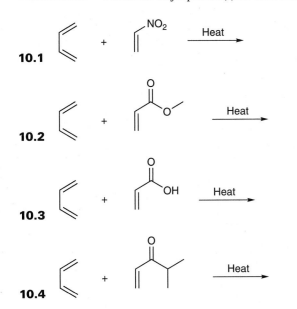

10.2 THE DIENOPHILE

In a Diels–Alder reaction, we refer to the reactants as the *diene* and the *dienophile*:

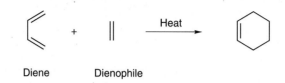

Diene   Dienophile

The former is called a diene because it has two π bonds, and the latter is called a dienophile because it reacts with the diene. In this section, we will explore a variety of dienophiles in Diels–Alder reactions. The next section will focus on dienes.

When a 1,2-disubstituted ethylene is used as the dienophile in a Diels–Alder reaction, the configuration of the dienophile is preserved in the product. For example, a *cis*-disubstituted dienophile will react with 1,3-butadiene to give a *cis*-disubstituted product:

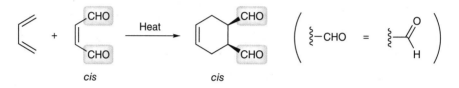

Similarly, a *trans*-disubstituted dienophile gives a *trans*-disubstituted product:

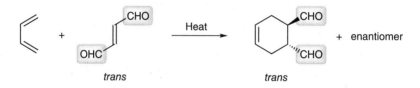

In each of the previous examples, the configuration of the dienophile was retained in the product.

Thus far, we have only seen alkenes (substituted ethylenes) functioning as dienophiles. There are, however, other functional groups that can function as dienophiles. The following example shows an alkyne functioning as a dienophile:

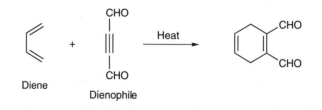

In this case, the dienophile has an extra π bond (compared with an alkene). This extra π bond "comes along for the ride," so to speak, which explains the extra π bond in the product (in between the two electron-withdrawing groups). Notice also that the product has no chiral centers:

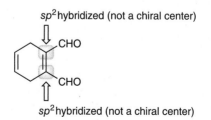

Therefore, we do not draw any wedges or dashes in this case.

PROBLEMS   Predict the major product for each of the following reactions:

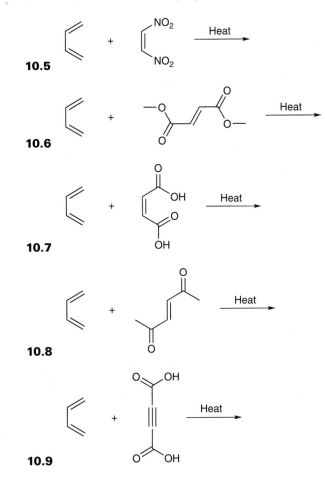

10.3 THE DIENE

In the previous section, we focused on dienophiles for Diels–Alder reactions. Now let's focus our attention on the other reactant in Diels–Alder reactions: the diene. As its name indicates, a diene has two π bonds. There are several conditions that must be met in order for a diene to react with a dienophile in a Diels–Alder reaction. First and foremost, the two π bonds of the diene must be *conjugated*:

A conjugated diene

In a conjugated diene, the two π bonds are separated by exactly one sigma bond, highlighted here:

A Diels–Alder reaction will NOT occur if the two π bonds of the diene are too close to each other (as in cumulated dienes) or too far apart from each other (as in isolated dienes):

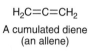
H$_2$C=C=CH$_2$
A cumulated diene
(an allene)

An isolated diene

Now let's explore another important condition that must be met in order for a diene to react in a Diels–Alder reaction. Not only does it have to be conjugated, but it must also be capable of adopting an *s-cis* conformation. To illustrate what this means, consider the following two conformations of 1,3-butadiene:

s-trans   *s-cis*

In each of these conformations, all four *p* orbitals are aligned so that they can overlap to form a single, extended π system. An equilibrium is established between these two conformations, although the *s-trans* conformation is favored because it is more stable. At any moment in time, only 2% of the molecules are in an *s-cis* conformation. The Diels–Alder reaction only occurs when the diene is in an *s-cis* conformation. When the compound adopts an *s-trans* conformation, the ends of the diene are too far apart to react with the dienophile.

Some dienes cannot adopt an *s-cis* conformation, and as a result, they cannot react with dienophiles in a Diels–Alder reaction. As an example, consider the following compound:

The geometry of the six-membered ring effectively locks this diene (permanently) in an *s-trans* conformation, so this compound will not react with a dienophile in a Diels–Alder reaction.

In contrast, a Diels–Alder reaction will be extremely rapid if the diene is permanently locked in an *s-cis* conformation, as in the following example, called cyclopentadiene:

Cyclopentadiene

When cyclopentadiene is used as the diene in a Diels–Alder reaction, the reaction occurs very rapidly, because all of the molecules of the diene spend 100% of the time (rather than just 2% of the time) in the conformation necessary for a Diels–Alder reaction.

Now let's try to draw the product that we expect when cyclopentadiene is used as the diene in a Diels–Alder reaction. We expect the extra carbon atom (highlighted below) to "come along for the ride," so to speak. We expect the following:

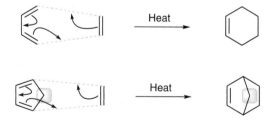

Heat

Heat

When cyclopentadiene is used as the diene, the product is bicyclic, and is more appropriately drawn in the following way:

This type of 3D drawing is very useful, and you should practice drawing it, using the following steps:

STEP ONE	STEP TWO	STEP THREE	STEP FOUR	STEP FIVE	STEP SIX
Draw two lines in the shape of a shallow roof.	Starting at each end of the roof, draw a line going up and to the right. These two new lines must be parallel to each other.	Draw another roof that connects the tops of the two lines that you just drew. But only draw the right side of this new roof. We will connect the other side last.	Draw two new lines that will form the bridge of our structure, as shown.	Complete the bicyclic structure with a broken line, to indicate it is behind the vertical line.	Don't forget the pi bond.

Practice these steps on a blank piece of paper, and repeat them until you feel comfortable drawing the skeleton without instructions. This skeleton will be formed whenever cyclopentadiene is used as the diene in a Diels–Alder reaction. If substituents are present, then we must place them in the correct locations, using the following numbering system:

For example, consider the following reaction:

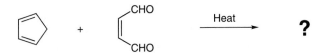

This is a Diels–Alder reaction involving cyclopentadiene, so we expect a bicyclic structure. To draw this bicyclic product, we must consider where the two aldehyde groups will be connected in the product:

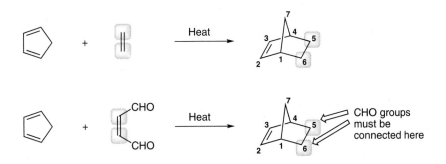

When we draw the product, we must place the aldehyde groups at positions C5 and C6, as shown above. When drawing these substituents in the product, remember that the configuration of the dienophile must be preserved in the product. In this case, the dienophile has a *cis* configuration, so the two aldehyde groups must be *cis* to each other in the product, as shown in each of the following two structures:

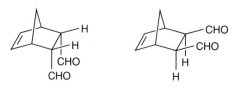

These are the two structures that we expect, however, the reaction does not actually produce both of them. In fact, one of these products is highly favored—the structure in which both aldehyde groups occupy *endo* positions:

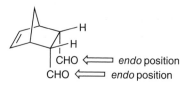

The *endo* position is defined as the position that is *syn* to the larger bridge, as illustrated here:

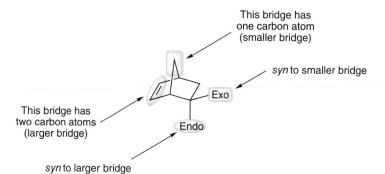

We observe a clear preference for the substituents (connected to the dienophile) to be located in *endo* positions in the bicyclic product, rather than in *exo* positions. What is the source of this preference? During formation of the *endo* product, there is a stabilizing interaction that occurs between the electron-withdrawing substituents (on the dienophile) and the developing π bond (on the diene). This stabilizing effect is absent during formation of the *exo* product. As such, formation of the *endo* product has a lower-energy transition state (and therefore a lower energy of activation) and so it occurs more rapidly than formation of the *exo* product.

If the dienophile is *trans*-disubstituted, then both substituents cannot occupy *endo* positions in the bicyclic product, because the *trans* configuration of the dienophile must be preserved. Two products are obtained:

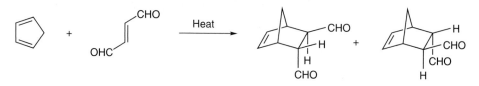

In each product, one aldehyde group occupies an *endo* position and the other aldehyde group occupies an *exo* position. These two products represent a pair of enantiomers. In order to see that they are enantiomers, imagine placing one in front of the other:

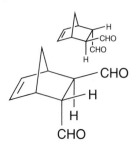

This more clearly shows that they are mirror images (the plane of the mirror is in between the two compounds).

If the dienophile is a monosubstituted ethylene, as in the following case, then the following products are obtained:

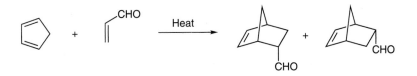

Notice that the aldehyde group occupies an *endo* position in each product. And once again, we can see that these products represent a pair of enantiomers (mirror images that are nonsuperimposable) if we place one compound in front of the other:

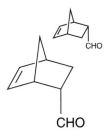

WORKED PROBLEM 10.10   Predict the products that are expected when cyclopentadiene is treated with nitroethylene:

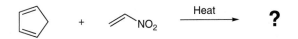

Answer   The reactants are a diene and a dienophile, so we expect a Diels–Alder reaction. The diene is cyclopentadiene, so we expect a bicyclic skeleton, which we can draw using the following steps

STEP ONE	STEP TWO	STEP THREE	STEP FOUR	STEP FIVE	STEP SIX
Draw two lines in the shape of a shallow roof.	Starting at each end of the roof, draw a line going up and to the right. These two new lines must be parallel to each other.	Draw another roof that connects the tops of the two lines that you just drew. But only draw the right side of this new roof. We will connect the other side last.	Draw two new lines that will form the bridge of our structure, as shown.	Complete the bicyclic structure with a broken line, to indicate it is behind the vertical line.	Don't forget the pi bond.

The dienophile bears one substituent (a nitro group), which will end up in one of the following two locations:

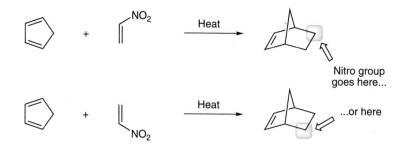

Nitro group
goes here...

...or here

At each of these locations, we expect the nitro group to occupy an *endo* position, rather than an *exo* position, giving the following two products:

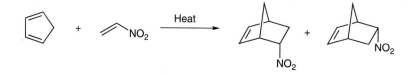

These products represent a pair of enantiomers.

PROBLEMS   Predict the major product(s) for each of the following reactions:

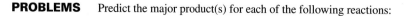

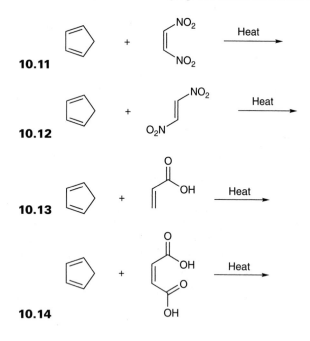

10.11

10.12

10.13

10.14

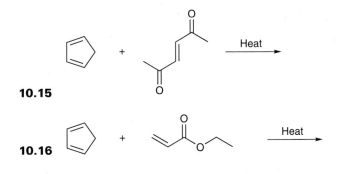

10.15

10.16

10.4 OTHER PERICYCLIC REACTIONS

Diels–Alder reactions belong to one category of pericyclic reactions, called cycloaddition reactions. However, there are two other major categories of pericyclic reactions, called electrocyclic reactions and sigmatropic rearrangements. The three major categories of pericyclic reactions are shown here:

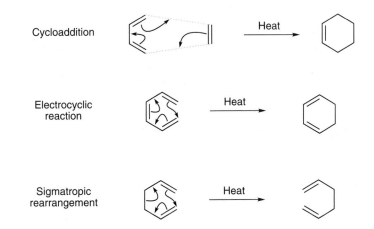

The first category (cycloaddition) was introduced in this chapter. If we compare the other two categories to cycloaddition, we find some similarities. In each case, the curved arrows are moving around in a ring, and the entire process occurs in one concerted step. Indeed, these are some of the features that place these reactions under the larger umbrella of pericyclic reactions. But if we inspect the reactions above more carefully, we will begin to see some important differences. For example, cycloaddition reactions require two reactants (an *inter*molecular process), while electrocyclic and sigmatropic reactions have only one reactant (*intra*molecular processes). Also, if we count the number of bonds being broken and formed in each category, we will see a key difference between the categories. In a cycloaddition process, two π bonds are replaced with two new σ bonds. In an electrocyclic reaction, one π bond is replaced with one new σ bond. In sigmatropic rearrangements, the number of π bonds and σ bonds does not change (only their locations change).

You should consult your textbook and/or lecture notes to see if you are responsible for any examples of electrocyclic reactions or sigmatropic rearrangements (such as the Cope rearrangement or the Claisen rearrangement).

END-OF-CHAPTER PROBLEMS

PRACTICE PROBLEMS *(Problems that involve only one skill)*

10.17 Identify starting materials that can be used to make the product below via a Diels-Alder reaction:

(racemic)

10.18 Identify starting materials that can be used to make the product below via a Diels-Alder reaction:

10.19 Identify starting materials that can be used to make the product below via a Diels-Alder reaction:

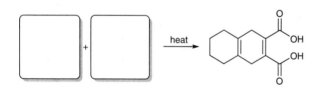

10.20 The following Diels–Alder reaction gives a racemic mixture of enantiomeric products. Draw one of the enantiomers:

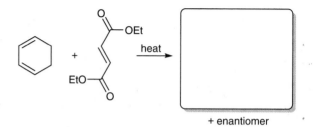

+ enantiomer

10.21 Predict the major product of the following Diels–Alder reaction:

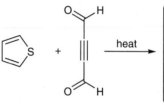

10.22 Identify starting materials that can be used to make the product below via a Diels–Alder reaction:

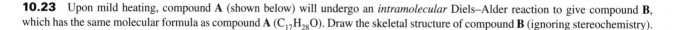

10.23 Upon mild heating, compound **A** (shown below) will undergo an *intramolecular* Diels–Alder reaction to give compound **B**, which has the same molecular formula as compound **A** ($C_{17}H_{28}O$). Draw the skeletal structure of compound **B** (ignoring stereochemistry).

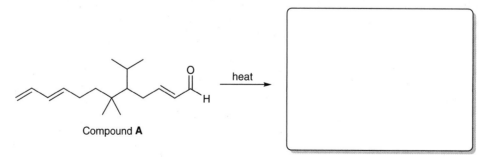

Compound **B**

10.24 Identify starting materials that can be used to make the product below via a Diels–Alter reaction:

10.25 Predict the major product of the following Diels–Alder reaction:

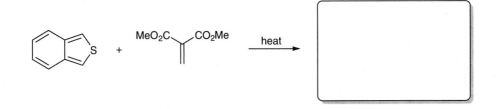

10.26 The triene below reacts with two equivalents of maleic anhydride to give a product with 14 carbon atoms. Draw the skeletal structure of the product (ignoring stereochemistry).

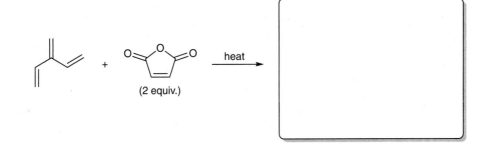

INTEGRATED PROBLEMS *(Problems that involve more than one skill)*

10.27 Propose an efficient synthesis for the following transformation:

10.28 Using acetaldehyde (CH₃CHO) as your only source of carbon atoms, show how you would make a racemic mixture of the following cyclic product:

(only source
of carbon)

+ enantiomer

10.29 Fill in the missing structures and reagents in the following reaction scheme:

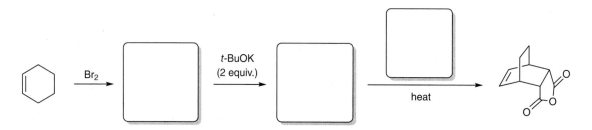

10.30 Fill in the missing structures and reagents in the following reaction scheme:

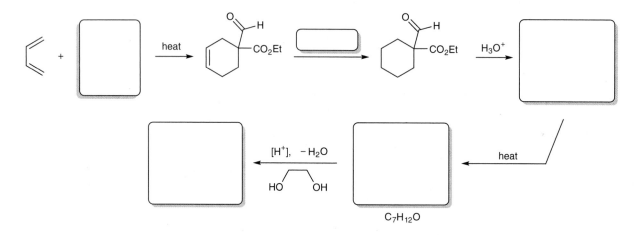

$C_7H_{12}O$

CHALLENGE PROBLEMS

10.31 In the following Diels–Alder reaction, the diene and dienophile react with each other *regioselectively* to give the product shown as the major product:

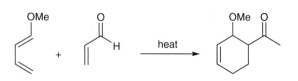

The reaction is said to be *regioselective* because there are two possible regiochemical outcomes for this reaction, one of which is favored (or "selected"). That is, this Diels–Alder reaction could theoretically produce two possible products, depending on the alignment of the diene and the dienophile during the reaction. One possible product is shown above, and that product is indeed favored. However, the other possible product, shown below, is obtained only in trace amounts (if at all):

(Trace amounts)

(a) On a separate sheet of paper, draw all significant resonance structures for the diene and for the dienophile, and then use those resonance structures to suggest a reason for the observed regioselectivity. (**Hint**: Try to align nucleophilic and electrophilic regions.)

(b) Use the logic developed in part (a) to predict the major product for the following Diels-Alder reaction, which is also regioselective:

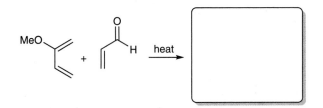

(c) Draw the major product of the following regioselective Diels–Alder reaction. The stereochemistry of this reaction is beyond the scope of our coverage, so it can be ignored.

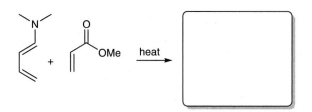

(d) Predict the major product for the following regioselective Diels–Alder reaction:

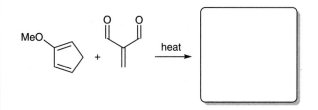

DETAILED SOLUTIONS

CHAPTER 1

1.2 This compound is a disubstituted benzene, and it can be named 1-methoxy-3-nitrobenzene. The methoxy group is assigned the lower number because it comes first alphabetically.

We can also name the compound as a monosubstituted derivative of anisole. Since we are using a common name (anisole) as the parent, we must assign numbers starting with the carbon atom connected to the methoxy group. This means that the methoxy group is connected to C1, by definition. In this case, we assign numbers in a clockwise fashion, so that the nitro group is at C3 rather than C5. Therefore, this compound can be named 3-nitroanisole or *meta*-nitroanisole.

1.3 This compound is a trisubstituted benzene, and it can be named 2-methyl-1,3-dinitrobenzene. When naming in this way, the numbers are assigned to give the lowest possible numbers to all three substituents (C1, C2, and C3). In the name, the methyl group appears first because "m" precedes "n" alphabetically.

We can also name the compound as a disubstituted derivative of toluene. Since we are using a common name (toluene) as the parent, we must assign numbers starting with the carbon atom connected to the methyl group. This means that the methyl group is connected to C1, by definition. In this case, the numbers can be assigned either clockwise or counterclockwise; either way, there will be a nitro group at C2 and another nitro group at C6. Therefore, this compound can be named 2,6-dinitrotoluene.

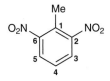

1.4 This compound is a disubstituted benzene, in which the two substituents are *para* to each other. This compound can be called 1,4-dibromobenzene or *para*-dibromobenzene.

1.5 This compound can be named as a trisubstituted derivative of benzoic acid. Since we are using a common name (benzoic acid) as the parent, we must assign numbers starting with the carbon atom connected to the carboxylic acid group. This means that the carboxylic acid group is connected to C1, by definition. In this case, the numbers can be assigned either clockwise or counterclockwise; either way, there will be three chloro substituents, at C2, C4, and C6. Therefore, this compound can be named 2,4,6-trichlorobenzoic acid.

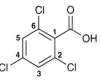

1.6 This compound is a disubstituted benzene, and it can be named 1-chloro-2-hydroxybenzene. The chlorine substituent is assigned the lower number because it comes first alphabetically.

Alternatively, we can name the compound as a monosubstituted derivative of phenol. Since we are using a common name (phenol) as the parent, we must assign numbers starting with the carbon atom connected to the hydroxy group. This means that the hydroxy group is connected to C1, by definition. In this case, we assign numbers in a counterclockwise fashion, so that the chloro substituent is at C2 rather than C6. Therefore, this compound can be named 2-chlorophenol or *ortho*-chlorophenol.

1.8 The first criterion for aromaticity is not satisfied, because the ring contains an sp^3 hybridized carbon atom (highlighted).

Since the ring lacks a continuous system of overlapping *p* orbitals, this anion is nonaromatic.

1.9 Recall that a carbocation represents an empty *p* orbital. Accordingly, the first criterion for aromaticity is satisfied (a ring with a continuous system of overlapping *p* orbitals), and the second criterion is also satisfied (there are 2 π electrons). Therefore, this cation is aromatic.

1.10 Recall that a carbocation represents an empty *p* orbital. Accordingly, the first criterion for aromaticity is satisfied (a ring with a continuous system of overlapping *p* orbitals), and the second criterion is also satisfied (there are 6 π electrons). Therefore, this cation is aromatic.

1.11 The first criterion for aromaticity is satisfied (a ring with a continuous system of overlapping *p* orbitals), but the second criterion is not satisfied (there are 8 π electrons in this case). As such, this structure is antiaromatic and is therefore very unstable.

Perhaps some of the instability could be reduced if the lone pair occupies an *sp³* hybridized orbital rather than a *p* orbital (thereby disrupting the continuous system of overlapping *p* orbitals), although this would render the lone pair localized (not stabilized by resonance) because a lone pair only participates in resonance when it occupies a *p* orbital. So, there is no way for this anion to be stable. Indeed, this anion is very unstable and is classified as antiaromatic.

1.12 The first criterion for aromaticity is satisfied (a ring with a continuous system of overlapping *p* orbitals), but the second criterion is not satisfied (there are 4 π electrons in this case). As such, this structure is antiaromatic.

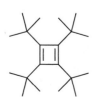

1.13 The first criterion for aromaticity is not satisfied, because the ring contains an *sp³* hybridized carbon atom (highlighted below). Since the ring lacks a continuous system of overlapping *p* orbitals, this structure is nonaromatic.

1.14 The lone pair highlighted here occupies a *p* orbital and is part of the aromatic system. So this lone pair is not available to function as a base:

In contrast, the other lone pair (highlighted below) occupies an *sp²* hybridized orbital and is not part of the aromatic system, so it is available to function as a base:

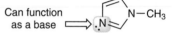

1.15 In the structure below, the sulfur atom has two lone pairs. One of these lone pairs occupies a *p* orbital, giving a continuous system of overlapping *p* orbitals with 6 π electrons. Therefore, this structure (called thiophene) is aromatic:

In the other structure (with two sulfur atoms), if each sulfur atom were to adopt *sp²* hybridization (and each were to place one lone pair in a *p* orbital), then there would be eight π electrons, which would be antiaromatic.

1.16 This compound has two rings, and we evaluate each of them separately. The ring on the left has an *sp³* hybridized carbon atom (highlighted) and therefore does not satisfy the first requirement for aromaticity (the ring does not have a continuous system of overlapping *p* orbitals). The ring on the right also has *sp³* hybridized carbon atoms (highlighted) and so does not satisfy the first requirement for aromaticity (this ring also does not have a continuous system of overlapping *p* orbitals). Therefore, neither of the rings is aromatic. This compound is nonaromatic.

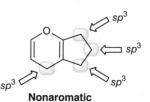

Nonaromatic

1.17 Recall that a carbocation is sp^2 hybridized, with an empty p orbital. Furthermore, the lone pair on the nitrogen atom can also occupy a p orbital, if doing so will establish aromaticity. Accordingly, the first criterion for aromaticity is satisfied (a ring with a continuous system of overlapping p orbitals). However, this system has four π electrons (two π electrons from the nitrogen atom, and two π electrons from the C=C bond).

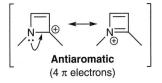

Antiaromatic
(4 π electrons)

A total of four π electrons fits the $4n$ formula, rather than $4n+2$, so the second criterion for aromaticity is not satisfied. Therefore, this cation is expected to be antiaromatic and unstable. The antiaromatic nature of this cation can be seen more readily in the following resonance structure, which resembles cyclobutadiene (where one of the carbon atoms has been replaced with a nitrogen atom):

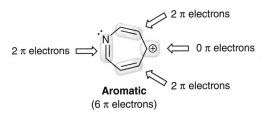

Antiaromatic
(4 π electrons)

1.18 Recall that a carbocation is sp^2 hybridized, with an empty p orbital. Accordingly, the first criterion for aromaticity is satisfied (a ring with a continuous system of overlapping p orbitals). The lone pair on the nitrogen atom occupies an sp^2 hybridized orbital and does not contribute to the count of π electrons, so the second criterion for aromaticity is also satisfied (there are six π electrons, which fits the $4n+2$ formula). Therefore, this cation is aromatic.

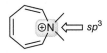

2 π electrons
0 π electrons
2 π electrons
2 π electrons

Aromatic
(6 π electrons)

1.19 The nitrogen atom is sp^3 hybridized. Accordingly, the first criterion for aromaticity is not satisfied (this ring lacks a continuous system of overlapping p orbitals). Therefore, this cation is nonaromatic.

Nonaromatic

1.20 Recall that a carbocation is sp^2 hybridized, with an empty p orbital. Furthermore, one of the lone pairs on the oxygen atom can also occupy a p orbital, if doing so will establish aromaticity. Accordingly, the first criterion for aromaticity is satisfied (a ring with a continuous system of overlapping p orbitals). However, this system has four π electrons (two π electrons from a lone pair on the oxygen atom, and two π electrons from the C=C bond).

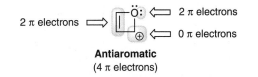

2 π electrons
2 π electrons
0 π electrons

Antiaromatic
(4 π electrons)

A total of four π electrons fits the $4n$ formula, rather than $4n+2$, so the second criterion for aromaticity is not satisfied. Therefore, this cation is expected to be antiaromatic and unstable. The antiaromatic nature of this cation can be seen more readily in the following resonance structure, which resembles cyclobutadiene (where one of the carbon atoms has been replaced with an oxygen atom):

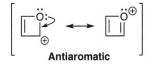

Antiaromatic

1.21 Recall that a carbocation is sp^2 hybridized, with an empty p orbital. Furthermore, one of the lone pairs on the oxygen atom and the lone pair on the nitrogen atom can also occupy p orbitals, if doing so will establish aromaticity. Accordingly, the first criterion for aromaticity is satisfied (a ring with a continuous system of overlapping p orbitals). The second criterion for aromaticity is also satisfied because this system has six π electrons (two π electrons from the oxygen atom, two π electrons from the nitrogen atom, and another two π electrons from the C=C bond), which fits the $4n+2$ formula. Therefore, this cation is aromatic.

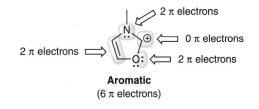

2 π electrons
2 π electrons
0 π electrons
2 π electrons

Aromatic
(6 π electrons)

The aromatic nature of this cation can be seen more readily in the following resonance structure, which resembles pyrrole (where a carbon atom has been replaced by an oxygen atom):

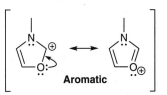

Aromatic

1.22 Recall that a carbocation is sp^2 hybridized, with an empty p orbital; and one of the lone pairs on the oxygen atom can also occupy a p orbital, if doing so will establish aromaticity. Accordingly, the first criterion for aromaticity is satisfied (a ring with a continuous system of overlapping p orbitals). Furthermore, this system has six π electrons (two π electrons from a lone pair on the oxygen atom, and another four π electrons from the two C=C bonds), which fits the $4n+2$ formula, so the second criterion for aromaticity is also satisfied. Therefore, this cation is aromatic.

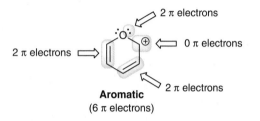

Aromatic
(6 π electrons)

The aromatic nature of this cation can be seen more readily in the following resonance structure, which resembles benzene or pyridine (but with an oxygen atom instead of carbon or nitrogen):

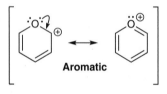

Aromatic

1.23 Recall that a carbocation is sp^2 hybridized, with an empty p orbital. Furthermore, one of the lone pairs on each oxygen atom can also occupy a p orbital, if doing so will establish aromaticity. Accordingly, the first criterion for aromaticity is satisfied (a ring with a continuous system of overlapping p orbitals). Furthermore, this system has six π electrons (two π electrons from each oxygen atom, and another two π electrons from the C=C bond). A total of six π electrons fits the $4n+2$ formula, so the second criterion for aromaticity is also satisfied. Therefore, this cation is aromatic.

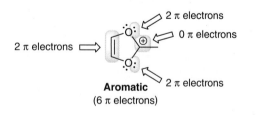

Aromatic
(6 π electrons)

The aromatic nature of this cation can be seen more readily in the following resonance structure, which resembles furan (but with an extra oxygen atom):

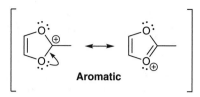

Aromatic

1.24 This ring has a carbon atom that is sp^3 hybridized (highlighted). Accordingly, the first criterion for aromaticity is not satisfied (this ring lacks a continuous system of overlapping p orbitals). Therefore, this anion is nonaromatic.

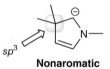

Nonaromatic

1.25 This compound has a ring with a continuous system of overlapping p orbitals, so the first criterion for aromaticity is satisfied. However, this system has sixteen π electrons (two π electrons from each of the eight C=C bonds). A total of sixteen π electrons fits the $4n$ formula, rather than $4n+2$, so the second criterion for aromaticity is not satisfied. Therefore, this structure would be antiaromatic if it were planar. However, much like we saw for cyclooctatetrene, this ring is large enough to avoid aromaticity entirely by puckering out of planarity. So, we expect this compound to adopt a nonplanar conformation, which is nonaromatic, rather than a planar conformation that would be antiaromatic.

16 π electrons

1.26 Let's begin with the ring on the left, which is a substituted pyridine ring. Just like in the case of pyridine, the nitrogen atom in this ring has a lone pair that occupies an sp^2 hybridized orbital (and therefore does not participate in resonance).

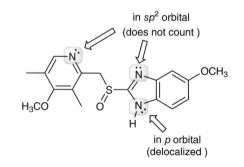

in sp^2 orbital
(does not count)

in p orbital
(delocalized)

This ring is aromatic, just like pyridine. The first criterion for aromaticity is satisfied (a ring with a continuous system of overlapping p orbitals). And the second criterion for aromaticity is also satisfied, because the lone pair on the nitrogen atom does not contribute to the count of π electrons, so there are six π electrons (from the three C=C bonds). Therefore, the ring on the left is aromatic.

The ring on the right is also aromatic, because it is a substituted benzene ring.

Now let's consider the final ring (in between the two rings that we have already analyzed). For this ring, the first criterion for aromaticity is satisfied (a ring with a continuous system of overlapping p orbitals), because the nitrogen atom at the bottom of the ring is sp^2 hybridized, with its lone pair in a p orbital. The nitrogen atom at the top of the ring is also sp^2 hybridized, but its lone pair occupies an sp^2 hybridized orbital (rather than a p orbital), so that lone pair does not contribute to the count of π electrons. This ring has six π electrons (two π electrons from each double bond, and another two π electrons from the nitrogen atom at the bottom of the ring), which fits the formula $4n+2$, so the second criterion for aromaticity is also satisfied. Therefore, this ring is aromatic.

In summary, all three rings are aromatic.

1.27 The strength of an acid is dependent on the stability of its conjugate base. If the conjugate base is more stable, then it is a weaker base with a stronger parent acid. Deprotonation of cyclopentadiene results in a conjugate base that is stabilized by aromaticity:

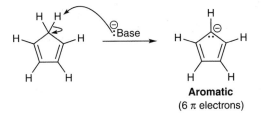

Aromatic
(6 π electrons)

This conjugate base is extremely stable, making it a relatively unreactive, weak base. As a result, cyclopentadiene is relatively acidic (for a hydrocarbon). In contrast, deprotonation of cycloheptatriene results in a conjugate base that is antiaromatic (an unstable, and therefore strong base), so cycloheptatriene is not acidic.

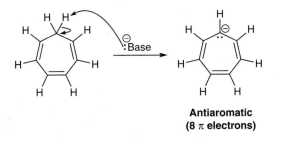

Antiaromatic
(8 π electrons)

1.28 The rate-determining step of an S_N1 reaction is the first step: loss of the leaving group to give a carbocation intermediate. One of the compounds shown will undergo an S_N1 reaction rapidly,

because loss of the leaving group (chloride) generates a carbocation that is aromatic (very stable).

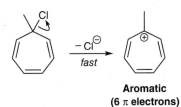

Aromatic
(6 π electrons)

The other compound will not undergo an S_N1 reaction rapidly, because loss of the leaving group generates a carbocation that is antiaromatic (unstable):

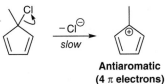

Antiaromatic
(4 π electrons)

1.29 We generally don't break C=C bonds to draw both C+ and C−, but this case represents an exception. In the resonance structure shown below, the ring on the left has six π electrons, and the ring on the right has two π electrons, so both rings are aromatic.

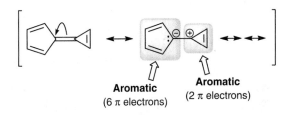

Aromatic   **Aromatic**
(6 π electrons)   **(2 π electrons)**

Since this resonance structure is so stable, it contributes significant character to the overall resonance hybrid, which explains the observed molecular dipole moment.

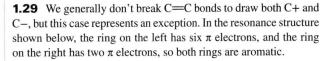

1.30 The rate-determining step of an S_N1 reaction is the first step: loss of the leaving group to give a carbocation intermediate. One of the compounds shown will undergo an S_N1 reaction rapidly, because loss of the leaving group (bromide) generates a carbocation that is aromatic (very stable).

Aromatic
(2 π electrons)

The other compound will not undergo an S_N1 reaction rapidly, because loss of the leaving group generates a carbocation that is antiaromatic (unstable):

Antiaromatic
(4 π electrons)

1.31 The first compound will readily undergo an E2 reaction, because the product is aromatic.

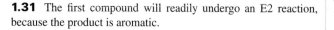

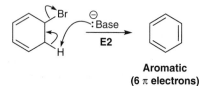

Aromatic
(6 π electrons)

In contrast, an E2 reaction with the second compound is unfavorable, because it would generate a product that is antiaromatic (unstable).

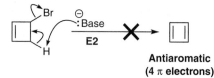

Antiaromatic
(4 π electrons)

1.32 This compound is aromatic, as can be seen in the third resonance structure shown below.

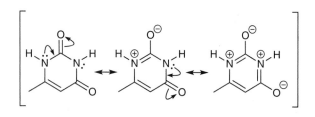

Inspection of the third resonance structure confirms that there is indeed a continuous system of overlapping *p* orbitals, so the first criterion for aromaticity is satisfied. And if we count the number of π electrons, there are a total of six π electrons in the ring, so the second criterion for aromaticity is also satisfied. This resonance structure contributes significant character to the overall resonance hybrid, rendering the compound aromatic.

1.33 Compound **A** is significantly more stable, because it has a resonance structure that is aromatic:

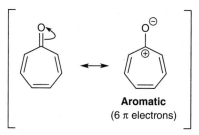

Aromatic
(6 π electrons)

This resonance structure contributes significant character to the overall resonance hybrid, giving compound **A** aromatic character. In contrast, compound **B** has a resonance structure that is antiaromatic, which contributes to the instability of the compound.

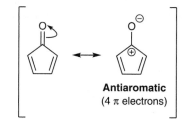

Antiaromatic
(4 π electrons)

1.34 Compound **B** can lose a bromide ion (loss of a leaving group) to give a carbocation that is highly stabilized, because it is aromatic.

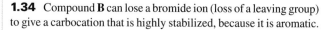

Aromatic
(6 π electrons)

As a result, compound **B** functions as a salt and is soluble in water. In contrast, when compound **A** loses a bromide ion, the resulting carbocation is not aromatic.

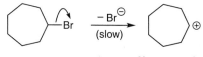

Nonaromatic

1.35 Each of these cations can be deprotonated to give an aromatic conjugate base:

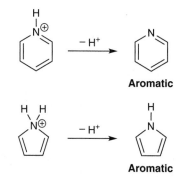

Aromatic

Aromatic

But only the first cation is itself aromatic. The second cation above is not aromatic, because it lacks a continuous system of overlapping *p* orbitals (the nitrogen atom lacks a *p* orbital).

Nonaromatic

As such, deprotonation of this cation will result in the generation of aromatic stabilization that did not previously exist. This process is highly favorable (as compared to deprotonation of the first cation, where aromaticity already existed and is not generated by the process), and as a result, the second cation more readily loses a proton. Therefore, this cation is the stronger acid:

Stronger acid

1.36 The rate-determining step of an S_N1 reaction is the first step: loss of the leaving group to give a carbocation intermediate. One of the compounds shown will undergo an S_N1 reaction rapidly, because loss of the leaving group generates a carbocation that is aromatic (very stable).

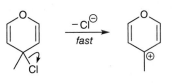

Aromatic
(6 π electrons)

The other compound will not undergo an S_N1 reaction rapidly, because loss of the leaving group generates a carbocation that is antiaromatic (unstable):

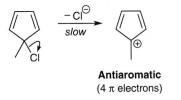

Antiaromatic
(4 π electrons)

1.37 The second compound will readily undergo an E2 reaction, because the resulting product is aromatic.

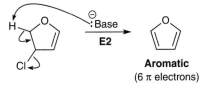

Aromatic
(6 π electrons)

In contrast, an E2 reaction with the first compound will generate a product that is not aromatic (although it should be noted that even this E2 reaction is expected to be fairly rapid in the presence of a strong base, because the product is a conjugated diene, which is stable).

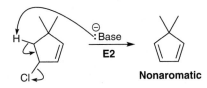

Nonaromatic

1.38 When treated with a strong acid, such as HBr, the compound shown below (which is similar to the structure in our problem statement, but without the phenyl groups) is protonated to give a cation that is aromatic and therefore highly stabilized:

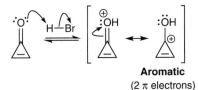

Aromatic
(2 π electrons)

In our case, the starting structure has two phenyl groups as well. These phenyl groups further stabilize the positive charge via resonance.

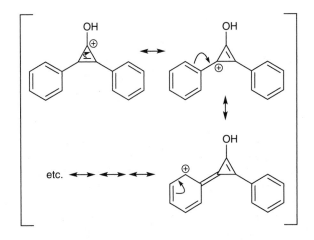

1.39 Boron is trivalent, which means that is has three valence electrons with which to form bonds. In borazine, each of the boron atoms forms three bonds and is sp^2 hybridized, with an empty p orbital (much like a carbocation, but without a positive charge). Each nitrogen atom has a lone pair that can occupy a p orbital. Since the ring is comprised of only boron and nitrogen atoms, the first criterion for aromaticity is satisfied (there is a continuous system of overlapping p orbitals). The second criterion for aromaticity is also satisfied, because the ring has six π electrons (two π electrons from each nitrogen atom, and no π electrons from each boron atom). As such, this compound is aromatic.

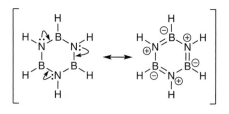

Aromatic
(6 π electrons)

The aromatic nature of borazine can be seen more clearly in the following resonance structure, which is reminiscent of benzene:

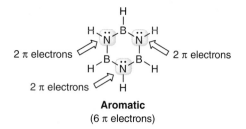

1.40 Imidazole has two nitrogen atoms, each of which has one lone pair. The nitrogen atom on the left has a delocalized lone pair that occupies a p orbital and contributes to aromaticity. As such, this lone pair is unavailable to function as a base.

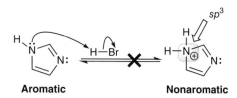

In contrast, the other nitrogen atom has a localized lone pair that occupies an sp^2 hybridized orbital and is not contributing to aromaticity. As such, this lone pair can function as a base to generate a conjugate acid that retains aromatic stabilization:

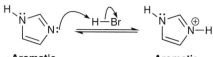

1.41 The lone pair on the nitrogen atom can occupy a p orbital, so the first criterion for aromaticity is satisfied (a continuous system of overlapping p orbitals). However, we count a total of four π electrons, which should be antiaromatic (unstable). This compound can partially reduce this instability if the nitrogen atom is sp^3 hybridized, but the compound is still expected to be very unstable.

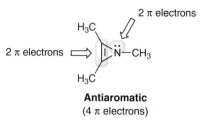

Antiaromatic
(4 π electrons)

1.42 When treated with a strong base, such as NaNH$_2$, indene is deprotonated to give the following anion, in which each ring is aromatic:

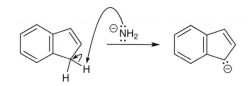

The ring on the left is a substituted benzene ring and is clearly aromatic. The ring on the right is also aromatic because it satisfies both criteria for aromaticity: it has a continuous system of overlapping p orbitals, and there are six π electrons:

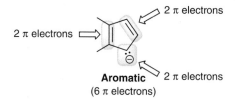

Aromatic
(6 π electrons)

Furthermore, if we treat the entire system as one entity (including both rings), both criteria for aromaticity are satisfied: there is a continuous system of overlapping p orbitals, and there are a total of ten π electrons (which fits the pattern, $4n+2$).

1.43 Each ring has one sp^3 hybridized carbon atom (highlighted below), so neither ring is aromatic (neither ring has a continuous system of overlapping p orbitals).

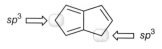

Nonaromatic

However, upon treatment with two equivalents of a strong base, such as NaH, the compound is deprotonated twice to give the following dianion.

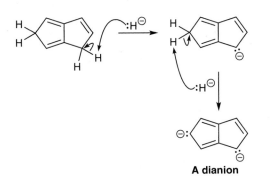

A dianion

This dianion is particularly stable, because each ring is aromatic. This can be seen more clearly in the resonance structure highlighted below:

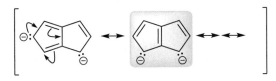

In this resonance structure, we can see that the ring on the left is aromatic, because it has a continuous system of overlapping p orbitals occupied by six π electrons,

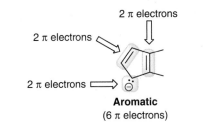

2 π electrons

2 π electrons

2 π electrons

Aromatic
(6 π electrons)

and the ring on the right is also aromatic because it also has a continuous system of overlapping p orbitals occupied by six π electrons.

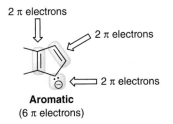

2 π electrons

2 π electrons

2 π electrons

Aromatic
(6 π electrons)

Furthermore, if we treat the entire system as one entity (including both rings), both criteria for aromaticity are satisfied: there is a continuous system of overlapping p orbitals, and there are a total of ten π electrons (which fits the pattern, $4n+2$).

1.44 The strength of an acid is dependent on the stability of its conjugate base. A stronger acid will generate a more stable conjugate base. Deprotonation of compound **A** results in a conjugate base that is resonance-stabilized:

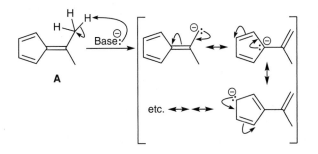

Furthermore, this conjugate base is not only resonance-stabilized, but it is also aromatic:

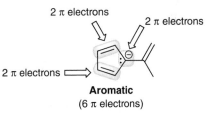

2 π electrons

2 π electrons

2 π electrons

Aromatic
(6 π electrons)

So this conjugate base is very stable, and as a result, compound **A** is a relatively strong acid.

In contrast, deprotonation of compound **B** would yield a conjugate base that is neither resonance-stabilized nor aromatic.

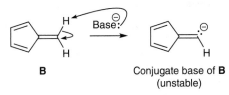

B

Conjugate base of **B**
(unstable)

As such, compound **B** is a much weaker acid than compound **A**.

1.45 Each of these compounds can be deprotonated to give an aromatic conjugate base,

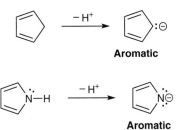

$- H^+$

Aromatic

$- H^+$

Aromatic

but we must also compare the nature of the starting materials as well. Cyclopentadiene is not aromatic, while pyrrole is aromatic:

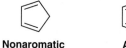

Nonaromatic   **Aromatic**

As such, deprotonation of cyclopentadiene will result in the generation of aromatic stabilization that did not previously exist. This process is highly favorable (as compared to deprotonation of pyrrole, where aromaticity already existed and is not generated by the process), and as a result, cyclopentadiene more readily loses a proton. Therefore, cyclopentadiene is more acidic than pyrrole.

1.46 The problem statement indicates that when pyrrole is treated with a strong acid, protonation occurs at the C2 position, rather than occurring at the nitrogen atom. This results in the following conjugate acid:

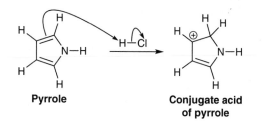

Pyrrole   **Conjugate acid of pyrrole**

This conjugate acid has the following resonance structures that delocalize the charge, and is therefore stabilized by resonance:

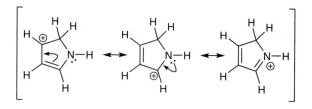

In contrast, if protonation were to occur at the nitrogen atom, the resulting conjugate acid would not be resonance-stabilized.

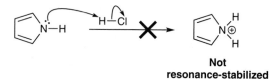

Not resonance-stabilized

While protonation at either position causes a loss of aromaticity, there is a difference in the stability of the two conjugate acids. In summary, protonation occurs at C2 rather than at the nitrogen atom, because protonation at C2 leads to a more stable conjugate acid than protonation at the nitrogen atom.

CHAPTER 2

2.1 Among the highlighted bonds, the C—H bond will produce the signal with the highest wavenumber, because hydrogen has the smallest mass. The other single bonds that are highlighted (C—O and C—Cl) will produce signals with low wavenumbers because they are single bonds. When comparing these two bonds to each other, oxygen has a smaller mass than chlorine, so a C—O bond will produce a signal with a higher wavenumber than a C—Cl bond. Therefore, the C—Cl bond will produce the signal with the lowest wavenumber, and the C—O bond will produce the signal with the second-lowest wavenumber. Among the remaining two bonds that are highlighted, the triple bond is stronger than the double bond and will therefore produce a signal with a higher wavenumber. In summary, the highlighted bonds are ranked below in order of decreasing wavenumber.

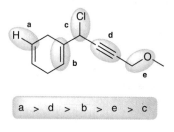

a > d > b > e > c

2.2 This compound has a double bond, but it is tetrasubstituted, so there are no C_{sp^2}—H bonds. Since the compound lacks C_{sp^2}—H and C_{sp}—H bonds, we do not expect there to be a signal above $3000 \, cm^{-1}$.

2.3 This compound has no double or triple bonds. In fact, it does not have any functional groups and is classified as an alkane. Since the compound lacks C_{sp^2}—H and C_{sp}—H bonds, we do not expect there to be a signal above $3000 \, cm^{-1}$.

2.4 This compound is a terminal alkyne. Since it has a C_{sp}—H bond (highlighted below), we expect there to be a signal above $3000 \, cm^{-1}$, likely near $3300 \, cm^{-1}$.

2.5 This compound is a disubstituted alkene. Since it has C_{sp^2}—H bonds (highlighted below), we expect there to be a signal above $3000 \, cm^{-1}$, likely near $3100 \, cm^{-1}$.

2.6 The carbonyl group "a" below is part of an ester, which generally produces a signal at 1740 cm^{-1}. The carbonyl group "b" below is part of a ketone, which generally produces a signal at 1720 cm^{-1}. The remaining carbonyl group is part of a conjugated unsaturated ketone, which is expected to produce a signal at a lower wavenumber than a saturated ketone (1680 cm^{-1}, rather than 1720 cm^{-1}).

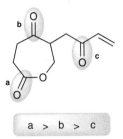

The conjugated unsaturated ketone ("c") produces a signal at a lower wavenumber than the saturated ketone ("b"), because of an additional resonance structure (shown below) that gives additional single-bond character to the C=O bond (see the highlighted bond below). This decreases the strength of the C=O bond, thereby resulting in a signal at a lower wavenumber.

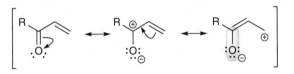

2.7 The strongest signal will be produced by the C=C bond with the greatest dipole moment. Among the three structures shown in the problem statement, the structure below has the C=C bond with the greatest dipole moment, because the two vinylic positions (the two carbon atoms of the C=C bond) are in different electronic environments.

One vinylic position (left) is electron-rich as a result of the electron-donating methyl groups, while the other vinylic position (right) is electron-poor as a result of the electron-withdrawing chlorine atoms. This results in the C=C bond having a relatively strong dipole moment for a C=C bond, giving rise to a relatively strong signal in an IR spectrum.

In contrast, each of the other two compounds has a C=C bond with no dipole moment, because in each case, both vinylic positions occupy identical electronic environments.

2.8 We begin by drawing all resonance structures and considering the location of the partial positive charge in the last resonance structure (highlighted).

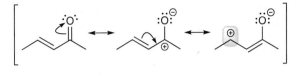

The last resonance structure contributes some of its character to the overall resonance hybrid. As a result, the highlighted carbon atom has partial positive character in the resonance hybrid, as shown here:

Since the two vinylic carbon atoms (the two carbon atoms of the C=C bond) occupy very different electronic environments, this C=C bond will have a relatively strong dipole moment, and as a result, this C=C bond will produce a relatively strong signal in an IR spectrum (compared with the relatively weak signals that are generally associated with C=C bonds).

2.9 This IR spectrum exhibits a broad signal between 3200 cm^{-1} and 3600 cm^{-1}. The presence of this signal in the IR spectrum is consistent with the structure of an alcohol.

2.10 This IR spectrum does not have any signals above 3000 cm^{-1}, so it is not consistent with the structure of an alcohol or a carboxylic acid.

2.11 This IR spectrum exhibits an extremely broad signal between 2200 cm^{-1} and 3600 cm^{-1}, as well as a strong signal just above 1700 cm^{-1}. The presence of these signals in the IR spectrum is consistent with the structure of a carboxylic acid.

2.12 This IR spectrum exhibits a broad signal between 3200 cm^{-1} and 3600 cm^{-1}. The presence of this signal in the IR spectrum is consistent with the structure of an alcohol.

2.13 This IR spectrum exhibits an extremely broad signal between 2200 cm^{-1} and 3600 cm^{-1}, as well as a strong signal just above 1700 cm^{-1}. The presence of these signals in the IR spectrum is consistent with the structure of a carboxylic acid.

2.14 This IR spectrum does not have any signals above 3000 cm^{-1}, so it is not consistent with the structure of an alcohol or a carboxylic acid.

2.15 This IR spectrum exhibits a broad signal between 3200 cm^{-1} and 3600 cm^{-1}. The presence of this signal in the IR spectrum is consistent with the structure of an alcohol.

2.16 This IR spectrum exhibits an extremely broad signal between 2200 cm^{-1} and 3600 cm^{-1}, as well as a strong signal just above 1700 cm^{-1}. The presence of these signals in the IR spectrum is consistent with the structure of a carboxylic acid.

2.17 This IR spectrum exhibits a pair of signals at 3350 cm^{-1} and 3450 cm^{-1}. The presence of these signals in the IR spectrum is consistent with the structure of a primary amine.

2.18 This IR spectrum exhibits a strong signal just above 1700 cm^{-1}. The presence of this signal in the IR spectrum is consistent with the structure of a ketone (the little bump at 3400 cm^{-1} can be ignored, as will be explained in the solution to Problem 2.21).

2.19 This IR spectrum exhibits a broad, weak signal at approximately 3400 cm^{-1}. The presence of this signal in the IR spectrum is consistent with the structure of a secondary amine.

2.20 This IR spectrum exhibits a broad signal between 3200 cm^{-1} and 3600 cm^{-1}. The presence of this signal in the IR spectrum is consistent with the structure of an alcohol.

2.22 Spectrum A exhibits a weak signal at around 1650 cm^{-1}, suggesting the presence of a C=C bond that is unsymmetrical. There is also a signal just above 3000 cm^{-1}, suggesting the presence of a C$_{sp^2}$—H bond. Both of these features are consistent with the following structure, which has an unsymmetrical C=C bond with at least one attached hydrogen atom:

A

Spectrum B exhibits an extremely broad signal between 2200 cm^{-1} and 3600 cm^{-1}, as well as a somewhat strong signal just above 1700 cm^{-1}. The presence of these signals in the IR spectrum is consistent with the structure of a carboxylic acid. There is also a signal at 1600 cm^{-1} suggesting the presence of C=C bonds. These features are consistent with the following structure:

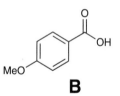

B

Spectrum C exhibits a strong signal just above 1700 cm^{-1}. The presence of this signal in the IR spectrum is consistent with the structure of a ketone (the C=O overtone signal at 3400 cm^{-1} can be ignored, as was explained in the solution to Problem 2.21).

C

Spectrum D exhibits a pair of signals at 3350 cm^{-1} and 3450 cm^{-1}. The presence of these signals in the IR spectrum is consistent with the structure of a primary amine.

D

Spectrum E exhibits a signal just above 2100 cm^{-1}, suggesting the presence of a C≡C bond that is unsymmetrical. There is also a narrow signal at 3300 cm^{-1}, suggesting the presence of a C$_{sp}$—H bond. Both of these features are consistent with the following structure, which is a terminal alkyne:

E

Finally, spectrum F exhibits a broad signal between 3200 cm^{-1} and 3600 cm^{-1}. The presence of this signal in the IR spectrum is consistent with the structure of an alcohol.

F

2.23 This compound is a ketone, so the IR spectrum of this compound is expected to have a signal at approximately 1720 cm^{-1} (C=O stretching).

2.24 This compound is an alcohol, so the IR spectrum of this compound is expected to have a broad signal that spans from 3200 cm^{-1} to 3600 cm^{-1} (O—H stretching).

2.25 This compound is a primary amine, so the IR spectrum of this compound is expected to have two signals at approximately 3350 and 3450 cm^{-1} (symmetric and asymmetric N—H stretching).

2.26 This compound is an alkene with an unsymmetrical C=C double bond, so the IR spectrum of this compound is expected to have a weak signal at approximately 1650 cm^{-1} (C=C stretching). Furthermore, this compound has a C$_{sp^2}$—H bond (a bond between an sp^2 hybridized carbon atom and a hydrogen atom), highlighted below:

Therefore, the IR spectrum of this compound is also expected to have a signal near 3100 cm^{-1} (C$_{sp^2}$—H stretching).

2.27 This compound is a secondary amine, so the IR spectrum of this compound is expected to have a signal at approximately 3400 cm^{-1} (N—H stretching).

2.28 This compound is a nitrile (it contains a C≡N group), so the IR spectrum of this compound is expected to have a signal at approximately 2200 cm^{-1} (C≡N stretching).

2.29 This compound is an internal alkyne (R—C≡C—R) with an unsymmetrical C≡C triple bond (the two R groups are not the same), so the IR spectrum of this compound is expected to have a signal at approximately 2100 cm^{-1} (C≡C stretching).

2.30 This compound has an unsymmetrical C≡C triple bond, so the IR spectrum is expected to have a signal at approximately 2100 cm^{-1} (C≡C stretching). Furthermore, this alkyne is terminal (R—C≡C—**H**), so it has a C$_{sp}$—H (a bond between an *sp* hybridized carbon atom and a hydrogen atom), highlighted below:

$$H_3C-C≡C-H$$

Therefore, the IR spectrum of this compound is also expected to have a signal near 3300 cm^{-1} (C$_{sp}$—H stretching).

2.31 This compound is an alkyne that has a symmetrical C≡C triple bond, so the IR spectrum of this compound will NOT have a signal in the region associated with triple bonds (2100–2300 cm^{-1}). However, this alkyne does have a C$_{sp}$—H bond (a bond between an *sp* hybridized carbon atom and a hydrogen atom):

$$H-C≡C-H$$

Therefore, the IR spectrum of this compound is expected to have a signal near 3300 cm^{-1} (C$_{sp}$—H stretching).

2.32 This compound is an ester, so the IR spectrum of this compound is expected to have a signal at approximately 1740 cm^{-1} (C=O stretching).

2.33 The first compound is an alcohol, so the IR spectrum of the first compound is expected to have a broad signal that spans from 3200 cm^{-1} to 3600 cm^{-1} (O—H stretching). The second compound is a primary amine, so the IR spectrum of the second compound is expected to produce two signals in the same region: one signal at 3350 cm^{-1} and another at 3450 cm^{-1} (symmetric and asymmetric N—H stretching). Therefore, these two compounds can be differentiated based on the number of signals above 3200 cm^{-1} in their IR spectra.

2.34 The first compound is a terminal alkyne (R—C≡C—**H**), so the IR spectrum of this compound is expected to have a signal at approximately 2100 cm^{-1} (for the C≡C triple bond) and another signal at approximately 3300 cm^{-1} (for the C$_{sp}$—H bond). In contrast, the second compound is an alkene. The IR spectrum of this compound will not have a signal at 2100 cm^{-1} because it lacks a triple bond. Instead, the IR spectrum of the second compound will have a signal at approximately 1650 cm^{-1} (for the C=C double bond), as well as another signal at approximately 3100 cm^{-1} (for the C$_{sp^2}$—H bond).

2.35 Both compounds are unsymmetrical alkenes, so both of these compounds will have signals at approximately 1650 cm^{-1} in their IR spectra (C=C stretching). These compounds can be differentiated from each other based on the presence or absence of

a signal at approximately 3100 cm^{-1}. The first compound will produce a signal near 3100 cm^{-1} from the stretching of the C$_{sp^2}$—H bond highlighted below,

while the second compound will not produce a signal near 3100 cm^{-1}, because the second compound lacks a C$_{sp^2}$—H bond.

2.36 Each of these compounds possesses both a C=C double bond and a C=O double bond, but these compounds differ from each other in the proximity of these double bonds. In the first compound, the C=C double bond and the C=O double bond are separated by exactly one sigma bond, so the double bonds are said to be conjugated. In contrast, in the second compound, the C=C double bond and the C=O double bond are separated by two sigma bonds, and are therefore not conjugated (they are considered to be isolated).

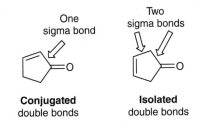

Conjugated
double bonds

Isolated
double bonds

As such, these compounds will produce C=C and C=O stretching signals at different wavenumbers of absorption in IR spectroscopy. The second compound will produce signals at 1650 and 1720 cm^{-1}, which are the characteristic signals that are expected for an isolated C=C double bond and an isolated ketone, respectively. In contrast, the first compound will produce signals at 1600 and 1680 cm^{-1}, which are the characteristic signals that are expected for a conjugated C=C double bond and a conjugated C=O double bond, respectively.

2.37 The starting material is a ketone, and the IR spectrum of a ketone is expected to have a strong signal at approximately 1720 cm^{-1} (C=O stretching). The product is an alcohol, and the IR spectrum of an alcohol is expected to have a broad signal that spans from 3200 cm^{-1} to 3600 cm^{-1} (O—H stretching). Therefore, this reaction can be monitored by observing the disappearance of the strong signal at 1720 cm^{-1} and the appearance of a broad signal from 3200 to 3600 cm^{-1}.

2.38 The starting material is a nitrile (R—C≡N), and the IR spectrum of a nitrile is expected to have a signal at approximately 2200 cm^{-1} (C≡N stretching). The product is a primary amine, and the IR spectrum of a primary amine is expected to have two signals above 3000 cm⁻¹ (one at 3350 cm⁻¹ and the other at 3450 cm⁻¹), due to symmetric and asymmetric stretching of the N—H bonds. Therefore, this reaction can be monitored by observing the disappearance of the signal at 2200 cm⁻¹ and the appearance of signals at both 3350 cm⁻¹ and 3450 cm⁻¹.

2.39 The starting material is a carboxylic acid, and the IR spectrum of a carboxylic acid is expected to have a very broad signal spanning from 2200 to 3600 cm⁻¹ (O—H stretching). The product is an ester, and the IR spectrum of an ester does NOT have a similar signal because it lacks an OH group. Therefore, this reaction can be monitored by observing the disappearance of the broad signal from 2200 to 3600 cm⁻¹.

2.40 In an IR spectrum, the region between 1600 and 1850 cm⁻¹ is associated with the stretching of double bonds (C=C, C=N, and C=O). Each of the compounds below has a double bond (highlighted) that will produce a signal in the region between 1600 and 1850 cm⁻¹.

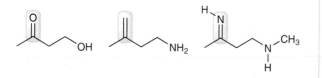

The strength of a signal is determined by the polarity of the bond that is stretching. A C=O double bond is more polar than a C=N double bond or a C=C double bond, and therefore, a C=O double bond will produce a stronger IR stretching signal than a C=N double bond or a C=C double bond. Indeed, the signal associated with a C=O double bond is often the strongest signal in an IR spectrum, if a C=O double bond is present.

2.41 Upon treatment with sodium borohydride, a ketone is reduced to give an alcohol.

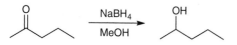

The starting ketone will have a strong signal near 1720 cm⁻¹ in its IR spectrum (C=O stretching). This signal will be absent in the IR spectrum of the alcohol product because it lacks a C=O double bond. Instead, the IR spectrum of the alcohol product will have a broad signal that spans from 3200 cm⁻¹ to 3600 cm⁻¹, characteristic of O—H stretching. Therefore, this reaction can be monitored by observing the disappearance of the strong signal near 1720 cm⁻¹, as well as the appearance of a broad signal from 3200 to 3600 cm⁻¹.

2.42 Upon treatment with a Grignard reagent, followed by aqueous acidic workup, a ketone is converted into a tertiary alcohol:

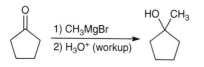

The starting ketone will have a strong signal near 1720 cm⁻¹ in its IR spectrum (C=O stretching). This signal will be absent in the IR spectrum of the alcohol product because it lacks a C=O double bond. Instead, the IR spectrum of the alcohol product will have a broad signal that spans from 3200 cm⁻¹ to 3600 cm⁻¹, characteristic of O—H stretching. Therefore, this reaction can be monitored by observing the disappearance of the strong signal near 1720 cm⁻¹, as well as the appearance of a broad signal from 3200 to 3600 cm⁻¹.

2.43 Upon treatment with PCC (pyridinium chlorochromate), a primary alcohol is oxidized to give an aldehyde.

The IR spectrum of the starting alcohol will have a broad signal that spans from 3200 cm⁻¹ to 3600 cm⁻¹, characteristic of O—H stretching. In contrast, this signal will be absent in the IR spectrum of the aldehyde product because it lacks an OH group. Instead, the IR spectrum of the aldehyde product will have a strong signal near 1720 cm⁻¹ (C=O stretching). Therefore, this reaction can be monitored by observing the disappearance of the broad signal between 3200 and 3600 cm⁻¹, as well as the appearance of a strong signal near 1720 cm⁻¹.

2.44 Upon treatment with sodium dichromate and aqueous sulfuric acid, a primary alcohol is oxidized to give a carboxylic acid.

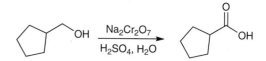

The IR spectrum of the starting alcohol will have a broad signal that spans from 3200 cm⁻¹ to 3600 cm⁻¹, characteristic of O—H stretching for alcohols. In contrast, the IR spectrum of the carboxylic acid product will have a broad signal that spans from 2200 cm⁻¹ to 3600 cm⁻¹, characteristic of O—H stretching for carboxylic acids. Furthermore, the IR spectrum of the carboxylic acid product is expected to have a signal near 1715 cm⁻¹, associated with stretching of the C=O double bond, while the IR spectrum of the starting alcohol lacks this signal (it does not have a C=O double bond). Therefore, this reaction can be monitored by observing a significant broadening of the broad O—H signal between 3200 and 3600 cm⁻¹, as well as the appearance of a strong signal near 1715 cm⁻¹.

2.45 Upon treatment with hydrogen gas in the presence of Lindlar's catalyst (a poisoned catalyst), a terminal alkyne is partially reduced to give a monosubstituted alkene:

The IR spectrum of the starting terminal alkyne is expected to have a signal at approximately $2100 \, \text{cm}^{-1}$ (for the $C \equiv C$ triple bond) and another signal at approximately $3300 \, \text{cm}^{-1}$ (for the C_{sp}—H bond). In contrast, the IR spectrum of the alkene product will NOT have these signals because the alkene product lacks a triple bond. Instead, the IR spectrum of the alkene product will have a signal at approximately $1650 \, \text{cm}^{-1}$ (for the $C \equiv C$ double bond), as well as another signal at approximately $3100 \, \text{cm}^{-1}$ (C_{sp2}—H stretching).

Therefore, this reaction can be monitored by observing the disappearance of the signals near $2100 \, \text{cm}^{-1}$ and $3300 \, \text{cm}^{-1}$, as well as the appearance of new signals in the regions of $1650 \, \text{cm}^{-1}$ and $3100 \, \text{cm}^{-1}$.

2.46 Upon treatment with sodium amide ($NaNH_2$), followed by methyl iodide, a terminal alkyne undergoes an alkylation process, thereby installing a methyl group to give an internal alkyne:

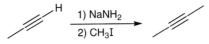

The starting material is an unsymmetrical (terminal) alkyne, so its IR spectrum is expected to have a signal at approximately $2100 \, \text{cm}^{-1}$ ($C \equiv C$ stretching) and another signal at approximately $3300 \, \text{cm}^{-1}$ (C_{sp}—H stretching). In contrast, the IR spectrum of the alkyne product will not have a signal near $2100 \, \text{cm}^{-1}$ because the triple bond in the product is symmetrical. Furthermore, the IR spectrum of the alkyne product will also not have a signal near $3300 \, \text{cm}^{-1}$ (no C_{sp}—H).

Therefore, this reaction can be monitored by observing the disappearance of the signals in both of those regions ($2100 \, \text{cm}^{-1}$ and $3300 \, \text{cm}^{-1}$).

2.47 The starting material is a tertiary alkyl halide, and hydroxide is a strong base, so we expect an E2 reaction. In this case, there are two regiochemical outcomes, and both products are expected, although the more substituted alkene (called the Zaitsev product) will be the major product.

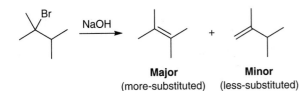

Major (more-substituted) **Minor** (less-substituted)

Both products are alkenes, but their IR spectra will look very different. The major product has a symmetrical $C \equiv C$ double bond, so the IR spectrum of this compound will not have a signal near $1650 \, \text{cm}^{-1}$; and since there is no C_{sp2}—H bond, there also won't be a signal near $3100 \, \text{cm}^{-1}$. In contrast, the IR spectrum of the minor product will have a signal near $1650 \, \text{cm}^{-1}$ (because the $C \equiv C$ double bond is unsymmetrical), as well as a signal near $3100 \, \text{cm}^{-1}$

(C_{sp2}—H stretching). Therefore, these products can be differentiated by looking for the presence or absence of signals near $1650 \, \text{cm}^{-1}$ and $3100 \, \text{cm}^{-1}$ in the IR spectra.

2.48 There are three constitutional isomers with the molecular formula C_3H_8O, shown here:

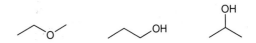

Two of these compounds have an OH group, so the IR spectrum for each of those compounds will have a broad signal from 3200 to $3600 \, \text{cm}^{-1}$. The first compound shown above lacks an OH group, and therefore, the IR spectrum of this compound will not have a signal above $3000 \, \text{cm}^{-1}$. As indicated in the problem statement, the first compound corresponds with compound **A**.

The problem statement also indicates that compound **B** is converted into a carboxylic acid upon treatment with $Na_2Cr_2O_7$ and aqueous sulfuric acid, so the primary alcohol shown above must be compound **B**. Only compound **C** remains, so compound **C** must be the secondary alcohol shown above.

Upon treatment with $Na_2Cr_2O_7$ and aqueous sulfuric acid, compound **C** is converted into a ketone (compound **D**, shown below). As is characteristic of ketones, compound **D** will have an IR spectrum with a signal at $1720 \, \text{cm}^{-1}$ ($C \equiv O$ stretching).

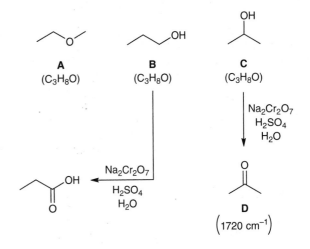

A (C_3H_8O) **B** (C_3H_8O) **C** (C_3H_8O)

$Na_2Cr_2O_7$ H_2SO_4 H_2O

$Na_2Cr_2O_7$ H_2SO_4 H_2O

D ($1720 \, \text{cm}^{-1}$)

2.49 There are only two compounds that have the molecular formula C_3H_6, shown here:

Propene (C_3H_6) Cyclopropane (C_3H_6)

The first compound (propene) has an unsymmetrical $C \equiv C$ double bond so we expect this compound to have an IR spectrum with a signal near 1650 cm⁻¹. Furthermore, propene has C_{sp^2}—H bonds, so we expect signals near 3100 cm⁻¹. According to the problem statement, compound **A** has both of these signals, so compound **A** must be propene. In contrast, cyclopropane lacks a C=C double bond or any C_{sp^2}—H bonds, so the IR spectrum of cyclopropane will NOT have signals near 1650 cm⁻¹ or 3100 cm⁻¹. Therefore, compound **B** is cyclopropane.

Compound **A**   Compound **B**

2.50 Compound **A** is an alkyl halide (RBr), and compound **B** is a nitrile (RC≡N).

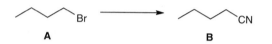

A   B

The conversion of **A** into **B** involves the installation of a C≡N group, which can be monitored with IR spectroscopy by looking for the appearance of a signal at approximately 2200 cm⁻¹ (associated with C≡N stretching).

Compound **B** is a nitrile, and compound **C** is a primary amine.

B   C

The conversion of **B** into **C** involves the loss of a C≡N triple bond and the formation of an NH₂ group. Therefore, this reaction can be monitored with IR spectroscopy by looking for the disappearance of a signal at approximately 2200 cm⁻¹ (associated with C≡N stretching) and for the appearance of two signals at approximately 3350 cm⁻¹ and 3450 cm⁻¹ (associated with symmetric and asymmetric N—H stretching of the NH₂ group).

Compound **C** is a primary amine, and compound **D** is an amide.

C   D

The conversion of **C** into **D** involves the installation of a C=O group. Therefore, this reaction can be monitored with IR spectroscopy by looking for the appearance of a signal at approximately 1650 cm⁻¹ (associated with C=O stretching of an amide). Furthermore, we expect the two signals at approximately 3350 cm⁻¹ and 3450 cm⁻¹ (associated with N—H stretching of the NH₂ group) to collapse into one signal, because the product has only one N—H bond.

2.51 In IR spectroscopy, a strong signal at 1720 cm⁻¹ is strongly suggestive of the presence of a C=O group. The molecular formula is C_4H_8O, so the compound must be either a ketone or an aldehyde

(an ester or carboxylic acid would require two oxygen atoms, and an amide would require a nitrogen atom). The lack of reactivity toward $Na_2Cr_2O_7$ indicates that the compound is a ketone, rather than an aldehyde (because ketones are unreactive toward $Na_2Cr_2O_7$, while aldehydes are oxidized by $Na_2Cr_2O_7$ to give carboxylic acids). There is only one ketone that has the molecular formula C_4H_8O, shown here:

C_4H_8O

2.52 In IR spectroscopy, a signal at 2100 cm⁻¹ is strongly suggestive of the presence of a triple bond. The molecular formula indicates that the compound has only carbon and hydrogen atoms. Therefore, the triple bond must be between two carbon atoms (C≡C), which means that the compound must be an alkyne. There are only two alkynes with the molecular formula C_4H_6, shown here:

C_4H_6 C_4H_6

The first option has a symmetrical C≡C triple bond, so we would not expect that compound to produce a stretching signal in IR spectroscopy. Yet, the problem statement indicates that there is a signal at 2100 cm⁻¹. Therefore, we conclude that the correct structure is the terminal alkyne:

C_4H_6

2.53 When a single functional group produces two nearby signals in an IR spectrum, it is highly suggestive that the functional group can undergo either symmetric or asymmetric stretching, much like we saw for the NH₂ group of a primary amine:

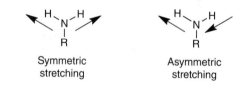

Symmetric   Asymmetric
stretching   stretching

Recall that a nitro group has resonance structures, shown here:

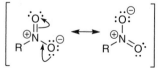

so the two N—O bonds have identical bond strength, as seen in this resonance hybrid:

As such, the N—O bonds of the nitro group can indeed undergo either symmetric or asymmetric stretching, giving rise to two signals:

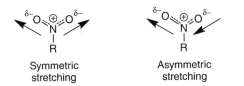

Symmetric   Asymmetric
stretching   stretching

Notice that the bond order of each N—O bond is 1.5 (exactly halfway between that of a single bond and a double bond), which explains the location of these signals, below $1600 \, cm^{-1}$. Double bonds typically produce signals above $1600 \, cm^{-1}$, but these N—O bonds have a bond order of 1.5, not 2, so they are weaker than typical double bonds. As such, they produce signals below $1600 \, cm^{-1}$, in the fingerprint region of the spectrum, where we typically observe signals from the stretching of bonds that are weaker than double bonds (i.e., single bonds, with the exception of single bonds to hydrogen).

2.54 Ketones (or aldehydes) are produced from the ozonolysis of alkenes. If ozonolysis of an alkene produces only one ketone (as the only product), then the alkene must have been symmetrical:

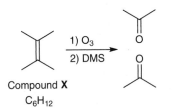

Compound **X**
C_6H_{12}

If ozonolysis produces two equivalents of acetone, compound **X** is 2,3-dimethyl-2-butene. This compound has a symmetrical C=C double bond, so we do not expect an IR signal near $1650 \, cm^{-1}$. Furthermore, this compound does not have any C_{sp^2}—H bonds, so the IR spectrum of 2,3-dimethyl-2-butene will not have any signals above $3000 \, cm^{-1}$.

2.55 Ozonolysis of a cycloalkene will cause cleavage of the C=C double bond to give an acyclic structure with two C=O groups, as shown here:

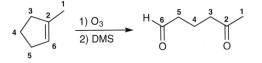

The product has two different C=O groups (a ketone and an aldehyde), so the IR spectrum of the product is expected to have two strong signals in the region of $1600–1850 \, cm^{-1}$. These two signals are likely to be very close to each other, at approximately $1720 \, cm^{-1}$.

2.56 Recall that methanol is more acidic than *tert*-butanol.

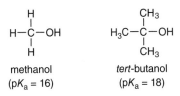

methanol   *tert*-butanol
($pK_a = 16$)   ($pK_a = 18$)

The difference in acidity can be explained by solvent effects. To review, let's compare the ability of solvent molecules to stabilize the conjugate base of each of these compounds. The conjugate base of methanol is methoxide, and the conjugate base of *tert*-butanol is *tert*-butoxide:

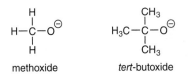

methoxide   *tert*-butoxide

Methoxide is less bulky than *tert*-butoxide, and as a result, methoxide can form a larger number of stabilizing interactions with the solvent molecules. In contrast, *tert*-butoxide is bulky (it is sterically hindered), and as a result, it forms fewer stabilizing interactions with the surrounding solvent molecules than methoxide does:

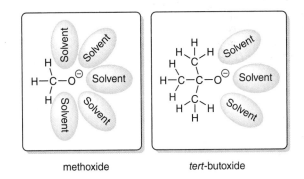

methoxide   *tert*-butoxide

Because of these solvent interactions, methoxide is more stable than *tert*-butoxide. Recall that a more stable (weaker) conjugate base indicates a stronger parent acid, so this explains why methanol is a stronger acid than *tert*-butanol.

These solvent interactions can also explain why the IR spectrum of methanol has an O—H stretching signal that is broader (wider) than the O—H stretching signal in the IR spectrum of *tert*-butanol. The OH group in methanol can form stronger hydrogen bonding interactions (with other molecules of methanol), as compared with the OH group in *tert*-butanol, which forms fewer hydrogen bonding interactions because of the bulkiness of the *tert*-butyl group. As a result, we expect the O—H stretching signal in methanol to be broader (extensive hydrogen bonding) compared with the O—H stretching signal of *tert*-butanol (fewer hydrogen bonding interactions).

2.57 Upon treatment with H_3O^+, the starting alkene will undergo a hydration reaction to give compound **A**, which is an alcohol:

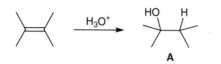

Upon treatment with OsO_4 and *t*-BuOOH, the starting alkene will undergo a dihydroxylation reaction to give compound **B**, which is a diol:

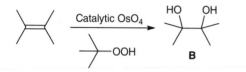

Since each of these products (**A** and **B**) has at least one hydroxyl (OH) group, we expect the IR spectrum of each product to have a broad O—H stretching signal from 3200 to 3600 cm^{-1}. Despite the expected similarity of the two spectra, we can still differentiate these compounds by acquiring an IR spectrum of each compound after it has been diluted in a solvent with which it cannot form hydrogen bonds (such as CCl_4). For compound **A**, the broad O—H signal is expected to collapse into a narrow signal at 3600 cm^{-1} (as a result of the absence of hydrogen bonding). In contrast, the IR spectrum of compound **B** should remain unchanged upon dilution. That is, the broad signal from 3200 to 3600 cm^{-1} will remain broad (it will NOT collapse to a narrow signal), because the diol can still undergo intramolecular hydrogen bonding, as shown here.

intramolecular hydrogen bonding

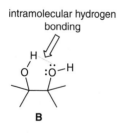

The presence of hydrogen bonding will cause the O—H stretching signal to remain broad, even when compound **B** is diluted.

2.58 We begin by drawing all resonance structures of each compound. Compound **A** has the following three resonance structures, typical of any ester:

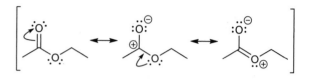

Because of this resonance, the C=O group of compound **A** has significant single-bond character.

Compound **B** is also an ester, so compound **B** also has three similar resonance structures, just like compound **A**:

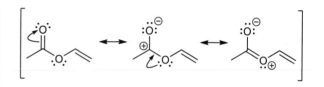

but compound **B** has an additional, fourth resonance structure, highlighted below:

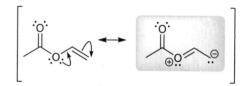

In the additional resonance structure, the carbonyl group (C=O) is drawn as a double bond. Therefore, the carbonyl group of **B** has more double-bond character than the carbonyl group of **A**. As such, the carbonyl group of **B** is a stronger bond than the carbonyl group of **A**, so we expect the carbonyl stretching signal for compound **B** to appear at a higher wavenumber of absorption than the carbonyl stretching signal for compound **A**.

CHAPTER 3

3.2 In this structure, all four methyl groups are interchangeable via symmetry, so the protons of these methyl groups (highlighted) are chemically equivalent:

$$H_3C\quad CH_3$$
$$H-C-C-H$$
$$H_3C\quad CH_3$$

The remaining two protons are interchangeable via symmetry and are therefore chemically equivalent:

$$H_3C\quad CH_3$$
$$H-C-C-H$$
$$H_3C\quad CH_3$$

so this compound will have a proton NMR spectrum with exactly two signals.

3.3 In this structure, the three methyl groups highlighted below are all chemically equivalent (in much the same way that the three protons of a single methyl group are equivalent to each other):

Furthermore, the remaining two methyl groups are interchangeable via symmetry, so they are chemically equivalent to each other:

$$H_3C\quad CH_3$$
$$H_3C-C-C-H$$
$$H_3C\quad CH_3$$

Note that there are a total of five methyl groups, and they are NOT all equivalent to each other. Rather, the three methyl groups on the left represent one type of proton, and the two methyl groups on the right represent a different type of proton.

Finally, there is one more proton, highlighted here:

$$H_3C\quad CH_3$$
$$H_3C-C-C-H$$
$$H_3C\quad CH_3$$

so this compound will have a proton NMR spectrum with exactly three signals.

3.4 In this structure, the three methyl groups on the left side of the structure are chemically equivalent (as we saw in the previous

problem), and the three methyl groups on the right side of the structure are also equivalent to each other (for the same reason).

$$H_3C\quad CH_3$$
$$H_3C-C-C-CH_3$$
$$H_3C\quad CH_3$$

In this case, there is additional symmetry that was absent in the previous problem. Here, the methyl groups on the left side ARE identical to the methyl groups on the right side, because they are interchangeable via symmetry. Therefore, all six methyl groups are chemically equivalent, so this compound will have a proton NMR spectrum with only one signal.

3.5 In this structure, the two methyl groups (highlighted) are interchangeable via symmetry, so the protons of these methyl groups are chemically equivalent:

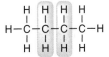

The remaining two methylene (CH$_2$) groups are interchangeable via symmetry, so all four of these protons (highlighted) are chemically equivalent:

$$\begin{array}{cccc} H & H & H & H \\ H-C-C-C-C-H \\ H & H & H & H \end{array}$$

In summary, this compound will have a proton NMR spectrum with exactly two signals.

3.6 The three protons of the methyl group are all equivalent to each other:

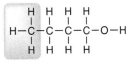

Furthermore, since this compound lacks a chiral center, the two protons of this methylene (CH$_2$) group are chemically equivalent to each other:

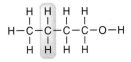

For a similar reason, the following two protons are also equivalent to each other:

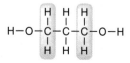

and these two protons are also equivalent to each other:

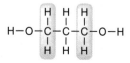

Note that the three CH₂ groups are NOT all equivalent to each other (they represent three different types of protons), because they do NOT occupy identical electronic environments (they are not interchangeable by symmetry). They differ from each other in their proximity to the oxygen atom.

Finally, there is one more signal from the proton connected to the oxygen atom:

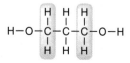

In summary, the proton NMR spectrum of this compound will have exactly five signals.

3.7 The protons connected to the oxygen atoms are interchangeable via symmetry and are therefore chemically equivalent:

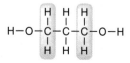

Furthermore, since this compound lacks a chiral center, the two protons of the central methylene (CH₂) group are chemically equivalent to each other:

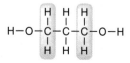

Similarly, the following four protons are also chemically equivalent to each other:

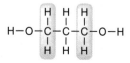

Note that the three CH₂ groups are NOT all equivalent to each other (they represent two different types of protons, not one type of proton), because they do NOT all occupy identical electronic environments. Each of the outer two CH₂ groups is connected directly to one of the oxygen atoms, while the central CH₂ group is NOT connected directly to an oxygen atom.

In summary, the proton NMR spectrum of this compound will have exactly three signals.

3.8 Each of the methyl groups produces its own signal:

These two methyl groups are not chemically equivalent to each other, because they occupy different electronic environments. The methyl group on the left is closer to the oxygen atom, while the methyl group on the right is more distant from the oxygen atom.

There is also a proton connected directly to the oxygen atom, and there is a CH proton (shown below), each of which produces its own signal:

So far, we have counted four signals (CH₃, CH₃, OH, and CH). Finally, we consider the methylene (CH₂) group. In this case, the structure contains a chiral center, and therefore, we expect the two protons of the methylene group to be different (not chemically equivalent):

In summary, the proton NMR spectrum of this compound will have six signals.

In practice, depending on the type of NMR spectrometer that is used, the two signals corresponding to the CH₂ protons could possibly appear as one signal (rather than two), especially if the spectrum is acquired with an NMR spectrometer that uses a relatively weak magnetic field. With an NMR spectrometer that uses a strong magnetic field, the two signals corresponding to the CH₂ protons can be easily distinguished.

3.9 The three protons of the methyl group produce one signal:

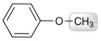

The five aromatic protons give rise to three distinct signals, corresponding to the highlighted protons shown here:

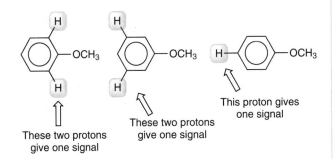

These two protons
give one signal

These two protons
give one signal

This proton gives
one signal

In summary, the proton NMR spectrum of this compound will have four signals.

3.10 In this structure, all four methyl groups are interchangeable via symmetry, so the protons of these methyl groups (highlighted) are chemically equivalent:

$$H_3C\ H\ H\ H\ CH_3$$
$$H-C-C-C-C-C-H$$
$$H_3C\ H\ H\ H\ CH_3$$

Similarly, the CH protons (highlighted below) are chemically equivalent to each other:

$$H_3C\ H\ H\ H\ CH_3$$
$$H-C-C-C-C-C-H$$
$$H_3C\ H\ H\ H\ CH_3$$

Furthermore, since this compound lacks a chiral center, the two protons of the central methylene (CH$_2$) group are chemically equivalent to each other:

$$H_3C\ H\ H\ H\ CH_3$$
$$H-C-C-C-C-C-H$$
$$H_3C\ H\ H\ H\ CH_3$$

Similarly, the following four protons are also chemically equivalent:

$$H_3C\ H\ H\ H\ CH_3$$
$$H-C-C-C-C-C-H$$
$$H_3C\ H\ H\ H\ CH_3$$

Note that the three CH$_2$ groups are NOT all equivalent to each other (they represent two different types of protons, not just one

type of proton), because they do NOT all occupy identical electronic environments.

In summary, the proton NMR spectrum of this compound will have exactly four signals.

3.11 Both of the structures below have the molecular formula C_9H_{18}, and in each case, all 18 protons are chemically equivalent. In the first structure, all six methyl groups are chemically equivalent (giving rise to just one signal); in the second structure, all nine CH$_2$ groups are chemically equivalent (once again, giving rise to just one signal).

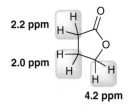

3.13 The proton NMR spectrum of this compound will have three signals, corresponding to the groups of protons highlighted below.

The methyl protons on the right side of the structure (benchmark value = 0.9 ppm) are alpha to the oxygen atom of an ester, which adds +3, so the predicted chemical shift = 0.9 + 3 = 3.9 ppm.

The methyl protons on the left side of the structure (benchmark value = 0.9 ppm) are beta to a carbonyl group, which adds +0.2, so the predicted chemical shift = 0.9 + 0.2 = 1.1 ppm.

And finally, the methine proton (benchmark value = 1.7 ppm) is alpha to a carbonyl group, which adds +1.0, so the predicted chemical shift = 1.7 + 1 = 2.7 ppm.

3.14 The proton NMR spectrum of this compound will have three signals, corresponding to the groups of protons highlighted below.

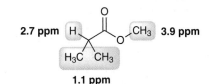

Starting at the top of the ring, and going counterclockwise around the ring, the first methylene group should give a signal near 2.3 ppm because a methylene group has a benchmark value of 1.2 ppm, and this methylene group is alpha to a carbonyl group (+1). Thus, 1.2 + 1 = 2.2 ppm.

The next methylene group should give a signal near 2.0 ppm because a methylene group has a benchmark value of 1.2 ppm, and this methylene group is beta to a carbonyl group (+0.2) and beta to the oxygen of an ester (+0.6). Thus, 1.2 + 0.2 + 0.6 = 2.0 ppm.

And finally, the last methylene group should give a signal near 4.2 ppm because a methylene group has a benchmark value of 1.2 ppm, and this methylene group is alpha to the oxygen of an ester (+3). Thus, 1.2 + 3 = 4.2 ppm.

3.15 The proton NMR spectrum of this compound will have four signals, corresponding to the groups of protons highlighted below.

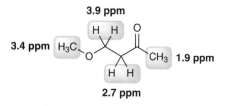

The methyl protons on the left side of the structure (benchmark value = 0.9 ppm) are alpha to the oxygen atom of an ether, which adds +2.5, so the predicted chemical shift = 0.9 + 2.5 = 3.4 ppm.

The methyl protons on the right side of the structure (benchmark value = 0.9 ppm) are alpha to a carbonyl group, which adds +1, so the predicted chemical shift = 0.9 + 1 = 1.9 ppm.

There are two other signals, both corresponding to methylene groups. Let's begin with the methylene group on the left. The benchmark value for a methylene group is 1.2 ppm, but this methylene group is alpha to the oxygen atom of an ether (+2.5) and beta to a carbonyl group (+0.2), so the predicted chemical shift = 1.2 + 2.5 + 0.2 = 3.9 ppm.

For the remaining methylene group, the benchmark value is 1.2 ppm, but this methylene group is alpha to a carbonyl group (+1) and beta to the oxygen atom of an ether (+0.5), so the predicted chemical shift = 1.2 + 1 + 0.5 = 2.7 ppm.

3.16 The proton NMR spectrum of this compound will have two signals, corresponding to the groups of protons highlighted below.

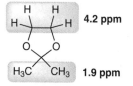

The methyl groups are chemically equivalent to each other, and they collectively produce one signal. The benchmark value for a methyl group is 0.9 ppm, and each of these methyl groups is beta to two ether groups (+0.5 + 0.5), which should give a signal near 0.9 + 0.5 + 0.5 = 1.9 ppm.

The methylene groups are chemically equivalent to each other, and they collectively produce one signal. The benchmark value for a methylene group is 1.2 ppm, and each of these methylene groups

is alpha to one ether group (+2.5) and beta to another ether group (+0.5), which should give a signal near 1.2 + 2.5 + 0.5 = 4.2 ppm.

3.17 The proton NMR spectrum of this compound will have four signals, corresponding to the groups of protons highlighted below.

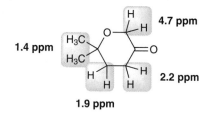

Going counterclockwise around the ring, we first encounter two equivalent methyl groups that collectively produce one signal. The benchmark value for a methyl group is 0.9 ppm, and each of these methyl groups is beta to an ether group (+0.5), which should give a signal near 0.9 + 0.5 = 1.4 ppm.

Moving counterclockwise through the structure, we next encounter a methylene group, which should give a signal near 1.9 ppm because a methylene group has a benchmark value of 1.2 ppm, and this methylene group is beta to a carbonyl group (+0.2) and beta to the oxygen of an ether (+0.5). Thus, 1.2 + 0.2 + 0.5 = 1.9 ppm.

Moving counterclockwise through the structure, we next encounter a methylene group, which should give a signal near 2.2 ppm because a methylene group has a benchmark value of 1.2 ppm, and this methylene group is alpha to a carbonyl group (+1). Thus, 1.2 + 1 = 2.2 ppm.

And finally, the last methylene group should give a signal near 4.7 ppm because a methylene group has a benchmark value of 1.2 ppm, and this methylene group is alpha to the oxygen of an ether (+2.5) and alpha to a carbonyl group (+1). Thus, 1.2 + 2.5 + 1 = 4.7 ppm.

3.18 The proton NMR spectrum of this compound will have five signals, corresponding to the groups of protons highlighted below.

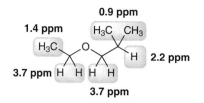

Moving from left to right through structure, we first encounter a methyl group (benchmark value = 0.9 ppm) that is beta to the oxygen atom of an ether, which adds +0.5, so the predicted chemical shift = 0.9 + 0.5 = 1.4 ppm.

Next, we encounter a methylene group (benchmark value = 1.2 ppm) that is alpha to the oxygen atom of an ether, which adds +2.5, so the predicted chemical shift = 1.2 + 2.5 = 3.7 ppm.

Next, we encounter another methylene group (benchmark value = 1.2 ppm) that is also alpha to the oxygen atom of an ether, which adds +2.5, so the predicted chemical shift = 1.2 + 2.5 = 3.7 ppm. Notice that we predict that two of the signals should be very close to each other, if not overlapping.

Next we encounter a methine proton (benchmark value = 1.7 ppm) that is beta to the oxygen atom of an ether, which adds +0.5, so the predicted chemical shift = 1.7 + 0.5 = 2.2 ppm.

Finally, there are two equivalent methyl groups that should appear at 0.9 ppm (all inductive effects are distant and practically negligible).

3.19 The proton NMR spectrum of this compound will have nine signals, corresponding to the groups of protons highlighted below. The values for these signals can be found in the table that immediately precedes the problem statement.

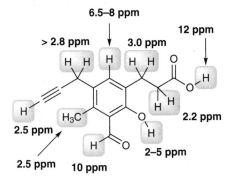

3.21 We identify the smallest integration value (in this case, 9.1) and we then divide all of the integration values by that number, giving the ratio 5:2:2:1. The molecular formula indicates that there are a total of 10 protons, so the numbers 5:2:2:1 represent not only relative values, but they also represent exact values (5 protons, 2 protons, 2 protons, and 1 proton).

3.22 We identify the smallest integration value (in this case, 10.8), and we then divide all of the integration values by that number, giving the ratio 1:6. The molecular formula indicates that there are a total of 14 protons, so the values 1 and 6 are just relative numbers. They actually represent 2 protons and 12 protons, in order for the total number of protons to be 14.

3.23 We identify the smallest integration value (in this case, 18.02), and we then divide all of the integration values by that number, giving the approximate ratio 1:1:1. The molecular formula indicates that there are a total of 6 protons, so the values 1, 1, and 1 are just relative numbers. They actually represent 2 protons, 2 protons, and 2 protons, in order for the total number of protons to be 6.

3.25 The proton NMR spectrum of this compound will have three signals, corresponding to the groups of protons highlighted below. In each case, the multiplicity is determined by the number of neighboring protons, using the $n + 1$ rule.

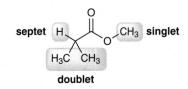

The methyl group on the right side of the structure has no neighbors ($n = 0$), so it will give a singlet ($n + 1 = 1$).

The methine proton has six neighbors ($n = 6$), so it will give a septet ($n + 1 = 7$).

And the two equivalent methyl groups have one neighbor ($n = 1$), so they will give a doublet ($n + 1 = 2$).

3.26 The proton NMR spectrum of this compound will have three signals, corresponding to the groups of protons highlighted below. In each case, the multiplicity is determined by the number of neighboring protons, using the $n + 1$ rule.

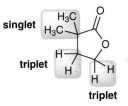

Moving counterclockwise around the structure, we first encounter two equivalent methyl groups, both of which are connected to a carbon atom that has no protons ($n = 0$). Therefore, these methyl groups collectively give rise to a singlet ($n + 1 = 1$).

The methylene groups are nonequivalent (because of their proximity to the oxygen atom), and they are neighboring each other, so each methylene group has two neighbors ($n = 2$). Therefore, each of these methylene groups will give rise to a triplet ($n + 1 = 3$).

3.27 The proton NMR spectrum of this compound will have four signals, corresponding to the groups of protons highlighted below. In each case, the multiplicity is determined by the number of neighboring protons, using the $n + 1$ rule.

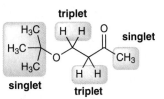

Moving from left to right, we first encounter a *tert*-butyl group, which has three equivalent methyl groups, all connected to a carbon atom that has no protons ($n = 0$). Therefore, these three methyl groups collectively give rise to a singlet ($n + 1 = 1$). This is characteristic of a *tert*-butyl group.

The nonequivalent methylene groups are neighboring each other, so each methylene group has two neighbors ($n = 2$).

Therefore, each of these methylene groups will give rise to a triplet ($n + 1 = 3$).

Finally, the methyl group on the right has no neighbors ($n = 0$), giving rise to a singlet ($n + 1 = 1$).

3.28 The proton NMR spectrum of this compound will have two signals, corresponding to the groups of protons highlighted below. In each case, the multiplicity is determined by the number of neighboring protons, using the $n + 1$ rule.

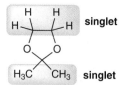

At the bottom of the structure, there are two equivalent methyl groups, each of which is connected to a carbon atom that has no protons ($n = 0$). Therefore, these two methyl groups collectively give rise to a singlet ($n + 1 = 1$).

The methylene groups are neighboring each other, but they don't split each other, because they are chemically equivalent protons. Instead, they collectively give rise to a singlet (chemically equivalent protons do NOT split each other).

3.29 The proton NMR spectrum of this compound will have four signals, corresponding to the groups of protons highlighted below. In each case, the multiplicity is determined by the number of neighboring protons, using the $n + 1$ rule.

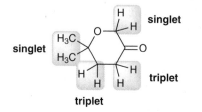

Moving counterclockwise around the structure, we first encounter two equivalent methyl groups, each of which is connected to a carbon atom that has no protons ($n = 0$). Therefore, these two methyl groups collectively give rise to a singlet ($n + 1 = 1$).

Next we encounter two nonequivalent methylene groups that are neighboring each other, so each methylene group has two neighbors ($n = 2$). Therefore, each of these methylene groups will give rise to a triplet ($n + 1 = 3$).

Finally, the remaining methylene group has no neighbors ($n = 0$), giving rise to a singlet ($n + 1 = 1$).

3.30 The proton NMR spectrum of this compound will have four signals, corresponding to the groups of protons highlighted below. In each case, the multiplicity is determined by the number of neighboring protons, using the $n + 1$ rule.

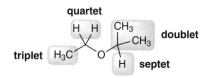

Moving from left to right, we first encounter a methyl group that has two neighbors ($n = 2$), giving rise to a triplet ($n + 1 = 3$).

Next we encounter a methylene group with three neighbors ($n = 3$), giving rise to a quartet ($n + 1 = 4$).

There is also a methine proton with six neighbors ($n = 6$), giving rise to a septet ($n + 1 = 7$).

Finally, there are two equivalent methyl groups, which are both connected to a carbon atom bearing only one proton ($n = 1$), so these methyl groups will collectively give rise to a doublet ($n + 1 = 2$).

3.31 This spectrum exhibits a septet with an integration of 1 and a doublet with an integration of 6. These two signals are characteristic of an isopropyl group.

3.32 We begin by identifying the smallest integration value (in this case, 13.9), and we then divide all of the integration values by that number, giving the ratio 1 : 1.5 : 1 : 1.5. There is no such thing as half of a proton, so we multiply these numbers by 2, to give the ratio 2 : 3 : 2 : 3. The molecular formula indicates that there are a total of 10 protons, so these values represent the actual number of protons giving rise to each signal (2 protons, 3 protons, 2 protons, and 3 protons = 10 protons).

Now that we know the integration values, we can see that this spectrum exhibits a quartet with an integration of 2, and a triplet with an integration of 3. These two signals are characteristic of an ethyl group.

3.33 We begin by identifying the smallest integration value (in this case, 10.8), and we then divide all of the integration values by that number, giving the ratio 1:6. Now that we know the integration values, we can see that this spectrum exhibits a septet with an integration of 1 and a doublet with an integration of 6. These two signals are characteristic of an isopropyl group.

The molecular formula indicates that there are a total of 14 protons, so the values 1 and 6 are just relative numbers. They actually represent 2 protons and 12 protons, in order for the total number of protons to be 14 (there are two equivalent isopropyl groups).

3.34 We begin by identifying the smallest integration value (in this case, 9.1), and we then divide all of the integration values by that number, giving the ratio 5:2:2:1. The molecular formula indicates that there are a total of 10 protons, so the numbers 5:2:2:1 represent not only relative values, but they also represent exact values (5 protons, 2 protons, 2 protons, and 1 proton).

Now that we know the integration values, we can see that this spectrum does not contain the characteristic signals for an isopropyl group, or an ethyl group, or a *tert*-butyl group. This structure does not have any of these three groups.

3.36 For purposes of calculating degrees of unsaturation, we inspect the molecular formula, and we can ignore the presence of oxygen atoms. So a compound with the molecular formula $C_6H_{10}O_4$ has the same HDI as a compound with the molecular formula C_6H_{10}. To be fully saturated, a compound with 6 carbon atoms would need to have 14 hydrogen atoms. This compound has only 10 hydrogen atoms, so it is missing 4 hydrogen atoms. Therefore, this compound has 2 degrees of unsaturation.

3.37 For purposes of calculating degrees of unsaturation, we inspect the molecular formula, and we can subtract one hydrogen atom for each nitrogen atom that is present. So a compound with the molecular formula $C_7H_{11}N$ has the same HDI as a compound with the molecular formula C_7H_{10}. To be fully saturated, a compound with 7 carbon atoms would need to have 16 hydrogen atoms. But the molecular formula C_7H_{10} has 10 hydrogen atoms, so 6 hydrogen atoms are missing. Therefore, this compound has 3 degrees of unsaturation.

3.38 For purposes of calculating degrees of unsaturation, we inspect the molecular formula, and we can ignore the presence of oxygen atoms. So a compound with the molecular formula $C_8H_{14}O_2$ has the same HDI as a compound with the molecular formula C_8H_{14}. To be fully saturated, a compound with 8 carbon atoms would need to have 18 hydrogen atoms. This compound has only 14 hydrogen atoms, so it is missing 4 hydrogen atoms. Therefore, this compound has 2 degrees of unsaturation.

3.39 For purposes of calculating degrees of unsaturation, we inspect the molecular formula, and we can ignore the presence of oxygen atoms. So a compound with the molecular formula $C_5H_{12}O_2$ has the same HDI as a compound with the molecular formula C_5H_{12}. To be fully saturated, a compound with 5 carbon atoms would need to have 12 hydrogen atoms. This compound does have 12 hydrogen atoms, so it is not missing any hydrogen atoms. Therefore, this compound is saturated and has 0 degrees of unsaturation.

3.40 For purposes of calculating degrees of unsaturation, we inspect the molecular formula, and we can subtract one hydrogen atom for each nitrogen atom that is present. So a compound with the molecular formula $C_6H_{15}N$ has the same HDI as a compound with the molecular formula C_6H_{14}, which is fully saturated. Therefore, this compound has 0 degrees of unsaturation.

3.41 For purposes of calculating degrees of unsaturation, we inspect the molecular formula, and we can ignore the presence of oxygen atoms. So a compound with the molecular formula $C_8H_{10}O$ has the same HDI as a compound with the molecular formula C_8H_{10}. To be fully saturated, a compound with 8 carbon atoms would need to have 18 hydrogen atoms. This compound has only 10 hydrogen atoms, so it is missing 8 hydrogen atoms. Therefore, this compound has 4 degrees of unsaturation.

3.43 The first step is to calculate the HDI. With a molecular formula of $C_5H_{10}O$, this compound has 1 degree of unsaturation. Therefore, the structure must contain either a ring or a double bond (but not both).

The next step is to consider the number of signals and the integration value of each signal. This spectrum has four signals, and we can determine the relative integration of each signal by dividing all of the given integration values by the smallest integration value (27.0). This process gives the following ratio:

$$1 : 1.5 : 1 : 1.5$$

There is no such thing as half of a proton, so we multiply these numbers by two, giving the same ratio in whole numbers:

$$2 : 3 : 2 : 3$$

The total of these integration values is 10, and the molecular formula indicates the presence of 10 protons in the structure. So these values are not only relative, but they also represent the *actual* number of protons associated with each signal (2H, 3H, 2H, and 3H).

The next step is to analyze each signal. Perhaps the easiest signal to analyze is the singlet just above 2 ppm. This signal has an integration of 3, which means that it is a methyl group, and the singlet indicates that there are no neighboring protons. A methyl group has a benchmark value of 0.9 ppm, but this signal has been shifted downfield by a little more than 1 ppm. This suggests that this methyl group is next to a carbonyl group:

$$H_3C-\overset{\overset{\displaystyle O}{\|}}{C}-\xi$$

Note that the double bond accounts for the 1 degree of unsaturation.

The signal at 2.4 ppm has an integration of 2, indicating that it is a CH_2 group, and its chemical shift is shifted downfield by a little more than 1 ppm (relative to the benchmark value of 1.2 ppm for a CH_2 group) indicating that this CH_2 group is likely adjacent to the carbonyl group. So we introduce a CH_2 group next to the carbonyl group:

$$H_3C-\overset{\overset{\displaystyle O}{\|}}{C}-\overset{\overset{\displaystyle H}{|}}{\underset{\underset{\displaystyle H}{|}}{C}}-\xi$$

Now let's analyze the last two signals. One signal (at 0.9 ppm) has an integration of 3, so this signal represents a methyl group. The other signal has an integration of 2, representing a CH_2 group. If we connect these pieces, we get the following structure:

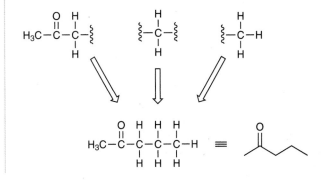

Notice the multiplicity of the signal associated with the CH_2 group that has neighbors on either side of it (the signal at 1.6 ppm). This CH_2 group has three neighbors on the right and two neighbors on the left, for a total of five neighbors. Since the J values are expected to be similar, the multiplicity will appear as if this CH_2 group has five equivalent neighboring protons, and according to the $n + 1$ rule, this gives a sextet, which is exactly what we see.

3.44 The first step is to calculate the HDI. With a molecular formula of $C_5H_{10}O_2$, this compound has 1 degree of unsaturation. Therefore, the structure must contain either a ring or a double bond (but not both).

The next step is to consider the number of signals and the integration value of each signal. This spectrum has four signals, and we can determine the relative integration of each signal by dividing all of the given integration values by the smallest integration value (33.2). This process gives the following ratio:

$$1 : 1.5 : 1 : 1.5$$

There is no such thing as half of a proton, so we multiply these numbers by 2, giving the same ratio in whole numbers:

$$2 : 3 : 2 : 3$$

The total of these integration values is 10, and the molecular formula indicates the presence of 10 protons in the structure. So these values are not only relative, but they also represent the *actual* number of protons associated with each signal (2H, 3H, 2H, and 3H).

The next step is to analyze each signal. Perhaps the easiest signal to analyze is the singlet just above 2 ppm. This signal has an integration of 3, which means that it is a methyl group, and the singlet indicates that there are no neighboring protons. A methyl group has a benchmark value of 0.9 ppm, but this signal has been shifted downfield by a little more than 1 ppm. This suggests that this methyl group is next to a carbonyl group:

$$H_3C-\overset{\overset{\displaystyle O}{\|}}{C}-\xi$$

Note that the double bond accounts for the 1 degree of unsaturation.

The signal at 4.0 ppm has an integration of 2, indicating that it is a CH_2 group, and its chemical shift is shifted downfield by approximately 3 ppm (relative to the benchmark value of 1.2 ppm for a CH_2 group) indicating that this CH_2 group is likely adjacent to an oxygen atom:

$$\xi-O-\overset{\overset{\displaystyle H}{|}}{\underset{\underset{\displaystyle H}{|}}{C}}-\xi$$

Now let's analyze the last two signals. One signal (at 0.9 ppm) has an integration of 3, so this signal represents a methyl group.

The other signal has an integration of 2, representing a CH_2 group. In summary, we have the following pieces:

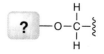

There is more than one way to connect these fragments. To help guide us, let's consider which fragment to connect to the oxygen atom:

We cannot connect either the CH_2 fragment or the CH_3 fragment directly to this oxygen atom, because then we would expect another signal around 4 ppm, and there is only one such signal. Therefore, this oxygen atom must be connected to the carbonyl group. This now gives only three fragments, which can only be connected in one way:

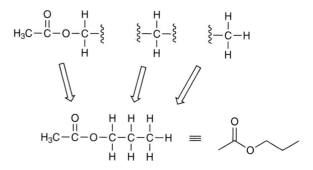

Notice the multiplicity of the signal associated with the CH_2 group that has neighbors on either side of it (the signal at 1.6 ppm). This CH_2 group has three neighbors on the right and two neighbors on the left, for a total of five neighbors. Since the J values are similar, which is usually the case, the multiplicity will appear as if this CH_2 group has five equivalent neighboring protons, and according to the $n+1$ rule, this gives a sextet, which is exactly what we see.

3.45 The first step is to calculate the HDI. With a molecular formula of $C_4H_6O_2$, this compound must have 2 degrees of unsaturation.

The next step is to consider the number of signals and the integration value of each signal. This spectrum has three signals, and we can determine the relative integration of each signal by dividing all of the given integration values by the smallest integration value (18.92). This process gives the following ratio:

$$1 : 1 : 1$$

The total of these relative integration values is 3, but the molecular formula indicates the presence of 6 protons in the structure. So we multiply all of the integration values by 2, to give the *actual* number of protons associated with each signal (2H, 2H, and 2H). So we conclude that each of the three signals is associated with a CH₂ group. Note the multiplicities of these three signals (triplet, triplet, and quintet). This indicates that all three CH₂ groups are connected as follows:

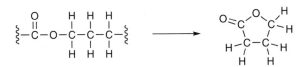

The central CH₂ group has four neighbors, so the signal for this CH₂ group is a quintet (*n* + 1 rule). And each of the other two remaining CH₂ groups is a triplet, because each has two neighbors.

Notice that one of the signals (4.3 ppm) appears shifted downfield by approximately +3 ppm (relative to the benchmark value of 1.2 ppm for a CH₂ group). This CH₂ groups must be next to an oxygen atom, likely of an ester group:

This accounts for one of the 2 degrees of unsaturation, but we still need one more degree of unsaturation, and we have already accounted for all atoms in the molecular formula ($C_4H_6O_2$). So the remaining degree of unsaturation must be associated with a ring, so we join the ends of the fragment above as follows:

If we consider the chemical shifts of the signals associated with all CH₂ groups, they are all consistent with expected predictions. The CH₂ group next to the carbonyl group is shifted a bit more than +1 ppm from its benchmark value of 1.2 ppm. And the central CH₂ group is beta to the carbonyl group (which adds +0.2 ppm) and also beta to the oxygen atom (which adds +0.6 ppm), so its signal is shifted downfield by almost +1 ppm.

3.46 The first step is to calculate the HDI. With a molecular formula of C_9H_{12}, this compound has 4 degrees of unsaturation, which is strongly suggestive of an aromatic ring.

The next step is to consider the number of signals and the integration value of each signal. This spectrum has three signals, and the integration values are 5H, 1H, and 6H, respectively.

The signal just above 7 ppm is characteristic of an aromatic ring, and the integration (5H) indicates that the ring is monosubstituted.

If we consider the other two signals, we will notice that these two signals are the characteristic pattern of an isopropyl group:

These two fragments account for all of the atoms in the molecular formula, and there is only one way to connect these fragments:

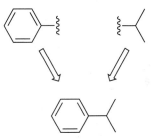

3.47 The first step is to calculate the HDI. With a molecular formula of $C_6H_{12}O_2$, this compound has 1 degree of unsaturation. Therefore, the structure must contain either a ring or a double bond (but not both).

The next step is to consider the number of signals and the integration value of each signal. This spectrum has four signals, and the integration values are 1H, 2H, 6H, and 3H, respectively.

Now we are ready to analyze each signal, one at a time, but in this case, it might be more efficient to look for patterns of signals. For example, the signal very far downfield (5 ppm) is a septet with an integration of 1H, and there is also a doublet with an integration of 6H. These two signals together suggest the presence of an isopropyl group:

Notice that the CH signal appears at 5 ppm, which is significantly downfield from the benchmark value for a CH group (1.7 ppm). Indeed the shift is more than +3 ppm, so we conclude that the CH group is connected directly to an oxygen atom (likely of an ester):

Now let's consider the remaining two signals. We have a quartet with an integration of 2H and a triplet with an integration of 3H. These two signals together suggest the presence of an ethyl group:

Notice that the CH₂ group produces a signal at 2.3 ppm, which is shifted approximately +1 ppm from the benchmark value for a CH₂ group (1.2 ppm). This suggests that the CH₂ group is connected to a carbonyl group:

So far, we have two fragments, and these two fragments account for all of the atoms in the molecular formula. There is only one way to connect these fragments, as shown:

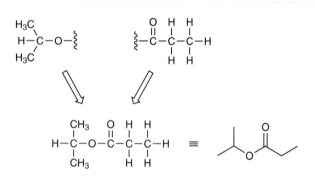

3.48 The first step is to calculate the HDI. With a molecular formula of $C_8H_{10}O$, this compound has 4 degrees of unsaturation, which is strongly suggestive of an aromatic ring.

The next step is to consider the number of signals and the integration value of each signal. This spectrum has four signals, and the integration values are 5H, 2H, 2H, and 1H, respectively.

The signal just above 7 ppm is characteristic of an aromatic ring, and the integration (5H) indicates that the ring is monosubstituted.

There are also two triplets, each of which has an integration of 2H, indicating that there are two CH₂ groups connected to each other (each being split into a triplet by the other):

These two fragments account for almost all of the atoms in the molecular formula, with the exception of one oxygen atom and one hydrogen atom. If we attached this hydrogen atom to either CH₂ group, it would turn the CH₂ group into a CH₃ group, which is not consistent with the spectrum. So the last proton must be connected directly to the oxygen atom, which is consistent with the broad 1H singlet around 2 ppm. This gives us three fragments, and there is only one way to connect these fragments:

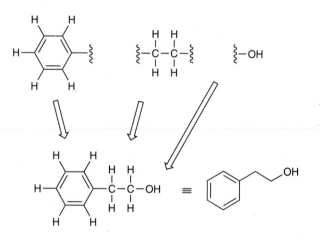

3.50 This compound has 10 carbon atoms, but it also has symmetry, so we expect fewer than 10 signals. Indeed, there are six different types of carbon atoms in this compound (highlighted below), so we expect a total of six signals in the ^{13}C NMR spectrum. Any carbon atom that is NOT highlighted is chemically equivalent to one of the highlighted carbon atoms. The expected location of each signal is indicated below.

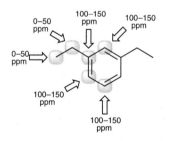

Notice that sp^2 hybridized carbon atoms are expected to produce signals between 100 and 150 ppm, while sp^3 hybridized carbon atoms are expected to produce signals in the range 0–50 ppm. The signal corresponding to the benzylic position is expected to be shifted downfield by its proximity to the aromatic ring, so perhaps it will appear closer to 50 ppm.

3.51 This compound lacks symmetry, and we expect one signal for each carbon atom, for a total of nine signals. The expected location of each signal is indicated below:

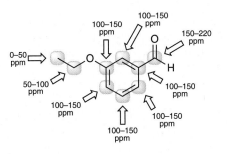

Notice that sp^2 hybridized carbon atoms are expected to produce signals between 100 and 150 ppm, while the carbonyl group is expected to produce a signal between 150 and 220 ppm. This structure has two sp^3 hybridized carbon atoms on the left side of the structure. One of them is connected to an oxygen atom and is expected to appear between 50 and 100 ppm. The other sp^3 hybridized carbon atom is expected to produce a signal in the range 0–50 ppm.

3.52 This compound has six carbon atoms, but two of them (the two methyl groups) are chemically equivalent (they occupy identical electronic environments), so we expect only five signals. The expected location of each signal is indicated below:

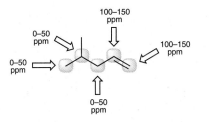

Notice that sp^2 hybridized carbon atoms are expected to produce signals between 100 and 150 ppm, while sp^3 hybridized carbon atoms are expected to produce signals in the range 0–50 ppm.

3.53 This compound has five carbon atoms, and there is no symmetry here, so we expect five signals. All five carbon atoms are sp^3 hybridized, although two of them are connected directly to the oxygen atom. These two carbon atoms will each produce a signal between 50 and 100 ppm (as a result of the electron-withdrawing effect from the oxygen atom), while the other three carbon atoms will produce signals between 0 and 50 ppm.

3.54 This compound has six carbon atoms, but there is symmetry here (the right half reflects the left half of the molecule), so we expect only three signals. The expected location of each signal is indicated below:

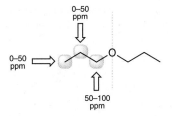

Notice that all three highlighted carbon atoms above are sp^3 hybridized, but one of them is connected directly to the oxygen atom. This carbon atom will produce a signal between 50 and 100 ppm (as a result of the electron-withdrawing effect from the oxygen atom), while the other two carbon atoms will produce signals between 0 and 50 ppm.

3.55 This compound has six carbon atoms, but two of the carbon atoms (the two methyl groups on the left side of the structure) are chemically equivalent because they occupy identical electronic environments. Accordingly, we expect only five signals. The expected location of each signal is indicated below:

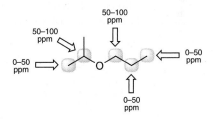

Notice that all highlighted carbon atoms above are sp^3 hybridized, but two of them are connected directly to the oxygen atom. These two carbon atoms will each produce a signal between 50 and 100 ppm (as a result of the electron-withdrawing effect from the oxygen atom), while the other three carbon atoms will produce signals between 0 and 50 ppm.

3.56 This compound has five different kinds of protons (labeled **a–e** below), so the ^{1}H NMR spectrum will have five signals,

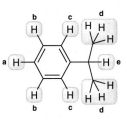

and there are six different kinds of carbon atoms (labeled **a–f** below), so the ^{13}C NMR spectrum will have six signals:

3.57 This compound has three different kinds of protons (labeled **a–c** below), so the ^{1}H NMR spectrum will have three signals,

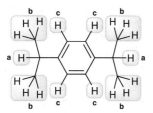

and there are four different kinds of carbon atoms (labeled **a–d** below), so the ^{13}C NMR spectrum will have four signals:

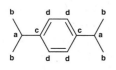

3.58 This compound has two different kinds of protons (labeled **a** and **b** below), so the ^{1}H NMR spectrum will have two signals,

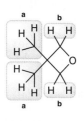

and there are three different kinds of carbon atoms (labeled **a–c** below), so the ^{13}C NMR spectrum will have three signals:

3.59 This compound has six different kinds of protons (labeled **a–f** below), so the ^{1}H NMR spectrum will have six signals,

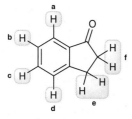

and there are nine different kinds of carbon atoms (labeled **a–i** below), so the ^{13}C NMR spectrum will have nine signals:

3.60 This compound has four different kinds of protons (labeled **a–d** below), so the ^{1}H NMR spectrum will have four signals,

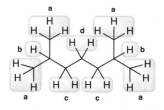

and there are four different kinds of carbon atoms (labeled **a–d** below), so the ^{13}C NMR spectrum will have four signals:

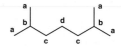

3.61 This compound has two different kinds of protons (labeled **a** and **b** below), so the ^{1}H NMR spectrum will have two signals,

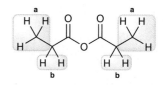

and there are three different kinds of carbon atoms (labeled **a–c** below), so the ^{13}C NMR spectrum will have three signals:

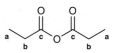

3.62 This compound has five different kinds of protons (labeled **a–e** below), so the ^{1}H NMR spectrum will have five signals,

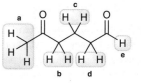

and there are six different kinds of carbon atoms (labeled **a–f** below), so the ^{13}C NMR spectrum will have six signals:

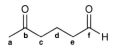

3.63 This compound has three different kinds of protons (labeled **a–c** below), so the ^1H NMR spectrum will have three signals,

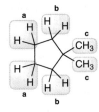

and there are four different kinds of carbon atoms (labeled **a–d** below), so the ^{13}C NMR spectrum will have four signals:

3.64 This compound has two different kinds of protons (labeled **a** and **b** below), so the ^1H NMR spectrum will have two signals,

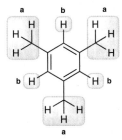

and there are three different kinds of carbon atoms (labeled **a–c** below), so the ^{13}C NMR spectrum will have three signals:

3.65 The molecular formula (C_4H_9Cl) indicates zero degrees of unsaturation, which means that the compound cannot have any double bonds, triple bonds, or rings. That is, the structure must be acyclic and cannot have any π bonds. With only one signal in the

^1H NMR spectrum, the structure must have a high degree of symmetry, such that all nine protons are equivalent. This can be achieved with three equivalent methyl groups. Three methyl groups will be chemically equivalent if they are all attached to the same carbon atom (i.e., if they are part of a *tert*-butyl group). A *tert*-butyl group has four carbon atoms and nine hydrogen atoms, which accounts for all atoms in the molecular formula except for the chlorine atom. So the structure must be *tert*-butyl chloride:

3.66 The molecular formula ($C_2H_3Br_3$) indicates zero degrees of unsaturation, which means that the compound cannot have any double bonds, triple bonds, or rings. That is, the structure must be acyclic and cannot have any π bonds. With only one signal in the ^1H NMR spectrum, all three protons must be chemically equivalent, indicative of a methyl group. This methyl group accounts for one of the carbon atoms and all three protons in the molecular formula. The molecular formula indicates that there is one other carbon atom and three bromine atoms, so the structure must be 1,1,1-tribromoethane:

3.67 The molecular formula ($C_2H_4Br_2$) indicates zero degrees of unsaturation, which means that the compound cannot have any double bonds, triple bonds, or rings. That is, the structure must be acyclic and cannot have any π bonds. With only one signal in the ^1H NMR spectrum, all four protons must be chemically equivalent, suggestive of two equivalent methylene (CH_2) groups. These two methylene groups account for both carbon atoms and all four protons in the molecular formula. The molecular formula indicates that there are also two bromine atoms, so the structure must be 1,2-dibromoethane:

3.68 The molecular formula ($C_4H_8O_2$) indicates one degree of unsaturation, which means that the compound must have either a double bond or a ring. With only one signal in the ^1H NMR spectrum, the structure must have a high degree of symmetry, such that all eight protons are equivalent. This can be achieved with four equivalent methylene (CH_2) groups. These four methylene groups account for all atoms in the molecular formula except for the two oxygen atoms. Since the structure is comprised of four methylene groups and two oxygen atoms, there is no way to incorporate a double bond into the structure. To account for the one degree of unsaturation, the structure must have a ring. In order for all four methylene groups to be equivalent, the two oxygen atoms must be incorporated on opposite sides of the ring, giving the following structure:

3.69 The molecular formula (C_5H_8) indicates two degrees of unsaturation, which means that the compound must have either a triple bond, or two double bonds, or two rings, or one ring and one double bond. With only one signal in the ^1H NMR spectrum, the structure must have a high degree of symmetry, such that all eight protons are equivalent. This can be achieved with four equivalent methylene (CH_2) groups. These four methylene groups account for all atoms in the molecular formula except for one carbon atom. Since the structure is comprised of four methylene groups and one carbon atom, there is no way to incorporate a double bond or triple bond into the structure. To account for the two degrees of unsaturation, the structure must have two rings. The following compound has two rings and four equivalent methylene groups:

3.70 The molecular formula (C_6H_{12}) indicates one degree of unsaturation, which means that the compound must have either a double bond or a ring. With only one signal in the ^1H NMR spectrum, the structure must have a high degree of symmetry, such that all twelve protons are equivalent. This can be achieved either with six equivalent methylene (CH_2) groups or with four equivalent methyl groups. The first option (six equivalent CH_2 groups) would account for all atoms in the molecular formula (C_6H_{12}), so we must account for the one degree of unsaturation by connecting the six CH_2 groups in a ring (giving the structure of cyclohexane, shown below). If, instead, there are four equivalent methyl groups, then we have accounted for all atoms except for two carbon atoms in the molecular formula. We must also account for the one degree of unsaturation, so these two carbon atoms must be connected by a double bond. In summary, each of the following two structures has the molecular formula C_6H_{12} and has a ^1H NMR spectrum with only one signal:

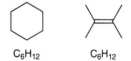

These two compounds can be differentiated from each other based on the number of signals in their ^{13}C NMR spectra. The first compound will have a ^{13}C NMR spectrum with only one signal, since all six carbon atoms are chemically equivalent. In contrast,

the second compound above (the alkene) has two different kinds of carbon atoms (labeled **a** and **b** below), so the ^{13}C NMR spectrum of the alkene will have two signals:

3.71 The molecular formula (C_9H_{18}) indicates one degree of unsaturation, which means that the compound must have either a double bond or a ring. With only one signal in the ^1H NMR spectrum, the structure must have a high degree of symmetry, such that all eighteen protons are equivalent. This can be achieved either with nine equivalent methylene (CH_2) groups or with six equivalent methyl groups.

The first option (nine equivalent CH_2 groups) would account for all atoms in the molecular formula (C_9H_{18}), so we must account for the one degree of unsaturation by connecting the nine CH_2 groups in a ring (giving the structure of cyclononane, shown below). If, instead, there are six equivalent methyl groups, then we have accounted for all atoms except for three carbon atoms in the molecular formula. We must account for these three atoms, as well as the one degree of unsaturation, and this can be accomplished by connecting the three carbon atoms in a ring, to give a cyclopropane ring with six methyl substituents, shown below.

In summary, each of the following two structures has the molecular formula C_9H_{18} and has a ^1H NMR spectrum with only one signal:

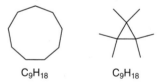

C_9H_{18} C_9H_{18}

These two compounds can be differentiated from each other based on the number of signals in their ^{13}C NMR spectra. The first compound will have a ^{13}C NMR spectrum with only one signal, since all nine carbon atoms are chemically equivalent. In contrast, the second compound has two different kinds of carbon atoms (labeled **a** and **b** below), so the ^{13}C NMR spectrum of this compound will have two signals:

3.72 The molecular formula ($C_{11}H_{24}$) indicates zero degrees of unsaturation, which means that the compound cannot have any double bonds, triple bonds, or rings. That is, the structure must be acyclic and cannot have any π bonds. With only two signals in the ^1H NMR spectrum, there must be a high degree of symmetry (there are only two types of protons, even though there are twenty-four protons in the structure). And we cannot simply draw a large ring of eleven carbon atoms, because the compound has zero degrees of unsaturation (no rings).

If we consider the structure of a chain with eleven carbon atoms, called undecane (shown below), there are six different kinds of protons (labeled **a–f** below), so the ^{1}H NMR spectrum of undecane will have six signals:

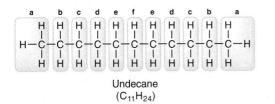

Undecane
($C_{11}H_{24}$)

Even though this compound has symmetry (there are only six signals, rather than eleven signals), nevertheless, we need a structure with even more symmetry, because we are looking for a compound that will have only two signals in its ^{1}H NMR spectrum. So, let's consider what happens if we introduce branching (on both sides of the compound, so as not to break the symmetry found in undecane), for example:

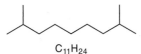

$C_{11}H_{24}$

This compound has only five signals, rather than six, so we are headed in the right direction. Further branching reduces the number of signals farther:

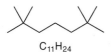

$C_{11}H_{24}$

This structure will have only three signals in its ^{1}H NMR spectrum. To arrive at a structure with even fewer signals, we can draw a shorter parent chain (pentane) with six methyl substituents:

$C_{11}H_{24}$

Indeed, this structure has only two signals in its ^{1}H NMR spectrum (labeled **a** and **b** below).

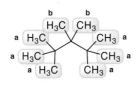

This compound has four different kinds of carbon atoms (labeled **a–d** below), so the ^{13}C NMR spectrum of this compound will have four signals:

3.73 Each of these compounds has three different CH$_2$ groups,

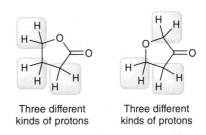

Three different   Three different
kinds of protons   kinds of protons

so each of these compounds is expected to have a ^{1}H NMR spectrum with three signals. Therefore, we cannot differentiate between these two compounds based on the number of signals in their ^{1}H NMR spectra. We also cannot differentiate between these two compounds based on the integration values of the signals, because in each ^{1}H NMR spectrum, all of the signals will have the same integration values.

If we want to use ^{1}H NMR spectroscopy to differentiate these compounds, we will have to find differences in the chemical shift values or multiplicities of the signals. For example, the first compound has only one CH$_2$ group adjacent to an oxygen atom, while the second compound has two CH$_2$ groups adjacent to an oxygen atom. So, the first compound will have a ^{1}H NMR spectrum with only one signal above 3 ppm, while the second compound will have a ^{1}H NMR spectrum with two signals above 3 ppm.

Predicted Chemical Shift Values

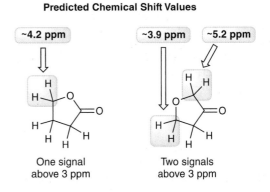

One signal   Two signals
above 3 ppm   above 3 ppm

Alternatively, these compounds can be differentiated based on the multiplicities of the signals. The second compound will have a singlet in its ¹H NMR spectrum, because one of the CH₂ groups has no neighboring protons. In contrast, the first compound will not have any singlets in its ¹H NMR spectrum.

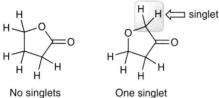

No singlets   One singlet

Now let's consider using ¹³C NMR spectroscopy. Both compounds will produce four signals in their ¹³C NMR spectra, so we cannot differentiate these compounds based on the number of signals in their ¹³C NMR spectra. Rather, we can differentiate them based on the location of those signals. The first compound will have only one signal in the region between 50 and 100 ppm, because only one CH₂ group is neighboring an oxygen atom. In contrast, the second compound will have two signals in the region between 50 and 100 ppm, because two CH₂ groups are neighboring the oxygen atom.

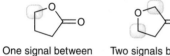

One signal between   Two signals between
50–100 ppm   50–100 ppm

3.74 These compounds can be differentiated based on the number of signals in their ¹H NMR spectra. The first compound has four different kinds of protons (labeled **a–d** below), so the ¹H NMR spectrum of this compound will have four signals.

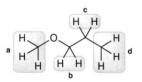

In contrast, the second compound has only two different kinds of protons (labeled **a** and **b** below), so the ¹H NMR spectrum of this compound will have only two signals.

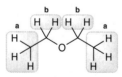

Similarly, these two compounds can also be differentiated by comparing the number of signals in their ¹³C NMR spectra. The

first compound has four different kinds of carbon atoms (labeled **a–d** below), so the ¹³C NMR spectrum of this compound will have four signals.

In contrast, the second compound has only two different kinds of carbon atoms (labeled **a** and **b** below) because of symmetry, so the ¹³C NMR spectrum of this compound will have only two signals.

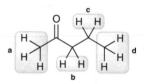

3.75 These compounds can be differentiated based on the number of signals in their ¹H NMR spectra. The first compound has four different kinds of protons (labeled **a–d** below), so the ¹H NMR spectrum of this compound will have four signals.

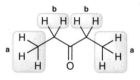

In contrast, the second compound has only two different kinds of protons (labeled **a** and **b** below) because of symmetry, so the ¹H NMR spectrum of this compound will have only two signals.

Similarly, these two compounds can also be differentiated by comparing the number of signals in their ¹³C NMR spectra. The first compound has five different kinds of carbon atoms (labeled **a–e** below), so the ¹³C NMR spectrum of this compound will have five signals.

In contrast, the second compound has only three different kinds of carbon atoms (labeled **a–c** below), so the ¹³C NMR spectrum of this compound will have only three signals.

3.76 These compounds can be differentiated based on the number of signals in their ^{1}H NMR spectra. The first compound has two different kinds of protons (labeled **a** and **b** below), so the ^{1}H NMR spectrum of this compound will have two signals. In contrast, the second compound has only one kind of proton, so the ^{1}H NMR spectrum of this compound will have only one signal.

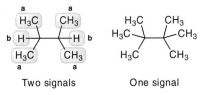

Two signals   One signal

These compounds cannot be differentiated easily with ^{13}C NMR spectroscopy, because both compounds will exhibit a ^{13}C NMR spectrum with two signals that appear below 50 ppm.

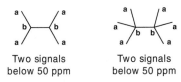

Two signals   Two signals
below 50 ppm   below 50 ppm

3.77 Each of these compounds has one isolated methyl group, and one ethyl group,

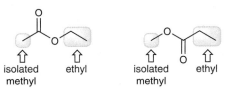

isolated   ethyl   isolated   ethyl
methyl   methyl

so the ^{1}H NMR spectra of these compounds will be quite similar. Both spectra will have three signals. And in both spectra, there will be a singlet for the isolated methyl group, as well as the characteristic signals of an ethyl group (a quartet with an integration of 2H, and a triplet with an integration of 3H). Furthermore, the integration values will be similar in each spectrum – in each case, the singlet will have an integration of 3H, the quartet will have an integration of 2H, and the triplet will have an integration of 3H.

If we want to use ^{1}H NMR spectroscopy to differentiate these compounds, we will have to find differences in the chemical shift values of the signals. We predict the following chemical shift values:

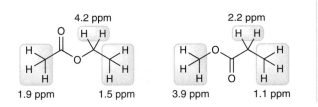

4.2 ppm   2.2 ppm

1.9 ppm   1.5 ppm   3.9 ppm   1.1 ppm

These compounds can be differentiated based on the location of the singlet (1.9 ppm vs. 3.9 ppm). Alternatively, we can focus on the location of the quartet (4.2 ppm vs. 2.2 ppm).

These compounds will be difficult to differentiate using ^{13}C NMR spectroscopy, because both compounds will produce four signals in their ^{13}C NMR spectra, and the locations of the signals will be similar (one signal near 200 ppm, one signal between 50 and 100 ppm, and two signals below 50 ppm.

3.78 These compounds can be differentiated based on the number of signals in their ^{1}H NMR spectra. The first compound has four different kinds of protons (labeled **a–d** below), so the ^{1}H NMR spectrum of this compound will have four signals.

In contrast, the second compound has only two different kinds of protons (labeled **a** and **b** below), so the ^{1}H NMR spectrum of this compound will have only two signals.

Similarly, these two compounds can also be differentiated by comparing the number of signals in their ^{13}C NMR spectra. The first compound has five different kinds of carbon atoms (labeled **a–e** below), so the ^{13}C NMR spectrum of this compound will have five signals.

In contrast, the second compound has only three different kinds of carbon atoms (labeled **a–c** below), so the ^{13}C NMR spectrum of this compound will have only three signals.

3.79 The molecular formula (C_3H_8O) indicates zero degrees of unsaturation, which means that the compound cannot have any double bonds, triple bonds, or rings. That is, the structure must be acyclic and cannot have any π bonds.

The IR spectrum has a broad signal from 3200 to 3600 cm^{-1}, characteristic of O—H stretching of an alcohol.

The ^1H NMR spectrum has four signals, and the integration values of those signals are 2H, 1H, 2H, and 3H, respectively. An integration of 2H suggests a methylene (CH_2) group, while an integration of 3H suggests a methyl group. So, based on the integration values alone, we expect the structure to have two methylene groups, one methyl group, and also a signal representing one proton (which is likely the proton of the OH group).

The signal just below 1 ppm with an integration of 3H (associated with the methyl group) is a triplet, indicating that it is adjacent to a methylene group. Similarly, the signal at 3.6 ppm with an integration of 2H (associated with one of the methylene groups) is a triplet, so it is also adjacent to a methylene group. The signal at 1.6 ppm (the second methylene group) is a sextet, indicating five neighbors (one methyl group and one methylene group). Putting this information together, we get the following fragment:

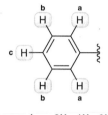

This accounts for all atoms in the molecular formula, except for one oxygen atom and one hydrogen atom. And we know from the IR spectrum that we have an alcohol, so the structure must be 1-propanol:

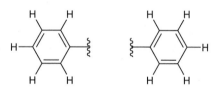

Notice that the signal at 3.6 ppm is shifted downfield (relative to the other signals in the ^1H NMR spectrum) because of the electron-withdrawing effects of the oxygen atom.

The ^{13}C NMR spectrum has three signals and is consistent with the structure that we deduced:

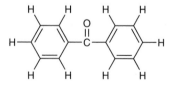

Notice that one signal in the ^{13}C NMR spectrum appears above 50 ppm. This signal (associated with the carbon atom labeled **a** above) is shifted downfield relative to the other signals because of the electron-withdrawing effects of the oxygen atom.

3.80 The molecular formula ($C_{13}H_{10}O$) indicates nine degrees of unsaturation, which is suggestive of two aromatic rings (each aromatic ring accounts for four degrees of unsaturation) plus one more site of unsaturation (either a ring or a double bond).

The IR spectrum has a group of weak signals just above 3000 cm^{-1}, as well as a group of signals around 1600 cm^{-1}, which suggest an aromatic ring (this is something we already suspected from our analysis of the molecular formula). There is also a strong signal above 1600 cm^{-1}, which looks like a carbonyl stretching signal, but the wavenumber of absorption is somewhat low for a carbonyl group (a ketone generally produces a signal at around 1720 cm^{-1}), so perhaps this is a conjugated carbonyl group.

The ^1H NMR spectrum has three signals, all of which are located between 7 and 8 ppm, suggesting that all of the protons in this compound are connected to aromatic rings. The relative integration values of the signals are 2:1:2, suggestive of a monosubstituted aromatic ring:

$$a : c : b = 2H : 1H : 2H$$

Since the compound has a total of ten protons, the actual integration values must be 4H, 2H, and 4H. This can be explained with two monosubstituted aromatic rings:

These two fragments account for all atoms in the molecular formula except for one carbon atom and one oxygen atom. If we draw a double bond between these two atoms, then this will account for the last site of unsaturation, and it will be consistent with the IR spectrum, which shows the presence of a carbonyl group. This carbonyl group is drawn in between the two monosubstituted rings (indeed, this is the *only* place where it can be drawn), to give the following compound:

Notice that the carbonyl group is conjugated with both aromatic rings, explaining why the carbonyl stretching signal is below 1720 cm^{-1}.

This compound is expected to have a ^{13}C NMR spectrum with five signals (labeled **a–e** below):

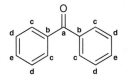

We expect one of those signals (labeled **a**) to be near 200 ppm on the spectrum, and the other four signals (**b–e**) will be between 100 and 150 ppm. This is indeed consistent with the ^{13}C NMR spectrum, which has five signals, in the locations predicted.

3.81 The molecular formula ($C_{11}H_{14}O_2$) indicates five degrees of unsaturation, which is suggestive of an aromatic ring (which accounts for four degrees of unsaturation) plus one more site of unsaturation (either a ring or a double bond).

The IR spectrum has a strong signal just above 1700 cm^{-1} and a very broad signal spanning from 2200 to 3600 cm^{-1}. These signals are characteristic of a carboxylic acid (which accounts for one degree of unsaturation):

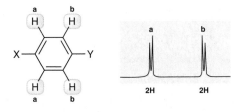

There is also a group of signals around 1600 cm^{-1}, which suggest an aromatic ring (this is something we already suspected from our analysis of the molecular formula).

The ^{1}H NMR spectrum has a broad signal between 11 and 12 ppm with an integration value of 1H, corresponding to the proton of the carboxylic acid group. There are also two signals between 7 and 8.5 ppm, corresponding to aromatic protons. These two signals have a combined integration value of 4H (each signal is 2H), indicating the presence of four aromatic protons (this aromatic ring is disubstituted). Notice that these two signals appear as a pair of doublets, which indicates that the two substituents must be on opposite sides of the ring (*para* substitution). This gives two kinds of aromatic protons, labeled **a** and **b** below (each has one immediate neighbor, and therefore, each signal is a doublet):

The signal at 1.5 ppm is a singlet with an integration of 9H, which is the characteristic signal for a *tert*-butyl group:

Thus far, we have identified three fragments (shown below) that collectively account for all atoms in the molecular formula. And there is only one way to assemble these three fragments:

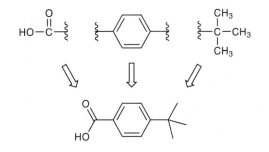

Now let's verify that this structure is consistent with the ^{13}C NMR spectrum provided. This compound is expected to have a ^{13}C NMR spectrum with seven signals (labeled **a–g** below):

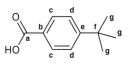

We expect one of those signals (labeled **a**) to be near 200 ppm on the spectrum, four signals (**b–e**) will be between 100 and 150 ppm, and two signals (**f** and **g**) will be below 50 ppm. This is indeed consistent with the given ^{13}C NMR spectrum, which has seven signals, in the locations predicted.

3.82 The molecular formula ($C_7H_{14}O$) indicates one degree of unsaturation, so the structure must have either a ring or a double bond.

The IR spectrum has a strong signal above 1700 cm^{-1}, suggesting the presence of a carbonyl group, which would account for the one and only degree of unsaturation:

The ^{1}H NMR spectrum has three signals, and the relative integration values of those signals are 2 : 2 : 3. An integration of 2 suggests a methylene (CH$_2$) group, while an integration of 3 suggests a methyl group. So, based on the integration values alone, we expect the structure to have two methylene groups and one methyl group. The signal just below 1 ppm with an integration of 3 (associated with the methyl group) is a triplet, indicating that it is adjacent to a methylene group. Similarly, the signal at 2.4 ppm with an integration of 2 (associated with one of the methylene groups) is a triplet, so it is also adjacent to a methylene group. The signal at 1.6 ppm (the second methylene group) is a sextet, indicating five neighbors (one methyl group and one methylene group). Putting this information together, we get the following fragment:

We have now analyzed all signals in the ^{1}H NMR spectrum, but we have not yet accounted for all atoms in the molecular formula. The structure must have seven carbon atoms, but we have only accounted for four carbon atoms (one from the carbonyl group, and three from the propyl group). Furthermore, the structure must have fourteen protons, but we have only accounted for seven protons. Clearly, the integration values (2 : 2 : 3) are only relative, and we must multiply by a factor of two in order to account for all protons in the structure. Therefore, the actual integration values are 4H, 4H, and 6H. This suggests that the structure has two chemically equivalent propyl groups.

By realizing that there are two propyl groups in the structure, we have now accounted for all atoms in the molecular formula. So, we can now connect the fragments (which can only be connected in one possible way) to give the following structure:

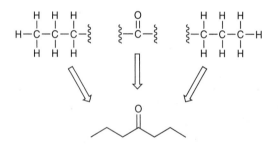

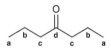

This compound is expected to have a ^{13}C NMR spectrum with four signals (labeled **a–d** below):

We expect one of those signals (labeled **d**) to be near 200 ppm on the spectrum, and the other three signals (**a–c**) will be below 50 ppm. This is indeed consistent with the given ^{13}C NMR spectrum, which has four signals, in the locations predicted.

3.83 The molecular formula ($C_7H_{14}O_2$) indicates one degree of unsaturation, so the structure must have either a ring or a double bond.

The IR spectrum has a strong signal at 1741 cm^{-1}, suggesting the presence of a carbonyl group of an ester, which would account for the one degree of unsaturation (and would also account for both oxygen atoms in the molecular formula):

The ^{1}H NMR spectrum has five signals. One of those signals (at 5.1 ppm) is a septet with an integration of 1H. This signal suggests a possible isopropyl group, so we look for a doublet with an integration of 6, and there is in fact such a signal at 1.3 ppm. These two signals indicate the presence of an isopropyl group:

The remaining three signals have integration values of 2H, 2H, and 3H. An integration of 2H suggests a methylene (CH_2) group, while an integration of 3H suggests a methyl group. So, in addition to the isopropyl group that we expect to be present, we also expect the structure to have two methylene groups and one methyl group. The signal at 1 ppm with an integration of 3H (associated with the methyl group) is a triplet, indicating that it is adjacent to a methylene group. Similarly, the signal at 2.3 ppm with an integration of 2H (associated with one of the methylene groups) is a triplet, so it is also adjacent to a methylene group. The signal at 1.7 ppm (the second methylene group) is a sextet, indicating five neighbors (one methyl group and one methylene group). Putting together the information from these three signals, we get the following fragment:

We have now analyzed all signals in the ^{1}H NMR spectrum, and we have the following three fragments, which account for all atoms in the molecular formula.

There are two different ways to connect these fragments, shown below. Notice that these two structures differ from each other based on the orientation of the ester group:

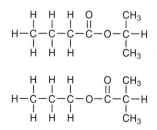

To differentiate these two possibilities, we can predict the following chemical shift values for the proton(s) on either side of each ester group:

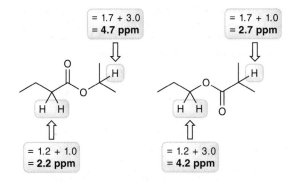

= 1.7 + 3.0
= **4.7 ppm**

= 1.7 + 1.0
= **2.7 ppm**

= 1.2 + 1.0
= **2.2 ppm**

= 1.2 + 3.0
= **4.2 ppm**

The given ^{1}H NMR spectrum is consistent with the first option above (in the spectrum, the CH septet is near 5 ppm, rather than near 3 ppm), so this must be our structure:

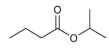

This compound is expected to have a ^{13}C NMR spectrum with six signals (labeled **a–f** below):

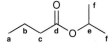

We expect one signal (labeled **d**) to be near 200 ppm on the spectrum, another signal (labeled **e**) to be between 50 and 100 ppm on the spectrum, and the remaining four signals (**a, b, c,** and **f**) will be below 50 ppm. This is indeed consistent with the given ^{13}C NMR spectrum, which has six signals, in the locations predicted.

3.84 The molecular formula ($C_8H_9NO_3$) indicates five degrees of unsaturation, which is suggestive of an aromatic ring (which accounts for four degrees of unsaturation) plus one more site of unsaturation (either a ring or a double bond).

The ^{1}H NMR spectrum has four signals. Two of these signals appear between 7 and 8.5 ppm, corresponding to aromatic protons. These two signals have a combined integration of 4H (each signal has an integration of 2H), indicating the presence of four aromatic protons (in other words, this aromatic ring is disubstituted). Notice that these two signals appear as a pair of doublets. This indicates that the two substituents must be on opposite sides of the ring (*para* substitution), which gives two kinds of aromatic protons, labeled **a**

and **b** below (each has one immediate neighbor, and therefore, each signal is a doublet):

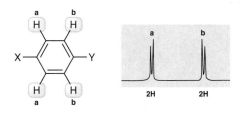

There are two other signals in the ^{1}H NMR spectrum: a quartet with an integration of 2H and a triplet with an integration of 3H. These two signals together represent the characteristic pattern for an ethyl group:

Notice that the quartet (representing the CH_2 group above) appears at 4.2 ppm, significantly downfield from the benchmark value of 1.2 ppm for a CH_2 group, so this CH_2 group must be next to an oxygen atom:

Thus far, we have accounted for all atoms in the molecular formula except for one nitrogen atom and two oxygen atoms. We also need to account for one more degree of unsaturation. This is suggestive of a nitro group:

We therefore have the following three fragments, which can only be connected to each other in one possible way, as shown below:

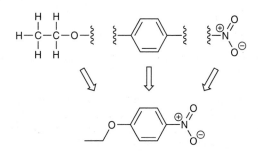

Now let's verify that this structure is consistent with the ^{13}C NMR spectrum provided. This compound is expected to have a ^{13}C NMR spectrum with six signals (labeled **a–f** below):

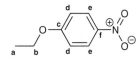

We expect one signal (labeled **a**) to appear below 50 ppm on the spectrum, one signal (labeled **b**) to appear between 50 and 100 ppm, and four signals (**c–f**) to appear above 100 ppm. This is indeed consistent with the given ^{13}C NMR spectrum, which has six signals, in the locations predicted.

3.85 The molecular formula ($C_6H_{14}O$) indicates zero degrees of unsaturation, which means that the compound cannot have any double bonds, triple bonds, or rings. That is, the structure must be acyclic and cannot have any π bonds.

The IR spectrum has no characteristic signals in the diagnostic region, other than the typical signals just below 3000 cm^{-1}, corresponding to C_{sp^3}–H stretching.

The ^1H NMR spectrum has two signals: a septet with a relative integration of 1, and a doublet with a relative integration of 6. These two signals are the characteristic signals for an isopropyl group:

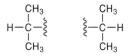

However, an isopropyl group has only seven protons, and the molecular formula shows fourteen protons. We have already analyzed all signals in the ^1H NMR spectrum, so, in order to account for all fourteen protons, there must be two chemically equivalent isopropyl groups:

These two isopropyl groups account for all atoms in the molecular formula, except for one oxygen atom. So, this oxygen atom must be in between the two isopropyl groups, giving the following compound:

$$H-\underset{\underset{CH_3}{|}}{\overset{\overset{CH_3}{|}}{C}}-O-\underset{\underset{CH_3}{|}}{\overset{\overset{CH_3}{|}}{C}}-H$$

Now let's verify that this structure is consistent with the ^{13}C NMR spectrum provided. This compound is expected to have a ^{13}C NMR spectrum with two signals (labeled **a** and **b** below):

We expect one signal (labeled **a**) to appear below 50 ppm on the spectrum, and the other signal (labeled **b**) will appear between 50 and 100 ppm. This is indeed consistent with the given ^{13}C NMR spectrum, which has two signals, in the locations predicted.

3.86 The molecular formula ($C_9H_{21}N$) indicates zero degrees of unsaturation, which means that the compound cannot have any double bonds, triple bonds, or rings. That is, the structure must be acyclic and cannot have any π bonds.

The ^1H NMR spectrum has three signals, and the relative integration values of those signals are 2 : 2 : 3. An integration of 2 suggests a methylene (CH_2) group, while an integration of 3 suggests a methyl group. So, based on the integration values alone, we expect the structure to have two methylene groups and one methyl group. The signal just below 1 ppm with an integration of 3 (associated with the methyl group) is a triplet, indicating that it is adjacent to a methylene group. Similarly, the signal at 2.4 ppm with an integration of 2 (associated with one of the methylene groups) is a triplet, so it is also adjacent to a methylene group. The signal at 1.5 ppm (the second methylene group) is a sextet, indicating five neighbors (one methyl group and one methylene group). Putting this information together, we get the following fragment:

$$\underset{\underset{H}{|}}{\overset{\overset{H}{|}}{C}}-\underset{\underset{H}{|}}{\overset{\overset{H}{|}}{C}}-\underset{\underset{H}{|}}{\overset{\overset{H}{|}}{C}}-H$$

We have now analyzed all signals in the ^1H NMR spectrum, but we have not yet accounted for all atoms in the molecular formula. The structure must have twenty-one protons, but we have only accounted for seven protons. Clearly, the integration values (2 : 2 : 3) are only relative, and we must multiply by three in order to account for all protons in the structure. Therefore, the actual integration values are 6H, 6H, and 9H. This suggests that the structure has three chemically equivalent propyl groups.

Three propyl groups collectively account for all atoms in the molecular formula except for the nitrogen atom. Nitrogen is trivalent (it forms three bonds), giving the following structure:

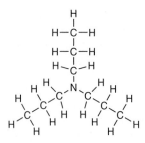

This compound is expected to have a ^{13}C NMR spectrum with three signals (labeled **a–c** below):

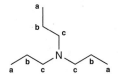

We expect one of those signals (labeled **c**) to be between 50 and 100 ppm on the spectrum, and the other two signals (**a** and **b**) will be below 50 ppm. This is indeed consistent with the given ^{13}C NMR spectrum, which has three signals, in the locations predicted.

3.87 The molecular formula ($C_{12}H_{10}$) indicates eight degrees of unsaturation, which is suggestive of two aromatic rings (each aromatic ring accounts for four degrees of unsaturation).

All of the signals in the ^{1}H NMR spectrum are located between 7 and 8 ppm, suggesting that all of the protons in this compound are connected to aromatic rings. The relative integration values of the signals are 2 : 3, suggestive of a ring with five protons (a monosubstituted aromatic ring):

Since the compound has a total of ten protons, the actual integration values must be 4H and 6H, which can be explained with two chemically equivalent, monosubstituted aromatic rings:

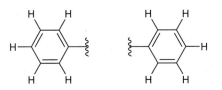

These two fragments account for all atoms in the molecular formula, so we simply connect the fragments, to give the following compound, called biphenyl.

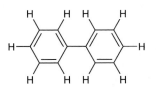

This compound is expected to have a ^{13}C NMR spectrum with four signals (labeled **a–d** below):

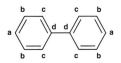

We expect all four signals to appear between 100 and 150 ppm in the ^{13}C NMR spectrum. This is indeed consistent with the ^{13}C NMR spectral data that was provided (four signals between 120 and 150 ppm).

3.88 The molecular formula ($C_9H_{10}O_2$) indicates five degrees of unsaturation, which is suggestive of an aromatic ring (which accounts for four degrees of unsaturation) plus one more site of unsaturation (either a ring or a double bond).

The ^{1}H NMR spectrum has two signals between 7 and 8.5 ppm, corresponding to aromatic protons. These two signals have a combined integration value of 4H (each signal has an integration of 2H), indicating the presence of four aromatic protons (this aromatic ring is disubstituted). Notice that these two signals appear as a pair of doublets, which indicates that the two substituents must be on opposite sides of the ring (*para* substitution). This gives two kinds of aromatic protons, labeled **a** and **b** below (each has one immediate neighbor, and therefore, each signal is a doublet):

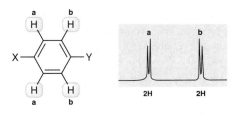

The signal at 3.8 ppm is a singlet with an integration of 3H. The integration value for this signal suggests that it is a methyl group, and the multiplicity (singlet) indicates that it is an isolated methyl group (no neighboring protons). The chemical shift is significantly downfield for a methyl group (the benchmark value for a methyl group is 0.9 ppm, if it is not next to any electron-withdrawing groups), so this indicates that the methyl group is next to an oxygen atom:

The last signal (at 2.5 ppm) is also a singlet, with an integration of 3H. The integration of this signal suggests that it is a methyl group, and the multiplicity (singlet) indicates that it is an isolated methyl group (no neighboring protons). The chemical shift suggests that it is adjacent to a carbonyl group (which would account for the final site of unsaturation):

Thus far, we have drawn the following fragments, which account for all atoms in the molecular formula. And there is only one way to connect these fragments:

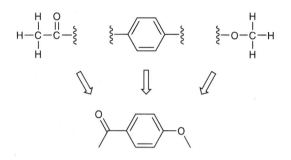

Now let's verify that this structure is consistent with the ^{13}C NMR spectrum provided. This compound is expected to have a ^{13}C NMR spectrum with seven signals (labeled **a–g** below):

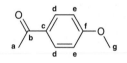

We expect one of those signals (labeled **b**) to be near 200 ppm on the spectrum, four signals (labeled **c–f**) will be between 100 and 150 ppm, one signal (labeled **g**) will be between 50 and 100 ppm, and one signal (labeled **a**) will be below 50 ppm. This is indeed consistent with the given ^{13}C NMR spectrum, which has seven signals, in the locations predicted.

3.89 The molecular formula ($C_5H_8O_4$) indicates two degrees of unsaturation, which means that the compound must have either a triple bond, or two double bonds, or two rings, or one ring and one double bond. With only one signal in the ^{1}H NMR spectrum, the structure must have a high degree of symmetry, such that all eight protons are equivalent. This can be achieved with four equivalent methylene (CH_2) groups. These four methylene groups account for all atoms in the molecular formula except for one carbon atom and the four oxygen atoms. Without the oxygen atoms, we could account for the two degrees of unsaturation (and the missing carbon atom) by drawing two rings, just like we saw in the solution to problem 3.36:

We must now insert the four oxygen atoms in such a way that it does not destroy the symmetry that makes all four methylene groups equivalent. There are two different ways to insert the oxygen atoms and preserve the symmetry:

Each of these compounds has the molecular formula $C_5H_8O_4$ and will have only one signal in its ^{1}H NMR spectrum.

These two compounds can be differentiated using ^{13}C NMR spectroscopy. The first compound above will have two signals in its ^{13}C NMR spectrum (labeled **a** and **b** below):

One signal (labeled **a**) will appear above 50 ppm (because of the electron-withdrawing effect of the oxygen atom), while the other signal (labeled **b**) will appear below 50 ppm in the spectrum (because that carbon atom is NOT connected directly to an oxygen atom).

The second possible structure will also have two signals in its ^{13}C NMR spectrum (once again, labeled **a** and **b** below),

but this time, both signals are expected to appear above 50 ppm, because every carbon atom in this structure is connected to at least one oxygen atom. Therefore, we can distinguish between the two possible structures by looking for a signal below 50 ppm in the ^{13}C NMR spectrum.

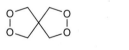

WILL	WILL NOT
have a signal below 50 ppm	have a signal below 50 ppm

3.90 The molecular formula ($C_{15}H_{24}$) indicates four degrees of unsaturation, which is suggestive of an aromatic ring.

The ^{1}H NMR spectrum has three signals. Among these signals, there is a septet with a relative integration of 1 and a doublet with a relative integration of 6. These two signals are the characteristic signals for an isopropyl group:

Notice that the relative integration values are 1 : 1 : 6, but the total number of protons is not 8. Rather, the molecular formula indicates that the total number of protons is 24. Therefore, the actual integration values must be 3H : 3H : 18H (and the compound must have a high degree of symmetry. That is, there must be three isopropyl groups, rather than just one.

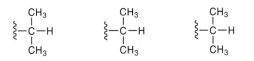

These isopropyl groups account for two of the three signals that appear in the spectrum. The final signal (appearing at 7.5 ppm) is a singlet with a relative integration of 1. The relative integration of this signal suggests that it is a methine (CH) group, and the multiplicity (singlet) indicates that it is isolated (no neighboring protons). The chemical shift suggests that it is a proton connected to an aromatic ring. While the relative integration is 1, we have seen that this signal must correspond with three equivalent protons. This information is consistent with the following fragment (a trisubstituted benzene ring, where all three substituents are identical, rendering all three protons identical as well):

This gives the following fragments:

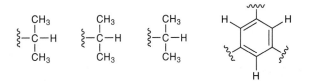

There is only one way to connect these four fragments together:

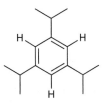

This compound is expected to have a ^{13}C NMR spectrum with four signals, labeled **a–d** below:

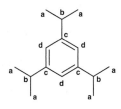

CHAPTER 4

4.2 Cl_2 reacts with $AlCl_3$ to generate a Lewis acid–Lewis base complex:

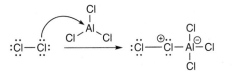

This complex can then serve as a delivery agent of Cl^+ (highlighted below):

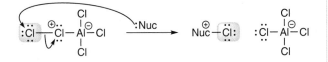

4.3 This mechanism has only two steps. In the first step, the ring functions as a nucleophile and attacks the Lewis acid–Lewis base complex (see the solution to the previous problem), thereby transferring Cl^+ to the aromatic ring, and generating a resonance-stabilized intermediate, called a sigma complex. In the second (and final) step of the mechanism, $AlCl_4^-$ removes a proton from the sigma complex to restore aromaticity, as shown.

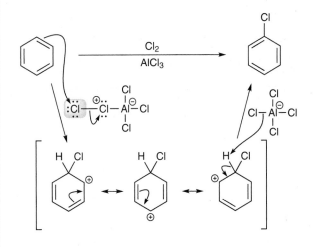

4.4 This mechanism has only two steps. In the first step, the ring functions as a nucleophile and attacks I^+, thereby generating a resonance-stabilized intermediate, called a sigma complex. In the second (and final) step of the mechanism, water functions as a base and removes a proton from the sigma complex to restore aromaticity.

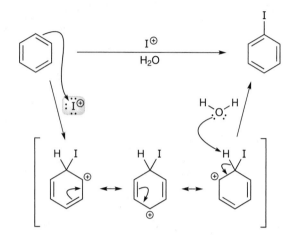

4.5 A bromine atom can be installed on the ring by treating benzene with Br_2 in the presence of a Lewis acid, such as $AlBr_3$:

4.6 A nitro group can be installed on the ring by treating benzene with a mixture of sulfuric acid and nitric acid:

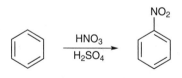

4.7 A chlorine atom can be installed on the ring by treating benzene with Cl_2 in the presence of a Lewis acid, such as $AlCl_3$:

4.10 This transformation requires the installation of a butyl group, without any rearrangements. This cannot be accomplished

directly via a Friedel–Crafts alkylation, because that would produce a mixture of products (as the result of rearrangement). The desired transformation can be achieved via a Friedel–Crafts acylation, followed by a Clemmensen reduction, as shown here:

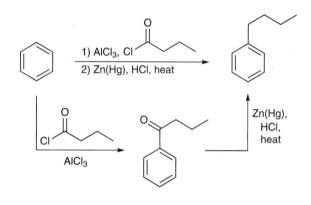

4.11 This alkylation can be achieved directly via a Friedel–Crafts alkylation process, without the concern of rearrangements (in this case, a secondary carbocation is involved, and there is no way for it to rearrange to become tertiary).

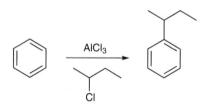

4.12 This alkylation can be achieved directly via a Friedel–Crafts alkylation process, without the concern of rearrangements (in this case, a tertiary carbocation is involved, and it cannot rearrange to become more stable).

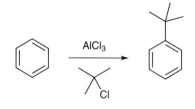

4.13 This alkylation can be achieved directly via a Friedel–Crafts alkylation process, without the concern of rearrangements (in this case, a secondary carbocation is involved, and there is no way for it to rearrange to become tertiary).

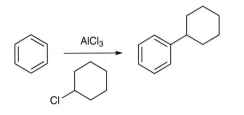

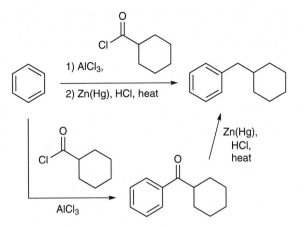

4.14 This transformation requires the installation of a primary alkyl group, without any rearrangements. This cannot be accomplished directly via a Friedel–Crafts alkylation, because that would produce a mixture of products (as the result of rearrangement). The desired transformation can be achieved via a Friedel–Crafts acylation, followed by a Clemmensen reduction, as shown here:

4.15 This is a Friedel–Crafts alkylation process in which a primary alkyl chloride is used. The mechanism involves a primary carbocation that can rearrange to a secondary carbocation, which can further rearrange to give a tertiary carbocation, as shown here:

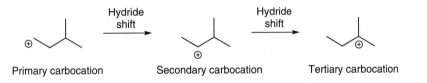

The aromatic ring can attack any one of the carbocations shown above, leading to a mixture of products:

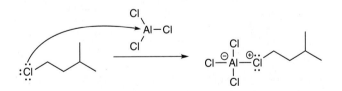

4.16 The first step is formation of a Lewis acid–Lewis base complex:

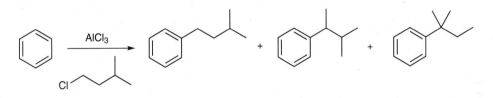

This complex can then function "as if" it were a primary carbocation:

The primary carbocation is very unstable, and it is unlikely that it is actually formed. Rather, the complex has the character of a primary carbocation (it can be attacked by a nucleophile or it can undergo rearrangement). If it is attacked by a nucleophile, without first rearranging, then a primary alkyl group is installed on the ring, as shown:

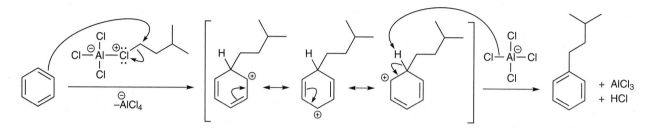

Now let's consider what happens if rearrangement occurs prior to nucleophilic attack, giving a secondary carbocation:

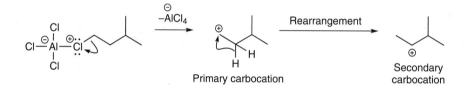

This secondary carbocation can rearrange further, but if instead, it is first attacked by the aromatic ring, then a secondary alkyl group is installed on the ring, as shown:

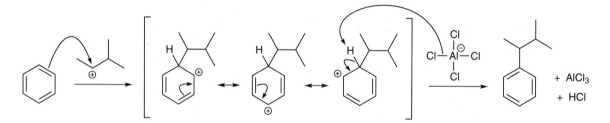

Now let's consider what happens if the secondary carbocation rearranges to give a tertiary carbocation prior to nucleophilic attack:

If this tertiary carbocation is attacked by the ring, then a tertiary alkyl group is installed on the ring, as shown:

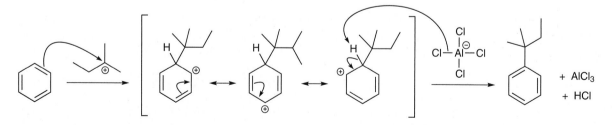

This product is likely the major product, although the exact product mixture is difficult to predict.

4.17 The first step is formation of the acylium ion, highlighted below:

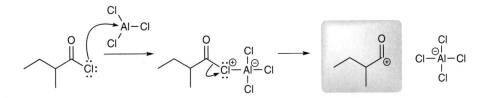

This acylium ion can function as an electrophile in an electrophilic aromatic substitution reaction, which has two steps. In the first step, the aromatic ring functions as a nucleophile and attacks the acylium ion to give a resonance-stabilized sigma complex. Then, in the second (and final) step, a proton is removed from the ring to restore aromaticity:

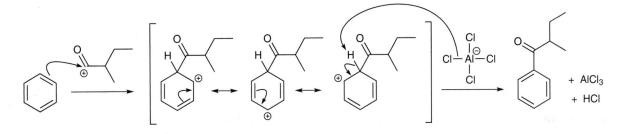

4.19 The desired transformation involves the installation of a sulfonate group on the aromatic ring, which can be achieved with concentrated fuming sulfuric acid.

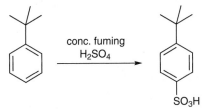

4.20 The desired transformation involves removing a sulfonate group from the aromatic ring, which can be achieved with dilute sulfuric acid.

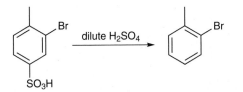

4.21 The desired transformation involves the installation of a sulfonate group on the aromatic ring, which can be achieved with concentrated fuming sulfuric acid.

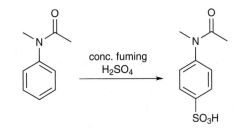

4.22 The desired transformation involves removing a sulfonate group from the aromatic ring, which can be achieved with dilute sulfuric acid.

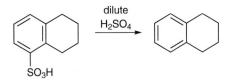

4.23 The desired transformation involves the installation of a bromine atom on the aromatic ring, which can be achieved with Br_2 and a Lewis acid, such as $AlBr_3$.

4.24 The desired transformation involves the installation of a nitro group on the aromatic ring, which can be achieved with a mixture of nitric acid and sulfuric acid.

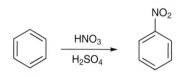

4.25 The desired transformation involves the installation of a chlorine atom on the aromatic ring, which can be achieved with Cl_2 and a Lewis acid, such as $AlCl_3$.

4.26 The desired transformation involves the installation of a methyl group on the aromatic ring, which can be achieved with a Friedel–Crafts alkylation.

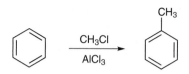

4.27 The desired transformation involves the installation of a primary alkyl group on the aromatic ring. This cannot be achieved efficiently with a Friedel–Crafts alkylation, because the possibility for carbocation rearrangement would generate a mixture of products. Instead, we can perform a Friedel–Crafts acylation (without carbocation rearrangements), and then reduce the resulting ketone with a Clemmensen reduction to give the desired product.

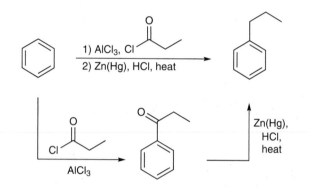

4.28 As indicated in the problem statement, this mechanism has three steps. In the first step of the mechanism, the aromatic ring functions as a nucleophile and attacks SO_3 (an electrophile) to give a resonance-stabilized intermediate, called a sigma complex. This sigma complex is then deprotonated (by water) to give a sulfonate anion, which is then protonated (under these acidic conditions) to give the product.

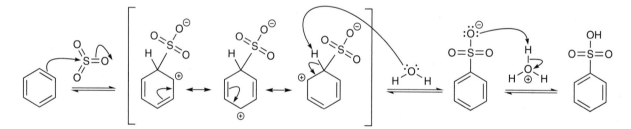

4.29 As indicated in the problem statement, this mechanism can be drawn in two steps, as shown below. In the first step, the aromatic ring is protonated to give a resonance-stabilized intermediate, called a sigma complex. This sigma complex is then deprotonated (by water) to give the product, as shown.

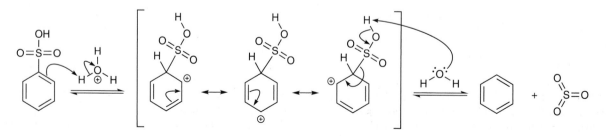

4.31 This ring bears a halogen substituent (Cl). Halogens are deactivators, but they are *ortho–para* directors.

4.32 This ring bears a hydroxyl group (OH), which is an activator and an *ortho–para* director.

4.33 This ring bears a nitro group (NO$_2$), which is a deactivator and a *meta* director.

4.34 The problem statement indicates that the substituent in this case is a deactivator. We have seen that all deactivators are *meta* directors, with the exception of the halogens. This substituent is not a halogen, so it is not an exception. If it is a deactivator, then it must be a *meta* director.

4.35 The problem statement indicates that the substituent in this case is an activator. We have seen that all activators are *ortho–para* directors. Therefore, this substituent must be an *ortho–para* director.

4.36 The problem statement indicates that the substituent in this case is a deactivator. We have seen that all deactivators are *meta* directors, with the exception of the halogens. This substituent is not a halogen, so it is not an exception. To be clear, this substituent does indeed contain halogens (three chlorine atoms), but none of those chlorine atoms are connected directly to the aromatic ring, so there are no resonance effects here. There are only inductive effects here, from each of the three chlorine atoms, which add together to give a powerful electron-withdrawing effect. Since this substituent (CCl$_3$) is a deactivator, it must be a *meta* director.

4.37 The problem statement indicates that the substituent in this case is an activator. We have seen that all activators are *ortho–para* directors. Therefore, this substituent must be an *ortho–para* director.

4.39 The reagents (nitric acid and sulfuric acid) indicate a nitration reaction, which means that a nitro group is installed on the aromatic ring. In order to draw the product(s), we must consider where the nitro group will be installed. That is, we must consider the existing substituent and its directing effects. In this case, the existing substituent is Br, which is an *ortho–para* director. Therefore, we expect the nitration reaction to occur at either the *ortho* position or the *para* position, giving two products, as shown:

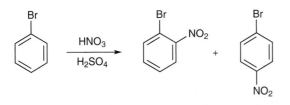

4.40 The reagents indicate a Friedel–Crafts alkylation reaction, so a methyl group is installed on the aromatic ring. In order to draw the product(s), we must consider where the methyl group will be installed. That is, we must consider the existing substituent and its

directing effects. In this case, the existing substituent is a methyl group, which is an *ortho–para* director. Therefore, we expect the methylation reaction to occur at either the *ortho* position or the *para* position, giving two products, as shown:

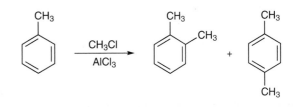

4.41 The reagents indicate a Friedel–Crafts acylation reaction, which means that an acyl group is installed on the aromatic ring. In order to draw the product(s), we must consider where the acyl group will be installed. That is, we must consider the existing substituent and its directing effects. In this case, the existing substituent is Br, which is an *ortho–para* director. Therefore, we expect the acylation reaction to occur at either the *ortho* position or the *para* position, giving two products, as shown:

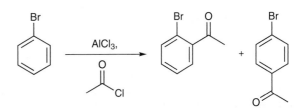

4.42 The reagents (nitric acid and sulfuric acid) indicate a nitration reaction, which means that a nitro group is installed on the aromatic ring. In order to draw the product(s), we must consider where the nitro group will be installed. That is, we must consider the existing substituent and its directing effects. The problem statement indicates that the existing substituent is a deactivator. Therefore, we expect that a nitro group will be installed at the *meta* position:

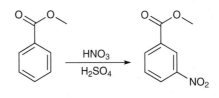

4.43 The reagents (Br$_2$ and AlBr$_3$) indicate a bromination reaction, which means that a bromine atom will be installed on the aromatic ring. In order to draw the product(s), we must consider where the bromine atom will be installed. That is, we must consider the existing substituent and its directing effects. The problem statement indicates that the existing substituent is a deactivator. Therefore, we expect that a bromine atom will be installed at the *meta* position:

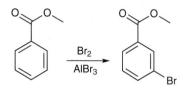

4.44 The reagents (Cl$_2$ and AlCl$_3$) indicate a chlorination reaction, which means that a chlorine atom will be installed on the aromatic ring. In order to draw the product(s), we must consider where the chlorine atom will be installed. That is, we must consider the existing substituent and its directing effects. The problem statement indicates that the existing substituent is an activator. Therefore, we expect that a chlorine atom will be installed either at the *ortho* position or at the *para* position:

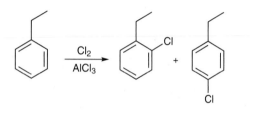

4.46 This aromatic ring has two groups: a strong deactivator and a strong activator. The deactivator is a *meta* director, while the activator is an *ortho-para* director. When a deactivator competes with an activator, the activator will control the directing effects. Therefore, we expect the following directing effects (*ortho* and *para* to the strong activator):

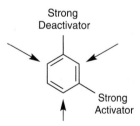

4.47 This aromatic ring has two groups: a strong activator and a weak activator. For electrophilic aromatic substitution reactions, the directing effects are controlled by the strongest activator. Therefore, we expect the directing effects to be *ortho* and *para* to the strong activator. In this case, the *para* position (relative to the strong activator) is already occupied, so we expect the following two locations to be the most reactive toward electrophilic aromatic substitution: (Note that substitution at either position gives the same product.)

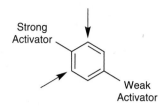

4.48 This aromatic ring has two groups: a strong activator and a strong deactivator. The activator is an *ortho–para* director, while the deactivator is a *meta* director. When an activator competes with a deactivator, the activator will control the directing effects (although there is no competition in this case, since they both direct to the same locations). Therefore, we expect the locations that are *ortho* to the strong activator to be the most reactive toward electrophilic aromatic substitution (the position that is *para* to the strong activator is already occupied). Note that substitution at either position gives the same product.

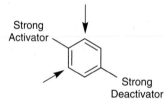

4.49 This aromatic ring has two groups: a strong activator and a strong deactivator. The activator is an *ortho–para* director, while the deactivator is a *meta* director. When an activator competes with a deactivator, the activator will control the directing effects (although in this case, they are not competing with each other because they both direct to the same locations). Therefore, we expect the locations that are *ortho* and *para* to the strong activator to be the most reactive toward electrophilic aromatic substitution (one of the positions that is *ortho* to the strong activator is already occupied):

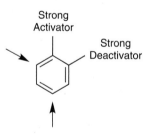

4.50 This aromatic ring has two groups: a strong activator and a weak activator. For electrophilic aromatic substitution reactions, the directing effects are controlled by the strongest activator. Therefore, we expect the directing effects to be *ortho* and *para* to the strong activator. In this case, one of the *ortho* positions (relative to the strong activator) is already occupied, so we expect the following two locations to be the most reactive toward electrophilic aromatic substitution:

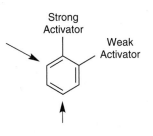

4.51 This aromatic ring has two groups: a weak activator and a strong deactivator. The activator is an *ortho–para* director, while the deactivator is a *meta* director. When an activator competes with a deactivator, the activator will control the directing effects (although there is no competition in this case, since they both direct to the same locations). Therefore, we expect the locations that are *ortho* and *para* to the activator to be the most reactive toward electrophilic aromatic substitution (one of the positions that is *ortho* to the activator is already occupied):

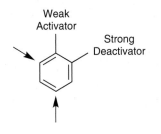

4.53 In this case, the substituent has a C=O bond connected directly to the aromatic ring, so this substituent is a moderate deactivator. Moderate deactivators are *meta* directors, so we expect the *meta* positions (indicated below) to be the most reactive toward electrophilic aromatic substitution.

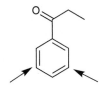

4.54 In this case, there is a halogen connected directly to the aromatic ring. We have seen that halogens are weak deactivators, but nevertheless, they are *ortho–para* directors. Therefore, we expect the *ortho* and *para* positions (indicated below) to be the most reactive toward electrophilic aromatic substitution.

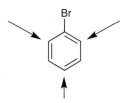

4.55 In this case, the substituent has a lone pair connected directly to the aromatic ring. This substituent is a strong activator and an *ortho–para* director. Therefore, we expect the *ortho* and *para* positions (indicated below) to be the most reactive toward electrophilic aromatic substitution.

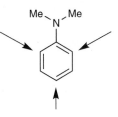

4.56 In this case, there is an alkyl group connected directly to the aromatic ring. Alkyl groups are weak activators and *ortho–para* directors. Therefore, we expect the *ortho* and *para* positions (indicated below) to be the most reactive toward electrophilic aromatic substitution.

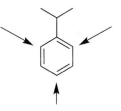

4.57 In this case, the substituent has a lone pair connected directly to the aromatic ring, but this lone pair is also participating in resonance outside of the ring. Therefore, this substituent is a moderate activator and an *ortho–para* director. As a result, we expect the *ortho* and *para* positions (indicated below) to be the most reactive toward electrophilic aromatic substitution.

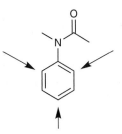

4.58 In this case, there is a nitro group connected directly to the aromatic ring, and nitro groups are strong deactivators. Strong deactivators are *meta* directors, so we expect the *meta* positions (indicated below) to be the most reactive toward electrophilic aromatic substitution.

4.59 In this case, there is a C≡N bond connected directly to the aromatic ring, so this substituent is a moderate deactivator. Moderate deactivators are *meta* directors, so we expect the *meta* positions (indicated below) to be the most reactive toward electrophilic aromatic substitution.

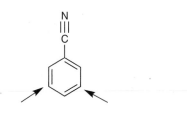

4.60 In this case, there is a C=N bond connected directly to the aromatic ring, so this substituent is a moderate deactivator. Moderate deactivators are *meta* directors, so we expect the *meta* positions (indicated below) to be the most reactive toward electrophilic aromatic substitution.

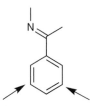

4.61 In this case, there is a CBr₃ group connected directly to the aromatic ring, and this group is a strong deactivator. Strong deactivators are *meta* directors, so we expect the *meta* positions (indicated below) to be the most reactive toward electrophilic aromatic substitution.

4.62 In this case, there is a C=O bond connected directly to the aromatic ring, so this substituent is a moderate deactivator. Moderate deactivators are *meta* directors, so we expect the *meta* positions (indicated below) to be the most reactive toward electrophilic aromatic substitution.

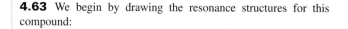

4.63 We begin by drawing the resonance structures for this compound:

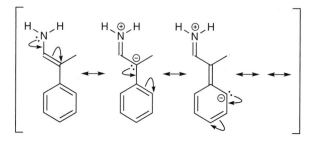

The resonance structures show that the lone pair on the nitrogen atom is delocalized, and the electron density is donated throughout the ring, which makes the ring electron-rich. The effect is similar to the effect observed when the lone pair is connected directly to the ring, so this substituent is a strong activator.

4.65 The reagents indicate a bromination reaction, so a bromine atom will be installed on the ring. The starting aromatic ring is monosubstituted, and the substituent is an alkyl group, which is a weak activator. Activators are *ortho–para* directors, so we expect that the bromine atom will be installed either at the *ortho* position or at the *para* position, giving the following products:

4.66 The reagents indicate a Friedel–Crafts alkylation reaction, so a methyl group will be installed on the ring. The starting aromatic ring is monosubstituted, and the substituent is a halogen, which is an *ortho–para* director. Therefore, we expect that the methyl group will be installed either at the *ortho* position or at the *para* position, giving the following products:

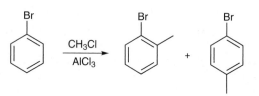

4.67 The reagents indicate a bromination reaction (a Lewis acid is not necessary here because the aromatic ring is moderately activated toward bromination), so a bromine atom will be installed on the ring. The starting aromatic ring is monosubstituted, and the substituent is an *ortho–para* director, so we expect that the bromine atom will be installed either at the *ortho* position or at the *para* position, giving the following products:

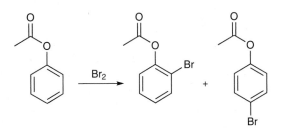

4.68 The reagents indicate a Friedel–Crafts acylation reaction, so an acyl group will be installed on the ring. The starting aromatic ring is monosubstituted, and the substituent is an ethyl group, which is an *ortho–para* director. Therefore, we expect that the acyl group will be installed either at the *ortho* position or at the *para* position, giving the following products:

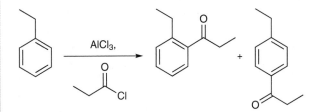

4.69 The reagents indicate a nitration reaction, so a nitro group will be installed on the ring. The starting aromatic ring is monosubstituted, and the substituent is a nitro group, which is a *meta* director. Therefore, we expect that the nitro group will be installed at the *meta* position, giving just one product:

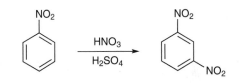

4.71 This aromatic ring has two substituents. The ester group is a moderate activator and an *ortho–para* director, while the methyl group is a weak activator and an *ortho–para* director. When a moderate activator competes with a weak activator during an electrophilic aromatic substitution, the more powerful activator (in this case, the moderate activator) controls the directing effects. The reagents indicate a nitration reaction, so we expect that a nitro group will be installed either at the position that is *ortho* to the ester or at the position that is *para* to the ester.

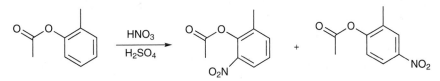

4.72 This aromatic ring has two substituents. The *tert*-butyl group is a weak activator and an *ortho–para* director, while the nitro group is a strong deactivator and a *meta* director. When an *ortho–para* director competes with a *meta* director during an electrophilic aromatic substitution, the *ortho–para* director controls the directing effects. The reagents indicate a chlorination reaction, so we expect that a chlorine atom will be installed either at a position that is *ortho* to the *tert*-butyl group (and there are two such positions) or at the position that is *para* to the *tert*-butyl group. This should give three products shown below, although we will see in the next section that one of these products (shown parenthetically) is actually a very minor product:

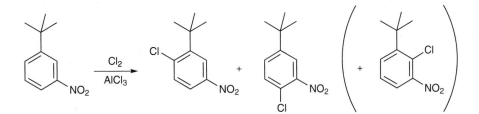

4.73 This aromatic ring has two substituents. The methoxy group is a moderate activator and an *ortho–para* director, while the chlorine atom is a weak deactivator and an *ortho–para* director. Since they are both *ortho–para* directors, the directing effects will be controlled by the more powerful activator (in this case, the methoxy group). The reagents indicate a Friedel–Crafts alkylation reaction, so we expect that a methyl group will be installed either at the position that is *ortho* to the methoxy group or at the position that is *para* to the methoxy group, giving the following two products:

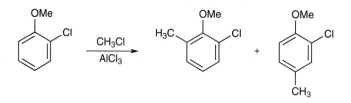

4.74   This aromatic ring has two substituents. One group is an alkoxy group (OR), which is a moderate activator and an *ortho–para* director. The other group is a carbonyl (C=O) group, which is a moderate deactivator and a *meta* director. The directing effects will be controlled by the *ortho–para* director (although in this case, there is not really a competition, because both substituents direct to the same two positions). The reagents indicate a Friedel–Crafts acylation reaction, so we expect that an acyl group will be installed either at the position that is *ortho* to the alkoxy group or at the position that is *para* to the alkoxy group:

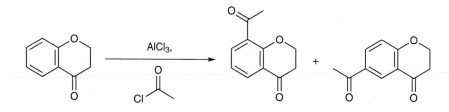

4.76   The starting material is a monosubstituted aromatic ring, and the substituent is a moderate activator and an *ortho–para* director. So, if we perform an electrophilic aromatic substitution reaction, we would expect the reaction to occur primarily at the *para* position. The desired transformation involves installing a group at the *ortho* position, so a blocking group will be necessary. First, we perform sulfonation to "block" the *para* position. Next, we perform the desired reaction (in this case, a nitration reaction), and finally, we remove the blocking group with dilute sulfuric acid to give the desired product.

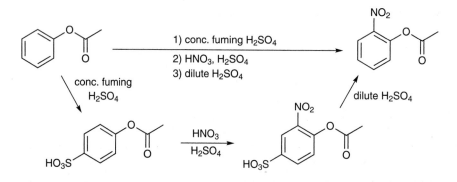

4.77   The starting material is a monosubstituted aromatic ring, and the substituent (a nitro group) is a strong deactivator and a *meta* director. So, if we perform an electrophilic aromatic substitution reaction, we would expect the reaction to occur primarily at the *meta* position. The desired transformation involves installing a group at the *meta* position, so a blocking group is not useful here. The bromine atom can be directly installed in the correct location by treating nitrobenzene with bromine and aluminum tribromide:

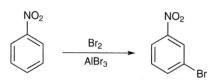

4.78 The starting material is a monosubstituted aromatic ring, and the substituent (an alkoxy group) is a moderate activator and an *ortho–para* director. So, if we perform an electrophilic aromatic substitution reaction, we would expect the reaction to occur primarily at the *para* position. The desired transformation involves installing a group at the *ortho* position, so a blocking group will be necessary. First, we perform sulfonation to "block" the *para* position. Next, we perform the desired reaction (in this case, a Friedel–Crafts acylation reaction), and finally, we remove the blocking group with dilute sulfuric acid to give the desired product.

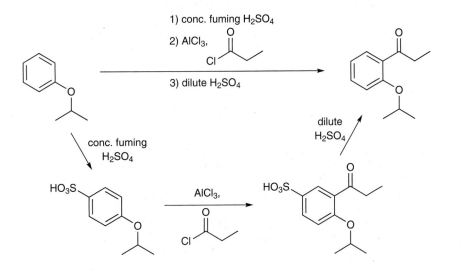

4.79 The starting material is a monosubstituted aromatic ring, and the substituent (an *n*-propyl group) is a weak activator and an *ortho–para* director. So, if we perform an electrophilic aromatic substitution reaction, we would expect the reaction to occur primarily at the *para* position. The desired transformation involves installing a group at the *para* position, so a blocking group is not needed here. A methyl group can be installed directly in the correct location via a Friedel–Crafts alkylation:

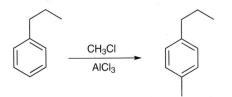

4.80 The starting material is a monosubstituted aromatic ring, and the substituent (an *n*-propyl group) is a weak activator and an *ortho–para* director. So, if we perform an electrophilic aromatic substitution reaction, we would expect the reaction to occur primarily at the *para* position. The desired transformation involves installing a group at the *ortho* position, so a blocking group will be necessary. First, we perform sulfonation to "block" the *para* position. Next, we perform the desired reaction (in this case, chlorination to install a chlorine atom at the *ortho* position), and finally, we remove the blocking group with dilute sulfuric acid to give the desired product.

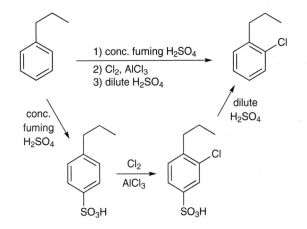

4.82 The reagents (nitric acid and sulfuric acid) indicate that this is a nitration reaction. The aromatic ring has one substituent (Cl), which is an *ortho–para* director. So we expect nitration to occur at either the *ortho* position or the *para* position, to give two products. Because of steric factors, the *para* product is the major product.

Major product

4.83 The reagents (nitric acid and sulfuric acid) indicate that this is a nitration reaction. The aromatic ring has one substituent, which is an *ortho–para* director. So we expect nitration to occur at either the *ortho* position or the *para* position, to give two products. Because of steric factors, the *para* product is the major product.

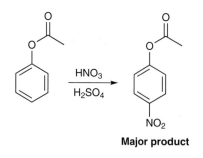

Major product

4.84 The reagents (Cl₂ and AlCl₃) indicate that a chlorine atom will be installed. The aromatic ring has two substituents, both of which are weak activators and *ortho–para* directors. So chlorination is favored at all positions (either *ortho* to the methyl group or *ortho* to the isopropyl group). Both products are expected, but because of steric factors, the major product results from chlorination at the position that is *ortho* to the methyl group. That position is more accessible than the position that is *ortho* to the larger isopropyl group.

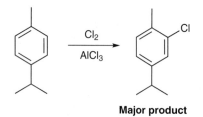

Major product

4.85 The reagent (concentrated fuming sulfuric acid) indicates a sulfonation reaction. The aromatic ring has two substituents, both of which are weak activators and *ortho–para* directors. Both groups direct to the same three locations.

When we compare these locations, the location directly between the alkyl groups is sterically less accessible, so the reaction does not

occur there rapidly. The reaction can occur at either of the other two locations, giving rise to two products. Because of steric factors, the major product results from sulfonation at the position that is *ortho* to the methyl group (as shown below). That position is more accessible than the position that is *ortho* to the larger isopropyl group.

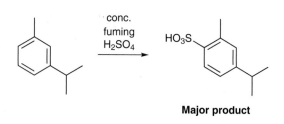

Major product

4.86 The reagents indicate a Friedel–Crafts alkylation reaction, so we expect that a methyl group will be installed. The aromatic ring has two substituents. One group is a carbonyl (C=O) group, which is a moderate deactivator and a *meta* director. The other group is an ester group, which is a moderate activator and an *ortho–para* director. The directing effects will be controlled by the *ortho–para* director, although in this case, there is not really a competition, because both substituents direct to the same two positions:

The reaction will occur more rapidly at the position on the left because it is sterically more accessible than the position on the right. Therefore, we expect the following major product:

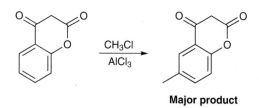

Major product

4.87 The reagents (Br₂ and AlBr₃) indicate that a bromine atom will be installed on the ring. This aromatic ring has four substituents, all of which are weak activators and *ortho–para* directors. These groups direct to the two available positions on the ring:

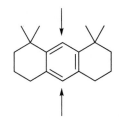

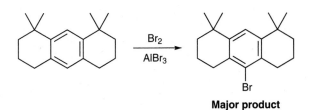

product results from bromination at the more sterically accessible position:

Major product

The upper position is sterically hindered (because of the methyl groups), while the lower position is more accessible. The major

4.89 We must install an isopropyl group and a chlorine atom on the aromatic ring. The isopropyl group can be installed via a Friedel–Crafts alkylation, and the chlorine atom can be installed by treating the aromatic ring with Cl_2 in the presence of catalytic $AlCl_3$.

Now that we know how to install each substituent individually, we must consider the order of events. Both substituents are *ortho–para* directors, so theoretically, either one can be installed first. But consider the fact that an isopropyl group is an activator, while chlorine is a deactivator. To maximize the efficiency of our synthesis, it makes more sense to install the isopropyl group first, so that we can then take advantage of the enhanced reactivity of isopropyl benzene (relative to benzene) to install Cl.

If we use the strategy described above (install the isopropyl group first and then try to install the chlorine atom next), we would expect to obtain both *ortho* and *para* products, although the reaction will occur primarily at the *para* position because of steric effects. The desired transformation involves installing a group at an *ortho* position, so a blocking group will be necessary here. After installing the isopropyl group, we perform sulfonation to "block" the *para* position. Next, we perform the desired reaction (installing Cl in the position that is *ortho* to the isopropyl group), and finally, we remove the blocking group with dilute sulfuric acid to give the desired product:

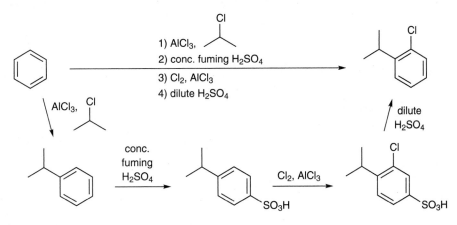

4.90 We must install a nitro group and a bromine atom on the aromatic ring. The nitro group can be installed with nitric acid and sulfuric acid, while the bromine atom can be installed with Br_2 in the presence of catalytic $AlBr_3$.

Now that we know how to install each substituent individually, we must consider the order of events. Nitro groups are *meta* directors, while Br is an *ortho–para* director. Since the two substituents must be installed so that they are *ortho* to each other, we conclude that we must install Br first, followed by the nitro group.

If we use the strategy described above (install Br first and then try to install the nitro group next), we would expect to obtain both *ortho* and *para* products, although the reaction will occur primarily at the *para* position because of steric effects. The desired transformation involves installing a group at an *ortho* position, so a blocking group will be necessary here. After installing the bromine atom, we perform sulfonation to "block" the *para* position. Next, we perform the desired reaction (installing a nitro group in the position that is *ortho* to Br), and finally, we remove the blocking group with dilute sulfuric acid to give the desired product:

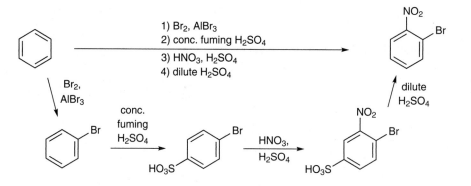

4.91 We must install a *tert*-butyl group and a methyl group on the aromatic ring. Each of these groups can be installed via a Friedel–Crafts alkylation.

Now that we know how to install each substituent individually, we must consider the order of events. Both substituents are *ortho–para* directors, so theoretically, either one can be installed first. But consider steric effects. If we install the methyl group first, any reaction we performed next will produce a mixture of products. The methyl group is not large enough to clearly favor one position over the other (*ortho* vs. *para*), and a mixture of products is undesirable. However, if we install the *tert*-butyl group first, there is a clear preference for the next reaction to occur at the *para* position, and we will exploit that preference to obtain the desired regiochemical outcome.

We first install the *tert*-butyl group, and we then use its steric bulk to perform sulfonation at the *para* position, thereby blocking that position. Next, we install Cl in the position that is *ortho* to the *tert*-butyl group, and finally, we remove the blocking group with dilute sulfuric acid to give the desired product:

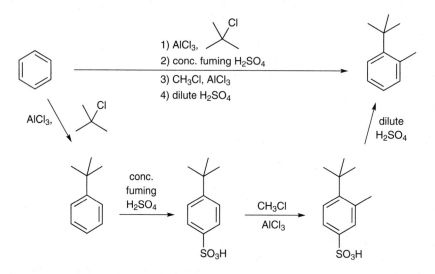

4.92 We must install an *n*-propyl group and a nitro group on the aromatic ring. A nitro group can be installed with nitric acid and sulfuric acid. The *n*-propyl group requires more than one step because it cannot be installed directly via a Friedel–Crafts alkylation, as that would produce a mixture of products (because of rearrangement). To install an *n*-propyl group, we must first perform a Friedel–Crafts acylation, and then reduce the resulting ketone with a Clemmensen reduction.

Now that we know how to install each substituent individually, we must consider the order of events. The nitro group is a *meta* director, while the *n*-propyl group is an *ortho–para* director, so we must install the *n*-propyl group first. If we try to install the nitro group first, it would direct the next reaction to the *meta* position, which is not the correct location.

If we use the strategy described above (install the *n*-propyl group first and then install the nitro group next), we would expect to obtain both *ortho* and *para* products, although the reaction will occur primarily at the *para* position because of steric effects. This is the desired product, so a blocking group is not needed here. The synthesis is summarized below. We perform a Friedel–Crafts acylation, followed by a Clemmensen reduction, followed by nitration:

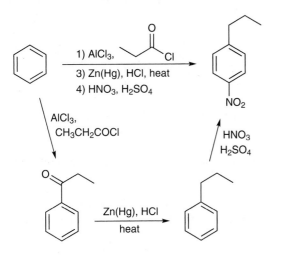

4.93 We must install a *tert*-butyl group and a nitro group on the aromatic ring. The *tert*-butyl group can be installed via a Friedel–Crafts alkylation, while the nitro group can be installed with nitric acid and sulfuric acid.

Now that we know how to install each substituent individually, we must consider the order of events. The *tert*-butyl group is an *ortho*–*para* director, while the nitro group is a *meta* director, so we must install the *tert*-butyl group first. If we try to install the nitro group first, it would direct the next reaction to the *meta* position, which is not the correct location.

If we use the strategy described above (install the *tert*-butyl group first and then install the nitro group next), we would expect to obtain primarily the *para* product because of steric effects. This is the desired product, so a blocking group is not needed here. The synthesis is summarized below. We perform a Friedel–Crafts alkylation, followed by nitration:

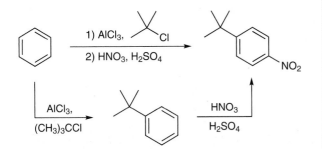

4.94 We must install an *n*-butyl group and a bromine atom on the aromatic ring. The bromine atom can be installed in one step by treating an aromatic ring with Br_2 and $AlBr_3$. But the *n*-butyl group cannot be installed directly via a Friedel–Crafts alkylation, as that would produce a mixture of products (because of rearrangement). To install an *n*-butyl group, we must first perform a Friedel–Crafts acylation, and then reduce the resulting ketone with a Clemmensen reduction.

Now that we know how to install each substituent individually, we must consider the order of events. An *n*-butyl group is an *ortho*–*para* director, and a bromine atom is also an *ortho*–*para* director. This is an issue, because we need to install the substituents so that they are ultimately *meta* to each other. How do we accomplish this?

Notice that the *n*-butyl group is installed in two steps, and if we pause after the first step (after acylation), we can capitalize on the *meta*-directing effects of the carbonyl group before we reduce it. This allows us to install the bromine atom at the *meta* position. This strategy gives the desired product, as shown below:

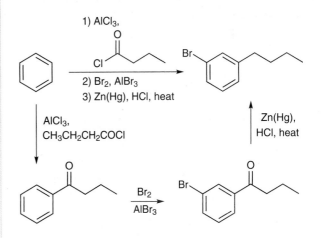

4.95 We must install a *tert*-butyl group and a bromine atom on the aromatic ring. A *tert*-butyl group can be installed via a Friedel–Crafts alkylation, and a bromine atom can be installed with Br_2 and $AlBr_3$.

Now that we know how to install each substituent individually, we must consider the order of events. Both substituents are *ortho*–*para* directors, so theoretically, either one can be installed first. But consider the fact that a *tert*-butyl group is an activator, while bromine is a deactivator. To maximize the efficiency of our synthesis, it makes more sense to install the a *tert*-butyl group first, so that we can then take advantage of the enhanced reactivity of *tert*-butylbenzene (relative to benzene) to install Br. Furthermore, the bulky *tert*-butyl group creates a large distinction between the *ortho* and *para* positions, which will now be exploited.

We first install the *tert*-butyl group, and we then use its steric bulk to perform sulfonation at the *para* position, thereby blocking that position. Next, we install Br in the position that is *ortho* to the *tert*-butyl group, and finally, we remove the blocking group with dilute sulfuric acid to give the desired product:

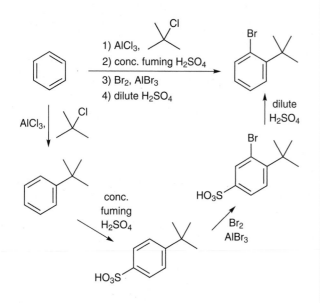

4.96 We must install a nitro group and a bromine atom on the aromatic ring. The nitro group can be installed with nitric acid and sulfuric acid, while the bromine atom can be installed with Br_2 in the presence of catalytic $AlBr_3$.

Now that we know how to install each substituent individually, we must consider the order of events. Nitro groups are *meta* directors, while Br is an *ortho–para* director. Since the two substituents must be installed so that they are *meta* to each other, we conclude that we must install the nitro group first, followed by Br:

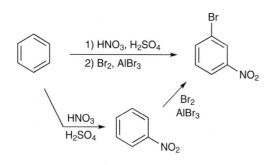

4.97 We must install an *n*-propyl group and a chlorine atom on the aromatic ring. A chlorine atom can be installed with Cl_2 and $AlCl_3$. The *n*-propyl group requires more than one step because it cannot be installed directly via a Friedel–Crafts alkylation, as that would produce a mixture of products (because of rearrangement). To install an *n*-propyl group, we must first perform a Friedel–Crafts

acylation, and then reduce the resulting ketone with a Clemmensen reduction.

Now that we know how to install each substituent individually, we must consider the order of events. Both substituents are *ortho–para* directors, so theoretically, either one can be installed first. But consider the fact that an *n*-propyl group is an activator, while chlorine is a deactivator. To maximize the efficiency of our synthesis, it makes more sense to install the *n*-propyl group first, so that we can then take advantage of the enhanced reactivity of *n*-propylbenzene (relative to benzene) to install Cl.

If we use the strategy described above (install the *n*-propyl group first and then install the chlorine atom next), we would expect to obtain both *ortho* and *para* products, although the reaction will occur primarily at the *para* position because of steric effects. The desired transformation involves installing a group at an *ortho* position, so a blocking group will be necessary here. After installing the *n*-propyl group via Friedel–Crafts acylation followed by a Clemmensen reduction, we perform sulfonation to "block" the *para* position. Next, we perform the desired reaction (installing a chlorine atom in the position that is *ortho* to the *n*-propyl group), and finally, we remove the blocking group with dilute sulfuric acid to give the desired product:

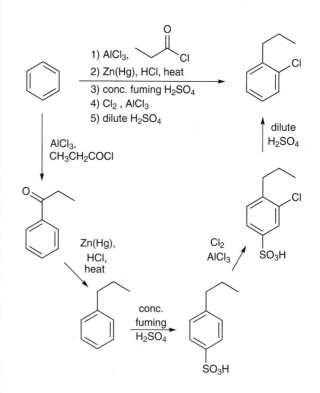

4.98 Three of the structures are activated toward electrophilic aromatic substitution (relative to benzene). A dimethyl amino [—$N(CH_3)_2$] group is a strong activator, a methoxy group is a moderate activator, and an isopropyl group is a weak activator.

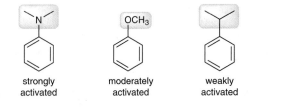

strongly activated   moderately activated   weakly activated

The other three structures are deactivated toward electrophilic aromatic substitution (relative to benzene). A nitro group is a strong deactivator, an aldehyde group is a moderate deactivator, and halogens (such as Cl) are weak deactivators.

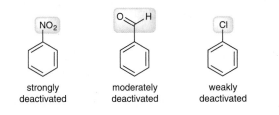

strongly deactivated   moderately deactivated   weakly deactivated

Based on these designations, all six structures are arranged below in order of reactivity toward electrophilic aromatic substitution (from least reactive to most reactive):

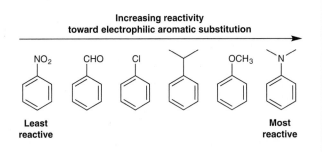

4.99 The starting material is toluene, which has a methyl group connected to an aromatic ring. Alkyl groups are activators, so toluene is activated toward electrophilic aromatic substitution (relative to benzene). Recall that all activators are *ortho-para* directors, so we expect this compound to undergo electrophilic aromatic substitution predominantly at the *ortho* and *para* positions:

The reagents (CH₃Cl and AlCl₃) indicate a Friedel–Crafts alkylation reaction, which will install a methyl group in the *ortho* or

para positions. The two *ortho* positions are equivalent, so substitution at either of those locations would generate the same product. We, therefore, expect the following two products:

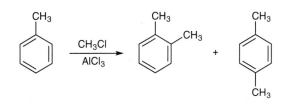

In this case (where the starting material is toluene, which has a very small alkyl group), there is not a strong preference for one product over the other, and you won't be able to predict which of these products will be the major product.

4.100 The reagents (Br₂ and AlBr₃) indicate a bromination reaction in which a bromine atom will be installed on an aromatic ring (via electrophilic aromatic substitution). The starting material has two aromatic rings, so we must decide which ring is more reactive. The ring on the left is deactivated by its one substituent (a carbonyl group),

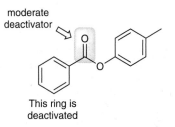

while the ring on the right is activated by both of its substituents:

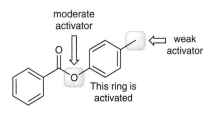

The reaction is expected to occur at the more activated ring (the ring on the right), so now we must determine the exact location where the reaction will occur. This ring has two substituents: a moderate activator and a weak activator. The directing effects will be controlled by the stronger activator (in this case, the moderate activator, rather than the weak activator), so we predict that the reaction will occur at the positions that are *ortho* or *para* to the moderate activator. The *para* position is already occupied (by the methyl group), so the reaction will occur at one of the *ortho* positions:

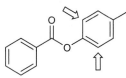

These two *ortho* positions are equivalent, so installation of a bromine atom in either position generates the same product:

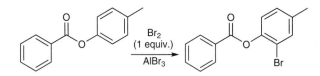

4.101 The reagents (nitric acid and sulfuric acid) indicate a nitration reaction, in which a nitro group is installed on the aromatic ring (electrophilic aromatic substitution). The starting material has two substituents: a moderate deactivator and a weak activator:

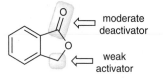

In this case, the directing effects are controlled by the weak activator (because an activator, even a weak one, always beats a deactivator), so we expect the reaction to occur at the positions that are *ortho* or *para* to the activator. One of these three positions is already occupied, leaving only two choices:

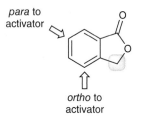

The reaction is expected to occur primarily at the *para* position because the *ortho* position is more sterically hindered.

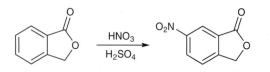

4.102 The reagents (nitric acid and sulfuric acid) indicate a nitration reaction, in which a nitro group is installed on the aromatic ring (electrophilic aromatic substitution). The starting material has three substituents: a weak activator, a strong deactivator, and a moderate activator:

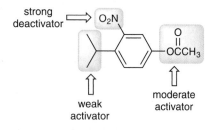

The directing effects are controlled by the stronger activator, which is the ester group, so we expect the reaction to occur at the positions that are *ortho* or *para* to the ester group. One of these three positions is already occupied, leaving only two choices:

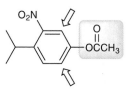

One of these two positions (the position in between the nitro group and the ester group) is too sterically hindered, and the reaction is less likely to occur there. Therefore, we expect the following major product:

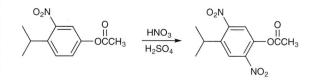

4.103 The reagents (Br$_2$ and AlBr$_3$) indicate a bromination reaction in which a bromine atom will be installed on the aromatic ring (via electrophilic aromatic substitution). The starting material has three substituents: a moderate activator, a strong deactivator, and a weak deactivator:

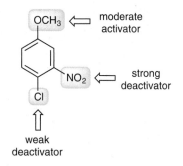

The directing effects are controlled by the activator (the methoxy group), so we expect the reaction to occur at the positions that are *ortho* or *para* to the methoxy group. One of these three positions is already occupied, leaving only two choices:

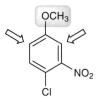

One of these two positions (the position in between the methoxy group and the nitro group) is too sterically hindered, and the reaction is less likely to occur there. Therefore, we expect the following major product:

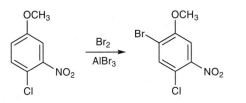

4.104 The reagents (CH_3Cl and $AlCl_3$) indicate a Friedel–Crafts alkylation reaction, which will install a methyl group on an aromatic ring. The starting material is a disubstituted benzene ring where both substituents are alkyl groups. Alkyl groups are activators, so this starting material is activated toward electrophilic aromatic substitution (relative to benzene). Recall that all activators are *ortho-para* directors, and all four unoccupied positions are either *ortho* or *para* to one of the alkyl groups, so we expect all four positions to be activated:

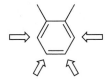

Two of these positions (the two positions that are closer to the methyl groups) are more sterically hindered, so it is less likely for the reaction to occur at those positions. The other two positions are equivalent. That is, installation of a methyl group in either position generates the same major product, shown here:

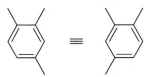

4.105 The reagents (CH_3Cl and $AlCl_3$) indicate a Friedel–Crafts alkylation reaction, which will install a methyl group on an aromatic ring. The starting material is a disubstituted benzene ring where

both substituents are alkyl groups. Alkyl groups are activators, so this starting material is activated toward electrophilic aromatic substitution (relative to benzene). Recall that all activators are *ortho-para* directors, so in this case, both substituents direct to the same three locations:

One of these positions (the position in between the two methyl groups) is more sterically hindered, so we don't expect the reaction to occur at that position. The other two positions are equivalent. That is, installation of a methyl group in either position generates the same major product, shown here:

4.106 The reagents (CH_3Cl and $AlCl_3$) indicate a Friedel–Crafts alkylation reaction, which will install a methyl group on an aromatic ring. The starting material is a disubstituted benzene ring where both substituents are alkyl groups. Alkyl groups are activators, so this starting material is activated toward electrophilic aromatic substitution (relative to benzene). Recall that all activators are *ortho-para* directors, and each of the four unoccupied positions is *ortho* to one of the substituents, so we expect all four positions to be activated:

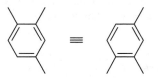

All four of these positions are equivalent. That is, installation of a methyl group in any one of these four positions will generate the same product, shown here:

4.107 The reagents (nitric acid and sulfuric acid) indicate a nitration reaction in which a nitro group will be installed on an aromatic ring (via electrophilic aromatic substitution). The starting material has two aromatic rings, so we must decide which ring is more reactive. The ring on the left is deactivated by its one substituent (a carbonyl group),

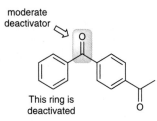

moderate
deactivator

This ring is
deactivated

while the ring on the right is deactivated by both of its substituents (two carbonyl groups):

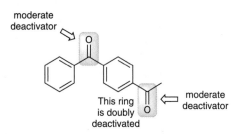

moderate
deactivator

This ring
is doubly
deactivated

moderate
deactivator

As such, both rings are deactivated. If we force the reaction using heat, we expect the reaction to occur on the ring that is less deactivated (the ring on the left, which is only deactivated by one group, rather than two groups): Since this ring has a deactivating substituent, we expect the reaction to occur at the *meta* position:

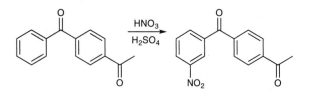

HNO₃
H₂SO₄

NO₂

Notice that the two *meta* positions are equivalent, so nitration at either position generates the same product.

4.108 The starting material has an aromatic ring and a halide leaving group. Upon treatment with catalytic AlCl₃, the starting material can undergo an intramolecular Friedel–Crafts reaction, giving the product shown below.

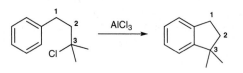

AlCl₃

4.109 The first set of reagents (nitric acid and sulfuric acid) indicate a nitration reaction in which a nitro group will be installed on the aromatic ring (via electrophilic aromatic substitution). The starting material is a disubstituted benzene ring where both substituents are alkyl groups. Alkyl groups are activators, so this starting material is activated toward electrophilic aromatic substitution (relative to benzene). Recall that all activators are *ortho-para* directors, so both substituents direct to the same three locations:

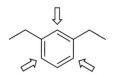

One of these positions (the position in between the two ethyl groups) is more sterically hindered, so we don't expect the reaction to occur at that position. The other two positions are equivalent. That is, installation of a nitro group in either position generates the same product, shown here:

NO₂   ≡   O₂N

This trisubstituted ring is then treated with molecular chlorine and aluminum trichloride to give a chlorination reaction. Recall that a nitro group is a strong deactivator, while alkyl groups are weak activators. Since activators (not deactivators) determine the directing effects, we expect chlorination to occur at the one of the positions that is *ortho* or *para* to the ethyl groups. There are three such positions, but one of those positions is occupied by the nitro group, leaving two positions to consider:

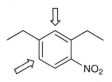

NO₂

One of these positions (the one in between the two ethyl groups) is more sterically hindered, so the reaction is less likely to occur at that position. We, therefore, expect the following major product:

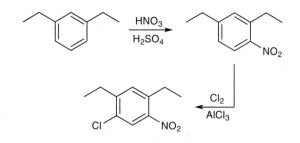

HNO₃
H₂SO₄

NO₂

Cl₂
AlCl₃

Cl   NO₂

4.110 We must install an acyl group, a *tert*-butyl group, and a bromine atom on an aromatic ring. The acyl group can be installed via a Friedel–Crafts acylation, the *tert*-butyl group can be installed via a Friedel–Crafts alkylation, and a bromine atom can be installed via bromination with molecular bromine and aluminum tribromide. Now we must consider the order of events.

Installation of the bromine atom can be the last step, because the *tert*-butyl group will direct the incoming bromine atom to the correct location.

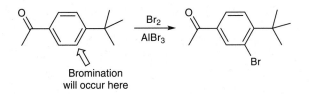

Bromination
will occur here

Now we must decide which of the other two groups to install first. If we install the *tert*-butyl group first, then that *tert*-butyl group will direct the subsequent reaction (Friedel–Crafts acylation) to occur in the correct location (*para* to the *tert*-butyl group). In contrast, installation of the acyl group first would lead to the wrong regiochemistry for the subsequent alkylation process (the acyl group would direct to the *meta* position). In addition, installation of the acyl group will deactivate the ring, making it difficult to achieve a Friedel–Crafts alkylation to install the *tert*-butyl group:

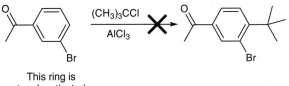

This ring is
too deactivated

In summary, we first install the *tert*-butyl group via a Friedel–Crafts alkylation, followed by a Friedel–Crafts acylation to install the acyl group in the *para* position. Finally, bromination gives the desired product:

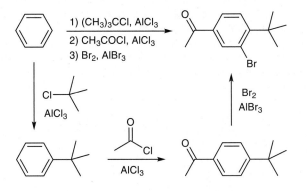

4.111 We must install an acyl group, a nitro group, and a bromine atom on an aromatic ring. The acyl group can be installed via a Friedel–Crafts acylation, the nitro group can be installed by treating with nitric acid and sulfuric acid, and a bromine atom can be installed via bromination with Br_2 and $AlBr_3$. Now let's consider the order of events.

All of the groups are *meta* to each other, which means that bromination must be the last step in our synthesis (otherwise, any subsequent reactions would be directed *ortho* and *para* to the bromine atom). In other words, we must first install the other two groups, both of which are *meta* directors. We cannot install the nitro group first, because then the ring would be too deactivated to perform a Friedel–Crafts reaction:

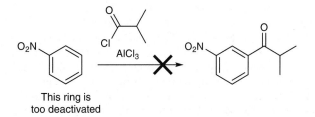

This ring is
too deactivated

Instead, we first install the acyl group, and then we use the directing effects of the acyl group to install the nitro group in the *meta* position (as desired). Finally, bromination gives the desired product:

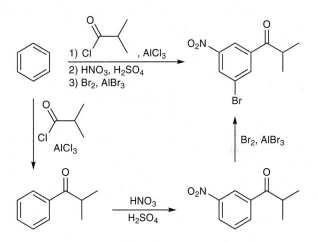

4.112 We must install a chlorine atom and an alkyl group on an aromatic ring. The chlorine atom can be installed by treating the aromatic ring with molecular chlorine and catalytic aluminum trichloride. The alkyl group cannot be installed directly with a Friedel–Crafts alkylation because a rearrangement is likely to occur:

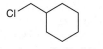

This primary alkyl halide **cannot**
be used for Friedel–Crafts alkylation
(rearrangement will occur)

Instead, the alkyl group must be installed via a two-step process: (1) installation of an acyl group via Friedel–Crafts acylation, followed by (2) Clemmensen reduction to reduce the carbonyl group.

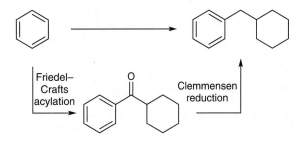

Now let's consider the order of events. That is, do we install the chlorine atom first, or do we install the alkyl group first? If we install the chlorine atom first, then that chlorine atom would direct the subsequent Friedel–Crafts reaction to occur at the wrong location (*ortho* or *para* to the chlorine atom, rather than *meta* to the chlorine atom). And if we install the alkyl group first, then once again, the alkyl group will direct chlorination to occur at an *ortho* or *para* position. In order to achieve the desired regiochemical outcome, we must take advantage of the two-step process for installing the alkyl group. After the first step of that process (installing an acyl group), the acyl group is a *meta* director, which will direct chlorination to the correct location.

In summary, we first install an acyl group, and then use the directing effects of that group to install a chlorine atom in the correct location. The last step of our synthesis is to reduce the acyl group with a Clemmensen reduction:

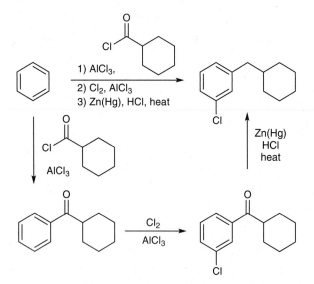

4.113 We must install two identical alkyl groups on the aromatic ring. Neither of these two groups can be installed directly with a Friedel–Crafts alkylation because a rearrangement is likely to occur:

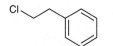

This primary alkyl halide ***cannot*** be used for Friedel–Crafts alkylation (rearrangement will occur)

Instead, each alkyl group must be installed via a two-step process: (1) installation of an acyl group via Friedel–Crafts acylation, followed by (2) Clemmensen reduction to reduce the carbonyl group.

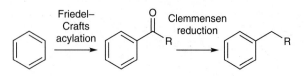

After the first alkyl group is installed, that alkyl group will direct the subsequent Friedel–Crafts reaction to occur at the desired location (*para* to the alkyl group). In summary, the following is a four-step synthesis of the desired product:

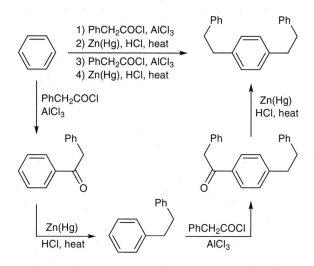

4.114 The starting material has both an aromatic ring and an isolated C=C bond. The isolated C=C bond (the alkene) is protonated in the presence of aqueous acid, giving a tertiary carbocation, shown below. This carbocation is an electrophile, while the aromatic ring can function as a nucleophile. So this intermediate has both an electrophilic center and a nucleophilic center, tethered to each other in one structure, which enables an intramolecular reaction in which the nucleophile attacks the electrophile (within the same molecule). This reaction is an electrophilic aromatic substitution reaction. The intermediate is a resonance-stabilized sigma complex, which is then deprotonated by the base that is present (H$_2$O) to give the product.

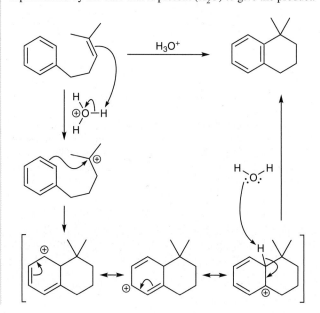

4.115 Trichloromethane will attack aluminum trichloride to give a Lewis acid–Lewis base complex, which can then lose aluminum tetrachloride to give a carbocation that can function as a very powerful electrophile. This carbocation then reacts with benzene in an electrophilic aromatic substitution reaction, which has two steps: (1) benzene attacks the carbocation to give a resonance-stabilized sigma complex, and (2) aluminum tetrachloride removes a proton from the sigma complex, generating $C_6H_5CHCl_2$.

This process is repeated to install a second and, finally, a third benzene ring on the central carbon atom, to give the product $(C_6H_5)_3CH$.

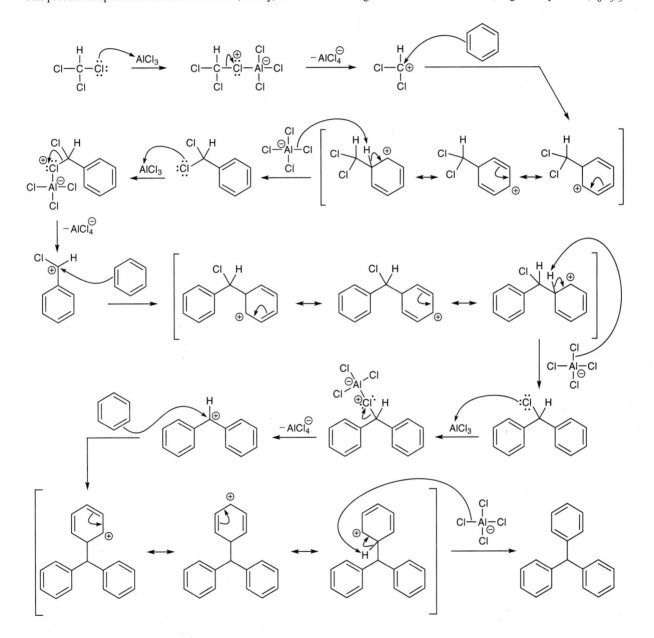

4.116 In the presence of sulfuric acid, a tertiary alcohol is protonated to give an oxonium ion, which then loses a leaving group (water) to give a tertiary carbocation intermediate. This carbocation is a powerful electrophile, and it will react with benzene in an electrophilic aromatic substitution reaction, which has two steps: (1) benzene attacks the carbocation to give a resonance-stabilized sigma complex, and (2) water (or HSO_4^-) functions as a base and removes a proton from the sigma complex, generating the product.

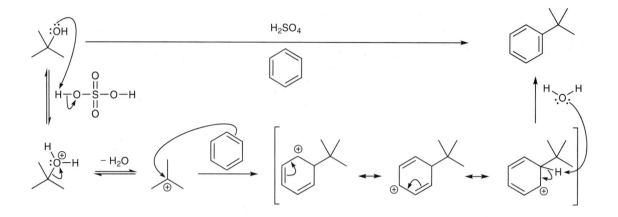

4.117 The starting material (benzenesulfonic acid) has a sulfonate (—SO₃H) group, which can be replaced by a proton upon treatment with dilute H₂SO₄ in H₂O:

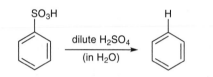

If instead, we treat benzenesulfonic acid with D₂SO₄ in D₂O, then we expect the sulfonate group to be replaced with a deuteron, rather than a proton:

4.118 The reagents indicate a bromination reaction, so we must decide where a bromine atom will be installed during this reaction. Of all the substituents, the OH group is the strongest activator, so the OH group will control the directing effects. Recall that all activators are *ortho-para* directors, so the OH group is directing to the following three locations:

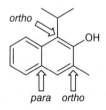

Note that all three of these locations are currently occupied, so the reaction cannot occur at any of these locations. However, there is a second aromatic ring, and the effects of the OH group extend to that ring as well. If we draw resonance structure of the starting material, we will see that the following five locations are electron-rich (δ−), as a result of resonance effects from the OH group:

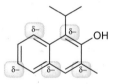

While three of these positions are occupied, two of these positions are available, and a bromination reaction can occur in either of these two locations.

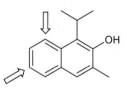

Of the two positions, we choose the one that is the least sterically hindered (no neighboring substituents), so we predict the following major product:

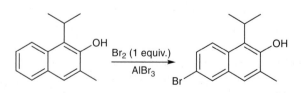

4.119 The desired product contains an acetal group, which can be made directly from the corresponding ketone, shown here:

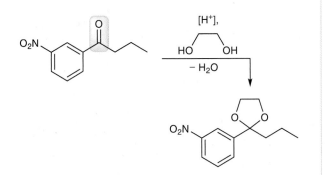

In order to make this ketone from benzene, we will need to install both a nitro group and an acyl group in a *meta* fashion. Both of these groups are *meta* directors, so we might be tempted to say that these two groups can be installed in either order. But recall that Friedel–Crafts reactions are not efficient on deactivated rings. If we install the nitro group first, the resulting ring (nitrobenzene) will be too deactivated to undergo a Friedel–Crafts acylation. So, we must first install the acyl group, and then we can install the nitro group. And finally, we convert the ketone into an acetal, as shown in the following synthesis:

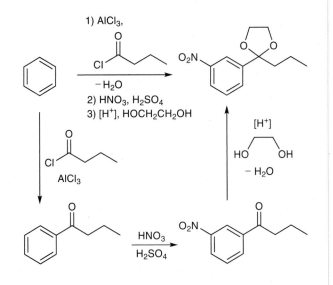

4.120 The desired product is an epoxide, which can be made by treating the corresponding alkene (called styrene) with a peroxy acid (RCO$_3$H). A commonly used peroxy acid is MCPBA

(*meta*-chloroperoxybenzoic acid) although others can also be used, including CH$_3$CO$_3$H.

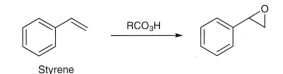

Styrene

We have not learned a direct method for converting benzene into styrene (directly installing a vinyl group on an aromatic ring), so we will have to accomplish this transformation in a number of steps:

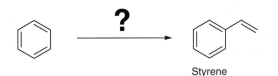

Styrene

First, we install an ethyl group on the ring, using a Friedel–Crafts alkylation, and then we use a radical bromination to install a bromine atom at the benzylic position. The resulting bromide is then treated with *tert*-butoxide to give an E2 reaction (*tert*-butoxide is used to disfavor the competing S$_N$2 reaction, although any strong base can be used). The resulting alkene (styrene) is then treated with a peroxy acid to give the desired product:

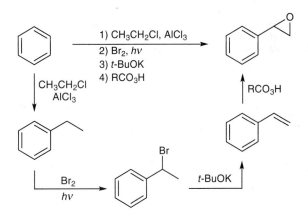

Alternatively, we can perform a Friedel–Crafts acylation, followed by reduction and then elimination to give styrene, and the last step of the synthesis would then be the same as above:

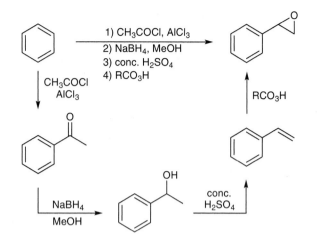

4.121 In compound **A**, the benzylic position (the position that is connected directly to the aromatic ring), highlighted below, is sp^3 hybridized:

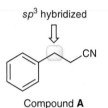

Compound **A**

As such, this substituent will behave much like an alkyl group—it is a weak activator. Since all activators are *ortho-para* directors, we expect compound **A** to undergo bromination at the *ortho* or *para* positions. Since the *ortho* position is sterically hindered, we expect bromination to occur predominantly at the *para* position:

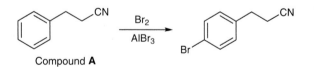

Compound **A**

In contrast, the benzylic position in compound **B** is sp^2 hybridized:

Compound **B**

In order to determine the effects of this substituent (whether it is an activator or a deactivator), we can draw resonance structures:

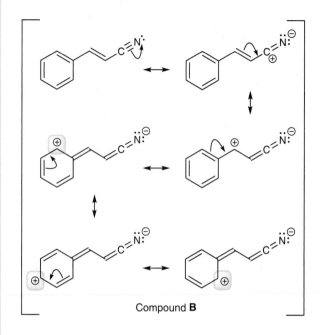

Compound **B**

Notice that this substituent removes electron density from the ring via resonance (many of the resonance structures have a positive charge in the ring), so this group is a deactivator, much like a cyano group is a deactivator:

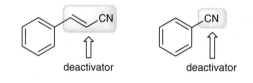

Indeed, in compound **B**, the C≡C bond serves to connect the aromatic ring and the cyano group (via conjugation), which enables the cyano group to have a deactivating effect on the aromatic ring (as if the cyano group were connected directly to the ring). And much like a cyano group is a *meta* director, so too, this substituent is also expected to be a *meta* director:

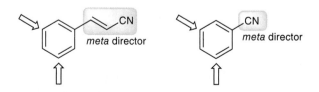

In compound **A**, the cyano group is not conjugated with the aromatic ring, because of the intervening sp^3 hybridized carbon atoms. As a result, the substituent in compound **A** functions much like an alkyl group (it is a weak activator, and an *ortho-para* director).

4.122

(a) The problem statement indicates that TCCA is an electrophilic chlorinating agent, so it functions much like Cl_2 and $AlCl_3$ (as a source of Cl^+). That is, we expect a chlorine atom to be installed on the aromatic ring. The methoxy group is an *ortho-para* director, so we expect chlorination to occur at the *ortho* and *para* positions, giving the following two products:

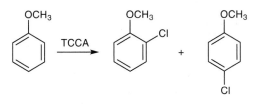

(b) The *para* product has more symmetry than the *ortho* product, so the *para* product will have fewer signals in its ^{13}C NMR spectrum. Specifically, the *para* product will have only five signals (labeled **a–e** below) in its ^{13}C NMR spectrum.

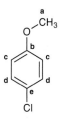

In contrast, the *ortho* product has seven unique carbon atoms (none of the carbon atoms are chemically equivalent), so the ^{13}C NMR spectrum of the *ortho* product will have seven signals, rather than five signals.

4.123 Treating benzene with Br_2 and $AlBr_3$ results in a bromination reaction, generating bromobenzene (compound **A**). When bromobenzene is treated with magnesium in diethyl ether, a magnesium atom is inserted between the C—Br bond, giving compound **B**, which is a Grignard reagent. This Grignard reagent will then attack a ketone (acetone in this case), followed by aqueous acidic workup, to generate a tertiary alcohol (compound **C**). Finally, treating compound **C** with concentrated sulfuric acid gives a dehydration reaction, thereby generating an alkene (compound **D**).

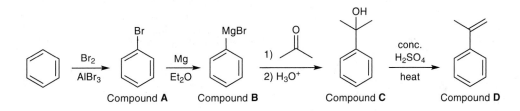

4.124 Treating benzene with an acyl chloride in the presence of catalytic $AlCl_3$ results in a Friedel–Crafts acylation. In the process, an acyl group is installed on the ring, giving compound **A**. When compound **A** is treated with a Wittig reagent, compound **A** is converted into an alkene (compound **B**). This alkene will then react with aqueous acid to give an acid-catalyzed hydration reaction, generating a tertiary alcohol (compound **C**). The final two steps represent a Williamson ether synthesis. NaH is a strong base, and it will deprotonate alcohol **C** to give the corresponding alkoxide ion, which will then attack methyl iodide in an S_N2 reaction to give the methyl ether shown below (compound **D**).

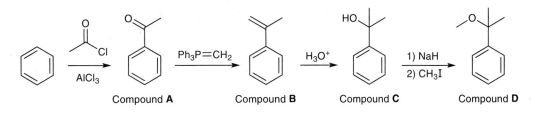

4.125 Recall that hydrohalogenation of an alkene occurs via the following two-step mechanism:

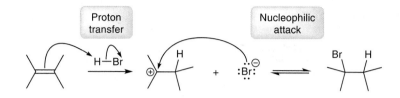

The first step of this process is protonation of the pi bond to from a carbocation, which is then attacked by a halide ion. Our starting material is an unsymmetrical alkene, so there are two regiochemical possibilities for the protonation step:

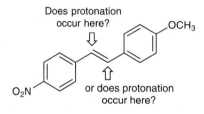

There is more than one way to rationalize why one position is preferentially protonated. We can either consider the electron density of the starting alkene (and determine which vinylic position is more electron-rich and therefore more susceptible to being protonated), or we can analyze the stability of the two possible intermediate carbocations and determine which one is more stable. We will now apply each of these methods, and we will see that they both give the same prediction.

Let's begin by considering the electron density of the starting alkene. Specifically, we must assess whether each of the vinylic positions is electron-rich or electron-poor. To do this, we will draw some of the resonance structures for this compound (the compound has other resonance structures that are not drawn below). The methoxy group is electron-donating via resonance, which places electron density on one of the vinylic positions, highlighted below:

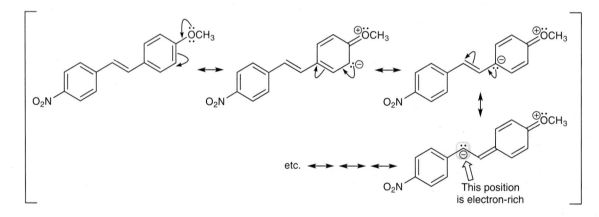

and the nitro group is electron-withdrawing via resonance, which removes electron density from one of the vinylic positions, highlighted below:

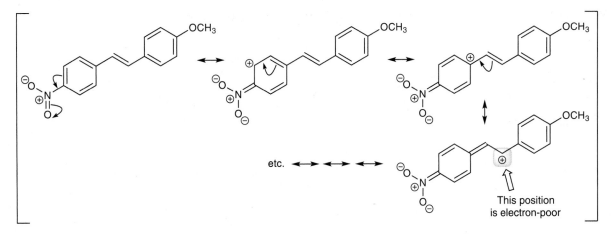

As a result, the starting alkene has the following electronic character:

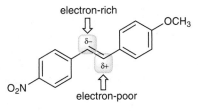

and we would expect protonation to occur at the site that is δ−, which gives the following addition product:

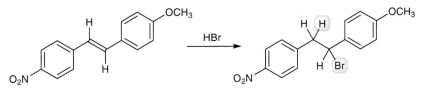

We can arrive at the same prediction if we instead consider the stability of the carbocation that would result from protonation at each location. Both possibilities are shown here:

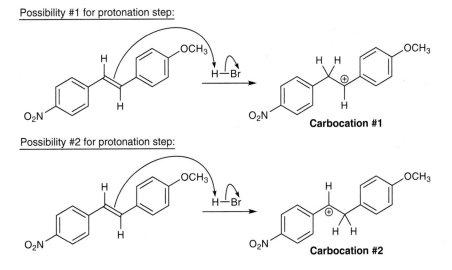

Notice that carbocation #1 is more stable, because of resonance effects from the methoxy group:

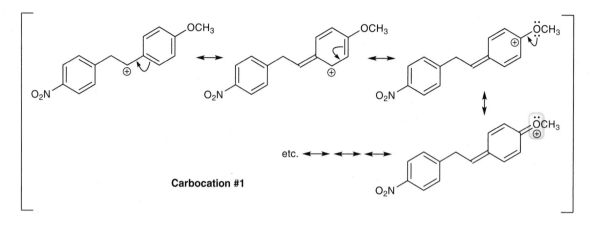

Carbocation #1

In carbocation #1, one of the resonance structures has the positive charge on the oxygen atom of the methoxy group (highlighted above), rather than on a carbon atom. In this resonance structure, all atoms have filled octets, so this resonance structure contributes stability to the overall resonance hybrid (all of the other resonance structures have C+, which lacks an octet of electrons).

Now let's consider the stability of carbocation #2. This carbocation is NOT stabilized by either the methoxy group or by the nitro group via resonance. We expect the reaction to proceed via the more stable carbocation intermediate (carbocation #1), so we predict the following product (note that this is the same prediction that we made by analyzing the electron density of the starting alkene):

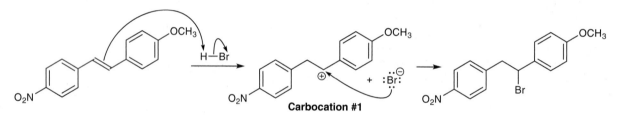

Carbocation #1

4.126 Compound **A** has the molecular formula C_9H_{12}, so it has four degrees of unsaturation, which is strongly suggestive of an aromatic ring.

Next, we consider the number of signals and the integration value of each signal. This spectrum has three signals, and the integration values are 5H, 1H, and 6H, respectively.

The multiplet just above 7 ppm is characteristic of an aromatic ring, and the integration (5H) indicates that the ring is monosubstituted.

The other two signals represent the characteristic pattern of an isopropyl group (a doublet with an integration of 6H, and a septet with an integration of 1H):

These two fragments (the aromatic ring and the isopropyl group) account for all of the atoms in the molecular formula, and there is only one way to connect these fragments, so compound **A** is isopropyl benzene:

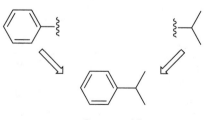

Compound **A**

Compound **B** has the molecular formula C$_{12}$H$_{18}$, so it also has four degrees of unsaturation (likely an aromatic ring). Indeed, the NMR spectrum of compound **B** looks very similar to The NMR spectrum of compound **A** (three signals in the same locations) except that the signal above 7 ppm has collapsed to a singlet, and the integration values of the three signals have changed. In the spectrum of compound **A**, the relative integration values were 5 : 1 : 6, but in the spectrum of compound **B**, the relative integration values are now 2 : 1 : 6. These values (in the spectrum of compound **B**) account for only half of the hydrogen atoms in this compound (9 out of the 18 protons in compound **B**). Since compound **B** has 18 protons, the actual integration values must be 4H, 2H, and 12H. The signal with an integration of 4H suggests that the aromatic ring in compound **B** is disubstituted (rather than monosubstituted), and the integration values indicate that there are two isopropyl groups in compound **B**. These two isopropyl groups must be *para* to each other, in order for the signal above 7 ppm to be a singlet:

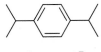

Compound **B**

If the two isopropyl groups were *ortho* or *meta* to each other, then the signal above 7 ppm would not be a singlet. So compound **B** must be *para*-diisopropyl benzene.

Compound **A** can be converted into compound **B** via a Friedel–Crafts alkylation, by treating compound **A** with isopropyl chloride in the presence of catalytic aluminum trichloride:

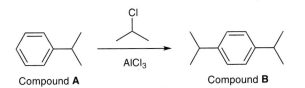

Compound **A**   Compound **B**

4.127 If we compare the structure of the starting material with the structure of the product, we can see that the product is formed from the reaction between two equivalents of the starting material, with the formation of two new sigma bonds, labeled **a** and **b** below:

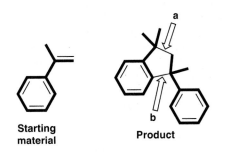

Starting material   **Product**

Let's begin by considering how the bond labeled **a** might be formed. The starting material has an aromatic ring, as well as a pi bond outside of the ring (conjugated with the ring). This pi bond can function as a nucleophile, but it can also function as a base (if it receives a proton from sulfuric acid) to generate a carbocation:

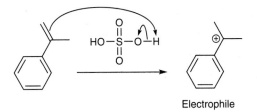

Electrophile

This carbocation is a powerful electrophile, and it has been formed in the presence of a nucleophile (the starting material). A reaction between the two can form the bond labeled **a**:

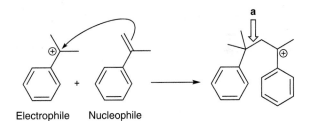

Electrophile   Nucleophile

The resulting intermediate has both an electrophilic center (C+) and a nucleophilic center (the aromatic ring):

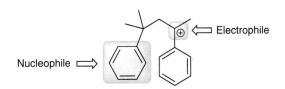

so a reaction can occur between these two centers (which results in the formation of the bond labeled **b**). In fact, the reaction between these two centers is just an electrophilic aromatic substitution reaction that is occurring in an intramolecular fashion, because the nucleophile and the electrophile are tethered to each other in one molecule.

The entire mechanism is shown below. First, the starting material is protonated to generate a carbocation intermediate, which is then attacked by a molecule of the starting material. The resulting intermediate undergoes an intramolecular electrophilic aromatic substitution reaction, which consists of two steps: (1) intramolecular attack to give an intermediate sigma complex and (2) deprotonation of the sigma complex to restore aromaticity. Note that for the last step of this mechanism, we can show either H$_2$O or HSO$_4^-$ as the base, because both are present.

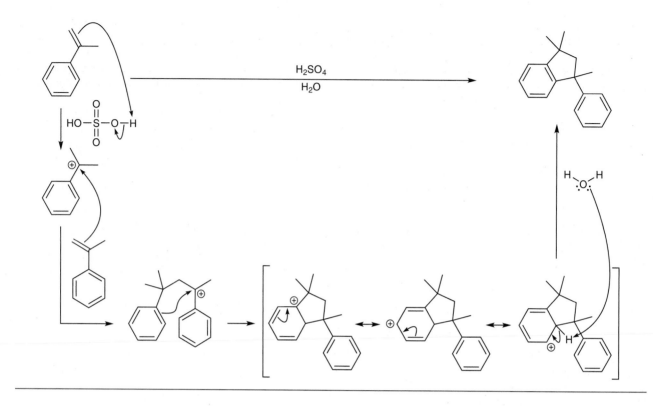

CHAPTER 5

5.2 This compound meets all three criteria for a nucleophilic aromatic substitution reaction: (1) it has an electron-withdrawing group (a nitro group), (2) it has a leaving group (bromide), and (3) the leaving group is *ortho* to the electron-withdrawing group. Therefore, this compound can serve as an electrophile in a nucleophilic aromatic substitution reaction.

5.3 This compound has an electron-withdrawing group (a nitro group) and a leaving group (chloride), but the leaving group is *meta* to the electron-withdrawing group. Therefore, this compound will NOT serve as a suitable electrophile in a nucleophilic aromatic substitution reaction.

5.4 This compound has an electron-withdrawing group (a nitro group) and a leaving group (a sulfonate group), but the leaving group is *meta* to the electron-withdrawing group. Therefore, this compound will NOT serve as a suitable electrophile in a nucleophilic aromatic substitution reaction.

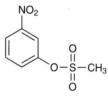

5.5 This compound meets all three criteria for a nucleophilic aromatic substitution reaction: (1) it has an electron-withdrawing group (a nitro group), (2) it has a leaving group (a sulfonate group), and (3) the leaving group is *para* to the electron-withdrawing group. Therefore, this compound can serve as an electrophile in a nucleophilic aromatic substitution reaction.

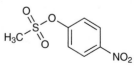

5.6 This compound has an electron-withdrawing group (a nitro group) but does NOT have a leaving group. The *tert*-butyl group cannot function as a leaving group, because a negative charge on a carbon atom is very unstable. Therefore, this compound will NOT serve as a suitable electrophile in a nucleophilic aromatic substitution reaction.

5.7 This compound has an electron-withdrawing group (a nitro group) but does NOT have a leaving group. Alkyl groups cannot function as leaving groups, because a negative charge on a carbon atom is very unstable. Therefore, this compound will NOT serve as a suitable electrophile in a nucleophilic aromatic substitution reaction.

5.9 The reagent in the first step is a strong nucleophile (hydroxide), and the starting compound meets all three criteria for an S_NAr mechanism: an electron-withdrawing group (NO_2) and a leaving group (iodide) that are *ortho* to each other. In a nucleophilic aromatic substitution reaction, the nucleophile (hydroxide) attacks at the position bearing the leaving group, and electrons are pushed up into the reservoir. The resulting intermediate is a Meisenheimer complex, and it has resonance structures that should be drawn. Then, in the second step of the mechanism, the leaving group is expelled to give the product:

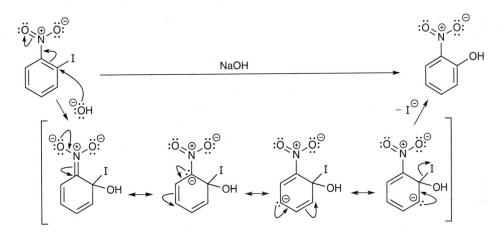

This might appear to be a complete mechanism. But remember that under basic conditions, the product is deprotonated.

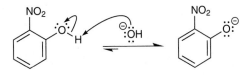

And that is why an acid source is necessary after the reaction is complete:

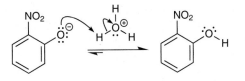

5.10 The reagent in the first step is a strong nucleophile (hydroxide), and the starting compound meets all three criteria for an $S_N Ar$ mechanism: an electron-withdrawing group (NO_2) and a leaving group (a sulfonate) that are *para* to each other. In a nucleophilic aromatic substitution reaction, the nucleophile (hydroxide) attacks at the position bearing the leaving group, and electrons are pushed up into the reservoir. The resulting intermediate is a Meisenheimer complex, and it has resonance structures that should be drawn. Then, in the second step of the mechanism, the leaving group is expelled to give the product:

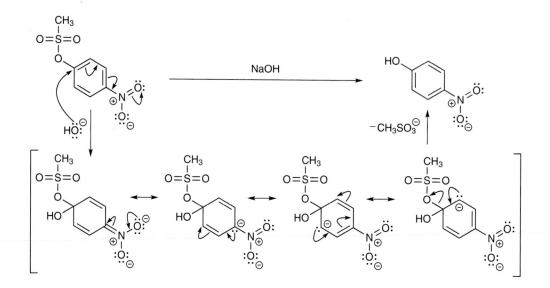

This might appear to be a complete mechanism. But remember that under basic conditions, the product is deprotonated.

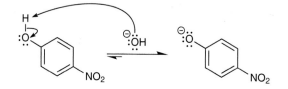

And that is why an acid source is necessary after the reaction is complete:

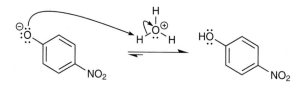

5.11 The reagent in the first step is a strong nucleophile (hydroxide), and the starting compound meets all three criteria for an $S_N Ar$ mechanism: an electron-withdrawing group (NO_2) and a leaving group (chloride) that are *ortho* to each other. In a nucleophilic aromatic substitution reaction, the nucleophile (hydroxide) attacks at the position bearing the leaving group, and electrons are pushed up into the reservoir. The resulting intermediate is a Meisenheimer complex, and it has resonance structures that should be drawn. Then, in the second step of the mechanism, the leaving group is expelled to give the product:

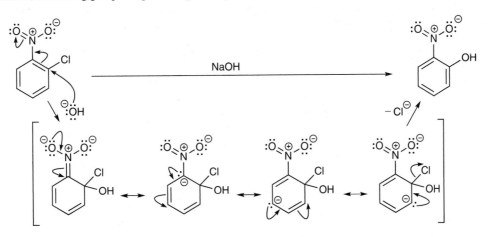

This might appear to be a complete mechanism. But remember that under basic conditions, the product is deprotonated.

And that is why an acid source is necessary after the reaction is complete:

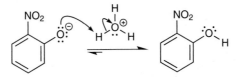

5.13 This compound has a leaving group, but it lacks a powerful electron-withdrawing group (such as a nitro group), so the reaction does NOT proceed via an $S_N Ar$ mechanism. Rather, the reaction must proceed via an elimination–addition process, as indicated by the high temperature. The reaction is expected to involve a benzyne intermediate (shown below), which can be attacked by hydroxide in either of two locations, leading to the two products shown.

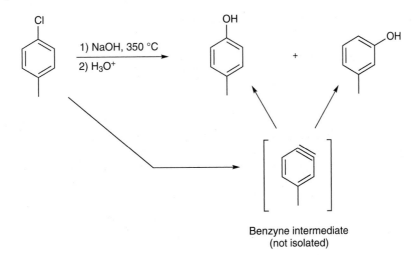

Benzyne intermediate
(not isolated)

5.14 This compound has a powerful electron-withdrawing group (a nitro group) and a leaving group (chloride), BUT they are not *ortho* or *para* to each other. They are *meta* to each other, so the reaction does NOT proceed via an S_NAr mechanism. Rather, the reaction must proceed via an elimination–addition process. The reaction is expected to involve one of two possible benzyne intermediates (shown below), which can be attacked by H_2N^- to give one of the three products shown.

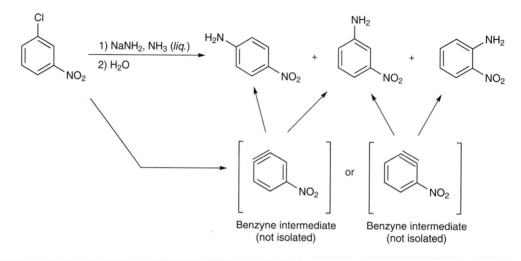

Benzyne intermediate
(not isolated)

Benzyne intermediate
(not isolated)

5.15 The reagent in the first step is a strong nucleophile (hydroxide), and the starting compound meets all three criteria for an S_NAr mechanism: an electron-withdrawing group (NO_2) and a leaving group (chloride) that are *ortho* to each other. We therefore expect an S_NAr reaction, in which the leaving group is replaced with a hydroxyl group:

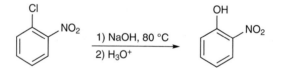

5.16 This compound has a leaving group, but it lacks a powerful electron-withdrawing group (such as a nitro group), so the reaction does NOT proceed via an S_NAr mechanism. Rather, the reaction must proceed via an elimination–addition process, as indicated by the high temperature. The reaction is expected to involve a benzyne intermediate (shown below). In this case, there is only one possible benzyne intermediate (there is only one hydrogen atom that is *ortho* to the leaving group). This benzyne intermediate is then attacked by methoxide in either of two locations, leading to the two products shown.

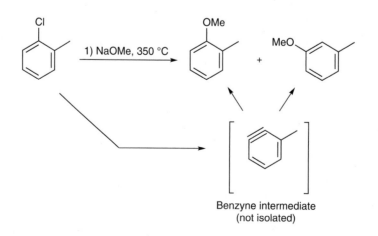

Benzyne intermediate
(not isolated)

5.17 This compound has a leaving group but lacks a powerful electron-withdrawing group (such as a nitro group). Therefore, the reaction does NOT proceed via an S_NAr mechanism. Rather, the reaction must proceed via an elimination–addition process, as indicated by the high temperature. The reaction is expected to involve one of two possible benzyne intermediates (shown below), which can be attacked by hydroxide to give one of the three products shown.

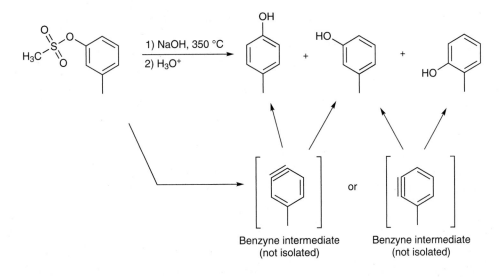

5.18 The reagent in the first step is a strong nucleophile (hydroxide), and the starting compound meets all three criteria for an S_NAr mechanism: an electron-withdrawing group (NO_2) and a leaving group (bromide) that are *para* to each other. We therefore expect an S_NAr reaction, in which the leaving group is replaced with a hydroxyl group:

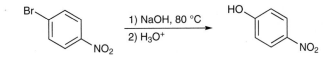

5.20 In this reaction, all three criteria for S_NAr have been met. For the mechanism, see the solution to Problem 5.11.

5.21 The reagent is a strong nucleophile (hydroxide), so we can rule out an electrophilic aromatic substitution reaction. The starting compound does NOT have all three criteria for an S_NAr mechanism. This compound has a leaving group (bromide), but it lacks a powerful electron-withdrawing group (such as a nitro group), so the reaction is NOT an S_NAr mechanism. Rather, the reaction must be an elimination–addition process (via a benzyne intermediate), as indicated by the high temperature.

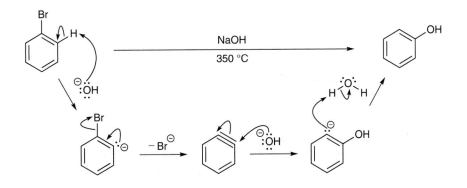

This might appear to be a complete mechanism. But remember that under basic conditions, the product is deprotonated.

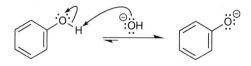

And that is why an acid source is necessary after the reaction is complete:

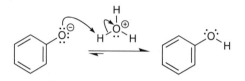

5.22 The reagent in the first step is a strong nucleophile (hydroxide), and the starting compound meets all three criteria for an S$_N$Ar mechanism: an electron-withdrawing group (NO$_2$) and a leaving group (fluoride) that are para to each other. In a nucleophilic aromatic substitution reaction, the nucleophile (hydroxide) attacks at the position bearing the leaving group, and electrons are pushed up into the reservoir. The resulting intermediate is a Meisenheimer complex, and it has resonance structures that should be drawn. Then, in the second step of the mechanism, the leaving group is expelled to give the product:

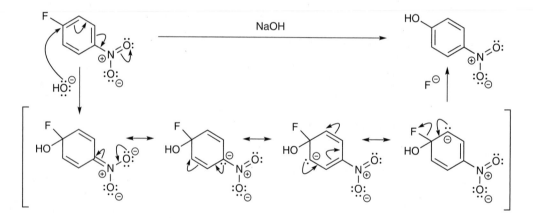

This might appear to be a complete mechanism. But remember that under basic conditions, the product is deprotonated.

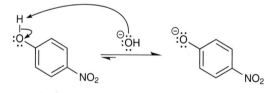

And that is why an acid source is necessary after the reaction is complete:

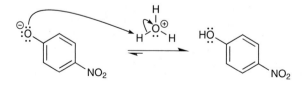

5.23 The reagent is a strong nucleophile (H$_2$N$^-$), so we can rule out an electrophilic aromatic substitution reaction. This compound has a leaving group (chloride), but it lacks a powerful electron-withdrawing group (such as a nitro group), so the reaction is NOT an S$_N$Ar mechanism. Rather, the reaction must be an elimination–addition process (via a benzyne intermediate).

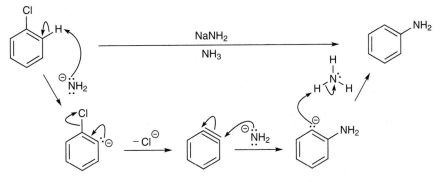

This might appear to be a complete mechanism. But, as we saw with phenol, aniline is deprotonated under these basic conditions:

And that is why a workup step is necessary after the reaction is complete:

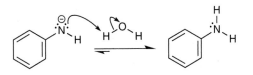

5.24 The reagent is sodium phenolate (also called sodium phenoxide), where sodium (Na$^+$) is a spectator ion (not involved in the reaction), and the phenolate (or phenoxide) ion can function as a nucleophile:

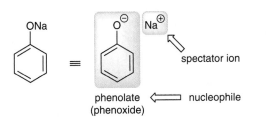

The starting material has an aromatic ring that meets all three criteria for a nucleophilic aromatic substitution reaction: (1) the ring has a strong electron-withdrawing group (a nitro group), (2) there is a leaving group (fluoride), and (3) the nitro group and the leaving group are *para* to each other.

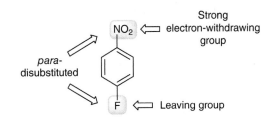

Therefore, we expect a nucleophilic aromatic substitution reaction in which the phenolate ion replaces the fluoride leaving group:

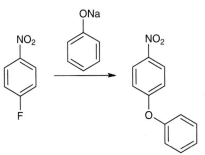

5.25 The reagent is CH₃CH₂SNa, where sodium (Na⁺) is a spectator ion (not involved in the reaction), and the thiolate ion is a very powerful nucleophile:

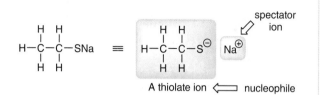

The starting material has an aromatic ring that meets all three criteria for a nucleophilic aromatic substitution reaction: (1) the ring has a strong electron-withdrawing group (a nitro group), (2) there is a leaving group (iodide), and (3) the nitro group and the leaving group are *para* to each other.

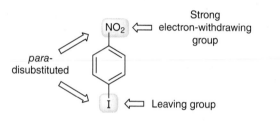

Therefore, we expect a nucleophilic aromatic substitution reaction in which the thiolate ion replaces the iodide leaving group:

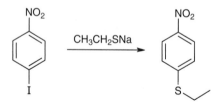

5.26 The reagent is NaCN, where sodium (Na⁺) is a spectator ion (not involved in the reaction), and the cyanide ion can function as a nucleophile:

The starting material has an aromatic ring that meets all three criteria for a nucleophilic aromatic substitution reaction: (1) the ring has a strong electron-withdrawing group (a nitro group), (2) there is a leaving group (bromide), and (3) the nitro group and the leaving group are *ortho* to each other.

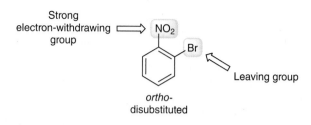

Therefore, we expect a nucleophilic aromatic substitution reaction in which the cyanide ion replaces the bromide leaving group:

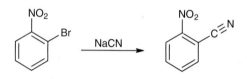

5.27 The reagent is NaOH, where sodium (Na⁺) is a spectator ion (not involved in the reaction), and the hydroxide ion can function as a nucleophile or as a base. The starting material is a trisubstituted aromatic ring that does NOT meet all criteria for a nucleophilic aromatic substitution reaction. Specifically, the ring does NOT have a strong electron-withdrawing group (a nitro group). However, a leaving group is present, so an elimination-addition reaction is possible. Indeed, the very high temperature (350°C) is suggestive of an elimination-addition process, which proceeds via a benzyne intermediate, shown below.

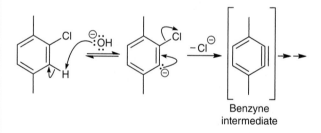

This benzyne intermediate is extremely high in energy (very short-lived) and is quickly attacked by a hydroxide ion, followed by protonation, to give the product:

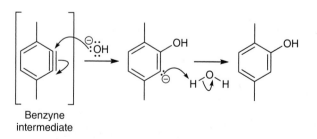

Under these basic conditions, the product is deprotonated as soon as it is formed, to give the conjugate base of the product (a substituted phenolate ion), which explains why an acidic workup is required. The workup step is used to regenerate the product:

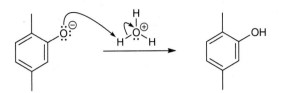

Note that the benzyne intermediate can be attacked by hydroxide at either carbon atom of the C≡C bond, but both options lead to the same product (because of the symmetry of the benzyne intermediate):

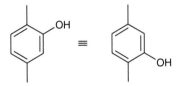

5.28 The reagent is NaOH, where sodium (Na$^+$) is a spectator ion (not involved in the reaction), and the hydroxide ion can function as a nucleophile or as a base. The starting material has an aromatic ring that meets all three criteria for a nucleophilic aromatic substitution reaction: (1) the ring has a strong electron-withdrawing group (a nitro group), (2) there is a leaving group (fluoride), and (3) the nitro group and the leaving group are *para* to each other.

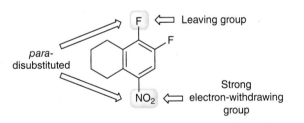

Note that there are two fluoride substituents in this compound, but one of them is *meta* to the nitro group, so this position is not suitable for nucleophilic attack and this fluoride will not be expelled as a leaving group in a nucleophilic aromatic substitution reaction.

In summary, we expect a nucleophilic aromatic substitution reaction in which the hydroxide ion replaces the fluoride leaving group that is *para* to the nitro group, giving the product shown below after aqueous acidic workup:

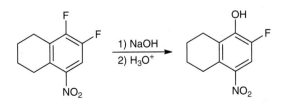

5.29 The reagent is NaNH$_2$, where sodium (Na$^+$) is a spectator ion (not involved in the reaction), and the amide ion can function as a nucleophile or as a base. The starting material is an aromatic compound that does NOT meet all the criteria for a nucleophilic aromatic substitution reaction. Specifically, the ring does NOT have a strong electron-withdrawing group (a nitro group). However, a leaving group (chloride) is present, so it is possible to have an elimination-addition reaction, which proceeds via a benzyne intermediate, shown below.

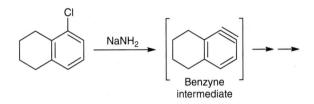

This benzyne intermediate is extremely high in energy (very short-lived) and is quickly attacked by an amide ion. The amide ion can attack either carbon atom of the C≡C bond of benzyne, ultimately leading to the two products shown below:

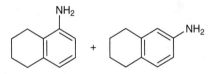

Note that under the strongly basic conditions of the reaction, these products are deprotonated to give their conjugate bases, which explains why a workup step is required (in order to regenerate the products shown).

5.30 Recall that there are three criteria for a nucleophilic aromatic substitution reaction: (1) the starting aromatic ring (which will function as an electrophile in this reaction) must have a strong electron-withdrawing group so that it is electron-poor (electrophilic), (2) there must be a leaving group, and (3) the electron-withdrawing group and the leaving group must be *ortho* or *para* to each other. If we consider the structure of the desired product, we see that there are two aromatic rings, one of which has nitro groups (powerful electron-withdrawing groups). In contrast, the other aromatic does not have any electron-withdrawing groups:

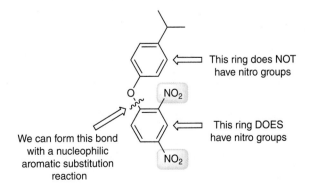

So, in order to meet all three criteria for a nucleophilic aromatic substitution reaction, the ring that has nitro groups must have been the electrophile in the reaction (which means that this ring must have had the leaving group). We, therefore, expect the product to be obtained from the reaction between the following two compounds:

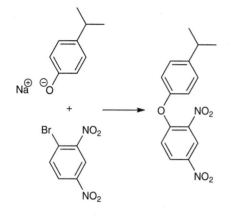

Note that the leaving group (shown above as bromide) could also have been iodide, chloride, fluoride, or a tosylate.

5.31 None of these compounds contains a nitro group, so the type of reaction occurring must be an elimination-addition reaction. Among the four choices, the compound below does not have a proton that is beta to the leaving group:

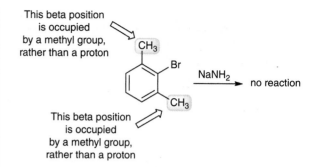

so this compound cannot undergo beta-elimination to form a benzyne intermediate. As such, this compound is unreactive toward NaNH₂ in liquid ammonia.

Each of the other three compounds has a proton that is beta to the leaving group, so treatment with NaNH₂ can result in elimination-addition:

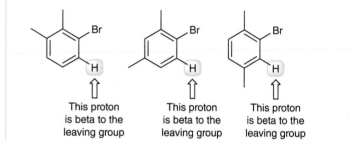

5.32

(a) If hydroxide attacks compound **A** at the location of the leaving group, the following Meisenheimer complex is formed as an intermediate:

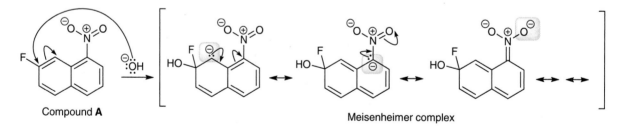

This intermediate is resonance-stabilized, with many resonance structures, a few of which have been drawn above (there are many other resonance structures that have not been drawn). Pay special attention to the third resonance structure above and compare it with the other resonance structures. This resonance structure has a negative charge on an oxygen atom (highlighted), whereas the other resonance structures have a negative charge on a carbon atom. This resonance structure is particularly good, and it contributes a lot of its character to the overall resonance hybrid of the Meisenheimer complex. Since this intermediate is stabilized, the reaction can occur, so hydroxide will react with compound **A** in a nucleophilic aromatic substitution reaction.

In contrast, if hydroxide attacks compound **B** at the location of the leaving group, the resulting Meisenheimer complex does not have a resonance structure that places the negative charge on the nitro group. The negative charge is delocalized exclusively over carbon atoms, but it is not stabilized by the nitro group. As a result, this Meisenheimer complex would not be sufficiently stabilized, so compound **B** will not react with hydroxide in a nucleophilic aromatic substitution reaction.

(b) As described above, compound **A** will react with hydroxide in a nucleophilic aromatic substitution reaction. The net result (after acidic workup) is the replacement of the fluoride leaving group with a hydroxy group:

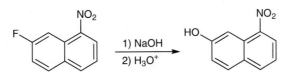

(c) As shown below (and as described above), the hydroxide ion functions as a nucleophile and attacks compound **A**, giving a resonance-stabilized, Meisenheimer complex. This intermediate has many resonance structures, some of which are shown (there are many other resonance structures that are not drawn below). This intermediate then loses a fluoride ion to regenerate aromaticity, giving the product. But under the conditions of its formation, this product is deprotonated by hydroxide to give a substituted phenolate ion:

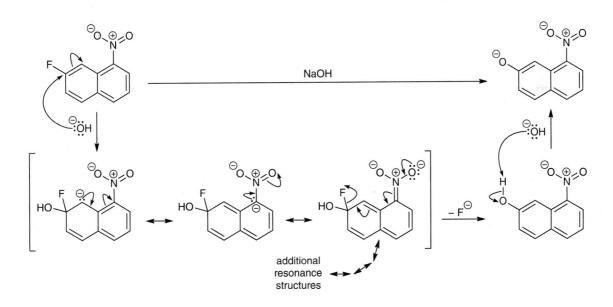

In order to protonate the resulting phenolate ion and regenerate the product, aqueous acidic workup is required:

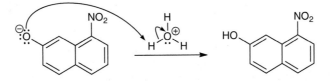

5.33 The reagent is NaSH (sodium hydrosulfide), where sodium (Na⁺) is a spectator ion (not involved in the reaction), and the hydrosulfide (HS⁻) ion is a strong nucleophile. The starting material has an aromatic ring that meets all three criteria for a nucleophilic aromatic substitution reaction: (1) the ring has a strong electron-withdrawing group (a nitro group), (2) there is a leaving group (tosylate), and (3) the nitro group and the leaving group are *para* to each other. Below is a mechanism showing a nucleophilic aromatic substitution reaction in which the hydrosulfide ion replaces the tosylate leaving group.

First, the hydrosulfide ion attacks the aromatic ring, generating a resonance-stabilized, Meisenheimer complex, which then loses the leaving group (tosylate) to regenerate aromaticity, giving the product. However, under the conditions of its formation, the product is deprotonated to give a thiolate ion (RS⁻):

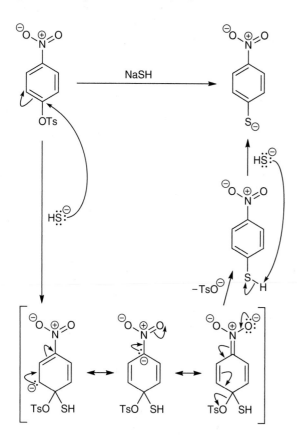

In order to protonate the thiolate ion and regenerate the product, aqueous acidic workup is required:

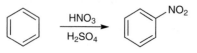

5.34 We must install a nitro group and a hydroxy group on an aromatic ring. Let's first consider how to install each of these groups individually. The nitro group can be installed by treating the aromatic ring with nitric acid and sulfuric acid:

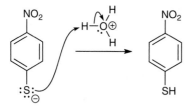

Installation of the hydroxy group will require more than one step. First, we install a leaving group on the ring, such as chloride, and then we can replace the chloride with a hydroxy group, using either nucleophilic aromatic substitution or elimination-addition.

If we choose to install the hydroxy group via elimination-addition, then it would be best to do so early in the synthesis. If we wait until later in the synthesis, when there is already another substituent on the ring, then elimination-addition is likely to produce a mixture of products (recall that a benzyne intermediate can be attacked at either carbon atom of the C≡C bond). So we will install the hydroxy group first, using the following two-step process:

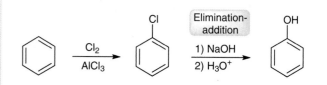

Once the hydroxy group has been installed, we can now install the nitro group. Note that a blocking group will be necessary in order to install the nitro group in the correct location (the *ortho* position), and this blocking group must be removed at the end of the synthesis:

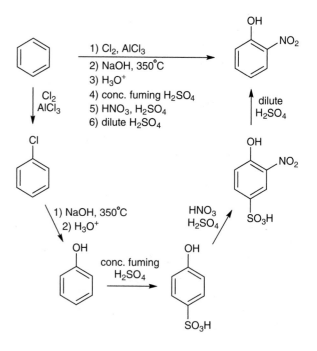

The synthesis above can be used to make the desired product. But now let's consider an alternate synthetic route, based on the possibility of installing the hydroxy group using a nucleophilic aromatic substitution reaction, rather than elimination-addition.

Recall that there are three criteria for a nucleophilic aromatic substitution: (1) the ring must have a strong electron-withdrawing group (such as a nitro group), (2) the ring must have a leaving group, and (3) the nitro group and the leaving group must be *ortho* or *para* to each other. Therefore, installation of the hydroxy group via nucleophilic aromatic substitution can be achieved like this:

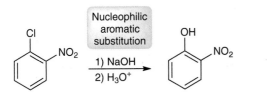

In order to make the starting chloride above, benzene can be chlorinated, and the product (chlorobenzene) can then be treated with concentrated fuming sulfuric acid to install a sulfonate blocking group in the *para* position. The nitro group can then be installed in the *ortho* position, followed by removal of the blocking group. And finally, a nucleophilic aromatic substitution reaction will give the desired product, after aqueous acidic workup:

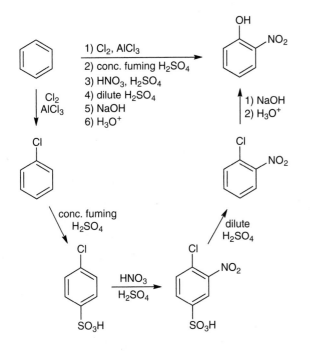

5.35 We must install a nitro group, a bromine atom, and a butoxy group on an aromatic ring. Let's first consider how to install each of these groups individually. The nitro group can be installed by treating the aromatic ring with nitric acid and sulfuric acid,

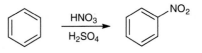

while a bromine atom can be installed via bromination with Br_2 and $AlBr_3$:

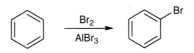

We have not seen a direct way (a one-step method) to install a butoxy group on a ring, but this chapter has provided us with several options for installing a butoxy group via more than one step. Below are two methods that each involve elimination-addition:

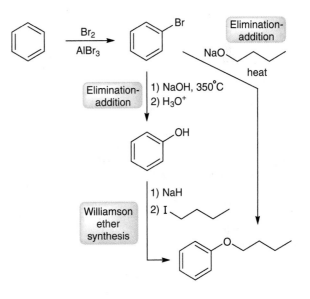

Alternatively, we have also seen a method for installing an alkoxy group via nucleophilic aromatic substitution, although let's first explore the possibilities above (elimination-addition), and then we will return to explore nucleophilic aromatic substitution as a method for installing a butoxy group.

If we choose to install the butoxy group via elimination-addition, then it would be best to do so early in the synthesis. If we wait until later in the synthesis, when there are already other substituents on the ring, then elimination-addition is likely to produce a mixture of products (recall that a benzyne intermediate can be attacked at either carbon atom of the C≡C bond). So we will install the butoxy group first, and then we can utilize its directing effects to install the nitro group next (in the *para* position), followed by bromination:

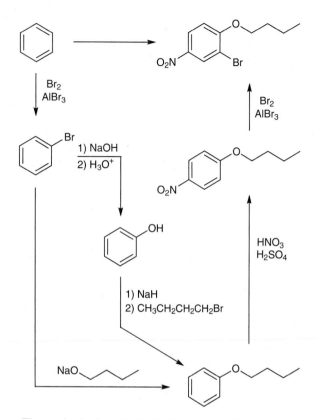

The synthesis above (in both of its variations) can be used to make the desired product. But now let's consider installing the butoxy group using a nucleophilic aromatic substitution reaction, rather than elimination-addition.

Recall that there are three criteria for a nucleophilic aromatic substitution: (1) the ring must have a strong electron-withdrawing group (such as a nitro group), (2) the ring must have a leaving group, and (3) the nitro group and the leaving group must be *ortho* or *para* to each other. Therefore, installation of the butoxy group via nucleophilic aromatic substitution can be achieved like this:

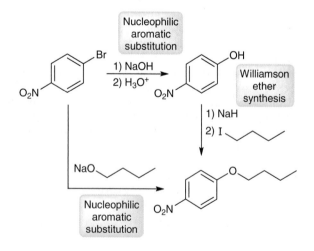

This process (whether the direct route, using sodium butoxide, or whether the less direct, multi-step route) would be the penultimate (second-to-last) process of the synthesis, with the final step being bromination. This would give the following synthesis, where the ring is first brominated (or chlorinated), followed by nitration, followed by nucleophilic aromatic substitution, followed by bromination:

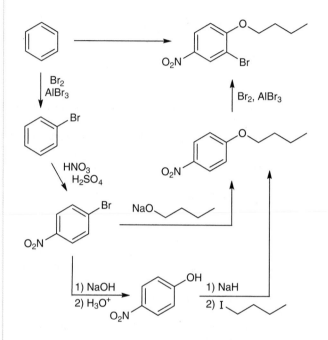

5.36 We must install a *tert*-butyl group, an ethoxy group, and two chlorine atoms on an aromatic ring. Let's first consider how to install each of these groups individually. The *tert*-butyl group can be installed via a Friedel–Crafts alkylation reaction:

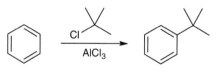

while chlorine atoms are installed with Cl_2 and $AlCl_3$:

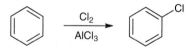

We have not seen a direct way (a one-step method) to install an ethoxy group on a ring, but this chapter has provided us with more than one multi-step method for installing an ethoxy group. Below are two similar methods that each involve elimination-addition:

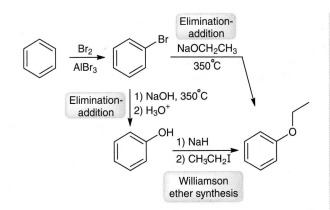

We have also seen that an alkoxy group can be installed on aromatic ring via nucleophilic aromatic substitution, although that route requires that the ring have a strong electron-withdrawing group, such as a nitro group. Our desired product does not have a nitro group or any other strong electron-withdrawing group, and we have not learned a method for removing a nitro group once it has been installed. So we can rule out nucleophilic aromatic substitution in this case. We must use elimination-addition.

When installing an ethoxy group via elimination-addition, it would be best to do so early in the synthesis. If we wait until later in the synthesis, when there are already other substituents on the ring, then elimination-addition is likely to produce a mixture of products (recall that a benzyne intermediate can be attacked at either carbon atom of the C≡C bond). So we will install the ethoxy group first, and then we can utilize its directing effects to install the *tert*-butyl group in the correct location (the *para* position). And finally, chlorination of such an activated ring will install a chlorine atom in every position that is *ortho* or *para* to the ethoxy group, even without the use of AlCl₃. The position that is *para* to the ethoxy group is already taken, so chlorination occurs at the two *ortho* positions, giving the desired product:

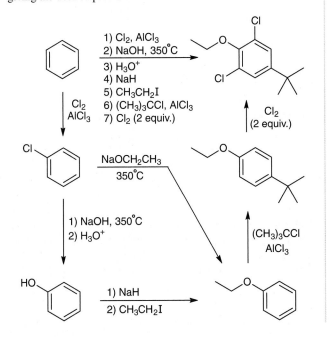

5.37

(a) The reagents and very high temperature indicate an elimination-addition process. Indeed, the starting material lacks a nitro group, so this cannot be a nucleophilic aromatic substitution reaction.

In elimination-addition, the base first deprotonates the aromatic ring, giving an anion, which then expels the leaving group to give a benzyne intermediate:

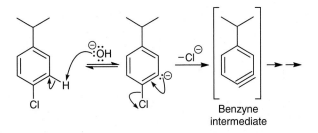

This intermediate can then be attacked by hydroxide at either carbon atom of the C≡C bond, ultimately giving rise to the following two products, after aqueous acidic workup:

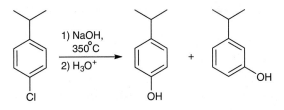

(b) The *para*-disubstituted product has more symmetry than the *meta*-disubstituted product, and, as a result, we expect that the ¹³C NMR spectrum of the *para*-disubstituted product will have only six signals (labeled **a–f** below):

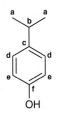

In contrast, the *meta*-disubstituted product has all carbon atoms in different electronic environments, with the exception of the two methyl groups (which are equivalent to each other). So the ¹³C NMR spectrum of the *meta*-disubstituted product will have eight signals (labeled **a–h** below):

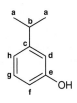

(c) The *para*-disubstituted product has more symmetry than the *meta*-disubstituted product, and, as a result, we expect that the ¹H NMR spectrum of the *para*-disubstituted product will have two doublets in the region between 7 and 8 ppm, corresponding to the two types of aromatic protons (labeled **a** and **b** below):

In contrast, in the *meta*-disubstituted product, all four aromatic protons occupy unique electronic environments (none of them are chemically equivalent), so the region between 7 and 8 ppm will have four signals that may be overlapping, rather than two clean doublets.

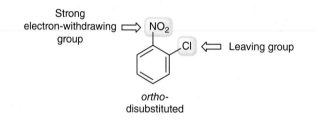

a = doublet, 2H
b = doublet, 2H

a = doublet, 1H
b = triplet, 1H
c = doublet, 1H
d = singlet, 1H

5.38 The reagent is NaC≡C—CH₃, where sodium (Na⁺) is a spectator ion (not involved in the reaction), and the alkynide ion can function as a nucleophile. The starting material has an aromatic ring that meets all three criteria for a nucleophilic aromatic substitution reaction: (1) the ring has a strong electron-withdrawing group (a nitro group), (2) there is a leaving group (chloride), and (3) the nitro group and the leaving group are *ortho* to each other:

Strong electron-withdrawing group ⟹ NO₂
Cl ⟸ Leaving group

ortho- disubstituted

In the first step of the mechanism, the alkynide ion attacks the aromatic ring, giving a resonance-stabilized Meisenheimer complex. This intermediate then loses the leaving group (chloride) to regenerate aromaticity, giving the product shown.

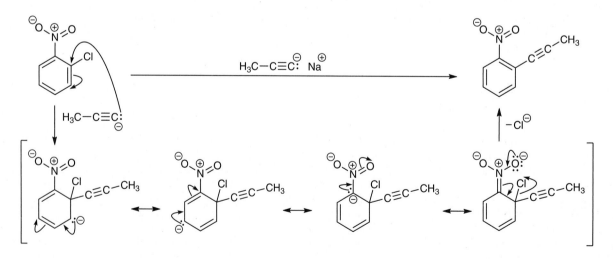

5.39 The reagent is NaH, where sodium (Na⁺) is a spectator ion (not involved in the reaction), and hydride (H⁻) is a strong, non-nucleophilic base. The starting material has an aromatic ring that meets all three criteria for a nucleophilic aromatic substitution reaction: (1) the ring has a strong electron-withdrawing group (a nitro group), (2) there is a leaving group (fluoride), and (3) the nitro group and the leaving group are *para* to each other:

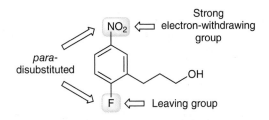

In the first step of the mechanism, the hydride ion deprotonates the OH group, giving an alkoxide ion. This alkoxide ion can then attack the aromatic ring in an intramolecular fashion (because the alkoxide nucleophile and the electrophilic aromatic ring are tethered to each other in the same compound), thereby generating a resonance-stabilized Meisenheimer complex. This intermediate then loses the leaving group (fluoride) to regenerate aromaticity, giving the product shown.

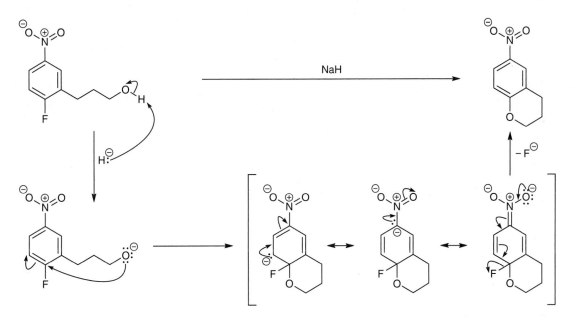

5.40 The reagent is sodium amide (NaNH$_2$), where sodium (Na$^+$) is a spectator ion (not involved in the reaction), and the amide ion (H$_2$N$^-$) is a strong base and a strong nucleophile. The starting material has two aromatic rings, neither of which has a nitro group, so neither of these rings is suitable for a nucleophilic aromatic substitution reaction. However, we can see that a bond is formed between the two aromatic rings, so we will consider the possibility of an elimination-addition process.

In the first step of the mechanism shown below, the amide ion deprotonates the substituted aniline, giving a negative charge on a nitrogen atom. Notice that two equivalents of NaNH$_2$ are used. The second equivalent can deprotonate the ring, giving an anion that can expel a leaving group to give a benzyne intermediate. If we draw a resonance structure showing the delocalized nature of the negative charge, we can see how this intermediate can undergo an intramolecular addition process. One aromatic ring (which has anionic character) acts as a nucleophile and attacks the other aromatic ring, which is a benzyne intermediate. The resulting anion is then protonated by NH$_3$. The resulting compound is a tautomer of the product. Notice that this tautomer has only one aromatic ring, rather than two. Tautomerization, via two proton transfers, will give the tautomer that has two aromatic rings (the more stable tautomer). An amide ion functions as a base to remove a proton and give a resonance-stabilized anion, and then during the workup step, water serves as a weak acid to protonate this anion and give the product:

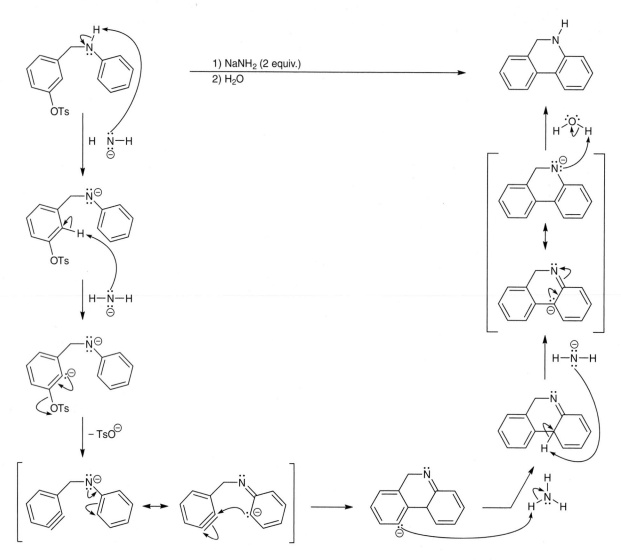

5.41 The reagent is NaH, where sodium (Na⁺) is a spectator ion (not involved in the reaction), and hydride (H⁻) is a strong, non-nucleophilic base. The starting material has two aromatic rings, one of which meets all three criteria for a nucleophilic aromatic substitution reaction: (1) the ring has at least one strong electron-withdrawing group (nitro group), (2) there is a group that can function as a leaving group (highlighted below), and (3) the nitro group and the leaving group are *ortho* to each other:

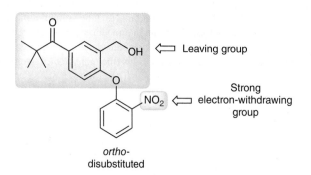

In the first step of the mechanism, the hydride ion deprotonates the alcohol group, giving an alkoxide ion. This alkoxide ion can then attack the electron-poor aromatic ring in an intramolecular fashion (because the alkoxide nucleophile and the electrophilic aromatic ring are tethered to each other in the same compound), thereby generating a resonance-stabilized Meisenheimer complex. This intermediate then loses the leaving group to regenerate aromaticity, giving a substituted phenolate ion. Finally, upon acidic workup, this phenolate ion is protonated to give the product:

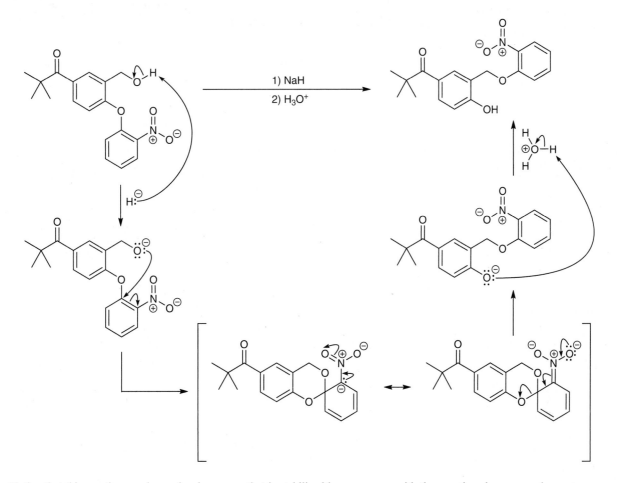

Notice that this reaction employs a leaving group that is stabilized by resonance, with the negative charge spread over two oxygen atoms (much like a carboxylate ion):

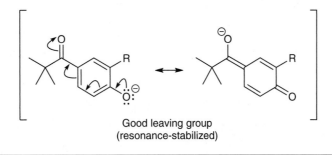

Good leaving group
(resonance-stabilized)

CHAPTER 6

6.2 When an alkene undergoes ozonolysis, the C=C bond is cleaved and is replaced by two C=O bonds. This compound has two C=C bonds, and both are cleaved, as shown.

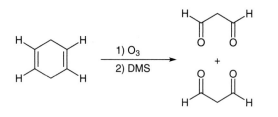

6.3 A secondary alcohol is converted into a ketone upon treatment with the Jones reagent:

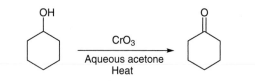

6.4 A secondary alcohol is converted into a ketone upon treatment with sodium dichromate and sulfuric acid:

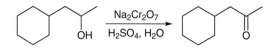

6.5 When an alkene undergoes ozonolysis, the C=C bond is cleaved and is replaced by two C=O bonds.

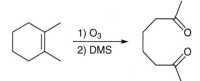

6.6 A primary alcohol is converted into a carboxylic acid upon treatment with sodium dichromate and sulfuric acid:

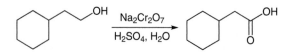

6.7 A primary alcohol is converted into an aldehyde upon treatment with PCC:

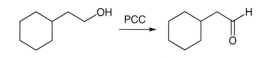

6.9 A primary alcohol can be converted into an aldehyde upon treatment with the mild oxidizing agent PCC.

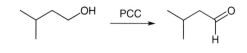

Treating this alcohol with a stronger oxidizing reagent would produce a carboxylic acid, not an aldehyde.

6.10 A primary alcohol can be converted into a carboxylic acid via a chromic acid oxidation, using either sodium dichromate and sulfuric acid, or the Jones reagent.

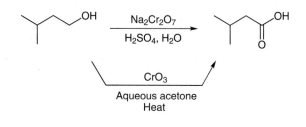

6.11 An alkene can be cleaved to give ketones via ozonolysis:

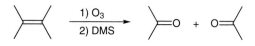

6.12 A secondary alcohol can be converted into a ketone with an oxidizing agent, such as sodium dichromate and sulfuric acid, or the Jones reagent, or even with PCC.

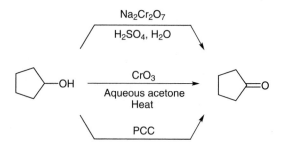

6.14 LiAlH$_4$ is a hydride reducing agent that will reduce a ketone to give a secondary alcohol:

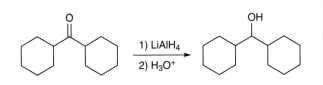

6.15 NaBH$_4$ is a hydride reducing agent that will reduce an aldehyde to give a primary alcohol:

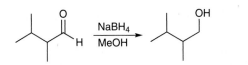

6.16 PCC is a mild oxidizing agent that will oxidize a primary alcohol to give an aldehyde, without further oxidizing the aldehyde to a carboxylic acid:

6.17 NaBH$_4$ is a hydride reducing agent that will reduce an aldehyde to give a primary alcohol:

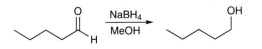

6.18 Ozonolysis of the alkene gives two equivalents of a ketone. This ketone is then reduced upon treatment with LiAlH$_4$ to give a secondary alcohol:

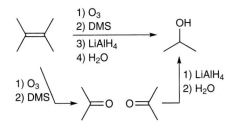

6.20 LiAlH$_4$ attacks the carbonyl group and functions as a "delivery agent" of a hydride ion, resulting in an alkoxide ion:

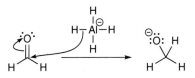

This alkoxide ion is then protonated upon treatment with a proton source, such as H$_3$O$^+$:

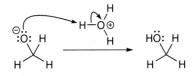

6.21 NaBH$_4$ attacks the carbonyl group and functions as a "delivery agent" of a hydride ion, resulting in an alkoxide ion:

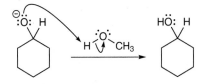

This alkoxide ion is then protonated by the proton source that is present in the reaction flask (in this case, CH$_3$OH):

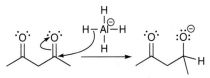

6.22 LiAlH$_4$ attacks one of the carbonyl groups and functions as a "delivery agent" of a hydride ion, resulting in an alkoxide ion:

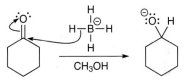

and then another equivalent of LiAlH$_4$ attacks the other carbonyl group:

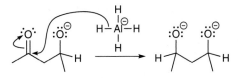

When the reaction is complete, the dianion above must be treated with a proton source, such as H$_3$O$^+$. When H$_3$O$^+$ is introduced into the reaction flask, each of the oxygen atoms is protonated. Make sure to draw each protonation step separately, and it doesn't matter which one is shown first. Here are the curved arrows for protonation of one of the oxygen atoms:

And here are the curved arrows for protonation of the other oxygen atom:

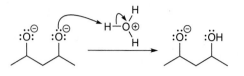

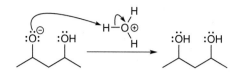

6.24 This mechanism has seven steps, described and shown below. Notice that the first three steps produce a hemiacetal, while the last four steps convert the hemiacetal into an acetal.

1. The ketone is protonated, thereby rendering the carbonyl group even more electrophilic.
2. The protonated carbonyl is then attacked by ethanol, giving a tetrahedral intermediate.
3. This intermediate is then deprotonated (by ethanol) to give a hemiacetal.
4. The hemiacetal is then protonated to generate an excellent leaving group (H₂O).
5. The leaving group (H₂O) is then expelled to regenerate the carbonyl group.
6. The carbonyl group is then attacked by ethanol to give an oxonium ion.
7. The oxonium ion is deprotonated (by ethanol) to give the product (an acetal):

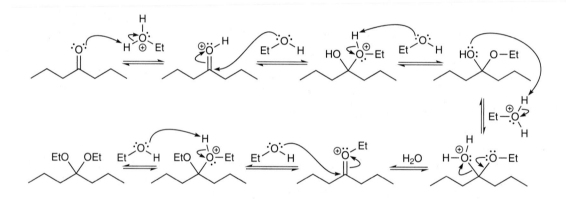

6.25 This mechanism has seven steps, described and shown below. Notice that the first three steps produce a hemiacetal, while the last four steps convert the hemiacetal into an acetal.

1. The ketone is protonated, thereby rendering the carbonyl group even more electrophilic.
2. The protonated carbonyl is then attacked by methanol, giving a tetrahedral intermediate.
3. This intermediate is then deprotonated (by methanol) to give a hemiacetal.
4. The hemiacetal is then protonated to generate an excellent leaving group (H₂O).
5. The leaving group (H₂O) is then expelled to regenerate the carbonyl group.
6. The carbonyl group is then attacked by methanol to give an oxonium ion.
7. The oxonium ion is deprotonated (by methanol) to give the product (an acetal):

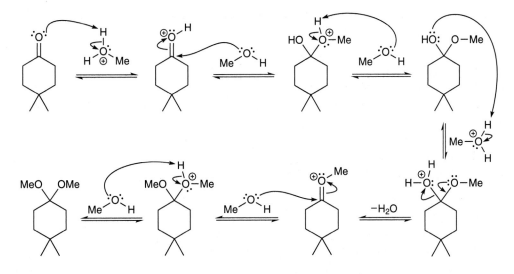

6.26 This mechanism has seven steps, described and then shown below. Notice that the first three steps produce a hemiacetal, while the last four steps convert the hemiacetal into an acetal.

1. The ketone is protonated, thereby rendering the carbonyl group even more electrophilic.

2. The protonated carbonyl is then attacked by the OH group (in an intramolecular fashion), giving a tetrahedral intermediate.

3. This intermediate is then deprotonated (by ethanol) to give a hemiacetal.

4. The hemiacetal is then protonated to generate an excellent leaving group (H_2O).

5. The leaving group (H_2O) is then expelled to regenerate the carbonyl group.

6. The carbonyl group is then attacked by ethanol to give an oxonium ion.

7. The oxonium ion is deprotonated (by ethanol) to give the product (an acetal):

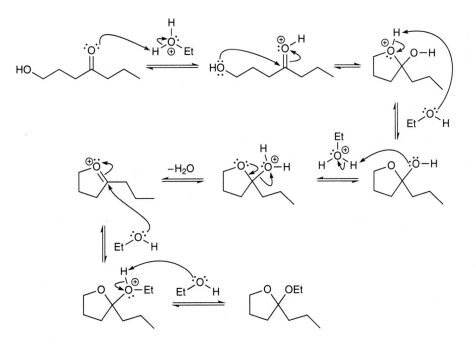

6.27 This process has seven steps, described and shown below. Notice that the order of the intermediates is the exact opposite order that we would expect for the reverse process (acetal formation).

1. The acetal is protonated, thereby converting the OEt group into a better leaving group (EtOH) that is consistent with acidic conditions (we cannot expel EtO⁻ in acidic conditions).
2. The leaving group (EtOH) is then expelled to generate a C=O bond.
3. The C=O bond is then attacked by water to give an oxonium ion.
4. The oxonium ion is deprotonated (by ethanol) to give a hemiacetal.
5. The hemiacetal is then protonated to convert the OEt group into a better leaving group (EtOH).
6. The leaving group is expelled to give a protonated ketone.
7. The protonated ketone is deprotonated (by ethanol) to give the product (a ketone):

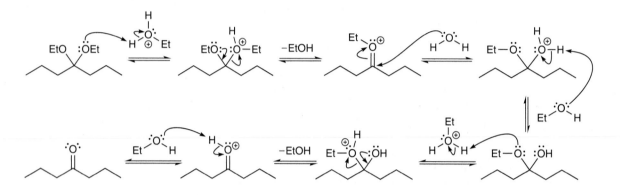

6.29 A cyclic acetal is obtained when a ketone is treated with ethylene glycol under acidic conditions and with removal of water via a Dean–Stark trap:

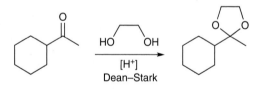

6.30 An acetal is obtained when a ketone is treated with excess alcohol under acidic conditions and with removal of water via a Dean–Stark trap:

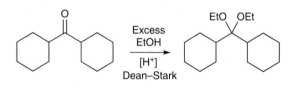

6.31 A ketone is converted into a cyclic acetal when treated with ethylene glycol under acidic conditions and with removal of water via a Dean–Stark trap. Note that the C=O bond of the ester is not reactive under these conditions.

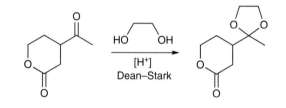

6.32 Each carbonyl group is converted into an acetal when treated with excess alcohol under acidic conditions and with removal of water via a Dean–Stark trap:

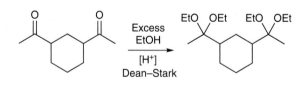

6.33 A ketone is converted into a cyclic thioacetal when treated with ethylene thioglycol under acidic conditions:

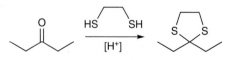

6.34 A ketone is converted into a cyclic thioacetal when treated with ethylene thioglycol under acidic conditions:

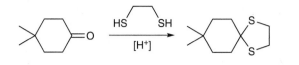

6.35 An aldehyde is converted into a cyclic thioacetal when treated with ethylene thioglycol under acidic conditions:

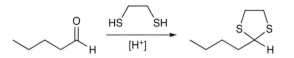

6.36 When a ketone is treated with ethylene thioglycol under acidic conditions, followed by Raney nickel, the carbonyl group is reduced to give a methylene (CH$_2$) group.

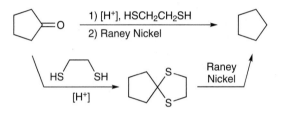

6.37 When an aldehyde is treated with ethylene thioglycol under acidic conditions, followed by Raney nickel, the carbonyl group is reduced to give a methyl group, as shown:

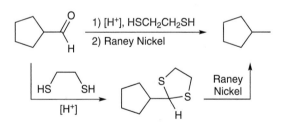

6.38 When a ketone is treated with ethylene thioglycol under acidic conditions, followed by Raney nickel, the carbonyl group is reduced to give a methylene (CH$_2$) group.

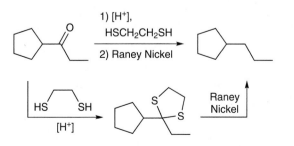

6.39 The starting material is propylene thioglycol (HSCH$_2$CH$_2$CH$_2$SH). This compound can be converted into the desired cyclic thioacetal upon treatment with formaldehyde under acidic conditions, as shown:

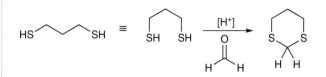

6.40 The desired transformation involves the reduction of the carbonyl group of a ketone to give a methylene (CH$_2$) group. This can be accomplished by converting the ketone into a cyclic thioacetal, and then reducing the cyclic thioacetal with Raney nickel, as shown below. Alternatively, the same transformation can be achieved with a Clemmensen reduction.

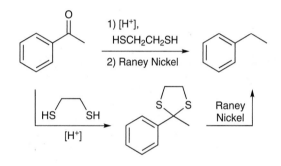

6.41 The desired transformation involves the complete reduction of the carbonyl group of an aldehyde to give a methyl group. This can be accomplished by converting the aldehyde into a cyclic thioacetal, and then reducing the cyclic thioacetal with Raney nickel, as shown below. Alternatively, the same transformation can be achieved with a Clemmensen reduction.

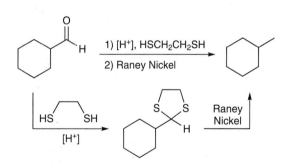

6.42 The desired transformation involves the complete reduction of the carbonyl group of an aldehyde to give an alkane. This can be accomplished by converting the aldehyde into a cyclic thioacetal, and then reducing the cyclic thioacetal with Raney nickel, as shown below. Alternatively, the same transformation can be achieved with a Clemmensen reduction.

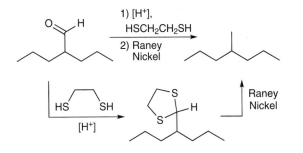

6.43 A ketone can be converted into a cyclic thioacetal upon treatment with ethylene thioglycol under acidic conditions, as shown here:

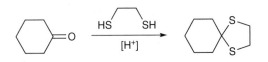

6.44 An aldehyde can be converted into a cyclic thioacetal upon treatment with propylene thioglycol under acidic conditions, as shown here:

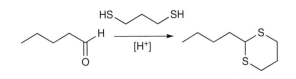

6.45 The desired transformation involves the reduction of the carbonyl group of a ketone to give a methylene (CH₂) group. This can be accomplished by converting the ketone into a cyclic thioacetal, and then reducing the cyclic thioacetal with Raney nickel, as shown below. Alternatively, the same transformation can be achieved with a Clemmensen reduction.

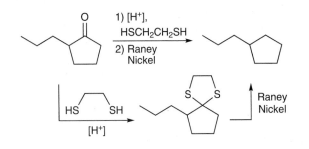

6.46 The desired transformation involves the complete reduction of the carbonyl group of an aldehyde to give an alkane. This can be accomplished by converting the aldehyde into a cyclic thioacetal, and then reducing the cyclic thioacetal with Raney nickel, as shown

below. Alternatively, the same transformation can be achieved with a Clemmensen reduction.

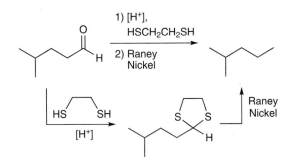

6.48 When a symmetrical ketone is treated with a primary amine under acidic conditions (and with removal of water), the ketone is converted into an imine:

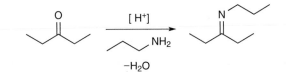

6.49 When a symmetrical ketone is treated with a secondary amine under acidic conditions (and with removal of water), the ketone is converted into an enamine:

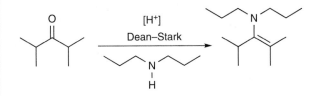

6.50 The starting material is a ketone, and the reagents indicate a Wolff–Kishner reduction. The net result of this process is to reduce the carbonyl group to give a methylene (CH₂) group:

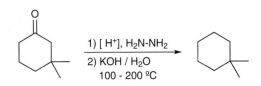

6.51 The reagent is hydroxylamine, and the starting material is an unsymmetrical ketone. Under these conditions, an unsymmetrical ketone is converted into a mixture of diastereomeric oximes:

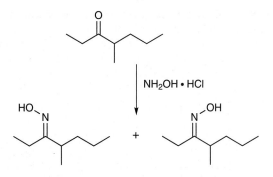

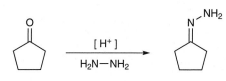

6.52 The starting material is a ketone, and the reagent is a secondary amine. A ketone is converted into an enamine upon treatment with a secondary amine under acidic conditions (and with removal of water). In this case, the ketone is unsymmetrical, so we expect a mixture of enamines, with the less-substituted enamine being favored:

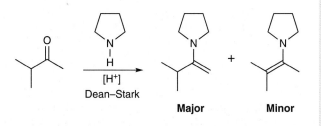

6.53 The starting material is a ketone, and the reagent is a primary amine. A ketone is converted into an imine upon treatment with a primary amine under acidic conditions (and with removal of water). In this case, the ketone is unsymmetrical, so we expect a mixture of diastereomeric imines:

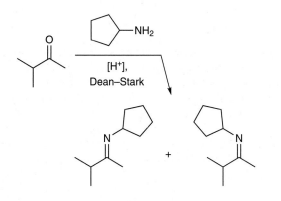

6.54 The starting material is a symmetrical ketone, and the reagent is hydrazine ($H_2N–NH_2$). A ketone is converted into a hydrazone upon treatment with hydrazine under acidic conditions.

6.55 The starting material is a symmetrical ketone, and the reagent is hydroxylamine (NH_2OH). A ketone is converted into an oxime upon treatment with hydroxylamine under acidic conditions.

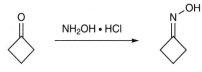

6.56 The starting material is a symmetrical ketone, and the reagent is a secondary amine. When a symmetrical ketone is treated with a secondary amine under acidic conditions (and with removal of water), the ketone is converted into an enamine:

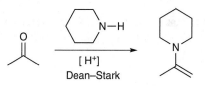

6.57 The starting material is a hydrazone, and the reagents indicate a Wolff–Kishner reduction, which gives an alkane (or in this case, a cycloalkane):

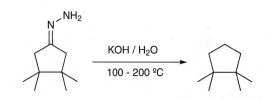

6.58 The starting material has both a primary amine and a ketone tethered together in the same molecule, allowing for an intramolecular reaction. Under acidic conditions (and with removal of water), a primary amine will react with a ketone to give an imine, as shown:

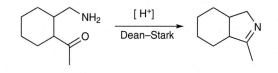

6.59 The starting material has both a secondary amine and an aldehyde tethered together in the same molecule, allowing for an intramolecular reaction. Under acidic conditions (and with removal of water), a secondary amine will react with an aldehyde to give an enamine (via the iminium ion intermediate shown):

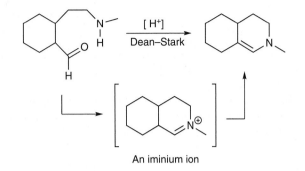

An iminium ion

6.60 The starting material has both a secondary amine and a ketone tethered together in the same molecule, allowing for an intramolecular reaction. Under acidic conditions (and with removal of water), a secondary amine will react with a ketone to give the less-substituted enamine (via the iminium ion intermediate shown):

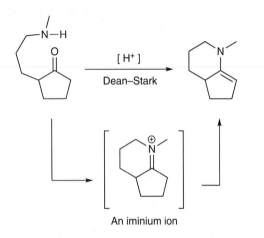

An iminium ion

6.62 When an aldehyde is treated with a Grignard reagent, followed by aqueous acidic workup, the aldehyde is reduced to an alcohol, accompanied by the installation of an R group from the Grignard reagent. In this case, the Grignard reagent is PhMgBr, so a phenyl group (Ph) is installed, as follows:

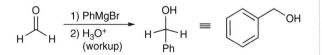

6.63 LiAlH$_4$ is a reducing agent. When an aldehyde is treated with LiAlH$_4$, followed by aqueous acidic workup, the aldehyde is reduced to give an alcohol:

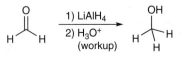

6.64 When a ketone is treated with a Grignard reagent, followed by aqueous acidic workup, the ketone is reduced to an alcohol, accompanied by the installation of an R group from the Grignard reagent. In this case, the Grignard reagent is cyclohexyl magnesium bromide, so a cyclohexyl group is installed, as follows:

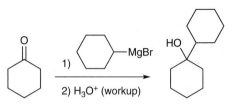

6.65 When a ketone is treated with a Grignard reagent, followed by aqueous acidic workup, the ketone is reduced to an alcohol, accompanied by the installation of an R group from the Grignard reagent. In this case, the Grignard reagent is PhMgBr, so a phenyl group (Ph) is installed, as follows:

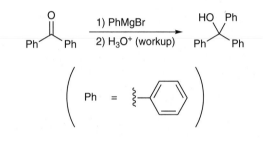

6.67 A ketone will react with a Wittig reagent to give an alkene. In this case, the Wittig reagent does not have any R groups (the carbon atom bearing the negative charge is connected to two hydrogen atoms and no R groups), so a methylene (CH$_2$) group is installed in place of the oxygen atom.

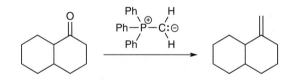

6.68 An aldehyde will react with a Wittig reagent to give an alkene. In this case, the Wittig reagent does not have any R groups (the carbon atom bearing the negative charge is connected to two hydrogen atoms and no R groups), so a methylene (CH$_2$) group is installed in place of the oxygen atom.

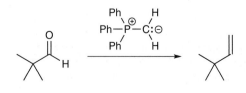

6.69 A ketone will react with a Wittig reagent to give an alkene. In this case, the Wittig reagent does not have any R groups (the carbon atom bearing the negative charge is connected to two hydrogen atoms and no R groups), so a methylene (CH$_2$) group is installed in place of the oxygen atom.

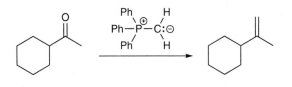

6.71 The starting material is a ketone, and the reagent is a sulfur ylide, which will convert the ketone into an epoxide, as shown:

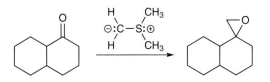

6.72 The starting material is an aldehyde, and the reagent is a sulfur ylide, which will convert the aldehyde into an epoxide, as shown:

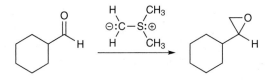

6.73 The starting material is formaldehyde, and the reagent is a sulfur ylide, which will convert formaldehyde into an epoxide, as shown:

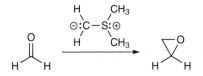

6.75 An aldehyde is being treated with a peroxy acid, so we expect a Baeyer–Villiger reaction to occur. Under these conditions, the aldehyde is oxidized to give a carboxylic acid:

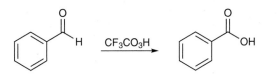

6.76 A ketone is being treated with a peroxy acid, so we expect a Baeyer–Villiger reaction to occur. We look closely at the starting ketone, and we see that it is unsymmetrical. So, we must predict where the oxygen atom will be inserted. We look at both sides, and we see that the left side is a phenyl group and the right side is tertiary (the carbon atom to the right of the carbonyl group is connected to three alkyl groups). Tertiary groups migrate faster than phenyl groups, so the oxygen atom will be inserted on the right side of the carbonyl group:

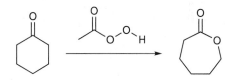

6.77 A ketone is being treated with a peroxy acid, so we expect a Baeyer–Villiger reaction to occur. The starting ketone is symmetrical, so we don't need to decide where the oxygen atom inserts itself (the left side generates the same product as the right side). The product is a cyclic ester:

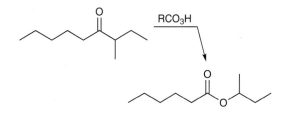

6.78 A ketone is being treated with a peroxy acid (RCO$_3$H), so we expect a Baeyer–Villiger reaction to occur. We look closely at the starting ketone, and we see that it is unsymmetrical. So, we must predict where the oxygen atom will be inserted. We look at both sides of the carbonyl group. The carbon atom to the left of the carbonyl group is connected to one alkyl group, while the carbon atom to the right of the carbonyl group is connected to two alkyl groups. The more-substituted group migrates faster, so the oxygen atom will be inserted on the right side of the carbonyl group:

6.79 An aldehyde is being treated with a peroxy acid (RCO$_3$H), so we expect a Baeyer–Villiger reaction to occur. Under these conditions, the aldehyde is oxidized to give a carboxylic acid:

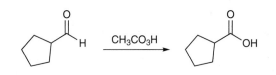

6.81 A ketone can be converted into an alkene (with the installation of a CH$_2$ group) via a Wittig reaction. This is achieved by treating the ketone with a phosphorus ylide (a Wittig reagent), shown here:

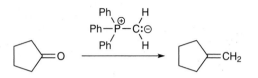

6.82 A symmetrical ketone can be converted into an ester (in this case, a cyclic ester) via a Baeyer–Villiger oxidation. This is achieved by treating the ketone with a peroxy acid, such as MCPBA:

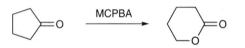

6.83 A secondary alcohol is oxidized to give a ketone upon treatment with a variety of oxidizing agents, including sodium dichromate, the Jones reagent, or even PCC:

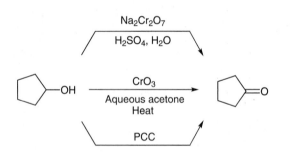

6.84 An acetal can be hydrolyzed to give a ketone upon treatment with water in the presence of an acid catalyst.

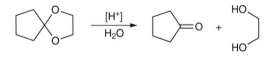

6.85 The starting material is a thioacetal (there is a carbon atom connected to two SR groups). This thioacetal can be reduced with Raney nickel to give the desired product:

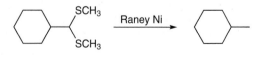

6.86 A ketone can be converted into an enamine upon treatment with a secondary amine and an acid catalyst (and with removal of water):

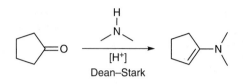

6.87 A ketone can be converted into a cyclic thioacetal upon treatment with ethylene thioglycol and an acid catalyst:

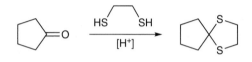

6.88 A primary alcohol is oxidized to give an aldehyde upon treatment with PCC:

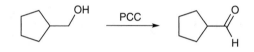

6.89 The starting material is a ketone and the product is an alcohol. That is, the carbonyl group is reduced, but together with the installation of a methyl group. This can be achieved by treating the ketone with a Grignard reagent (methyl magnesium bromide) followed by aqueous acidic workup:

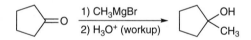

6.90 A ketone can be reduced to give a secondary alcohol upon treatment with lithium aluminum hydride, followed by aqueous acidic workup. Alternatively, sodium borohydride can be used (note that sodium borohydride is shown together with the proton source, not as a separate step):

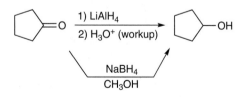

6.91 The starting material is a cyclic thioacetal, which can be reduced with Raney nickel to give the desired product:

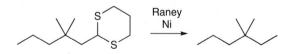

6.92 A symmetrical ketone can be converted into an imine upon treatment with a primary amine and an acid catalyst (and with removal of water):

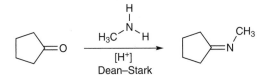

6.93 A symmetrical ketone can be converted into an oxime upon treatment with hydroxyl amine and an acid catalyst:

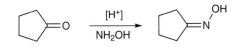

6.94 The starting material is a cyclic thioacetal, which can be reduced with Raney nickel to give the desired product:

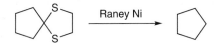

6.95 The starting material is a hydrazone, which can undergo a Wolff–Kishner reduction to give the desired product:

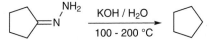

6.96 An aldehyde can be oxidized to give a carboxylic acid upon treatment with a strong oxidizing agent, such as sodium dichromate or the Jones reagent:

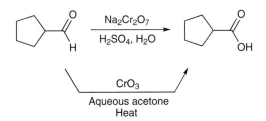

Alternatively, an aldehyde can be oxidized to give a carboxylic acid upon treatment with a peroxy acid (such as MCPBA), via a Baeyer–Villiger oxidation:

6.97 A ketone can be converted into a cyclic acetal upon treatment with ethylene glycol and an acid catalyst (and with removal of water):

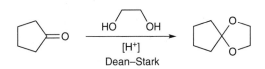

6.98 A ketone can be converted into an epoxide upon treatment with a sulfur ylide:

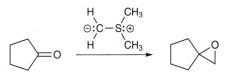

6.99 The starting material is an unsymmetrical ketone, and the product is an ester. The difference between the starting material and the product is an oxygen atom, which needs to be inserted between the aromatic ring and the carbonyl group. This can be accomplished via a Baeyer–Villiger oxidation reaction. Treating the starting ketone with a peroxy acid, such as MCPBA, will give the desired product.

6.100 A symmetrical ketone can be converted into a hydrazone upon treatment with hydrazine and an acid catalyst:

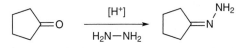

6.103 The starting material is an aldehyde, and the desired product is an alcohol. Notice also that we must install two methyl groups. We can use a Grignard reaction (with MeMgBr) to convert the aldehyde into an alcohol and simultaneously install one methyl group. The resulting alcohol is similar in structure to the desired product, but we are missing a methyl group. In order to install another methyl group, we need to oxidize the alcohol to a ketone, and then we can treat the ketone with another Grignard reagent, thereby installing another methyl group, as shown in the following synthesis:

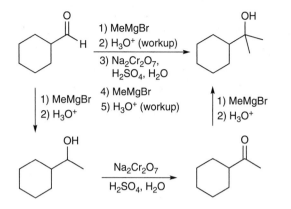

6.104 The starting material is an alcohol, and the desired product is an ester. We have not yet seen many ways to make esters, but we have seen that an ester can be made from a ketone via a Baeyer–Villiger oxidation process. So the last step of our synthesis might be a Baeyer–Villiger oxidation process, in which the product is made from the following ketone:

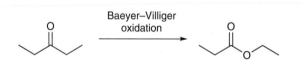

This ketone can be made directly from the starting material via oxidation. In summary, we have developed a two-step synthesis for the desired product. In the first step, the alcohol is treated with an oxidizing agent to give a ketone, and then the ketone is treated with a peroxy acid (such as MCPBA) to give the desired ester:

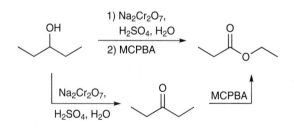

6.105 The starting material is an acetal, and the desired product is an epoxide. We have not yet seen many ways to make epoxides. But in this chapter, we have seen that an epoxide can be made by treating a ketone with a sulfur ylide. So the last step of our synthesis might be the conversion of the following ketone into the desired epoxide:

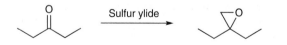

This ketone can be made directly from the starting acetal upon treatment with aqueous acid. In summary, we have developed a two-step synthesis for the desired product. In the first step, the acetal is treated with water and an acid catalyst to give a ketone, and then the ketone is treated with a sulfur ylide to give the desired epoxide:

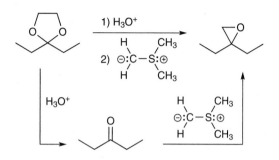

6.106 The starting material is an alkene, and the desired product is an alkane (actually, a cycloalkane). We have not seen any direct ways to remove a methylene (CH_2) group. But in this chapter, we have seen more than one way to convert a ketone into an alkane. For example, the following transformation can be achieved by converting the ketone below into a thioacetal, followed by desulfurization with Raney nickel:

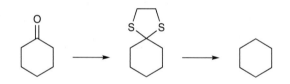

This ketone can be made directly from the starting alkene via ozonolysis. In summary, we have developed the following synthesis for the desired product. First, the alkene undergoes ozonolysis, and the resulting ketone is converted to a cyclic thioacetal, followed by desulfurization with Raney nickel, to give the desired product:

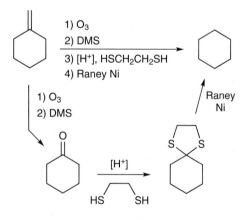

Alternatively, the ketone can be reduced to a CH_2 group via a Clemmensen reduction or via a Wolff–Kishner reduction.

6.107 The starting material is an alkene, and the desired product is an alcohol. We have not seen any direct ways to achieve this transformation. But in this chapter, we have seen that the desired alcohol can be made via reduction of the corresponding ketone:

This ketone can be made directly from the starting alkene via ozonolysis. In summary, we have developed the following synthesis for the desired product. First, ozonolysis is performed to convert the alkene into a ketone, and this ketone is then reduced to give the desired product. The reduction step can be achieved either with lithium aluminum hydride (as shown) or with NaBH$_4$:

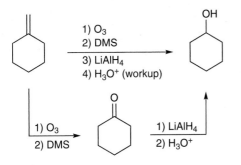

6.108 The starting material is an aldehyde, and the desired product is an alkene. We have not seen any direct ways to achieve this transformation. But in this chapter, we have seen that the desired alkene can be made via from a ketone via a Wittig reaction:

This ketone can be made from the starting aldehyde via a two-step process: (1) a Grignard reaction, followed by (2) oxidation. In summary, we have developed the following synthesis for the desired product. The starting aldehyde is treated with ethyl magnesium bromide, followed by aqueous acidic workup, to give an alcohol, which is then oxidized to give a ketone. And finally, the ketone is treated with a Wittig reagent to give the desired product:

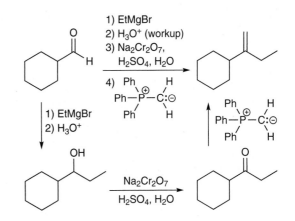

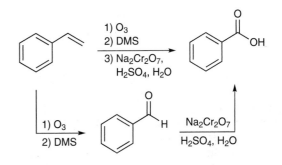

6.109 The starting material is an alkene, and the product is a carboxylic acid. Notice also that the product has one less carbon atom than the starting material. Removing a carbon atom requires breaking a bond between carbon atoms, and that can be accomplished via ozonolysis. The resulting aldehyde can then be oxidized to give the desired carboxylic acid:

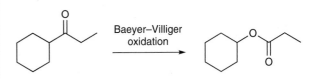

6.110 The starting material is an aldehyde, and the desired product is an ester. We have not yet seen many ways to make esters, but we have seen that an ester can be made from a ketone via a Baeyer–Villiger oxidation process. So the last step of our synthesis might be a Baeyer–Villiger oxidation process, in which the product is made from the following ketone:

This ketone can be made from the starting aldehyde via a two-step process: (1) a Grignard reaction, followed by (2) oxidation. In summary, we have developed the following synthesis for the desired product. The starting aldehyde is treated with ethyl magnesium bromide, followed by aqueous acidic workup, to give an alcohol, which is then oxidized to give a ketone. And finally, the ketone is treated with a peroxy acid (such as MCPBA) to give the desired ester:

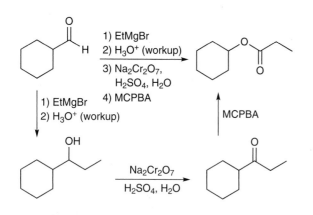

6.111 The starting material is a acetal, and the desired product is an ester. We have not yet seen many ways to make esters, but we have seen that an ester can be made from a ketone via a Baeyer–Villiger oxidation process. So the last step of our synthesis might be a Baeyer–Villiger oxidation process, in which the product is made from the following ketone:

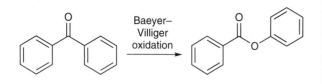

This ketone can be made directly from the starting acetal upon treatment with aqueous acid. In summary, we have developed a two-step synthesis for the desired product. In the first step, the acetal is treated with water and an acid catalyst to give a ketone, and the ketone is then treated with a peroxy acid (such as MCPBA) to give the desired ester:

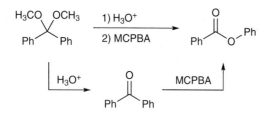

6.112 The desired product is an enamine, and we have only seen one way to make an enamine (from the reaction between a ketone and a secondary amine), so the last step of our synthesis must be the following:

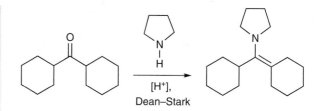

Now let's work our way forward, beginning with the starting material, which is a primary alcohol. We need to introduce a cyclohexyl group into the compound, which means that we need to make a C–C bond. This can be accomplished by the reaction between a Grignard reagent and a carbonyl group, so we first oxidize the alcohol with PCC to give an aldehyde, and then treat that aldehyde with cyclohexyl magnesium bromide, followed by aqueous acidic workup. The resulting alcohol can then be oxidized to give the ketone above, which can be converted into the desired product as shown above.

The entire synthesis is summarized here:

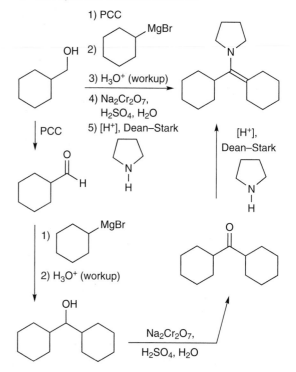

6.113 The starting material has three alcohol groups, and the reagent is PCC (an oxidizing agent). Under these conditions, the tertiary alcohol is unreactive, while the primary alcohol group is oxidized to give an aldehyde, and the secondary alcohol group is oxidized to give a ketone:

6.114 The reagents (ethylene glycol and catalytic acid) indicate formation of a cyclic acetal. The starting material has two carbonyl groups (a ketone and an ester). Under these conditions, the carbonyl group of a ketone is converted into a cyclic acetal, while the carbonyl group of an ester is unreactive, giving the following product:

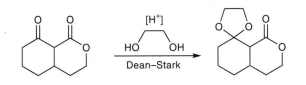

6.115 The starting material is an aldehyde, and the reagents (ethyl magnesium bromide, followed by aqueous acidic workup) indicate a Grignard reaction. The Grignard reagent (ethyl magnesium bromide) attacks the aldehyde, thereby installing an ethyl group, and subsequent workup gives the following, secondary alcohol:

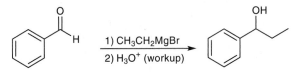

6.116 The product is a cyclic acetal. Recall that cyclic acetals are formed from the reaction between a diol and a ketone/aldehyde:

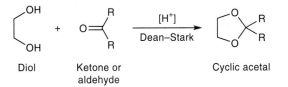

In this case, the structure of the starting diol has been given, and we must determine the structure of the missing reagent (which must be a ketone or aldehyde). If we look at the structure of the acetal product, we see that the central carbon atom of the acetal group (highlighted below) is connected to two methyl groups.

This carbon atom must have been a carbonyl group in the structure of the missing reagent, and since this carbonyl group must have been connected to two methyl groups, we conclude that missing reagent is acetone. Below is the complete reaction:

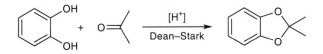

6.117 Recall that a thioacetal can be made from a ketone or aldehyde:

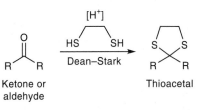

In this case, the starting material must have been an aldehyde (specifically, benzaldehyde) as shown below:

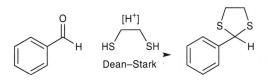

6.118

(a) The product is a cyclic acetal. Recall that cyclic acetals are formed from the reaction between a diol and a ketone/aldehyde:

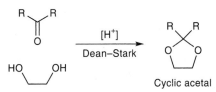

In this case, the structure of the starting aldehyde has been given, and the missing reagent must be ethylene glycol:

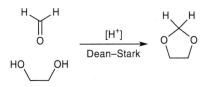

(b) Cyclic acetal formation occurs via a mechanism with seven steps, shown below. In the first step, the aldehyde is protonated to give a protonated carbonyl group, which is then attacked by ethylene glycol. The resulting oxonium ion then undergoes two successive proton transfer steps to give an intermediate oxonium ion with an excellent leaving group (water). Water then leaves as a leaving group, giving an oxonium ion, which is then attacked by a nucleophile in an intramolecular process. The resulting oxonium ion is then deprotonated to give the cyclic acetal product.

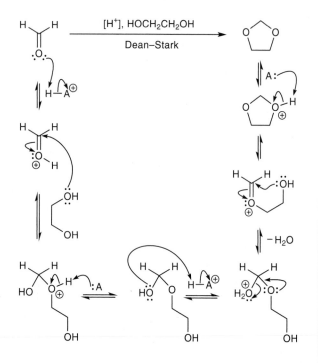

In the mechanism shown above, HA⁺ is indicated as the acid, and A is indicated as the base. The likely identities of HA⁺ and A are shown below:

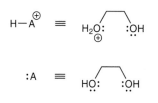

6.119   A Grignard reaction can be used to convert a ketone into a tertiary alcohol:

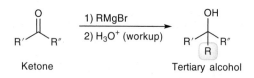

Notice that a tertiary alcohol has three alkyl groups (R, R′, and R″), and any one of these three groups can be installed via the process above, giving three different ways to make the desired alcohol using a Grignard reaction:

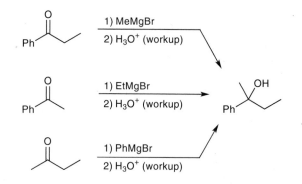

6.120   The product is an imine. Recall that an imine can be made from the reaction between a primary amine and a carbonyl group (of a ketone or an aldehyde):

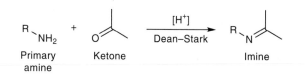

In this case, the primary amine and the carbonyl group must be tethered to each other to give a cyclic imine (the reaction must be intramolecular):

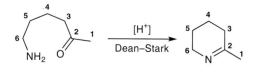

Notice that we use numbers to help us count the carbon atoms. This is a very a useful technique to use whenever drawing intramolecular reactions.

6.121   The starting material is a ketone, and the desired product is an alkene that has one more carbon atom than the ketone. This transformation can be accomplished directly (in one step), using a Wittig reaction:

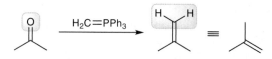

Alternatively, the extra carbon atom can be installed with a Grignard reaction, which converts the ketone into a tertiary alcohol. The resulting tertiary alcohol can then be converted into the desired product via dehydration, by treating the alcohol with concentrated sulfuric acid (an E1 process):

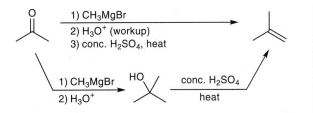

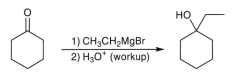

Note that there are other acceptable ways to achieve the desired transformation. For example, shown below is a variation on the second approach (the Grignard approach). Rather than using an E1 process to convert the tertiary alcohol into the desired alkene, we can use an E2 process. This requires converting the alcohol into a tosylate (which has a better leaving group), and then using a strong base (such as NaOH, NaOMe, or NaOEt) to give an E2 reaction:

Now we must remove the OH group. We have not learned a direct way to reduce an alcohol to an alkane, although this can be accomplished in two steps. Dehydration of the alcohol gives an alkene, which can then be reduced (via hydrogenation) to give the desired product. This process is shown below:

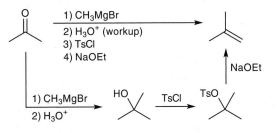

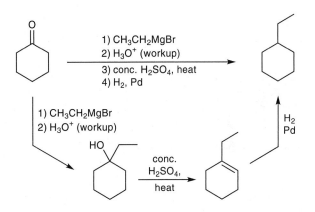

6.122 The starting material (a ketone) has six carbon atoms, while the product (an alkane) has eight carbon atoms. We must install an ethyl group in the location of the carbonyl group, and in the process, the carbonyl group must be reduced to an alkane.

Installing an ethyl group requires formation of a carbon–carbon bond, and in this chapter, we have learned at least two ways to form a carbon–carbon bond: via a Grignard reaction or via a Wittig reaction. Let's first explore a Wittig reaction.

Using a Wittig reaction, we first treat the starting ketone with a Wittig reagent that will install two carbon atoms, and then we reduce the resulting alkene (via hydrogenation) to give the desired product:

6.123 The desired product is a cyclic thioacetal, and we have seen only one way to make a thioacetal (from a ketone), so this must be the last step of our synthesis:

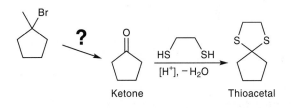

Now we must bridge the gap between the starting alkyl halide and the ketone above. Notice that the ketone has only five carbon atoms, while the alkyl halide has six carbon atoms (a methyl group plus a five-membered ring). In order to synthesize the ketone, we will need to break a carbon–carbon bond. This can be achieved via ozonolysis, provided that we can first create a C=C double bond at the desired location. This can be achieved via elimination (E2), using a strong base:

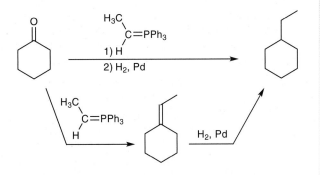

Alternatively, we can form the necessary carbon–carbon bond via a Grignard reaction. The ketone is first treated with ethyl magnesium bromide, followed by aqueous acidic workup, thereby installing an ethyl group.

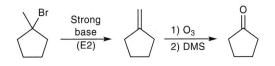

Notice that in the elimination step, we will need to form the Hoffmann product (a disubstituted alkene), rather than the Zaitsev product (a trisubstituted alkene). So we must use a sterically hindered base, such as *tert*-butoxide.

The complete synthesis is shown below:

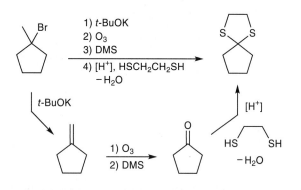

6.124 The desired product is a cyclic acetal, and we have seen only one way to make a cyclic acetal (from a diol reacting with a ketone or aldehyde), so this must be the last step of our synthesis:

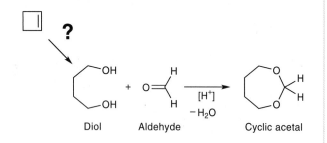

Now we must bridge the gap between the starting cycloalkene and the diol above. This requires opening the ring, which can be achieved with ozonolysis, giving a dialdehyde. This dialdehyde can then be reduced to the necessary diol with two equivalents of lithium aluminum hydride (or sodium borohydride):

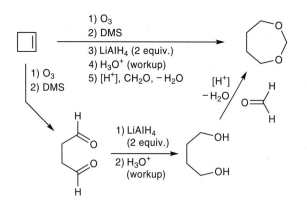

6.125 The product is a *trans* diol, which can be made by *anti*-dihydroxylation of the corresponding alkene (shown below). This can be achieved by treatment with a peroxy acid (RCO_3H), such as MCPBA, followed by aqueous acid to open the epoxide ring:

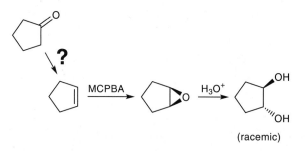

Now we must bridge the gap between the starting ketone and the alkene. We have not seen a direct way to convert a ketone into an alkene, although this can be accomplished in two steps: (1) reduction of the ketone with hydride to give an alcohol, followed by (2) dehydration with concentrated sulfuric acid to give the alkene:

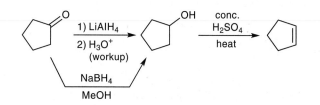

The complete synthesis is shown below:

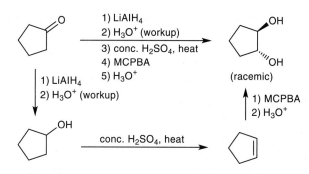

6.126 The product is an epoxide, which can be made by treating the corresponding alkene (shown below) with a peroxy acid, such as MCPBA, so this can be the last step of our synthesis:

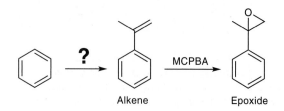

Now we must bridge the gap between the starting material (benzene) and the alkene. Continuing to work backwards, the alkene might be prepared from a ketone via a Wittig reaction:

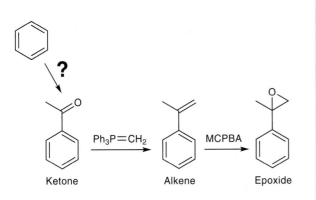

Ketone   Alkene   Epoxide

This is useful, because we have seen a direct way to convert benzene into the ketone required. Specifically, we can perform a Friedel–Crafts acylation:

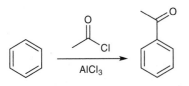

Now that we have worked backwards to the starting material, we can draw the entire synthesis, from the beginning to the end:

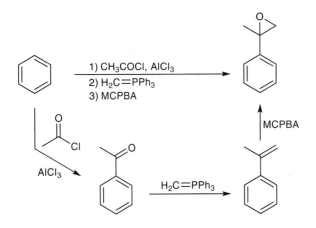

There are certainly other acceptable syntheses. For example, the variation below replaces the Wittig reaction for a Grignard reaction, followed by elimination. This synthesis is also reasonable:

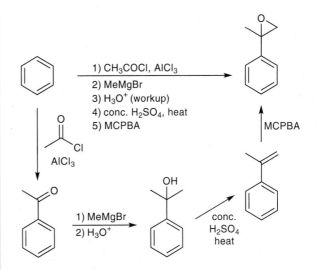

6.127 The product is an enamine, which can be made by treating the corresponding ketone (shown below) with a secondary amine, in the presence of catalytic acid:

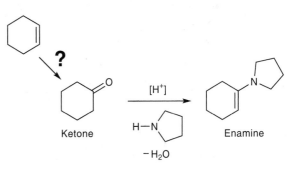

Ketone   Enamine

Now we must bridge the gap between the starting alkene and the ketone shown above. We have not seen a direct (one-step) method to convert an alkene into a ketone, although we have seen a way to achieve this transformation in two steps: acid-catalyzed hydration of the alkene to give an alcohol, followed by oxidation of the alcohol with an oxidizing agent, such as PCC. The complete synthesis is shown below:

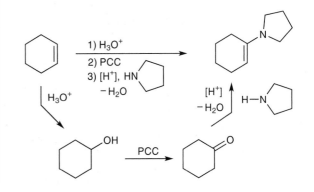

6.128 The product is an imine, which can be made by treating the corresponding aldehyde (shown below) with methylamine, in the presence of catalytic acid:

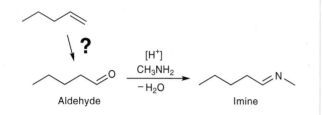

Aldehyde   Imine

Now we must bridge the gap between the starting alkene (1-pentene) and the aldehyde shown above (1-pentanal). We have not seen a direct (one-step) method to convert an alkene into an aldehyde, although this transformation can be achieved, if we use more than one reaction. As shown in the synthesis below, hydroboration-oxidation will convert the alkene into a primary alcohol (*anti*-Markovnikov addition of H_2O), which can then be oxidized with PCC to give the necessary aldehyde. Treating the aldehyde with methylamine and catalytic acid completes the synthesis:

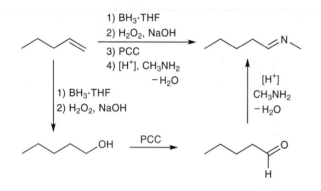

Alternatively, the alkene can be converted into an alkyne, which can then be converted into the necessary aldehyde, as shown in the following synthesis:

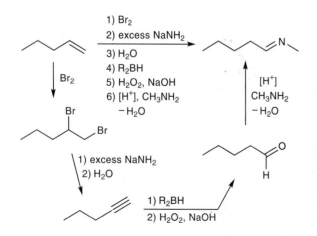

6.129 The starting material is an alkane, and the product is an alkene with one additional carbon atom (compared with the starting alkane):

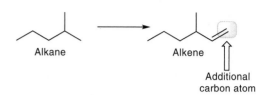

Alkane   Alkene

Additional
carbon atom

The alkene can be made via a Wittig reaction, which can be used to install this additional carbon atom:

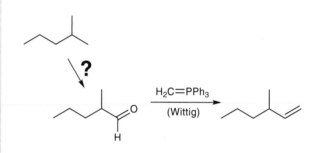

Now we must bridge the gap between the starting alkane and the aldehyde. The first step of our synthesis must be the installation of a functional group on the alkane, and we have only seen one way to do that. Specifically, we can install a bromine atom selectively at the tertiary position of an alkane via radical halogenation:

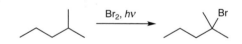

This gives us an alkyl bromide, so now we must bridge the gap by converting this alkyl bromide into the aldehyde necessary for the last step of our synthesis:

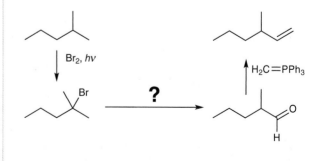

This requires that we move the location and the identity of the functional group.

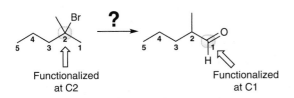

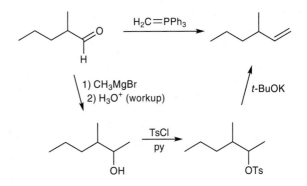

There are many ways to accomplish this. Two such methods are shown below:

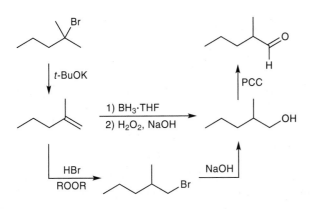

The following is a complete synthesis (one of many possibilities):

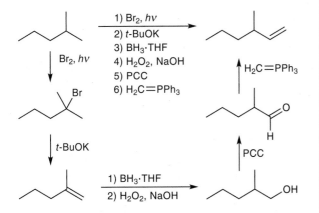

Note that there are other acceptable ways to convert the starting alkane into the desired alkene. For example, the last step of the synthesis (the Wittig reaction) can be replaced with a Grignard reaction, followed by tosylation and a subsequent E2 process (with a sterically hindered base), as shown below:

6.130 The starting material is cyclic, and the desired product is acyclic, so our synthesis must involve opening the ring (breaking a carbon–carbon bond). This can be accomplished via ozonolysis, if we first convert the alcohol to an alkene via an elimination process:

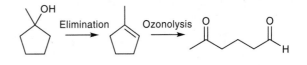

The resulting acyclic compound can be converted into the product upon treatment with a strong oxidizing agent, thereby oxidizing the aldehyde group to a carboxylic acid:

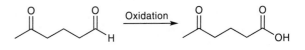

The complete synthesis is shown below:

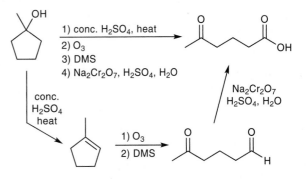

6.131 The starting material is a five-membered ring, while the desired product is a six-membered lactone (a cyclic ester). This type of transformation requires the installation of an oxygen atom within the ring, giving a larger ring (ring expansion). This can be accomplished with a Baeyer–Villiger oxidation, which can be the last step of our synthesis (conversion of a ketone to an ester):

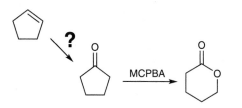

Now we must bridge the gap between the starting cycloalkene and the necessary ketone. This can be accomplished in just two steps: acid-catalyzed hydration to give an alcohol, followed by oxidation. The complete synthesis is shown below:

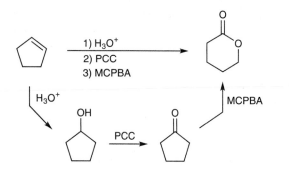

6.132 The starting material is cyclic, and the desired product is acyclic, so our synthesis must involve opening the ring (breaking a carbon–carbon bond), which can be accomplished with an ozonolysis process. But this is not the only change that we must make to the carbon skeleton. Notice that the starting material has five carbon atoms, while the desired product has seven carbon atoms, so we will need to make new carbon–carbon bonds as well. This can be accomplished with Wittig reactions. So the last two steps of our synthesis could be ozonolysis, followed by treatment with excess Wittig reagent, as shown:

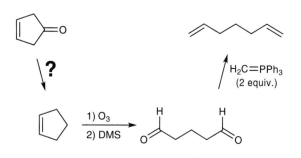

Now we must bridge the gap between the starting material and the necessary alkene, which requires complete reduction of the ketone group that is present in the starting material. We have

learned three different ways to achieve this type of transformation: (1) Clemmensen reduction, or (2) conversion of the ketone to a thioacetal, followed by desulfurization of the thioacetal, or (3) Wolff–Kishner reduction:

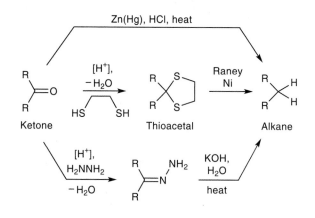

We should avoid the first of these three approaches (Clemmensen reduction) because it requires the use of HCl, which can react with the alkene group of the starting material to give an undesired addition reaction. Therefore, we should use one of the other two approaches above.

The following is a synthesis in which the starting ketone is converted into a thioacetal, followed by desulfurization. We then open the resulting cycloalkene with ozonolysis, and then employ Wittig conditions to give the desired product:

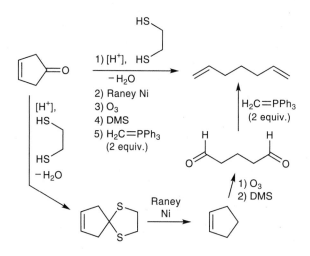

Alternatively, the first two steps above can be replaced with imine formation, followed by Wolff–Kishner reduction.

6.133 The starting material is an alcohol, and the product is an alkane. We have not yet learned a way to achieve this transformation in one step, so we will have to identify a multistep method. We have seen that a secondary alcohol can be oxidized to give a ketone, and we have seen that a ketone can be reduced to give an alkane:

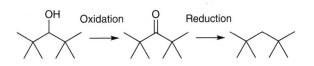

The complete synthesis is shown here:

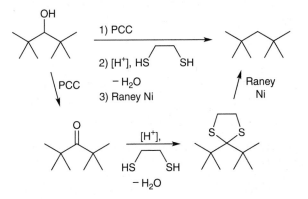

Note that it is not possible to dehydrate the starting alcohol (to give an alkene), because the starting alcohol lacks beta protons.

6.134 The desired product is an oxime, which can be made from the corresponding ketone:

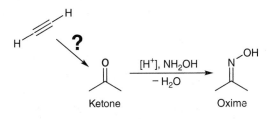

Now we must bridge the gap between the starting material (acetylene) and the ketone above. Notice that acetylene has two carbon atoms, and the ketone has three carbon atoms. The additional carbon atom can be installed via alkylation of acetylene, and then we can install the correct functional group via acid-catalyzed hydration (Markovnikov addition):

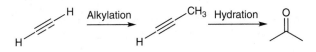

The complete synthesis is shown here:

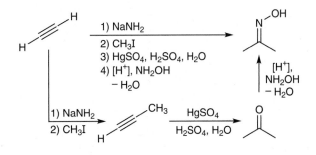

6.135 The starting material and the product have the same carbon skeleton. This transformation requires that we change the location and the identity of the functional group. The starting material has an OH group at C2, while the product has a carbonyl group at C1:

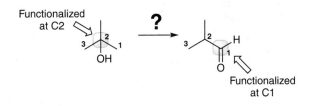

In order to move the location of the functional group, we first perform an elimination reaction (E1) to give an alkene, which is now functionalized at both C1 and C2. Next, we can add H and OH across this bond in an *anti*-Markovnikov fashion, using hydroboration-oxidation:

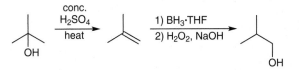

And finally, this new alcohol, with the OH group at C1 (rather than C2) can be oxidized with PCC to give the product:

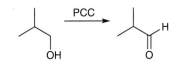

The complete synthesis is shown here:

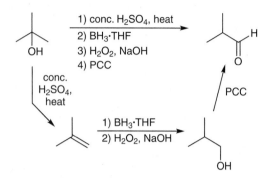

6.136 The product is an acetal, and we have seen that an acetal can be made from the corresponding aldehyde or ketone (in this case, aldehyde):

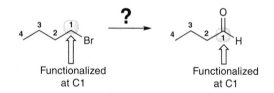

Now we must bridge the gap between the starting alkyl bromide and the aldehyde above. These compounds have the same carbon skeleton, and the location of the functional group is the same in both compounds (in each case, the functional group is at C1):

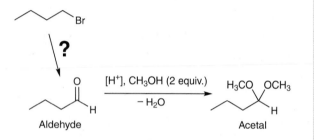

We must change only the identity of the functional group. This can be accomplished by converting the alkyl halide into an alcohol (via S_N2), and then oxidizing the alcohol with PCC to give an aldehyde. The complete synthesis is shown here:

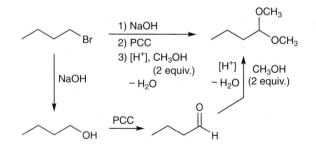

6.137 The product is an acetal:

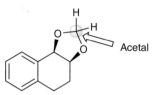

We have seen that a cyclic acetal can be made from the reaction between an aldehyde (or ketone) and a diol. So this can be the last step of our synthesis:

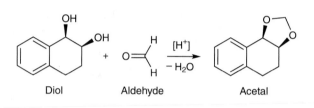

Recall that a diol can be made from the corresponding alkene:

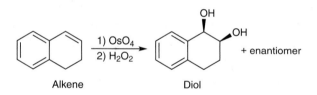

So now we must bridge the gap between the starting ketone and this alkene:

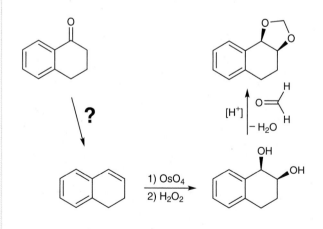

This can be accomplished in two steps: (1) reduction to an alcohol, followed by (2) acid-catalyzed dehydration:

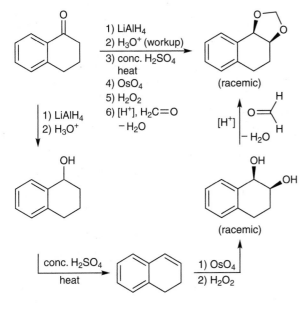

The complete synthesis is shown here:

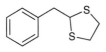

6.138 The product is a thioacetal:

Thioacetal

We have seen that a thioacetal can be made from an aldehyde (or ketone). So this can be the last step in our synthesis:

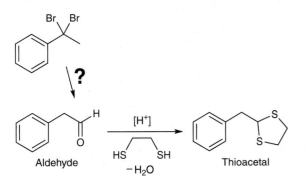

Now we must bridge the gap between the starting material and the aldehyde above. This can be accomplished via elimination to give an alkyne, followed by hydroboration-oxidation to give the necessary aldehyde:

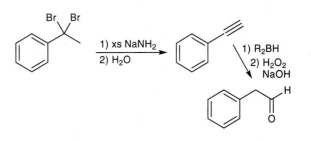

The complete synthesis is shown here:

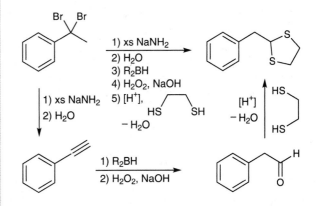

6.139

(a) The starting material is a ketone, and it is being treated with methyl magnesium bromide (a Grignard reagent), followed by aqueous acidic workup. This results in the installation of a methyl group, giving a tertiary alcohol:

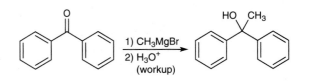

(b) The product has protons that the starting material lacks. Specifically, the product has a methyl group and an OH group, each of which will produce a signal in the ^{1}H NMR spectrum:

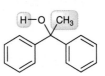

The signal for the OH group is expected to appear between 2 and 5 ppm, while the signal for the methyl group is expected to appear near 2 ppm (rather than the benchmark value of 0.9 ppm) because of the effects of the two, nearby aromatic rings. These two signals will be absent in the ^{1}H NMR spectrum of the starting material.

(c) The starting material and the product have different carbon skeletons. The product has an additional carbon atom (the methyl group) that is absent in the starting material.

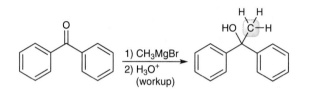

Accordingly, the ^{13}C NMR spectrum of the product will have one additional signal, corresponding to this methyl group. Furthermore, the starting material will have a weak signal near 200 ppm, corresponding to the carbon atom of the carbonyl group. In the product, this carbon atom is no longer a carbonyl group, but is rather next to an OH group. So this carbon atom will produce a signal between 50 and 100 ppm. So compared to the starting ketone, the alcohol product will be missing the carbonyl signal around 200 ppm, and will have two signals below 100 ppm, corresponding to sp^3 hybridized carbon atoms. The starting material has no signals below 100 ppm.

6.140

(a) Compound **A** has the molecular formula C_8H_8O, which has five degrees of unsaturation. We are therefore on the lookout for an aromatic ring (accounting for four degrees of unsaturation), and one more degree of unsaturation (either a pi bond or a ring). The strong signal at 1680 cm^{-1} suggests that we have a carbonyl group that is conjugated to the aromatic ring. The multiplet at 7.4 ppm confirms the presence of an aromatic ring, and the integration value of this signal (5H) indicates that the ring is monosubstituted:

The ^{1}H NMR spectrum has one other signal: a singlet with an integration of 3H. This signal suggests that there is a methyl

group that has no neighbors, so we propose the following structure:

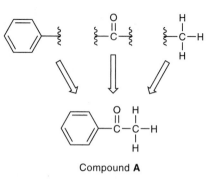

Compound **A**

When this compound is treated with lithium aluminum hydride, followed by aqueous acidic workup, the carbonyl group is reduced to an alcohol, giving compound **B**:

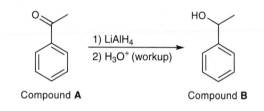

Compound **A**   Compound **B**

(b) Compounds **A** and **B** have the same carbon skeleton, and we expect their ^{13}C NMR spectra to have the same number of signals (six signals in each spectrum). In the ^{13}C NMR spectrum of compound **A**, we expect one signal near 200 ppm, corresponding to the carbonyl group at the benzylic position. In contrast, the ^{13}C NMR spectrum of compound **B** will have a signal between 50 and 100 ppm, corresponding to the benzylic position (which is connected to the OH group). Therefore, compounds **A** and **B** can be differentiated by the location of the signals associated with the benzylic positions:

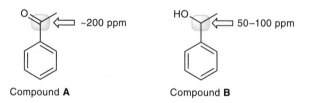

Compound **A**   Compound **B**

6.141 Compound **A** is an alcohol, which is converted into the corresponding tosylate upon treatment with tosyl chloride and pyridine. This tosylate then undergoes an E2 reaction upon treatment with a non-nucleophilic base such as DBU, giving an alkene (Compound **B**). Ozonolysis of compound **B** gives compound **C**, which is an aldehyde (note that one carbon atom has been removed in the process). When compound **C** is treated with lithium aluminum hydride, followed by aqueous acidic workup, the carbonyl group is reduced to give an alcohol (Compound **D**). Compound **D** can be converted back into compound **C** upon treatment with a mild oxidizing agent, such as PCC (a stronger oxidizing agent, such as chromic acid, would oxidize this primary alcohol to a carboxylic acid). Compound **C** can be converted back into compound **B** by installing a carbon atom via a Wittig reaction. And finally, the conversion of compound **B** into compound **A** requires an *anti*-Markovnikov addition of H and OH across the alkene, which can be accomplished via hydroboration-oxidation.

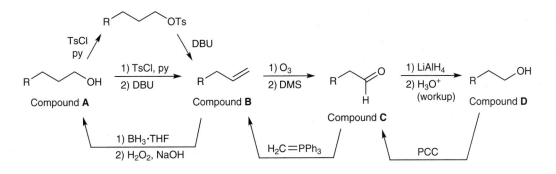

6.142 Compound **A** is an aldehyde, while compound **B** is an alcohol with an extra methyl group. So conversion of compound **A** to compound **B** requires installation of a methyl group and conversion of the carbonyl group to an alcohol. Both of these changes can be accomplished with a Grignard reaction in which methyl magnesium bromide is used as the Grignard reagent. The conversion of compound **B** to compound **C** is an oxidation process, and it can be achieved with a number of oxidizing agents, including PCC. Alternatively, we could use sodium dichromate and sulfuric acid (which gives chromic acid, a strong oxidizing agent). Compound **C** is a ketone, while compound **D** is the corresponding cyclic acetal, so compound **C** can be converted into compound **D** upon treatment with ethylene glycol in catalytic acid (with removal of water). This acetal group can be removed, converting compound **D** back into compound **C**, upon treatment with aqueous acid.

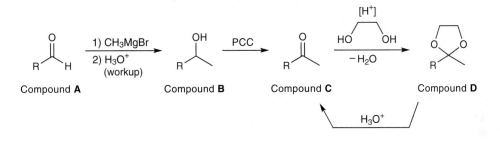

6.143 Recall that an acetal is made from the reaction between an aldehyde (or ketone) and a diol:

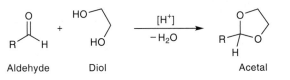

Note that the carbon atom connected to two oxygen atoms (in the structure of the acetal) is the same carbon atom that was the carbonyl group in the starting aldehyde:

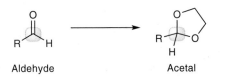

Also note that the highlighted carbon atom (in the structure of the acetal) is connected to two oxygen atoms, and each of these C—O bonds is formed during acetal formation:

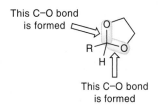

With this in mind, let's analyze the specific acetal in our case. We identify the carbon atom connected to two oxygen atoms (highlighted below), and we determine that this carbon atom must have been the carbonyl group in the acyclic starting material.

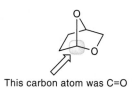

This carbon atom was C=O

Furthermore, we can determine which two C—O bonds are formed in the process (highlighted below):

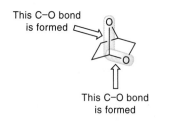

This C—O bond is formed

This C—O bond is formed

Working backwards, we break these two C—O bonds, and we redraw the highlighted carbon atom as a carbonyl group, giving the following compound:

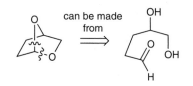

can be made from

This is our starting material for the reaction:

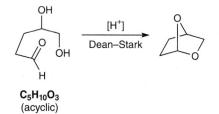

[H⁺]

Dean–Stark

$C_5H_{10}O_3$
(acyclic)

6.144

(a) Recall that a cyclic acetal is made from the reaction between an aldehyde/ketone and a diol:

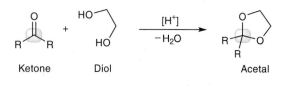

Ketone   Diol   Acetal

Note that, in the structure of the acetal, the highlighted carbon atom connected to two oxygen atoms is the same carbon

atom that was the carbonyl group in the starting ketone, and that each of these C—O bonds is formed during acetal formation:

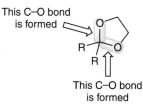

This C—O bond is formed

This C—O bond is formed

With this in mind, let's analyze the specific acetal in our case. We identify the carbon atom connected to two oxygen atoms (highlighted below), and we determine that this carbon atom must be a carbonyl group in the acyclic starting material.

This carbon atom was C=O

Furthermore, we can determine which two C—O bonds are formed in the process:

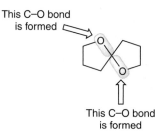

This C—O bond is formed

This C—O bond is formed

Working backwards, we break these C—O bonds, and we redraw the highlighted carbon atom as a carbonyl group, giving the following compound:

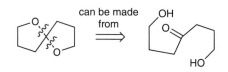

can be made from

This is our starting material for the reaction:

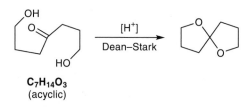

[H⁺]

Dean–Stark

$C_7H_{14}O_3$
(acyclic)

(b) The acyclic starting material has symmetry, with four different types of carbon atoms (labeled **a–d** below), so we expect four signals in the ^{13}C NMR spectrum of this compound.

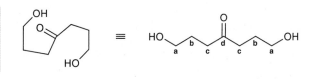

CHAPTER 7

7.2 An acid halide is being treated with an excess of Grignard reagent. First, a Grignard reagent attacks the carbonyl group of the acid halide, giving a tetrahedral intermediate, which then re-forms the carbonyl group by expelling chloride as a leaving group. The resulting ketone is then attacked by another equivalent of the Grignard reagent to give an alkoxide ion. This alkoxide ion cannot re-form the carbonyl group because it has no leaving groups, so this is the end of the reaction. After the reaction is complete, aqueous acid (H_3O^+) is added to the reaction flask, thereby protonating the alkoxide ion to give an alcohol:

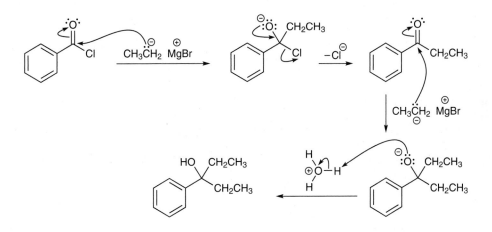

7.3 An acid halide is being treated with an alcohol in the presence of pyridine. First, the alcohol attacks the carbonyl group of the acid halide, giving a tetrahedral intermediate. This intermediate then re-forms the carbonyl group by expelling chloride as a leaving group. And finally, the resulting cation is then deprotonated by pyridine to give an ester:

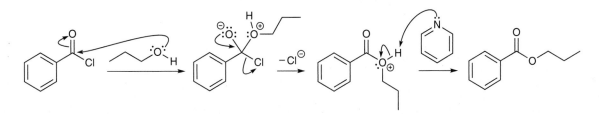

7.4 An acid halide is being treated with excess amine. First, the amine attacks the carbonyl group of the acid halide, giving a tetrahedral intermediate. This intermediate then re-forms the carbonyl group by expelling chloride as a leaving group. And finally, the resulting cation is then deprotonated by a second equivalent of the amine to give an amide:

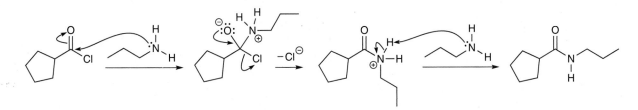

7.5 An acid halide is converted into a carboxylic acid upon treatment with thionyl chloride. First, the carboxylic acid functions as a nucleophile and attacks the S=O bond. The S=O bond is then re-formed (by expelling a chloride ion), followed by a proton transfer step. The net result of these first three steps is the conversion of an OH group into a better leaving group. A chloride ion (formed in the second step) then attacks the carbonyl group, giving a tetrahedral intermediate, which can then re-form the carbonyl group by expelling SO₂ gas and a chloride ion. The product is an acid halide:

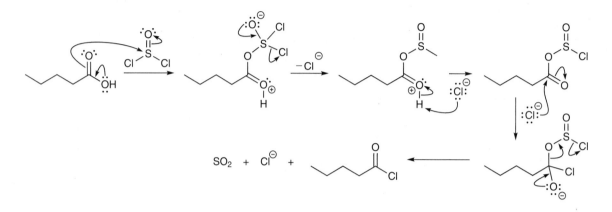

7.6 An acid halide is being treated with an excess of LiAlH₄, which is a hydride reducing agent. First, LiAlH₄ functions as a delivery agent of hydride (H⁻) and attacks the carbonyl group of the acid halide, giving a tetrahedral intermediate, which then re-forms the carbonyl group by expelling chloride as a leaving group. The resulting aldehyde is then attacked by another equivalent of LiAlH₄ to give an alkoxide ion. This alkoxide ion cannot re-form the carbonyl group because it has no leaving groups, so this is the end of the reaction. After the reaction is complete, aqueous acid (H₃O⁺) is added to the reaction flask, thereby protonating the alkoxide ion to give an alcohol:

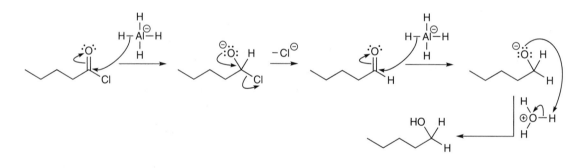

7.8 Upon treatment with thionyl chloride (SOCl₂), a carboxylic acid is converted into an acid halide. When this acid halide is treated with an excess of ethyl magnesium bromide (a Grignard reagent), followed by aqueous acid (H₃O⁺), the acid halide reacts with two equivalents of the Grignard reagent to give an alcohol. Notice that two ethyl groups are installed.

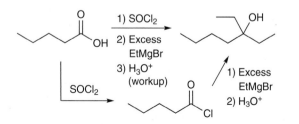

7.9 Upon treatment with thionyl chloride (SOCl₂), a carboxylic acid is converted into an acid halide. When this acid halide is treated with lithium diethyl cuprate (a tame carbon nucleophile), the acid halide reacts with only one equivalent of the nucleophile to give a ketone.

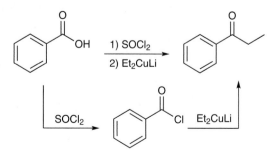

7.10 Upon treatment with excess ammonia (NH₃), a carboxylic acid is converted into an amide.

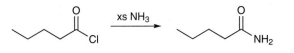

7.11 When an acid halide is treated with an excess of a hydride reducing agent, such as NaBH₄ or LiAlH₄, the acid halide will react with two equivalents of the reducing agent to give an alcohol.

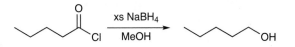

7.12 An acid halide is converted into an ester upon treatment with an alcohol in the presence of pyridine.

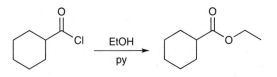

7.14 The desired ester can be made from an acid halide, which can be made from the starting carboxylic acid. This gives the following two-step synthesis:

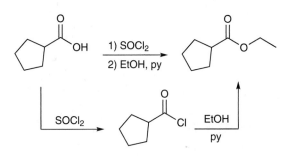

7.15 The desired ketone can be made by treating an acid halide with lithium dimethyl cuprate. The acid halide can be made from the starting carboxylic acid. This gives the following two-step synthesis:

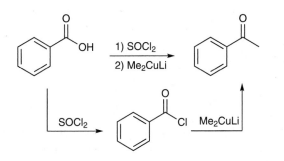

7.16 The starting material is a carboxylic acid, and the product is an alcohol. In making the alcohol, we must install one ethyl and one methyl group. This cannot be achieved by treating an acid halide with an excess of a Grignard reagent, because we need to install two different alkyl groups. Rather, the desired transformation can be achieved by converting the carboxylic acid into an acid halide, and then treating the acid halide with a lithium dialkyl cuprate. This results in the installation of only one alkyl group to give a ketone. And this ketone can then be treated with a Grignard reagent.

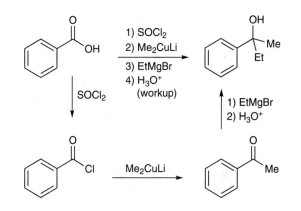

In the synthesis shown, we used Me₂CuLi followed by EtMgBr. Alternatively, we could have used Et₂CuLi, followed by MeMgBr, which would have accomplished the same goal.

7.17 The starting material is a carboxylic acid, and the product is an alcohol. In making the alcohol, we must install one ethyl. This can be achieved by converting the carboxylic acid into an acid halide, and then treating the acid halide with lithium diethyl cuprate. This results in the installation of only one alkyl group to give a ketone. And this ketone can then be treated with a reducing agent to give the desired alcohol.

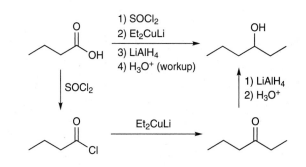

7.18 An acid halide can be converted directly into an acid anhydride upon treatment with a carboxylic acid in the presence of pyridine. The carboxylic acid functions as a nucleophile that attacks the carbonyl group of the acid halide, and pyridine functions as a base to remove the acidic protons generated during this process.

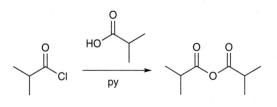

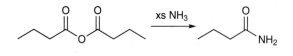

7.20 An anhydride is converted into an amide upon treatment with excess ammonia:

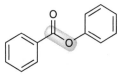

7.21 An anhydride is converted into an alcohol when treated with excess reducing agent:

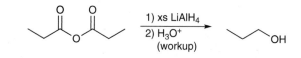

7.22 An anhydride is converted into a carboxylic acid when treated with aqueous acid:

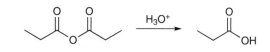

7.23 This is an intramolecular Fischer esterification process, and the mechanism should have six steps, like any Fischer esterification. There are two core steps (attack the carbonyl group, and re-form the carbonyl group), both highlighted below, and the remaining four steps are proton transfer steps that occur in order to facilitate the two core steps. In the first step of the mechanism, the carbonyl group is protonated, rendering it more electrophilic and subject to attack by a weak nucleophile. In this case, the nucleophile is an alcohol that is tethered to the carbonyl group (in the same molecule), allowing for an intramolecular attack. The resulting cyclic oxonium ion is then deprotonated by water and protonated by a hydronium ion, to give an intermediate that can re-form the carbonyl group via expulsion of a leaving group (H_2O). The resulting intermediate is then deprotonated by water to give the product:

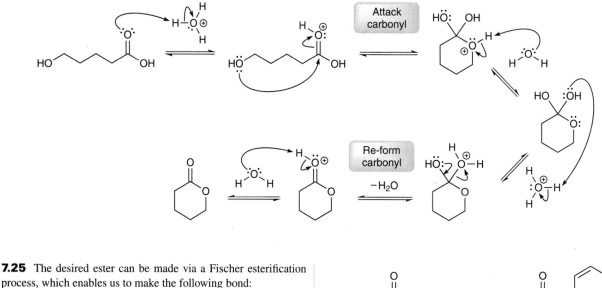

7.25 The desired ester can be made via a Fischer esterification process, which enables us to make the following bond:

The required starting materials (a carboxylic acid and an alcohol) are shown:

7.26 The desired ester can be made via a Fischer esterification process, which enables us to make the following bond:

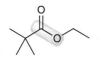

The required starting materials (a carboxylic acid and an alcohol) are shown below:

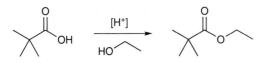

7.27 The desired ester can be made via an intramolecular Fischer esterification process, which enables us to make the following bond:

The required starting material (which is both a carboxylic acid and an alcohol) is shown:

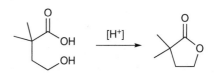

7.28 The desired ester can be made via an intramolecular Fischer esterification process, which enables us to make the following bond:

The required starting material (which is both a carboxylic acid and an alcohol) is shown below:

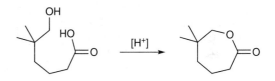

7.30 Under aqueous acidic conditions, an ester can be hydrolyzed to give a carboxylic acid and an alcohol. This process is the reverse of a Fischer esterification process, and the mechanism should have six steps. There are two core steps (attack the carbonyl group, and re-form the carbonyl group), both highlighted below, and the remaining four steps are proton transfer steps that occur in order to facilitate the two core steps. In the first step of the mechanism, the carbonyl group is protonated, rendering it more electrophilic and subject to attack by a weak nucleophile (H$_2$O). Water attacks the carbonyl group to give a tetrahedral intermediate. This intermediate is then deprotonated by water and then protonated by a hydronium ion, to give an intermediate that can re-form the carbonyl group via expulsion of a leaving group (ROH). Notice that after the leaving group leaves, the alcohol and the carboxylic acid are tethered to each other in one molecule (because the starting ester was cyclic). The resulting intermediate is then deprotonated by water to give the product:

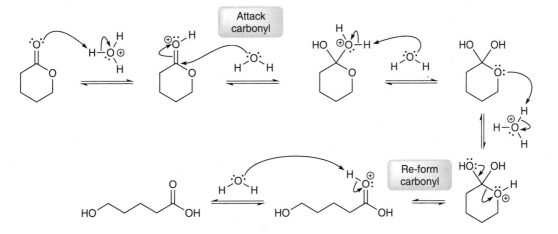

7.31 Under aqueous acidic conditions, an ester can be hydrolyzed to give a carboxylic acid and an alcohol. This process is the reverse of a Fischer esterification process, and the mechanism should have six steps. There are two core steps (attack the carbonyl group, and re-form the carbonyl group), both highlighted below, and the remaining four steps are proton transfer steps that occur in order to facilitate the two core steps. In the first step of the mechanism, the carbonyl group is protonated, rendering it more electrophilic and subject to attack by a weak nucleophile (H_2O). Water attacks the carbonyl group to give a tetrahedral intermediate. This intermediate is then deprotonated by water and then protonated by a hydronium ion, to give an intermediate that can re-form the carbonyl group via expulsion of a leaving group (EtOH). The resulting intermediate is then deprotonated by water to give a carboxylic acid:

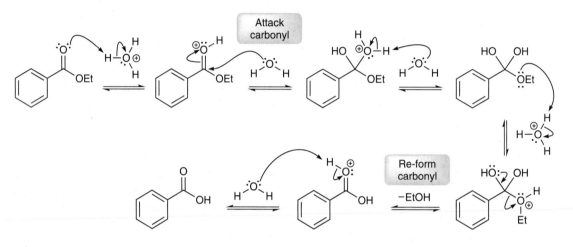

7.33 Upon treatment with aqueous acid (H_3O^+), an ester is hydrolyzed to give a carboxylic acid and an alcohol. This is the reverse of a Fischer esterification:

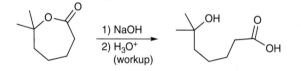

7.34 Upon treatment with hydroxide, followed by a proton source, an ester is hydrolyzed to give a carboxylic acid and an alcohol. This process is called saponification:

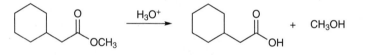

7.35 Upon treatment with an alcohol in the presence of catalytic acid, a carboxylic acid is converted to an ester. This process is called Fischer esterification:

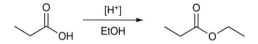

7.36 Upon treatment with hydroxide, followed by a proton source, an ester is hydrolyzed to give a carboxylic acid and an alcohol. This process is called saponification:

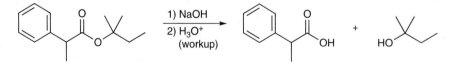

7.37 In the first step of the mechanism, hydroxide functions as a nucleophile and attacks the cyano group. The resulting anion is then protonated by water, followed by a deprotonation step to give a resonance-stabilized anion. This anion can receive a proton (from water, because there are no stronger acids present) to give the product:

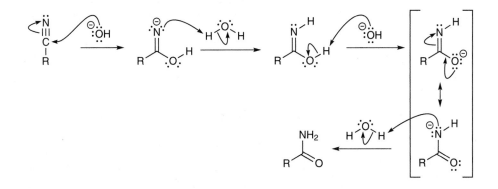

7.39 An amide is hydrolyzed under aqueous acidic conditions. This process is similar to the acid-catalyzed hydrolysis of an ester, and the mechanism should have six steps. There are two core steps (attack the carbonyl group, and re-form the carbonyl group), both highlighted below, and the remaining four steps are proton transfer steps that occur in order to facilitate the two core steps. In the first step of the mechanism, the carbonyl group is protonated, rendering it more electrophilic and subject to attack by a weak nucleophile (H_2O). Water attacks the carbonyl group to give a tetrahedral intermediate. This intermediate is then deprotonated by water and then protonated by a hydronium ion, to give an intermediate that can re-form the carbonyl group via expulsion of a leaving group (R_2NH). Notice that after the leaving group leaves, the amine and the carboxylic acid are tethered to each other in one molecule (because the starting amide was cyclic). The resulting intermediate is then deprotonated by water to give the product:

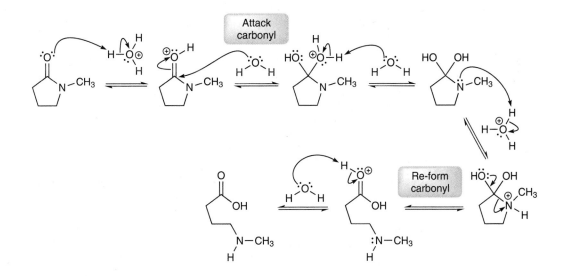

7.40 In the first step of the mechanism, hydroxide functions as a nucleophile and attacks the cyano group. The resulting anion is then protonated by water, followed by a deprotonation step to give a resonance-stabilized anion. This anion can receive a proton (from water, because there are no stronger acids present) to give the product:

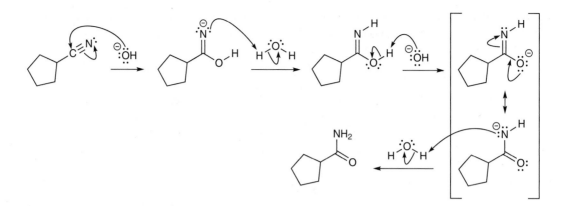

7.41 Hydroxide will attack the carbonyl group of an amide to give a tetrahedral intermediate, which can then re-form the carbonyl group. Under these basic conditions, the resulting anion then undergoes two proton transfer steps, giving a carboxylate ion:

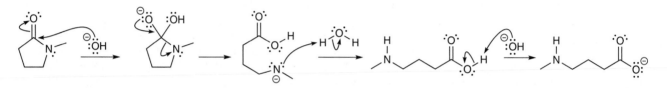

7.42 Under acidic conditions, the carbonyl group is first protonated, which renders it more electrophilic. Water then attacks the carbonyl group, giving a tetrahedral intermediate. In order to re-form the carbonyl group and break a C–N bond, we must first protonate a nitrogen atom (so that it is a better leaving group), and we can only do this after first removing a proton (we must avoid drawing an intermediate with two positive charges). So, the intermediate undergoes two successive proton transfer steps. Then, the carbonyl group is re-formed by breaking a C–N bond, and the resulting cation is then deprotonated. In order to break the other C–N bond, the nitrogen atom must first be protonated, rendering it a better leaving group. Finally, deprotonation (and loss of carbon dioxide) gives the products. Note that this final step can be alternatively drawn as two separate steps: (1) deprotonation to make the carboxylate ion, followed by (2) expulsion of the amine and carbon dioxide.

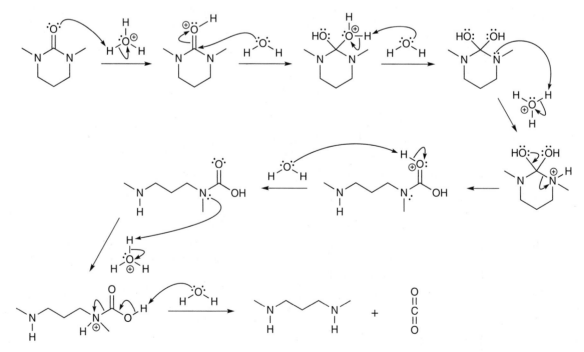

7.44 The starting material is an amide, and the reagent is aqueous acid. We expect the amide to undergo acid-catalyzed hydrolysis to give a carboxylic acid and an amine:

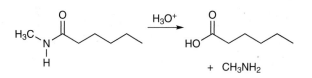

Actually, under these acidic conditions, the amine is protonated to give an ammonium ion, $CH_3NH_3^+$, (the reason will be explored in Chapter 9).

7.45 The starting material is a primary (benzylic) halide, and the cyanide is a good nucleophile that can replace bromide in an S_N2 reaction. The resulting nitrile will then react with hydroxide to give an amide, as shown:

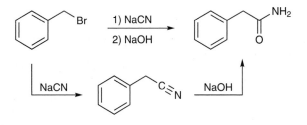

7.46 The starting material is a cyclic amide, and the reagent is aqueous acid. We expect the amide to undergo acid-catalyzed hydrolysis to give a carboxylic acid and an amine:

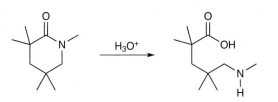

Actually, under these acidic conditions, the amine is protonated to give an ammonium ion (the reason will be explored in Chapter 9):

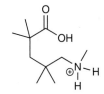

7.48 We must convert an ester into an acid halide, but we have not learned a way to do this directly in one step (esters are less reactive than acid halides). Instead, we can first convert the ester into a carboxylic acid, and then we can convert the carboxylic acid into the desired acid halide. To achieve this transformation, we can use the following reagents:

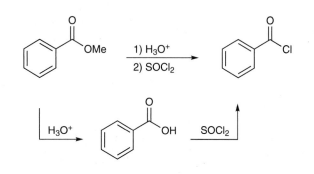

7.49 An acid chloride is more reactive than an ester, so an acid halide can be converted directly into an ester. This transformation can be achieved with an alcohol and pyridine. In this case, the desired product is an ethyl ester, so we use ethanol as the alcohol:

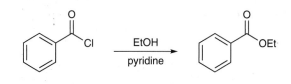

7.50 We must convert an amide into an acid halide, but we have not learned a way to do this directly in one step (amides are less reactive than acid halides). Instead, we can first convert the amide into a carboxylic acid, and then we can convert the carboxylic acid into the desired acid halide. To achieve this transformation, we can use the following reagents:

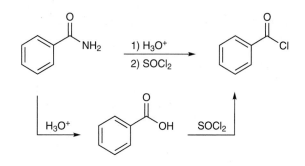

7.51 We must convert an ester into an acid anhydride, but we have not learned a way to do this directly in one step (esters are less reactive than acid anhydrides). Instead, we can first convert the ester into a carboxylic acid, and then convert the carboxylic acid into an acid halide. Finally, the acid halide can be converted into the desired acid anhydride. To achieve this transformation, we can use the following reagents:

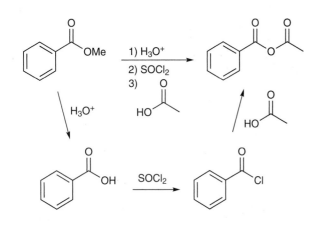

7.52 We must convert an amide into an ester, but we have not learned a way to do this directly in one step (amides are less reactive than esters). Instead, we can first convert the amide into a carboxylic acid, and then we can convert the carboxylic acid into the desired methyl ester. To achieve this transformation, we can use the following reagents:

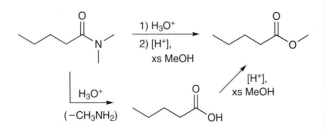

7.54 The product is an acetal, which can be made from the corresponding ketone, as shown here:

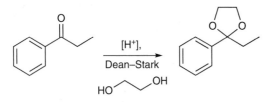

This can be the last step of our synthesis, if we can convert the starting material (a carboxylic acid) into the ketone above. Indeed, this ketone can be made in two steps from the starting carboxylic acid. The carboxylic acid is first converted into an acid halide upon treatment with thionyl chloride (SOCl$_2$), and this acid halide is then converted into the ketone upon treatment with lithium diethyl cuprate. Finally, the ketone is converted into the desired acetal upon treatment with ethylene glycol in the presence of an acid catalyst (with removal of water):

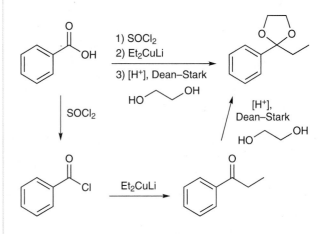

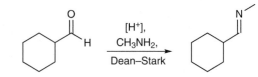

7.55 The product is an imine, which can be made from the corresponding aldehyde, as shown here:

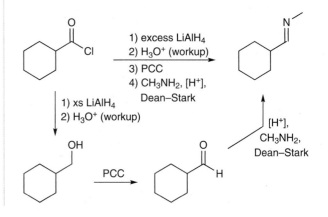

This can be the last step of our synthesis, if we can convert the starting material (an acid halide) into the aldehyde above. Indeed, this aldehyde can be made in two steps from the starting acid halide. The acid halide is first reduced to an alcohol upon treatment with excess reducing agent (such as LiAlH$_4$ or NaBH$_4$), and this alcohol is then oxidized to give the aldehyde upon treatment with PCC. Finally, the aldehyde is converted into the desired imine upon treatment with methyl amine in the presence of an acid catalyst (with removal of water):

7.56 The product is an acid halide, which can be made from the corresponding carboxylic acid, as shown here:

This can be the last step of our synthesis, if we can convert the starting material (a ketone) into the carboxylic acid above. Indeed, this carboxylic acid can be made in two steps from the starting ketone, as shown below. The ketone is converted into an ester upon treatment with a peroxy acid, such as MCPBA, and this ester is then hydrolyzed with aqueous acid to give the carboxylic acid. Finally, the carboxylic acid is converted into the desired acid halide upon treatment with thionyl chloride ($SOCl_2$):

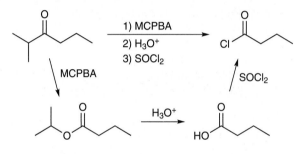

7.57 The product is an acid halide, which can be made from the corresponding carboxylic acid, as shown here:

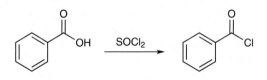

This can be the last step of our synthesis, if we can convert the starting material (an alcohol) into the carboxylic acid above. Indeed, this carboxylic acid can be made directly from the starting alcohol upon treatment with a strong oxidizing agent, giving the following synthesis:

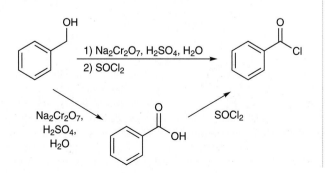

7.58 The product is an enamine, which can be made from the corresponding ketone, as shown here:

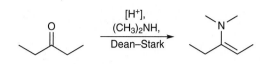

This can be the last step of our synthesis, if we can convert the starting material (an amide) into the ketone above. Indeed, this ketone can be made in a few steps from the starting amide. The amide is first hydrolyzed to a carboxylic acid upon treatment with aqueous acid. This acid is then converted into an acid halide, which is then converted into ketone above upon treatment with lithium diethyl cuprate. Finally, the ketone is converted into the desired enamine upon treatment with dimethyl amine in the presence of an acid catalyst (with removal of water):

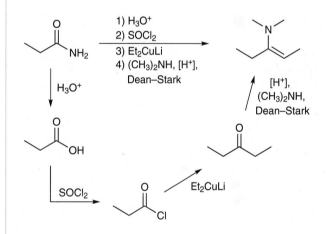

7.59 The product is a cyclic thioacetal, which can be made from the corresponding aldehyde, as shown here:

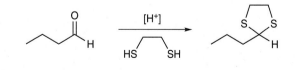

This can be the last step of our synthesis, if we can convert the starting material (an ester) into the aldehyde above. Indeed, this aldehyde can be made in two steps from the starting ester. The ester is first reduced to an alcohol upon treatment with excess $LiAlH_4$, and this alcohol is then oxidized to an aldehyde upon treatment with PCC. Finally, the aldehyde is converted into the desired thioacetal upon treatment with ethylene thioglycol (HSCH₂CH₂SH) in the presence of an acid catalyst:

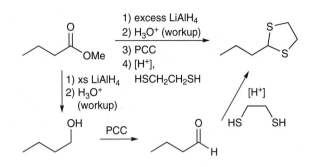

7.60 The starting material is an alcohol, and it can be oxidized upon treatment with an oxidizing agent. The resulting ketone can then be converted into the desired carboxylic acid in just two steps. The ketone is first oxidized via a Baeyer–Villiger oxidation to give an ester, and the ester is then hydrolyzed with aqueous acid to give the desired product:

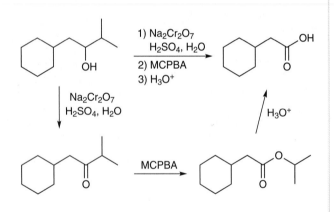

7.61 The product is an alkene, which can be made from a ketone via a Wittig reaction. This can be the last step of our synthesis:

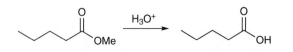

The starting material is an ester, so the first step of our synthesis could be hydrolysis of the ester to give a carboxylic acid:

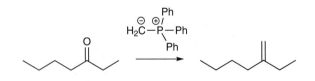

The synthesis can be completed if we can convert the carboxylic acid into the ketone:

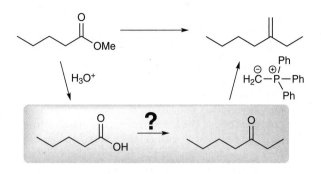

This can be accomplished in just two steps. The carboxylic acid is first converted into an acid halide, which is then converted into the ketone upon treatment with lithium diethyl cuprate.

The entire synthesis is shown here:

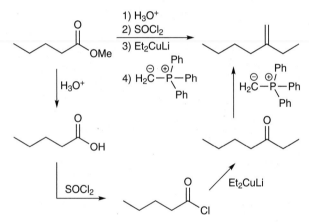

7.62 The product is a cyclic acetal, which can be made from the corresponding aldehyde, as shown here:

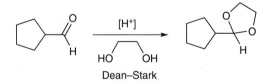

This can be the last step of our synthesis, if we can convert the starting material (a carboxylic acid) into the aldehyde above. Indeed, this aldehyde can be made in just a few steps from the starting carboxylic acid. The carboxylic acid is first converted into an acid halide, which is then reduced to an alcohol upon treatment with excess LiAlH₄. This alcohol is then oxidized to an aldehyde upon treatment with PCC. Finally, the aldehyde is converted into the desired acetal upon treatment with ethylene glycol (HOCH₂CH₂OH) in the presence of an acid catalyst (with removal of water):

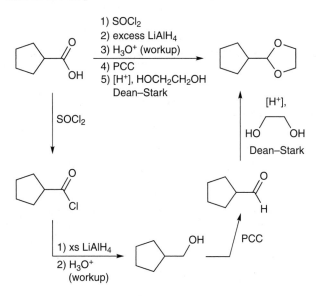

7.63 The product is an acid halide, which can be made from the corresponding carboxylic acid, so this can be the last step of our synthesis:

And the starting material is an alcohol, so the first step of our synthesis might be oxidation to give a ketone:

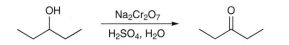

We can complete the synthesis if we can find a way to convert the ketone into a carboxylic acid:

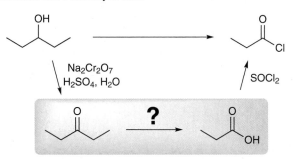

Conversion of the ketone into the carboxylic acid above can be accomplished in two steps. First the ketone is treated with a peroxy acid to give an ester (via a Baeyer–Villiger oxidation), and this ester is then hydrolyzed with aqueous acid to give the carboxylic acid.

The entire synthesis is shown here:

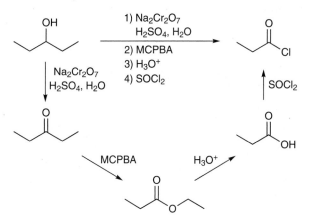

7.64 The product is an amide, which can be made from the corresponding acid halide, as shown here:

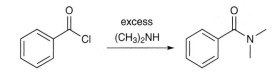

This can be the last step of our synthesis, if we can convert the starting material (an alcohol) into the acid halide above. Indeed, this acid halide can be made from the starting alcohol in just two steps. First the alcohol is oxidized to give a carboxylic acid, which is then converted into an acid halide upon treatment with thionyl chloride (SOCl₂). Finally, the acid halide is converted into the desired amide upon treatment with excess dimethyl amine:

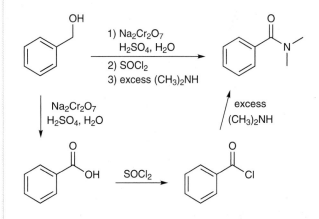

7.65 Recall that an ester is hydrolyzed when treated with aqueous acid to give a carboxylic acid and an alcohol:

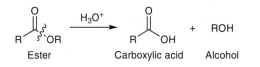

The same process occurs for cyclic esters (called lactones), to give a single product containing both a carboxylic acid group and an alcohol group (tethered together in the same molecule):

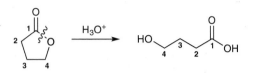

In our case, the starting material has two rings, and each ring is classified as a lactone (a cyclic ester). So upon treatment with aqueous acid, each ring will open to give a carboxylic acid group and an alcohol group, as follows:

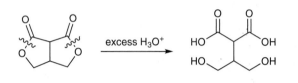

7.66 Recall that in the presence of an acid catalyst, an ester will react with a nucleophile in a nucleophilic acyl substitution reaction:

A similar process can occur for a lactone (a cyclic ester), with the ring being opened in the process:

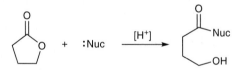

In our case, the nucleophile is an alcohol, so the lactone is converted into an acyclic ester, as shown below:

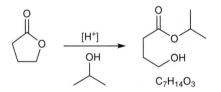

7.67 Recall that an ester will react with two equivalents of a Grignard reagent, thereby installing two alkyl groups (highlighted below) and converting the ester group into an alcohol:

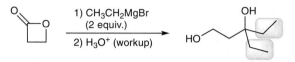

In our case, the starting material is a cyclic ester (a lactone), and it will react with two equivalents of the Grignard reagent to give an acyclic product:

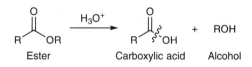

7.68 Recall that an ester is hydrolyzed when treated with aqueous acid to give a carboxylic acid and an alcohol:

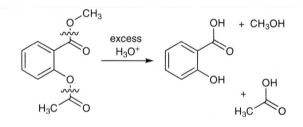

Our starting material has two ester groups, and each of them is hydrolyzed upon treatment with aqueous acid to give the following products:

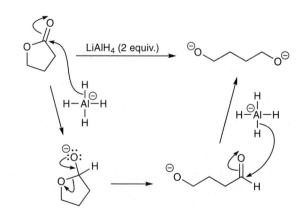

7.69 Recall that an ester will react with two equivalents of lithium aluminum hydride, as shown in the mechanism below. First the starting ester is attacked by lithium aluminum hydride, which serves as a delivery agent of nucleophilic hydride. This gives a tetrahedral intermediate that can reform the carbonyl group by expelling an alkoxide ion (thereby opening up the ring). The resulting intermediate has an aldehyde group, which can then be attacked by another equivalent of lithium aluminum hydride, giving a dianion:

When the reaction is complete, aqueous acid is introduced into the reaction flask to workup the reaction (to protonate the dianion and give the diol product). When drawing the workup steps, note that each alkoxide ion is protonated separately, so we must draw two separate proton transfer steps. Also note that each proton transfer step requires two curved arrows:

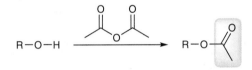

7.70 Recall that an alcohol will react with acetic anhydride to install an acyl group, thereby converting the alcohol into an ester:

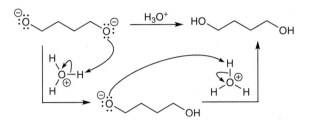

In our case, the starting material has four alcohol groups. When treated with excess acetic anhydride, we expect each of the four alcohol groups to be converted into an ester group. In other words, we expect the installation of an acyl group (a process called acylation) at each of the four alcohol groups, giving the following product:

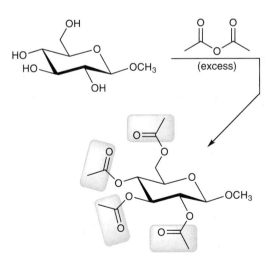

7.71 The starting material is a secondary amine, and it will react with an acyl halide to install an acyl group, thereby giving an amide as the product:

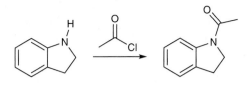

To remove the amide group and regenerate the secondary amine, the amide group can be hydrolyzed upon treatment with aqueous acid and heating.

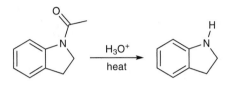

7.72 Recall that a Fischer esterification is a reaction between a carboxylic acid and an alcohol, in the presence of catalytic acid, giving an ester as the product:

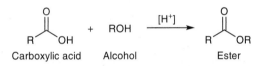

In our case, the desired product is a lactone (an ester that is part of a ring), which means that the starting carboxylic acid and the starting alcohol must be tethered together in the same molecule, so that the reaction can occur in an intramolecular fashion. Note the use of numbers to help us draw the starting material (for example, the product has two methyl groups at C6, so we draw two methyl groups at C6 in the starting material):

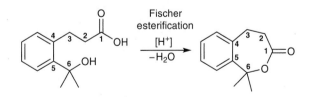

7.73 Recall that an alcohol will react with an acyl halide to give an ester:

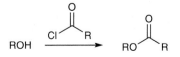

Similarly, we would expect phosgene (COCl$_2$, shown below) to react with two equivalents of an alcohol to give the product below, called a carbonate:

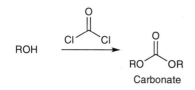

Carbonate

In our case, phosgene is being converted into a carbonate, and notice that the carbonate product is cyclic, which means that the missing reagent must be a diol (the two alcohol groups must be tethered to each other in the same molecule), as shown below:

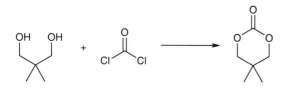

7.74 The product is a cyclic acetal, and we have only seen one way to make a cyclic acetal (from a diol):

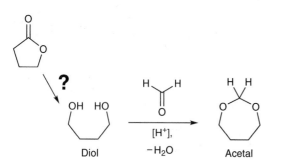

Now we must bridge the gap between the starting material (which is a cyclic ester) and the diol above. This can be accomplished by treating the ester with two equivalents of lithium aluminum hydride, followed by aqueous acidic workup. The complete synthesis is shown here:

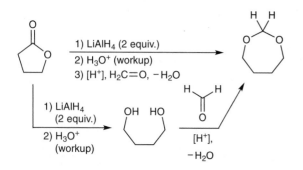

7.75 The starting material is cyclic, and the product is acyclic, so we will need to break a carbon–carbon bond at some point during our synthesis. We have not seen many ways to break a carbon–carbon bond, so let's focus on how we might accomplish this. One way to break a carbon–carbon bond is with ozonolysis, which requires forming an alkene first. This strategy would require that we reduce the ketone to an alcohol, and then use E1 conditions to give an alkene, which could then be converted into an acyclic product upon ozonolysis:

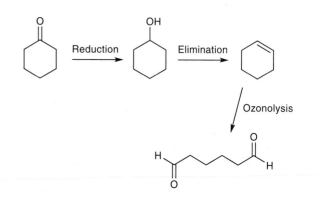

However, at this point, we encounter an insurmountable problem. Specifically, we will need to differentiate between two identical groups so that a reaction will occur only at one of them. This is not possible. If, for example, we treat this dialdehyde with one equivalent of a Grignard reaction, we won't obtain one clean product. Instead, some molecules will get attacked twice (at both functional groups), others will get attacked once, and others will not get attacked at all (because the Grignard reagent will have been consumed). Similarly, if we oxidize one aldehyde group, the other will also be oxidized under those conditions. We cannot do anything to one aldehyde without affecting the other aldehyde. So we are at a dead end, and we must go back to the drawing board and find a different way to break open the ring.

Another method for breaking a carbon–carbon bond is to perform a Baeyer–Villiger oxidation, which converts a ketone into an ester by inserting an oxygen atom between two carbon atoms. If we perform a Baeyer–Villiger oxidation on our starting material, the result is a lactone (a cyclic ester), which can then be opened in a variety of ways:

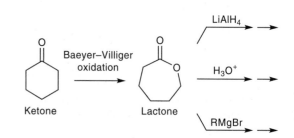

To determine the best way to open this lactone, note that the product has two additional methyl groups (highlighted below) that are not present in the lactone:

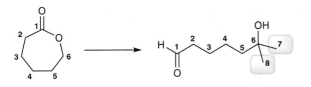

These methyl groups can be installed if we open the lactone with two equivalents of methyl magnesium bromide, to give the following diol:

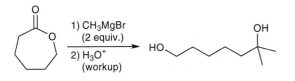

Then, to complete the synthesis, we treat the diol with an oxidizing agent, such as PCC, to give the product:

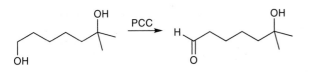

The complete mechanism is shown here:

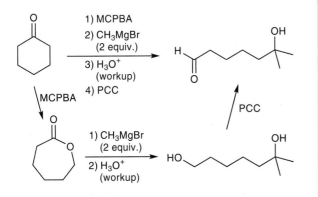

7.76 The starting material has one aromatic ring, while the desired product has two aromatic rings. The second aromatic ring can be installed by converting the starting carboxylic acid into an acyl halide, and then treating the acyl halide with lithium diphenyl cuprate:

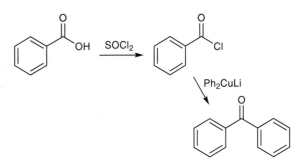

Now we just need to bridge the gap between this ketone and the desired product:

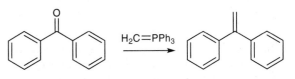

This can be achieved in one step, with a Wittig reaction. The complete synthesis is shown here:

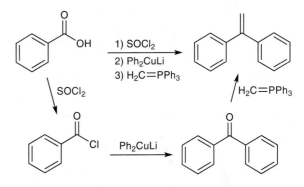

Alternatively, the last step (a Wittig reaction) can be replaced with a Grignard reaction, followed by dehydration with concentrated sulfuric acid, as shown here:

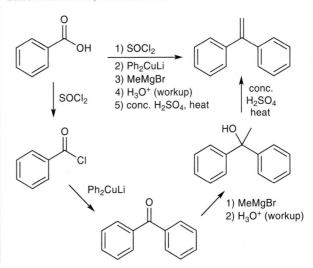

7.77 The desired product is an ester, which can be made from the corresponding carboxylic acid via a Fischer esterification:

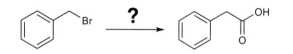

Now we must bridge the gap between the starting material (benzyl bromide) and this carboxylic acid:

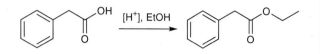

There are certainly many ways to achieve this transformation. Below is one such method, where the starting bromide is treated with cyanide to give a nitrile, which can then be hydrolyzed to give the necessary carboxylic acid. The complete synthesis is shown here:

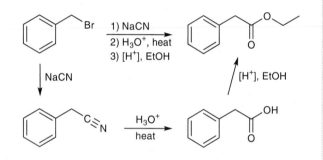

Alternatively, the starting bromide can be treated with magnesium to give a Grignard reagent, which can then be treated with carbon dioxide, followed by aqueous acidic workup to give the necessary carboxylic acid. The complete synthesis is shown here:

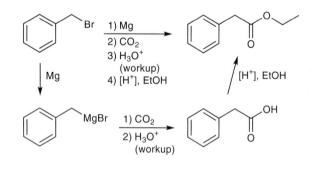

7.78 The product has many more carbon atoms than the starting material:

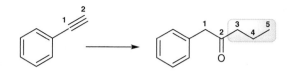

We can install a propyl group by alkylating the starting terminal alkyne:

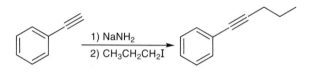

However, it will be difficult to control the regiochemistry of the next step in our synthesis, because we now have an internal (disubstituted) alkyne. For example, hydroboration-oxidation of this alkyne is likely to give a mixture of products:

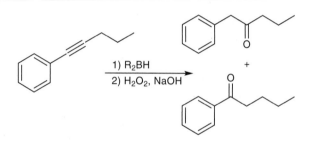

So we must install the propyl chain in another way. One way to do this is to convert the starting terminal alkyne into an aldehyde via hydroboration-oxidation, followed by a Grignard reaction to install the propyl chain:

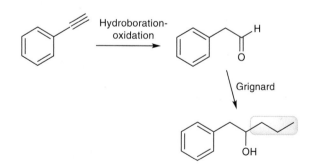

Indeed, the resulting alcohol can be converted into the desired product upon treatment with an oxidizing agent, such as PCC. The complete synthesis is shown here:

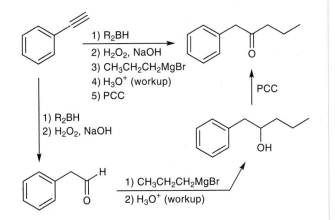

Alternatively, after hydroboration-oxidation, the aldehyde can be oxidized to give a carboxylic acid, which can be converted into an acyl halide. At that point, treatment with lithium dipropyl cuprate will give the desired product:

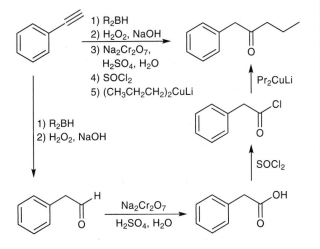

The two methods shown above are certainly not the only ways to convert the starting material into the desired product. There are other acceptable syntheses not shown here.

7.79

(a) The starting material is a carbonate, which has two leaving groups, highlighted below:

A carbonate

Compare the structure of a carbonate with the structure of a carboxylic acid derivative, which has only one leaving group rather than two leaving groups:

A carbonate   A carboxylic acid derivative

We have seen that a carboxylic acid derivative (with one leaving group) will react with two equivalents of a Grignard reagent. But a carbonate has an additional leaving group, so it can react with one additional equivalent of the Grignard reagent. That is, a carbonate will react with three equivalents of the Grignard reagent (according to the mechanism shown below) to give *tert*-butanol and ethylene glycol:

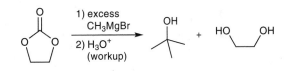

(b) One equivalent of the Grignard reagent functions as a nucleophile and attacks the carbonyl group of the carbonate. The resulting intermediate then reforms the carbonyl group to expel an alkoxide leaving group (thereby opening the ring). The resulting ester is then attacked by another equivalent of the Grignard reagent, to give an intermediate that can once again reform the carbonyl group to expel an alkoxide leaving group. The resulting ketone (acetone) is then attacked by a third equivalent of the Grignard reagent to give *tert*-butoxide.

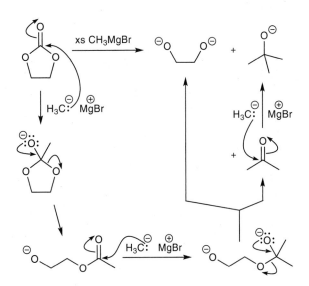

Then, upon workup with aqueous acid, each anion is protonated. Note that we must draw each proton transfer step separately, showing two curved arrows for each proton transfer step. Below are the two proton transfer steps that produce ethylene glycol,

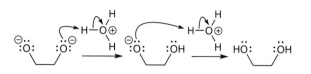

and below is the proton transfer step that produces *tert*-butanol:

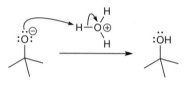

7.80

(a) In the presence of catalytic acid, an alcohol will react with a carboxylic acid to give an ester, in a process called Fischer esterification:

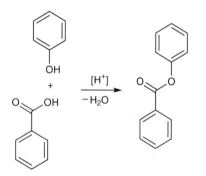

(b) The IR spectrum of the starting alcohol will have a broad signal between 3200 and 3600 cm^{-1} (characteristic of alcohols), while the IR spectrum of the starting carboxylic acid will have a very broad signal between 2200 and 3600 cm^{-1} (characteristic of carboxylic acids). In contrast, the product is an ester, and it will not have either of these broad signals in its IR spectrum. Therefore, the success of the reaction can be measured by the lack of these broad signals.

(c) The ^{1}H NMR spectrum of the starting carboxylic acid will have a signal near 12 ppm, corresponding to the proton of the carboxylic acid. In contrast, the product (an ester) will lack such a signal. Indeed, the ^{1}H NMR spectrum of the product should not have any signals downfield of the signals for the aromatic protons (there should be no signals appearing to the left of the signals near 7 ppm).

7.81

(a) In the ^{1}H NMR spectrum of compound **B**, the signal at 12 ppm (with an integration of 1H) is characteristic of a carboxylic acid. The other two signals (a septet with an integration of 1H, and a doublet with an integration of 6H) represent the characteristic pattern for an isopropyl group, so we can draw the following structure for compound **B**:

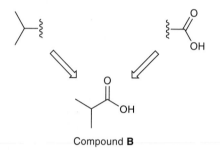

Compound **B**

(b) The IR spectrum of compound **A** has a signal at 1740 cm^{-1}, which suggests an ester group. Indeed, an ester is hydrolyzed to give a carboxylic acid upon treatment with aqueous acid, and this is exactly what is happening in this case. Compound **A** has the molecular formula $C_6H_{12}O_2$, and we have already accounted for four of the carbon atoms in compound **A**, because the acyl group in compound **B** is the same as the acyl group in compound **A** (four carbon atoms):

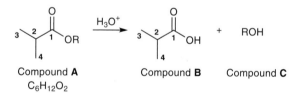

Compound **A**　　　　Compound **B**　　Compound **C**
$C_6H_{12}O_2$

We are missing two carbon atoms, so the R group above must be an ethyl group. That is, compound **A** must be an ethyl ester:

Compound **A**
$C_6H_{12}O_2$

(c) Hydrolysis of compound **A** gives compound **B** and ethanol, so compound **C** is ethanol:

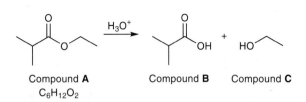

Compound **A**
$C_6H_{12}O_2$

Compound **B**   Compound **C**

(d) The ^{1}H NMR spectrum of ethanol will have three signals, corresponding to the three types of protons, labeled **a–c** below:

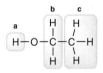

(e) The ^{13}C NMR spectrum of ethanol will have two signals, corresponding to the two types of carbon atoms, labeled **a** and **b** below:

(f) The ^{13}C NMR spectrum of compound **B** will have three signals, corresponding to the three types of carbon atoms, labelled **a–c** below:

Compound **B**

7.82

(a) The starting material has two degrees of unsaturation (one ring and one double bond), while the molecular formula of the product ($C_9H_{14}O_2$) indicates three degrees of unsaturation. This is suggestive of the formation of another ring. Indeed, the starting material has both an ester group and a built-in nucleophile (an OH group), allowing for an intramolecular nucleophilic acyl substitution reaction, in which the OH group functions as a nucleophile and displaces the methoxy group. This process is intramolecular, giving the following cyclic ester (called a lactone) as the product:

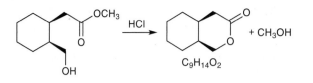

$C_9H_{14}O_2$

(b) The complete mechanism is shown below. First, the ester group is protonated by the acid catalyst to give a protonated ester, which is then attacked by the OH group in an intramolecular fashion. The resulting oxonium ion then undergoes two successive proton transfer steps to give a new oxonium ion that can expel water as a leaving group to give a protonated lactone. This intermediate is then deprotonated to give the product:

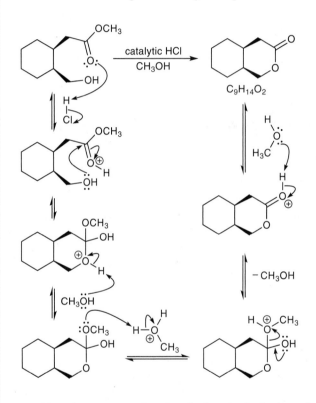

Note that we use methanol as the base in the last step (it is generally best to avoid using chloride as a base if there is a better option present, because HCl is a very strong acid, which makes chloride a very poor base).

(c) The starting material has one additional carbon atom (highlighted) that the product lacks:

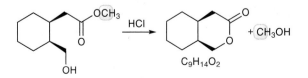

$C_9H_{14}O_2$

Therefore, we can differentiate the starting material from the product based on the number of signals in their ^{13}C NMR spectra. The starting material will have ten signals in its ^{13}C NMR spectrum, while the product will have only nine signals in its ^{13}C NMR spectrum.

More specifically, the ^{13}C NMR spectrum of the starting material will have two signals that appear between 50 and 100 ppm (for the two carbon atoms connected directly to oxygen atoms), while the ^{13}C NMR spectrum of the product will have only signal that appears between 50 and 100 ppm:

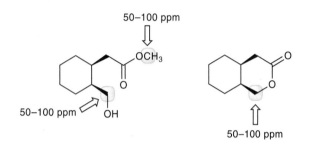

7.83 The starting material is a carboxylic acid derivative, where Br_3C^- is the leaving group. Therefore, after the starting material is attacked by hydroxide to give a tetrahedral intermediate, this intermediate can then expel Br_3C^- as a leaving group. Recall that a leaving group is, by definition, a weak base (while a strong base is a poor leaving group). Since Br_3C^- is a relatively weak base (certainly relative to other carbanions, but even relative to hydroxide), it can be expelled as a leaving group, thereby giving a carboxylic acid. Under these basic conditions, the carboxylic acid does not survive and is quickly deprotonated to give the carboxylate ion. Notice that Br_3C^- can function as the base that deprotonates the acid, because Br_3C^- is a stronger base than a carboxylate ion.

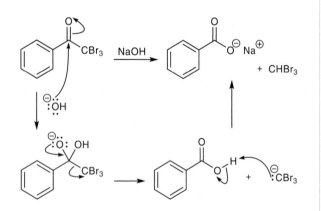

7.84 This process involves an ester reacting with an alcohol in intramolecular fashion:

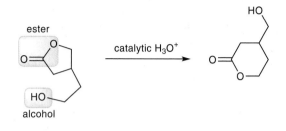

This entire process occurs under acid catalysis. The first step is protonation of the carbonyl group of the ester to give a protonated ester, which is then attacked by the OH group in an intramolecular fashion. The resulting oxonium ion then undergoes two successive proton transfer steps to give another oxonium ion that can expel a leaving group to give an intermediate that resembles the product. Indeed, deprotonation of this intermediate (with water functioning as the base) gives the product.

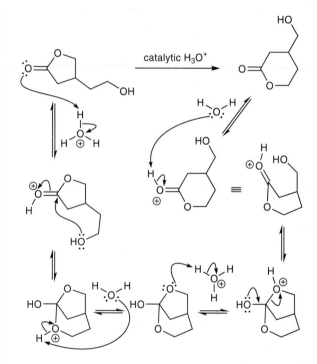

7.85 The lactone is first attacked by hydroxide to give a tetrahedral intermediate, which can reform a carbonyl group by expelling an alkoxide ion. The resulting carboxylic acid group is likely to be deprotonated under these conditions, and this step is shown as well. The resulting intermediate is a dianion because it has two negative charges (a carboxylate ion and an alkoxide ion). The alkoxide ion is a strong nucleophile, and it can attack another molecule of the lactone. This gives a tetrahedral intermediate, which can reform a carbonyl group by expelling an alkoxide ion. This process can repeat, thereby growing the polymer shown.

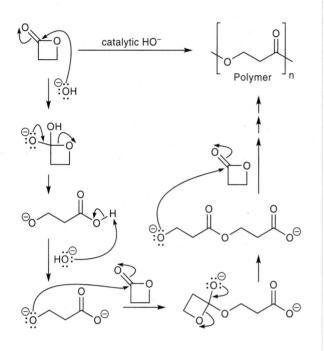

7.86

(a) The reagent is acetic anhydride, which can react with a variety of nucleophiles. Acetic anhydride can be used to install an acyl group on an amine (to give an amide),

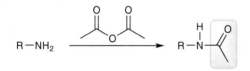

or to install an acyl group on an alcohol (to give an ester),

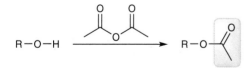

In our case, the starting material has both functional groups (a hydroxyl group and an amino group). But only one equivalent of acetic anhydride is used in this case, so we must decide which group is acylated: the hydroxyl group or the amino group? To answer this question, we note that the product of this reaction has an IR spectrum with a broad signal between 3200 and 3600 cm^{-1}, characteristic of an alcohol. This means that the OH group survives the reaction, (it is not acylated), so we conclude

that it is the amino group that is acylated during this process, giving the product shown (acetaminophen):

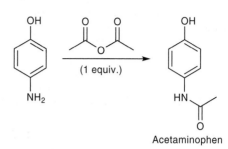

Acetaminophen

(b) The lone pair on the nitrogen atom functions as a nucleophile and attacks one of the carbonyl groups of acetic anhydride. The resulting intermediate then reforms a carbonyl group by expelling a leaving group (an acetate ion). Finally, a proton transfer step (where acetate functions as the base) gives acetaminophen and acetic acid:

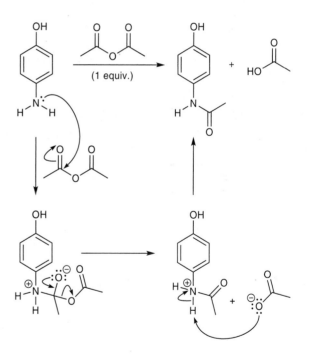

7.87 Recall that the reactivity of a carboxylic acid is related to the stability of the leaving group:

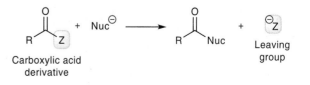

The more stable the leaving group, the more reactive the carboxylic acid derivative. With this in mind, we must evaluate the stability of the leaving groups (highlighted below) in order to predict whether an amide or an imidazolide will be more reactive toward a nucleophilic acyl substitution reaction.

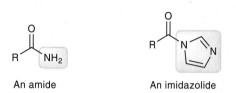

An amide   An imidazolide

In the case of an amide, the leaving group is H_2N^-, which is a very strong base, and therefore a very poor leaving:

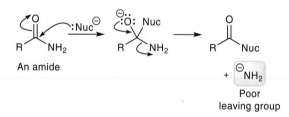

An amide

+ $^{\ominus}NH_2$
Poor
leaving group

In contrast, an imidazolide has a leaving group that is resonance-stabilized (and aromatic) and therefore extremely stable. As such, it is a good leaving group (as compared with H_2N^-, which is a bad leaving group):

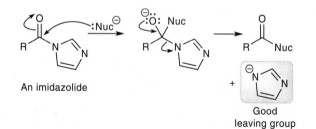

An imidazolide

+ Good
leaving group

For this reason, imidazolides are more reactive than amides toward nucleophilic acyl substitution.

7.88 The starting material has three functional groups, all of which will be hydrolyzed under these conditions. The acetal group is hydrolyzed to give a diol and a ketone; the ester is hydrolyzed to give a carboxylic acid and an alcohol; and the imine is hydrolyzed to give a ketone and ammonia:

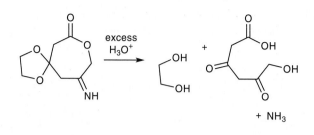

+ NH_3

7.89 Cocaine has an amino group and two ester groups. Upon treatment with aqueous acid, each ester group is hydrolyzed to give a carboxylic acid and an alcohol:

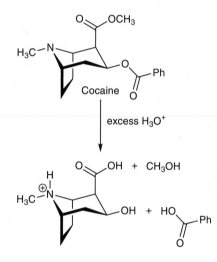

Note that the amino group is protonated under these conditions to give an ammonium ion. This will be discussed more in an upcoming chapter.

7.90 Notice that the starting material has an acetal group, and we know that a cyclic acetal is hydrolyzed upon treatment with aqueous acid to give a diol and a ketone:

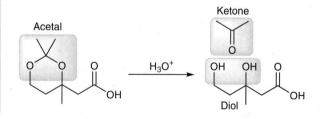

One of the OH groups of the diol can then react with the carboxylic acid group in an intramolecular Fischer esterification process, to give the product:

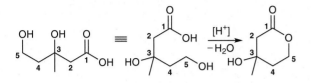

Notice the use of numbers to help us compare the diol and the product (it is very helpful to use numbers in this way).

Now let's draw the mechanism for each part of this process, beginning with hydrolysis of the acetal. First, one of the oxygen atoms of the acetal is protonated, and then expelled as a leaving group to give a C=O bond. This carbonyl group is then attacked by water to give an oxonium ion. Two successive proton transfers give a new oxonium ion, which can reform a carbonyl group to expel the diol:

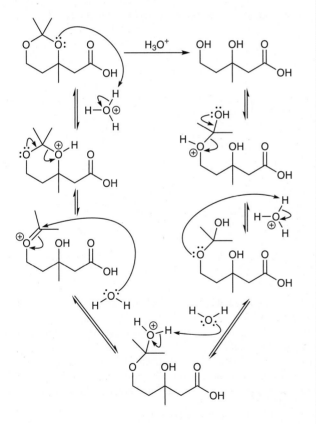

Next, we draw the steps for an intramolecular Fischer esterification. The carboxylic acid group is first protonated to give an intermediate, which is then attacked by an OH group in an intramolecular fashion, giving a six-membered ring. The resulting

oxonium ion then undergoes two successive proton transfer steps to give another oxonium ion that can expel water as a leaving group. The resulting intermediate is then deprotonated by water to give the product.

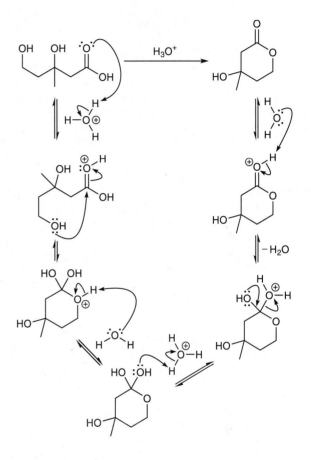

7.91 When we see an ester reacting with a Grignard reagent, we might be tempted to explore what happens when the Grignard reagent functions as a nucleophile and attacks the ester:

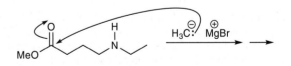

However, we can see from the structure of the product that the nitrogen atom (rather than a Grignard reagent) must attack the carbonyl group of the ester. So the Grignard reagent must be performing some other function in this case. Recall that Grignard reagents are both strong nucleophiles and strong bases, so we consider the Grignard reagent functioning as a base in this case, rather than a nucleophile. We look at the starting material and we find that it has a proton that is likely to be removed by the Grignard reagent (because this proton transfer step effectively transfers the negative charge from a carbon atom to a more electronegative nitrogen atom):

In general, a proton transfer step will occur more rapidly than a nucleophilic attack, which would explain why the Grignard reagent functions as a base in this case. This deprotonation step renders the nitrogen atom a very strong nucleophile, which can now attack

the ester in a nucleophilic acyl substitution reaction. The complete mechanism is shown below.

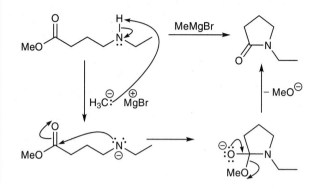

7.92   The starting material is an alkene, and the reagents indicate a hydroboration-oxidation, which accomplishes *anti*-Markovnikov hydration of the alkene to give a primary alcohol (compound **A**). Then, upon treatment with sodium dichromate and sulfuric acid (which gives a very strong oxidizing agent called chromic acid), the alcohol group in compound **A** is oxidized to give a carboxylic acid (compound **B**). This carboxylic acid is then converted into the corresponding acyl halide (compound **C**) upon treatment with thionyl chloride. And finally, compound **C** will react with hydrazine (a strong nucleophile) in a nucleophilic acyl substitution reaction to give compound **D**:

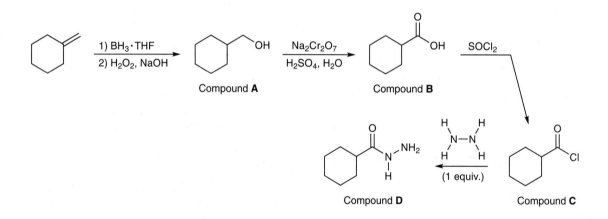

7.93   The starting material is a ketone, and the reagent is a peroxy acid (RCO_3H), which will convert the ketone into an ester, by inserting an oxygen atom next to the carbonyl group. This gives compound **A**, which is a cyclic ester (called a lactone). Upon treatment with excess methyl magnesium bromide, the ester will react with two equivalents of the Grignard reagent, thereby opening the ring, and giving a diol product (compound **B**) after aqueous acidic workup. Compound **B** is then oxidized upon treatment with sodium dichromate and sulfuric acid (which gives chromic acid, a strong oxidizing agent). Note that the tertiary alcohol is not oxidized. Only the primary alcohol is oxidized, and under these strong oxidizing conditions, the primary alcohol is converted into a carboxylic acid (compound **C**). Under acid catalysis, compound **C** can undergo an intramolecular Fischer esterification to give compound **D**. Compound **D** can be converted back into compound **C** via acid-catalyzed hydrolysis of the ester group:

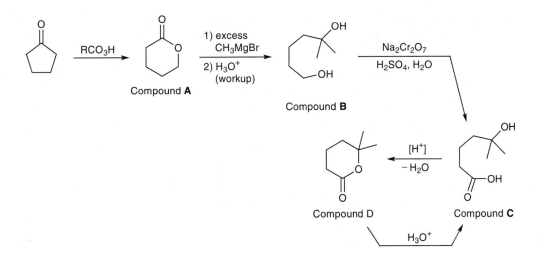

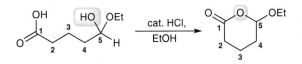

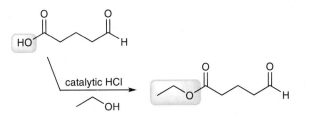

7.94 If we inspect the starting material (which has a carboxylic acid group) and we consider the function of the reagents, we might be tempted to conclude that a Fischer esterification is occurring, thereby replacing the OH of the carboxylic acid with an OR group:

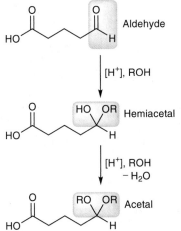

Yet, it is hard to see how this process gets us any closer to the product. The structure of the product is cyclic, which means that we must form a ring at some point. So we consider what else can happen, other than a Fischer esterification.

Notice that the starting material has an aldehyde group, and we have learned that an aldehyde will react with catalytic acid and an alcohol to give an acetal (via a hemiacetal intermediate, also shown below):

Acetal formation was covered in the previous chapter. Notice that the hemiacetal has an OH group. This is important, because with this OH group (and the carboxylic acid group in the same molecule), we can now have an intramolecular Fischer esterification that will form a ring and give the desired product:

Now let's draw all of the steps of this process. Conversion of the aldehyde into a hemiacetal occurs via three mechanistic steps, shown below. First the aldehyde is protonated by the acid catalyst, giving an intermediate that is then attacked by a molecule of ethanol. The resulting oxonium ion is then deprotonated by ethanol to give the hemiacetal:

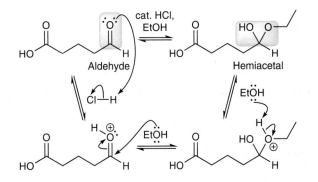

Next, this hemiacetal undergoes an intramolecular Fischer esterification to give the product. The mechanism for this process is similar to the mechanism for any Fischer esterification process. First, the carboxylic acid group is protonated, and the resulting intermediate is attacked by the OH group of the hemiacetal (in an intramolecular fashion). The resulting cyclic oxonium ion then undergoes two successive proton transfer steps to give a new oxonium ion that can expel water as a leaving group. The resulting intermediate is then deprotonated by ethanol to give the product:

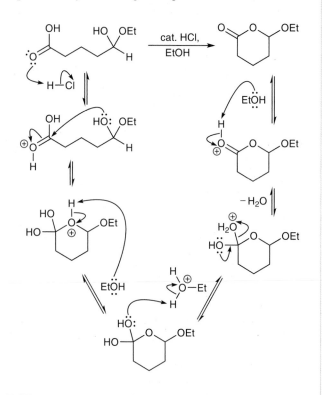

7.95 Recall that an alcohol will react with acetic anhydride to install an acyl group, thereby giving an ester:

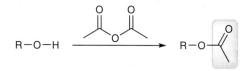

In this case, the starting material is not an alcohol, but it is an oxime, which has a nucleophilic OH group (much like an alcohol). So, it is reasonable to consider the possibility that the OH group

of the oxime might attack acetic anhydride. An alcohol reacts with acetic anhydride via a mechanism with three steps, and we can draw the same three steps here. First the OH group (of the oxime) attacks one of the carbonyl groups of acetic anhydride to give an intermediate, which is then deprotonated by whatever base is present (the bases that are present are the acetate ion, the starting oxime, and even acetic anhydride can function as a base). The resulting anion then reforms a carbonyl group by expelling an acetate ion:

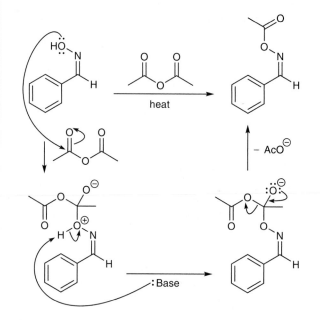

And finally, in the presence of an acetate ion, which can function as a base, an E2 process can occur, giving the product and generating another acetate ion (as a leaving group):

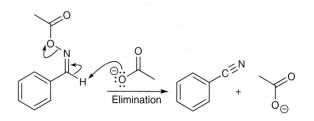

CHAPTER 8

8.2 This compound has only one alpha carbon atom (to the right of the carbonyl group), and that carbon atom is connected to one alpha proton (highlighted below). Note that the proton to the left of the carbonyl group is not an alpha proton, because it is not connected to an alpha carbon (it is connected directly to the carbonyl group):

8.3 This compound has two alpha carbon atoms. The alpha carbon atom to the right of the carbonyl group is not connected directly to any protons. The alpha carbon atom to the left of the carbonyl group is connected to one alpha proton (highlighted below).

8.4 This compound has two alpha carbon atoms, but neither of them is connected to any protons, so this compound has no alpha protons.

8.5 This compound has two alpha carbon atoms. The alpha carbon atom to the right of the carbonyl group is connected to three alpha protons (highlighted), and the alpha carbon atom to the left of the carbonyl group is connected to one alpha proton (highlighted).

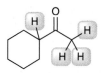

8.6 This compound has two alpha carbon atoms. The alpha carbon atom to the right of the carbonyl group is connected to two alpha protons (highlighted), while the alpha carbon atom to the left of the carbonyl group is not connected to any alpha protons.

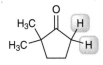

8.7 This compound does not have any alpha carbon atoms, because the carbonyl group is not connected to any carbon atoms. Therefore, this compound (called formaldehyde) does not have any alpha protons.

8.9 In this case, a ketone is converted into an enol, so this is a keto-enol tautomerization, which has two steps (both of which are proton transfer steps). Under basic conditions (hydroxide), we first show deprotonation to give a resonance-stabilized anion, and then we show protonation to give the product. Notice that hydroxide (rather than H$_2$O) functions as the base during the deprotonation step, and water (rather than H$_3$O$^+$) must be shown as the proton source during the protonation step, in order to stay consistent with basic conditions:

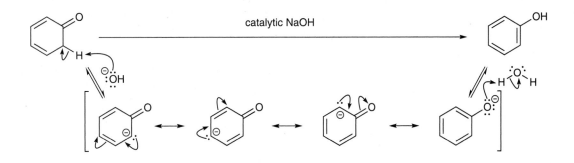

8.10 In this case, a ketone is converted into an enol, so this is a keto-enol tautomerization, which has two steps (both of which are proton transfer steps). Under acidic conditions (H$_3$O$^+$), we first show protonation to give a resonance-stabilized cation, and then we show deprotonation to give the product. Notice that H_3O^+ functions as the acid during the protonation step, and water (rather than hydroxide) must be shown as the base during the deprotonation step, in order to stay consistent with acidic conditions:

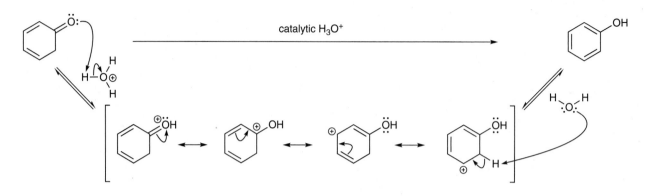

8.11 In this case, an enol is converted into a ketone, so this is a keto-enol tautomerization, which has two steps (both of which are proton transfer steps). Under basic conditions (hydroxide), we first show deprotonation to give a resonance-stabilized enolate anion, and then we show protonation to give the product. Notice that hydroxide (rather than H_2O) functions as the base during the deprotonation step, and water (rather than H_3O^+) must be shown as the proton source during the protonation step, in order to stay consistent with basic conditions:

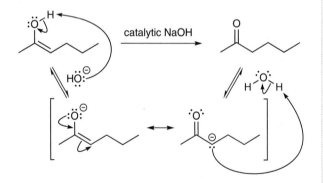

8.12 In this case, an enol is converted into a ketone, so this is a keto-enol tautomerization, which has two steps (both of which are proton transfer steps). Under acidic conditions (H_3O^+), we first show protonation to give a resonance-stabilized cation, and then we show deprotonation to give the product. Notice that H_3O^+ functions as the acid during the protonation step, and water (rather than hydroxide) must be shown as the base during the deprotonation step, in order to stay consistent with acidic conditions:

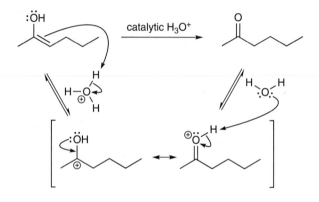

8.13 In this case, an enol is converted into a ketone, so this is a keto-enol tautomerization, which has two steps (both of which are proton transfer steps). Under acidic conditions (H_3O^+), we first show protonation to give a resonance-stabilized cation (notice that this intermediate has three resonance structures), and then we show deprotonation to give the product. Notice that H_3O^+ functions as the acid during the protonation step, and water (rather than hydroxide) must be shown as the base during the deprotonation step, in order to stay consistent with acidic conditions:

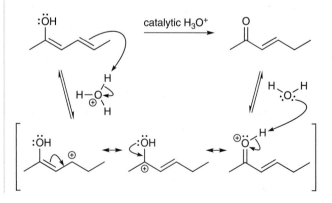

8.15 The starting material is a ketone, and the reagents are Br$_2$ and a mild acid. Under these conditions, a ketone will undergo alpha halogenation. This ketone has two alpha positions, but only one of these positions has protons, so the reaction occurs at this location. One of the alpha protons is replaced with Br:

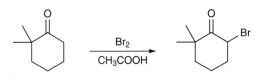

8.16 The starting material is a carboxylic acid, and the reagents indicate a Hell–Volhard–Zelinsky reaction. The result of this reaction is to install Br at the alpha position:

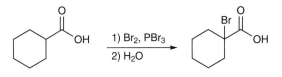

8.17 The starting material is a carboxylic acid, and the reagents indicate a Hell–Volhard–Zelinsky reaction. The result of this reaction is to install Br at the alpha position:

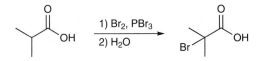

8.18 The starting material is a ketone, and the reagents are Br$_2$ and H$_3$O$^+$. Under these conditions, a ketone will undergo alpha halogenation. This ketone has two alpha positions, but only one of these positions has protons, so the reaction occurs at this location. One of the alpha protons is replaced with Br:

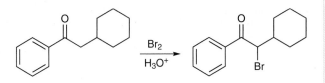

8.20 Hydroxide is a strong base, and it will remove a proton from the position in between the two carbonyl groups (this is where the most acidic proton is), to give an enolate:

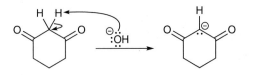

This enolate is highly stabilized by resonance. The negative charge is delocalized over two oxygen atoms:

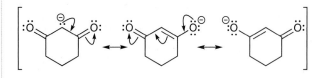

8.21 Hydroxide is a strong base, and it will remove a proton from the position in between the two carbonyl groups (this is where the most acidic proton is), to give an enolate:

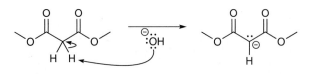

This enolate is highly stabilized by resonance. The negative charge is delocalized over two oxygen atoms:

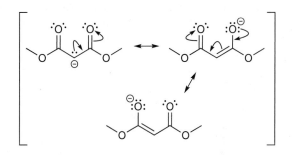

8.22 Hydroxide is a strong base, and it will remove a proton from an alpha position, resulting in an enolate ion. In this case, the starting ketone is symmetrical, so the two alpha positions are the same (deprotonation at either location leads to the same enolate):

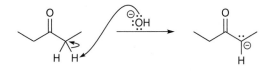

Enolate ions are stabilized by resonance:

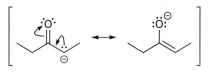

8.23 Hydroxide is a strong base, and it will remove a proton from an alpha position, resulting in an enolate ion. In this case, the starting ketone is symmetrical, so the two alpha positions are the same (deprotonation at either location leads to the same enolate):

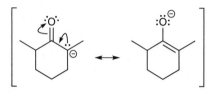

Enolate ions are stabilized by resonance:

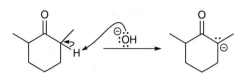

8.25 The starting material is a methyl ketone, and the reagents indicate a haloform reaction. Under these conditions, the methyl ketone is converted into a carboxylic acid, and bromoform is produced as a by-product:

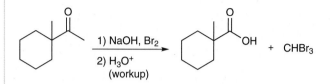

8.26 The starting material is a methyl ketone, and the reagents indicate a haloform reaction. Under these conditions, the methyl ketone is converted into a carboxylic acid, and bromoform is produced as a by-product:

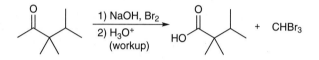

8.27 Hydroxide is a strong base, and it will deprotonate the alpha position to give an enolate ion. This enolate ion then functions as a nucleophile and attacks Br_2, thereby installing a bromine atom at the alpha position. These two steps (deprotonation, followed by nucleophilic attack) are repeated two more times, thereby installing three bromine atoms at the alpha position. Thus far, the methyl (CH_3) group has been converted into a tribromomethyl (CBr_3) group. Hydroxide now functions as a nucleophile and attacks the carbonyl group, giving a tetrahedral intermediate, which expels Br_3C^- as a leaving group to give a carboxylic acid. Under these conditions, the carboxylic acid is deprotonated to give a carboxylate ion.

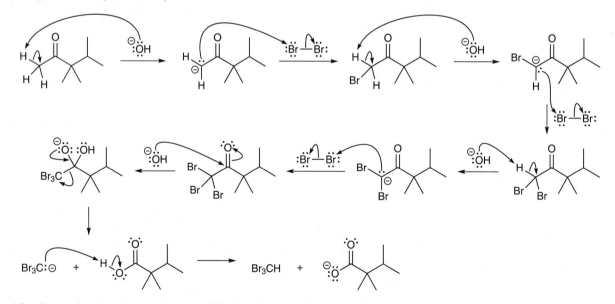

After the reaction above is complete, an acid is introduced into the reaction flask to protonate the carboxylate ion, and the carboxylic acid is obtained:

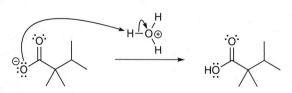

8.29 The starting material is an unsymmetrical ketone, and when treated with LDA, this ketone is deprotonated at the less-substituted position to give the kinetic enolate, shown here:

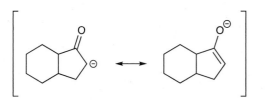

This enolate will react with methyl iodide in an S_N2 reaction, thereby installing a methyl group at the less-substituted alpha carbon, to give the product shown:

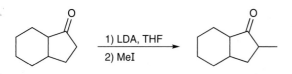

8.30 The starting material has only one alpha proton. Upon treatment with LDA, the starting material is deprotonated to give an enolate, shown here:

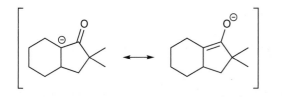

This enolate will react with ethyl bromide in an S_N2 reaction, thereby installing an ethyl group, to give the product shown:

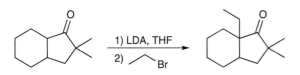

8.31 The starting material is an unsymmetrical ketone, and when treated with LDA, this ketone is deprotonated at the less-substituted position to give the kinetic enolate, shown here:

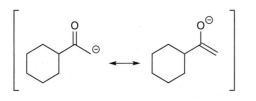

This enolate will react with *n*-propyl chloride in an S_N2 reaction, thereby installing a propyl group at the less-substituted alpha carbon, to give the product shown:

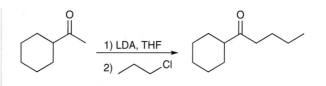

8.32 The starting material is an unsymmetrical ketone, and when treated with LDA, this ketone is deprotonated at the less-substituted position to give the kinetic enolate, shown here:

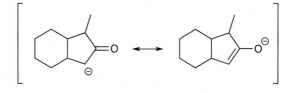

This enolate will react with methyl iodide in an S_N2 reaction, thereby installing a methyl group at the less-substituted alpha carbon, to give the product shown:

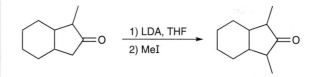

8.34 If we look at the difference between the starting material and the product, we see that there is an extra methyl group that was introduced at an alpha position. This methyl group was installed at the less-substituted alpha position, which can be achieved by treating the ketone with LDA followed by methyl iodide:

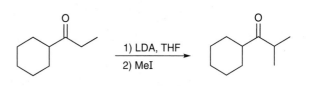

8.35 If we look at the difference between the starting material and the product, we see that there is an extra *n*-propyl group that was introduced at an alpha position. This *n*-propyl group was installed at the less-substituted alpha position, which can be achieved by treating the ketone with LDA followed by *n*-propyl iodide:

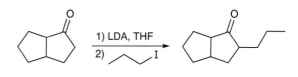

8.36 If we look at the difference between the starting material and the product, we see that there is an extra alkyl group that was introduced at an alpha position. This alkyl group was installed at the less-substituted alpha position, which can be achieved by treating the ketone with LDA followed by an alkyl halide, shown below:

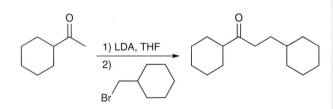

8.38 We start by drawing two molecules of the ketone, so that the oxygen atom of one ketone is pointing directly at the alpha protons of the other ketone. Then, we erase the two alpha protons and the oxygen atom (highlighted), and we push the fragments together, connecting them with a double bond. This is NOT a mechanism, but it is a useful method for drawing the product of an aldol condensation.

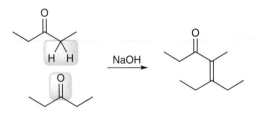

8.39 We start by drawing two molecules of the ketone, so that the oxygen atom of one ketone is pointing directly at the alpha protons of the other ketone. Then, we erase the two alpha protons and the oxygen atom (highlighted), and we push the fragments together, connecting them with a double bond. This is NOT a mechanism, but it is a useful method for drawing the product of an aldol condensation.

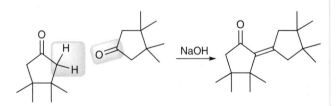

8.40 We start by drawing two molecules of the aldehyde, so that the oxygen atom of one aldehyde is pointing directly at the alpha protons of the other aldehyde. Then, we erase the two alpha protons and the oxygen atom (highlighted), and we push the fragments together, connecting them with a double bond. In this case, two stereoisomeric products are obtained, and the first one shown is likely the major product (the aldehydic group has lone pairs and is likely larger than a methyl group, so that aldehydic group will be *trans* to the ethyl group).

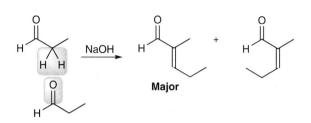

8.42 Below is a retrosynthetic analysis for the desired product, showing the C=C bond that is made during an aldol condensation. To draw the starting materials for this aldol condensation, consider the C=C bond in the product (the alpha and beta positions). The alpha position must have originally been the location of two alpha protons, and the beta position must have originally been the location of a carbonyl group, as shown here:

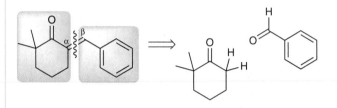

Upon treatment with NaOH, these starting materials will react with each other via an Aldol condensation to give the desired product.

8.43 Below is a retrosynthetic analysis for the desired product, showing the C=C bond that is made during an aldol condensation. To draw the starting materials for this aldol condensation, consider the C=C bond in the product (the alpha and beta positions). The alpha position must have originally been the location of two additional alpha protons, and the beta position must have originally been the location of a carbonyl group, as shown here:

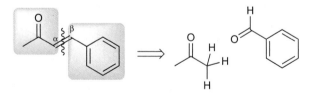

Upon treatment with NaOH, these starting materials will react with each other via an Aldol condensation to give the desired product.

8.44 Below is a retrosynthetic analysis for the desired product, showing the C=C bond that is made during an aldol condensation. To draw the starting materials for this aldol condensation, consider the C=C bond in the product (the alpha and beta positions). The alpha position must have originally been the location of two alpha protons, and the beta position must have originally been the location of a carbonyl group, as shown here:

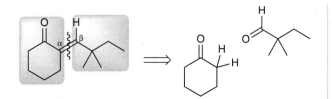

Upon treatment with NaOH, these starting materials will react with each other via an Aldol condensation to give the desired product.

8.45 Below is a retrosynthetic analysis for the desired product, showing the C=C bond that is made during an aldol condensation. To draw the starting materials for this aldol condensation, consider

the C=C bond in the product (the alpha and beta positions). The alpha position must have originally been the location of two alpha protons, and the beta position must have originally been the location of a carbonyl group, as shown here:

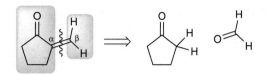

Upon treatment with NaOH, these starting materials will react with each other via an Aldol condensation to give the desired product.

8.47 In the first step, hydroxide functions as a base and removes an alpha proton from the ketone. The resulting enolate ion then functions as a nucleophile and attacks the carbonyl group of formaldehyde. The resulting alkoxide ion is then protonated by H₂O (this is the proton source that is present under these conditions) to give a β-hydroxy ketone. Under these conditions, the β-hydroxy ketone can undergo elimination to give an α,β-unsaturated ketone. This elimination process occurs via two steps. First, hydroxide functions as a base and removes an alpha proton, and then hydroxide is expelled as a leaving group to give the product:

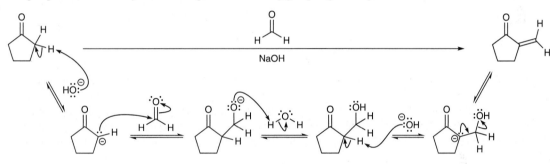

8.48 In the first step, hydroxide functions as a base and removes an alpha proton from the ketone bearing alpha protons. The resulting enolate ion then functions as a nucleophile and attacks the carbonyl group of the other ketone. The resulting alkoxide ion is then protonated by H₂O (this is the proton source that is present under these conditions) to give a β-hydroxy ketone. Under these conditions, the β-hydroxy ketone can undergo elimination to give an α,β-unsaturated ketone. This elimination process occurs via two steps. First, hydroxide functions as a base and removes an alpha proton, and then hydroxide is expelled as a leaving group to give the product:

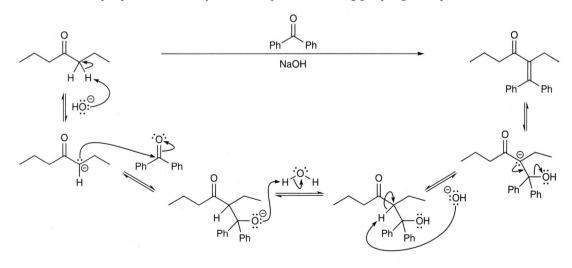

8.49 In the first step, hydroxide functions as a base and removes an alpha proton from the ketone. The resulting enolate ion then functions as a nucleophile and attacks the carbonyl group of another molecule of the same ketone. The resulting alkoxide ion is then protonated by H$_2$O (this is the proton source that is present under these conditions) to give a β-hydroxy ketone. Under these conditions, the β-hydroxy ketone can undergo elimination to give an α,β-unsaturated ketone. This elimination process occurs via two steps. First, hydroxide functions as a base and removes an alpha proton, and then hydroxide is expelled as a leaving group to give the product:

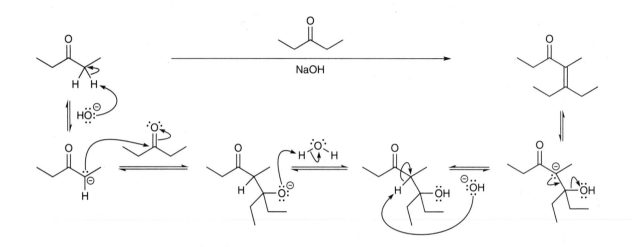

8.50 In the first step, hydroxide functions as a base and removes an alpha proton from the ketone bearing alpha protons. The resulting enolate ion then functions as a nucleophile and attacks the carbonyl group of the aldehyde. The resulting alkoxide ion is then protonated by H$_2$O (this is the proton source that is present under these conditions) to give a β-hydroxy ketone. Under these conditions, the β-hydroxy ketone can undergo elimination to give an α,β-unsaturated ketone. This elimination process occurs via two steps. First, hydroxide functions as a base and removes an alpha proton, and then hydroxide is expelled as a leaving group to give the product:

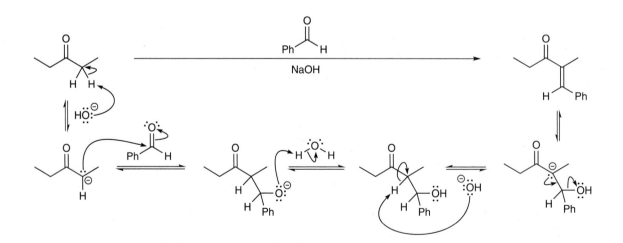

8.52 To draw the product of this Claisen condensation, we start by drawing two molecules of the ester, so that the alkoxy of one ester is pointing directly at the alpha protons of the other ester. Then, we erase the alkoxy group and one of the alpha protons, and we push the fragments together, as follows:

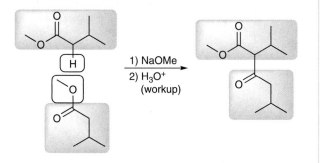

8.53 To draw the product of this Claisen condensation, we start by drawing two molecules of the ester, so that the alkoxy of one ester is pointing directly at an alpha proton of the other ester. Then, we erase the alkoxy group and the alpha proton, and we push the fragments together, as follows:

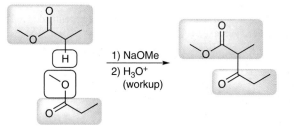

8.54 To draw the product of this Claisen condensation, we start by drawing two molecules of the ester, so that the alkoxy of one ester is pointing directly at the alpha protons of the other ester. Then, we erase the alkoxy group and one of the alpha protons, and we push the fragments together, as follows:

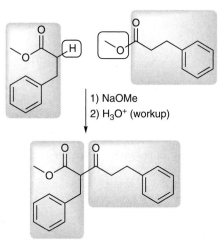

8.55 To draw the product of this Claisen condensation, we start by drawing two molecules of the ester, so that the alkoxy of one ester is pointing directly at the alpha protons of the other ester. Then, we erase the alkoxy group and one of the alpha protons, and we push the fragments together, as follows:

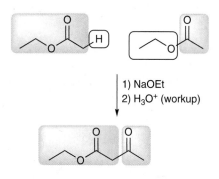

8.57 Below is a retrosynthetic analysis for the desired product, showing the C–C bond that is made during a Claisen condensation. To draw the starting materials for this Claisen condensation, consider the C–C bond in the product (between the alpha and beta positions). The alpha position must have originally been the location of an additional alpha proton, and the beta position must have originally been connected to an alkoxy group (specifically, a methoxy group, so that both esters have the same alkoxy group), as shown here:

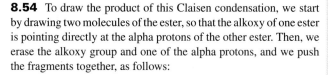

The two esters above can be used to make the desired product via a Claisen condensation reaction, as shown below. Notice that we use methoxide as our base (to avoid transesterification):

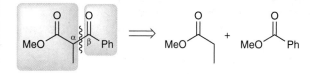

8.58 Below is a retrosynthetic analysis for the desired product, showing the C–C bond that is made during a Claisen condensation. To draw the starting materials for this Claisen condensation, consider the C–C bond in the product (between the alpha and beta positions). The alpha position must have originally been the location of an additional alpha proton, and the beta position must have originally been connected to an alkoxy group (specifically, an ethoxy group, so that both esters have the same alkoxy group), as shown here:

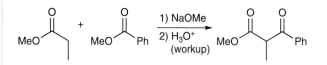

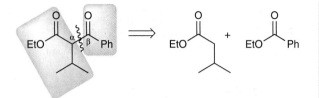

The two esters above can be used to make the desired product via a Claisen condensation reaction, as shown below. Notice that we use ethoxide as our base (to avoid transesterification):

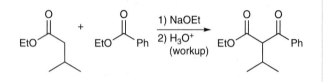

8.59 Below is a retrosynthetic analysis for the desired product, showing the C–C bond that is made during a Claisen condensation. To draw the starting materials for this Claisen condensation, consider the C–C bond in the product (between the alpha and beta positions). The alpha position must have originally been the location of an additional alpha proton, and the beta position must have originally been connected to an alkoxy group (specifically, a methoxy group, so that both esters have the same alkoxy group), as shown here:

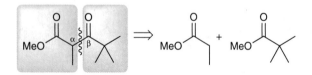

The two esters above can be used to make the desired product via a Claisen condensation reaction, as shown below. Notice that we use methoxide as our base (to avoid transesterification):

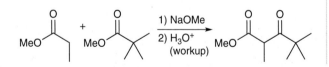

8.60 Below is a retrosynthetic analysis for the desired product, showing the C–C bond that is made during a Claisen condensation. To draw the starting materials for this Claisen condensation, consider the C–C bond in the product (between the alpha and beta positions). The alpha position must have originally been the location of an additional alpha proton, and the beta position must have originally been connected to an alkoxy group (specifically, a methoxy group, so that both esters have the same alkoxy group), as shown here:

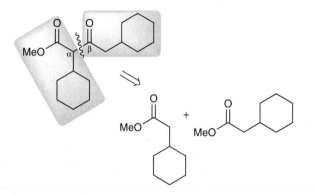

Notice that, in this case, the two esters are identical. This ester can be used to make the desired product via a Claisen condensation reaction, as shown below. Notice that we use methoxide as our base (to avoid transesterification):

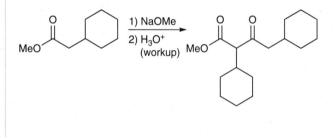

8.62 Ethoxide functions as a base and removes a proton from the alpha position of the starting ester, giving an enolate ion. This enolate ion then functions as a nucleophile and attacks another molecule of the ester (that has not been deprotonated) to give a tetrahedral intermediate. This intermediate then re-forms the carbonyl group by expelling an ethoxide ion. The resulting product (which is our desired product) is deprotonated under these basic conditions to give a stabilized enolate ion. The formation of this particularly stable anion is a driving force for the reaction. When the reaction is complete, we must then introduce a source of protons (H_3O^+) into the reaction flask to protonate the anion and regenerate the desired product.

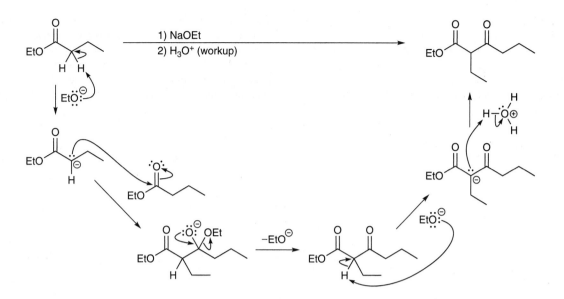

8.63 Methoxide functions as a base and removes a proton from the alpha position of the starting ester, giving an enolate ion. This enolate ion then functions as a nucleophile and attacks another molecule of the ester (that has not been deprotonated) to give a tetrahedral intermediate. This intermediate then re-forms the carbonyl group by expelling a methoxide ion. The resulting product (which is our desired product) is deprotonated under these basic conditions to give a stabilized enolate ion. The formation of this particularly stable anion is a driving force for the reaction. When the reaction is complete, we must then introduce a source of protons (H_3O^+) into the reaction flask to protonate the anion and regenerate the desired product.

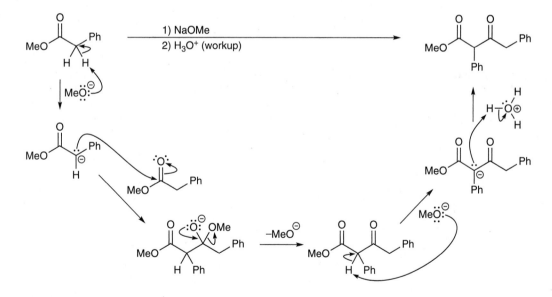

8.64 Let's begin by redrawing the starting diester:

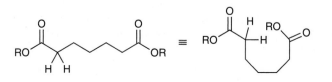

In the first step of the mechanism, an alkoxide ion functions as a base and removes a proton from one of the alpha positions of the starting diester, giving an enolate ion. This enolate ion then functions as a nucleophile and attacks the other carbonyl group (in an intramolecular fashion) to give a tetrahedral intermediate. This intermediate then re-forms the carbonyl group by expelling an alkoxide ion. The resulting product (which is our desired product) is deprotonated under these basic conditions to give a stabilized enolate ion. The formation of this particularly stable anion is a driving force for the reaction. When the reaction is complete, we must then introduce a source of protons (H_3O^+) into the reaction flask to protonate the anion and regenerate the desired product.

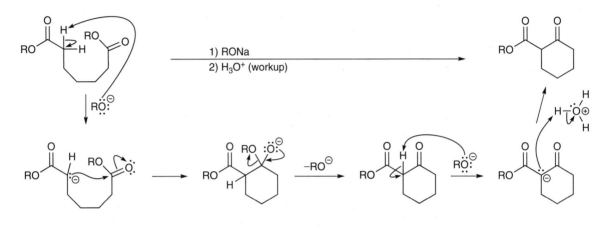

8.66 We begin by identifying an alkyl halide that can be used to make the desired product via an acetoacetic ester synthesis:

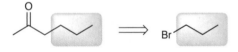

Now we can draw the forward process. In the first step, ethyl acetoacetate is treated with sodium ethoxide followed by *n*-propyl bromide, thereby installing an *n*-propyl group at the alpha position. The ester group is then hydrolyzed, and the resulting β-ketoacid is then heated to give a decarboxylation reaction, affording the desired product.

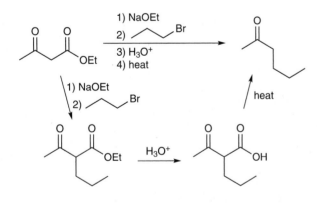

Now we can draw the forward process. In the first step, ethyl acetoacetate is treated with sodium ethoxide followed by isobutyl bromide, thereby installing an isobutyl group at the alpha position. The ester group is then hydrolyzed, and the resulting β-ketoacid is then heated to give a decarboxylation reaction, affording the desired product.

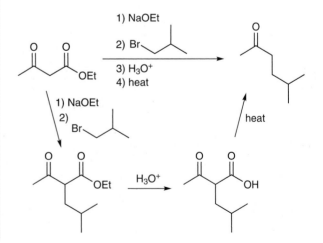

8.67 We begin by identifying an alkyl halide that can be used to make the desired product via an acetoacetic ester synthesis:

8.68 We begin by identifying an alkyl halide that can be used to make the desired product via an acetoacetic ester synthesis:

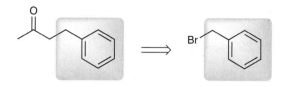

Now we can draw the forward process. In the first step, ethyl acetoacetate is treated with sodium ethoxide followed by benzyl bromide, thereby installing a benzyl group at the alpha position. The ester group is then hydrolyzed, and the resulting β-ketoacid is then heated to give a decarboxylation reaction, affording the desired product.

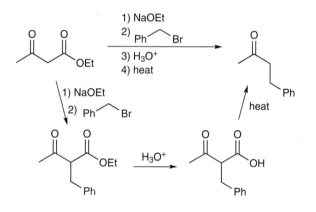

8.69 Consider the following proposed acetoacetic ester synthesis:

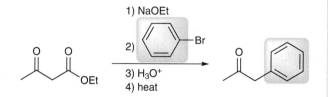

This process will not work because it would require performing an S_N2 reaction with phenyl bromide as the substrate. This will not work because S_N2 reactions cannot be performed with a substrate in which the leaving group is connected to an sp^2 hybridized carbon atom.

8.70 The desired product is a derivative of acetone containing two R groups, highlighted below, and can be made via an acetoacetic ester synthesis, starting with the two alkyl halides shown below:

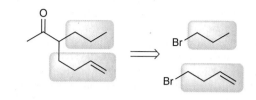

The forward process is shown here. In the first step, ethyl acetoacetate is treated with sodium ethoxide followed by propyl bromide, thereby installing an *n*-propyl group at the alpha position. This compound is then treated again with sodium ethoxide, followed by the other alkyl halide, installing the second alkyl group (note that the two alkyl groups can be installed in either order). The ester group is then hydrolyzed, and the resulting β-ketoacid is then heated to give a decarboxylation reaction, affording the desired product.

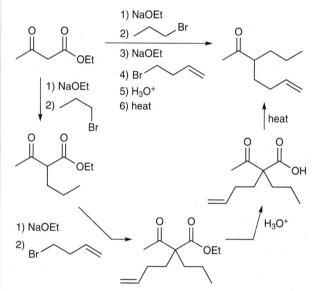

8.71 The desired product is a derivative of acetone containing two R groups, highlighted below, and can be made via an acetoacetic ester synthesis, starting with the two alkyl halides shown below:

You might wonder why one alkyl halide is a bromide and the other is an iodide. In general, alkyl bromides are less expensive than alkyl iodides (even though iodides are more reactive), and they get the job done, so bromides are often used. However, methyl bromide is not easy to use because it is a gas at room temperature. When using a methyl halide, it is best to use methyl iodide, which is a liquid at room temperature, rather than methyl bromide.

The forward process is shown here. In the first step, ethyl acetoacetate is treated with sodium ethoxide followed by methyl iodide, thereby installing a methyl group at the alpha position. This compound is then treated again with sodium ethoxide, followed by *n*-propyl bromide, thereby installing the *n*-propyl group (note that the two alkyl groups can be installed in either order). The ester group is then hydrolyzed, and the resulting β-ketoacid is then heated to give a decarboxylation reaction, affording the desired product.

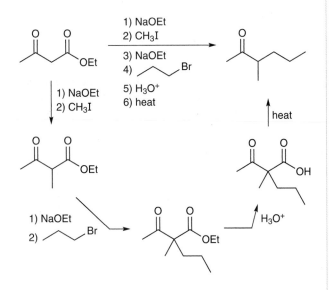

8.72 The desired transformation can be achieved using the same strategy employed in an acetoacetic acid synthesis. Three steps are involved: 1) alkylation, 2) hydrolysis, and 3) decarboxylation:

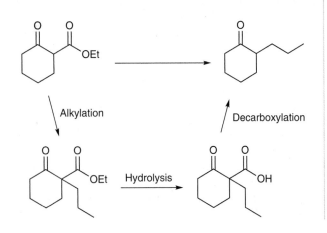

The reagents for this process are shown here:

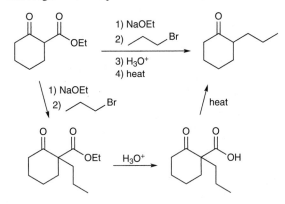

8.74 We begin by identifying an alkyl halide that can be used to make the desired product via a malonic ester synthesis:

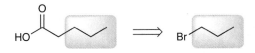

Now we can draw the forward process. In the first step, diethyl malonate is treated with sodium ethoxide followed by *n*-propyl bromide, thereby installing an *n*-propyl group at the alpha position. The ester groups are then both hydrolyzed, and the resulting diacid is then heated to give a decarboxylation reaction, affording the desired product.

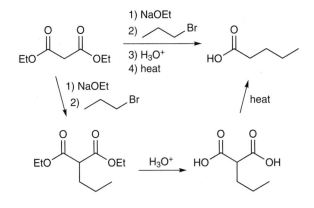

8.75 The desired product is a derivative of acetic acid containing two R groups, highlighted below. This product can be made via an acetoacetic ester synthesis, starting with the two alkyl halides shown below:

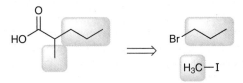

You might wonder why one alkyl halide is a bromide and the other is an iodide. You will find the explanation in the solution to Problem 8.71.

The forward process is shown here. In the first step, diethyl malonate is treated with sodium ethoxide followed by methyl iodide, thereby installing a methyl group at the alpha position. This compound is then treated again with sodium ethoxide, followed by *n*-propyl bromide, thereby installing the *n*-propyl group (note that the two alkyl groups can be installed in either order). The ester groups are then hydrolyzed, and the resulting diacid is then heated to give a decarboxylation reaction, affording the desired product.

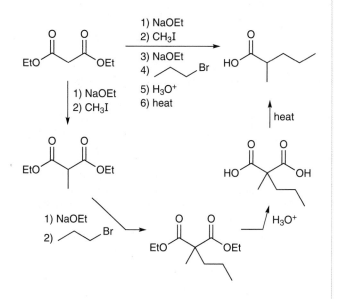

8.76 The desired product is a derivative of acetic acid containing two R groups, highlighted below. This product can be made via an acetoacetic ester synthesis, starting with the two halides shown below:

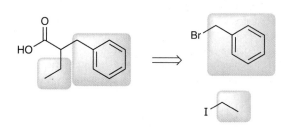

The forward process is shown here. In the first step, diethyl malonate is treated with sodium ethoxide followed by ethyl iodide, thereby installing an ethyl group at the alpha position. This compound is then treated again with sodium ethoxide, followed by benzyl bromide, thereby installing the benzyl group (note that the two R groups can be installed in either order). The ester groups are then hydrolyzed, and the resulting diacid is then heated to give a decarboxylation reaction, affording the desired product.

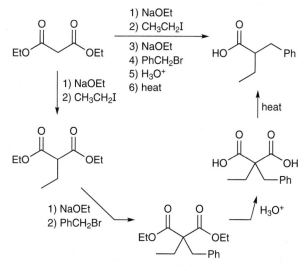

8.78 The starting material is an α,β-unsaturated ketone, which can function as a Michael acceptor. The reagent is a stabilized nucleophile, which can function as a Michael donor. So in this case, we do expect a Michael reaction in which the nucleophile is installed at the beta position:

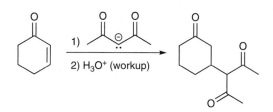

8.79 In this case, we do not expect a clean Michael reaction, because a Grignard reagent is not a good Michael donor. It is too reactive. In this case, a 1,2-addition is expected.

8.80 The starting material is an α,β-unsaturated ketone, which can function as a Michael acceptor. The reagent is lithium dimethyl cuprate (Me₂CuLi), which is a stabilized nucleophile that will attack the β position of the α,β-unsaturated ketone in a 1,4-addition reaction. In this case, a methyl group is installed at the beta position:

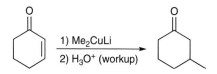

8.82 The starting material is an α,β-unsaturated nitrile, which can function as a Michael acceptor. When we compare the starting material and the desired product, we see that we must install a group at the beta position (highlighted below):

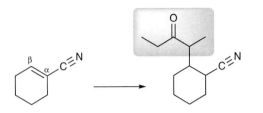

Now let's consider the nucleophile that we would need to install this group:

This is an enolate, which is not sufficiently stable to function as a Michael donor. Therefore, we will need to use a Stork enamine synthesis, in which a ketone is first converted into an enamine (rather than an enolate), because an enamine can function as a Michael donor. The enamine group is then removed (converted back to a ketone) with aqueous acid at the end of the synthesis:

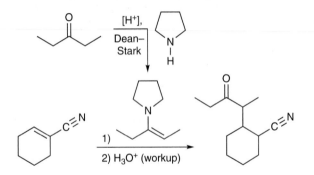

8.83   The starting material is an α,β-unsaturated ester, which can function as a Michael acceptor. When we compare the starting material and the desired product, we see that we must install a group at the beta position (highlighted below):

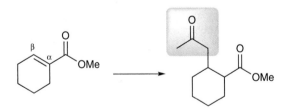

Now let's consider the nucleophile that we would need to install this group:

This is an enolate, which is not sufficiently stable to function as a Michael donor. Therefore, we will need to use a Stork enamine

synthesis, in which a ketone is first converted into an enamine (rather than an enolate), because an enamine can function as a Michael donor. The enamine group is then removed (converted back to a ketone) with aqueous acid at the end of the synthesis:

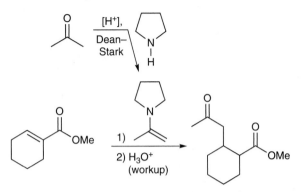

8.84   The starting material is an α,β-unsaturated aldehyde, which can function as a Michael acceptor. When we compare the starting material and the desired product, we see that we must install a group at the beta position (highlighted below):

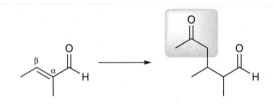

Now let's consider the nucleophile that we would need to install this group:

This is an enolate, which is not sufficiently stable to function as a Michael donor. Therefore, we will need to use a Stork enamine synthesis, in which a ketone is first converted into an enamine (rather than an enolate), because an enamine can function as a Michael donor. The enamine group is then removed (converted back to a ketone) with aqueous acid at the end of the synthesis:

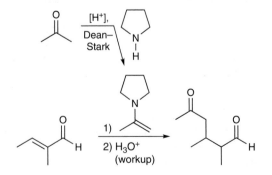

8.85 The starting material is an α,β-unsaturated aldehyde, which can function as a Michael acceptor. When we compare the starting material and the desired product, we see that we must install a group at the beta position (highlighted below):

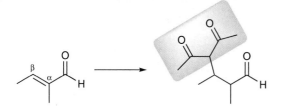

Now let's consider the nucleophile that we would need to install this group:

This is a stabilized enolate, which IS sufficiently stable to function as a Michael donor. Therefore, we will NOT need to use a Stork enamine synthesis in this case. Instead, we can directly perform a Michael reaction to obtain the desired product:

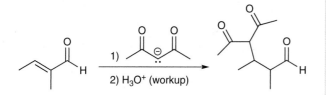

8.86 The starting material is a ketone, and the reagents indicate enamine formation:

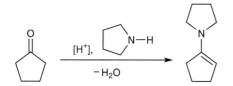

This enamine can function as a Michael donor, and it will react with an α,β-unsaturated aldehyde in a Michael addition, followed by aqueous acidic workup, to give the following product:

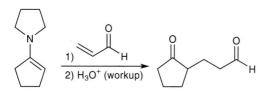

8.87 The desired product is an α,β-unsaturated ketone, which can be made via an aldol condensation. The carbon–carbon bond between the α and β positions is the bond that is formed during an aldol condensation:

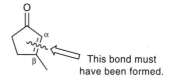

This bond (between the α and β positions) is part of a ring, which means that we will need to use an intramolecular aldol condensation (thereby forming a ring). That is, the two starting ketones must be tethered together in the same starting material. To draw this starting material, it is helpful to use a numbering scheme so that we can keep track of all carbon atoms:

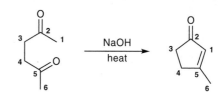

8.88 The starting material is an α,β-unsaturated ketone, which can function as a Michael acceptor. Lithium diphenyl cuprate (Ph_2CuLi) can function as a Michael donor, and it will react with the Michael acceptor, followed by aqueous acidic workup, to give a Michael addition (a 1,4-addition). This initially gives an enol, shown below, but under the acidic workup conditions, the enol quickly tautomerizes to give a ketone:

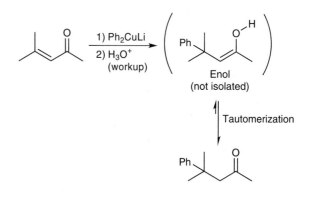

8.89 The desired product is a β-keto ester, which can be made via a Claisen condensation. The carbon–carbon bond between the α and β positions is the bond that is formed during a Claisen condensation:

This bond must
have been formed.

This bond (between the α and β positions) is part of a ring, which means that we will need to use an intramolecular Claisen condensation (thereby forming a ring). That is, the two starting esters must be tethered together in the same starting material. To draw this starting material, it is helpful to use a numbering scheme so that we can keep track of all carbon atoms:

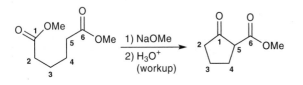

8.90

(a) In order to draw the conjugate base of compound **A**, we must identify the most acidic proton (the removal of which will lead to the most stable conjugate base). Compound **A** has carbonyl groups, and the protons connected to the alpha positions will be the most acidic protons in the compound. There are three alpha positions, identified below:

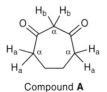

Compound **A**

Two of these alpha positions are equivalent, because of symmetry, and the protons at both of those locations have been labeled H_a. The remaining alpha position (at the top of the structure) also has alpha protons, and those protons have been labeled H_b.

Now we must determine whether H_a or H_b is more acidic. To do this, let's consider deprotonation at each location, and draw the resulting conjugate base in each case. Let's start with H_a. Deprotonation of H_a gives the following conjugate base, which is a resonance-stabilized enolate:

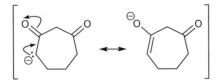

Conjugate base via deprotonation of H_a

Deprotonation of H_b gives the following conjugate base, which is also resonance-stabilized, but notice that this enolate has more resonance structures than the previous enolate:

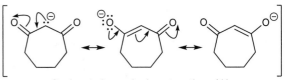

Conjugate base via deprotonation of H_b

This enolate has an additional resonance structure, and the negative charge is now delocalized over two oxygen atoms (rather than just one oxygen atom). Therefore, this conjugate base is more stable than the previous conjugate base, which makes H_b the most acidic proton.

When compound **A** is treated with a base, H_b is removed to give the following resonance-stabilized enolate:

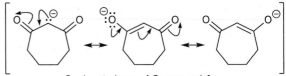

Conjugate base of Compound **A**

(b) We have already drawn the conjugate base of compound **A**, so now we must draw the conjugate base of compound **B**. Compound **B** has carbonyl groups, and the protons connected to the alpha positions will be the most acidic protons in the compound. There are four alpha positions, identified below:

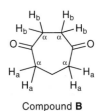

Compound **B**

Two of these alpha positions are equivalent because of symmetry, and the protons at both of those locations have been labeled H_a. The remaining two alpha positions (at the top of the structure) are also equivalent because of symmetry, and the protons at those locations have been labeled H_b.

Now we must determine whether H_a or H_b is more acidic. To do this, let's consider deprotonation at each location, and draw the resulting conjugate base in each case. Let's start with H_a. Deprotonation of H_a gives the following conjugate base, which is a resonance-stabilized enolate:

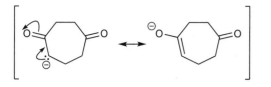

Deprotonation of H_b gives the following conjugate base, which is also a resonance-stabilized enolate:

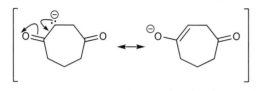

Notice that deprotonation of either H_a or H_b results in enolates that are similar in stability. Either way, the conjugate base will have a negative charge that is delocalized over one oxygen atom and one carbon atom. We cannot draw a conjugate base that has the negative charge delocalized over *two* oxygen atoms (as we saw for the conjugate base of compound **A**). As such, the conjugate base of compound **B** is less stable than the conjugate base of compound **A**.

This can be seen more clearly if we compare the resonance hybrids for the conjugate bases of compounds **A** and **B**:

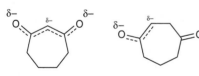

| Conjugate base of compound **A** (resonance hybrid) | Conjugate base of compound **B** (resonance hybrid) |

Notice that the first enolate above (the conjugate base of compound **A**) has the negative charge delocalized over two oxygen atoms. In contrast, the second enolate (the conjugate base of compound **B**) has the negative charge delocalized over only one oxygen atom. As such, the conjugate base of compound **A** is more stable than the conjugate base of compound **B**.

(c) We have seen that the conjugate base of compound **A** is more stable than the conjugate base of compound **B**. The conjugate base of compound **A** is the more stable, weaker conjugate base. The weaker conjugate base has the stronger parent acid. Therefore, compound **A** is more acidic than compound **B**.

8.91 In each compound, we consider the proximity of the ketone group to the carboxylic acid group, using Greek letters (α, β, etc.):

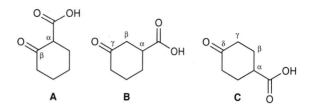

In order for decarboxylation to occur, the ketone group must be beta (β) to the carboxylic acid group. Only compound **A** has this relationship. Upon heating, compound **A** will undergo decarboxylation to give an enol, which quickly tautomerizes to give a ketone (cyclohexanone) as the product:

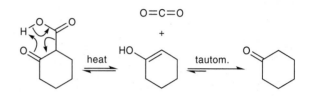

8.92 In the presence of catalytic hydroxide, the starting material is deprotonated at the alpha position to give a resonance-stabilized enolate ion. This enolate ion is then protonated (by water, thereby regenerating the catalyst, hydroxide) to give the product:

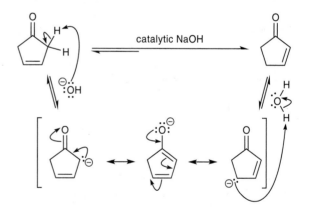

8.93

(a) Notice that the starting material has ten carbon atoms, and the product also has ten carbon atoms. If this is an aldol condensation, then it must be an intramolecular process (otherwise, the product would have twenty carbon atoms). When drawing the product of an intramolecular aldol condensation, it is always helpful to number the starting material (this will help us keep track of all carbon atoms when drawing the product):

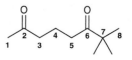

In the presence of a strong base, this starting material can be deprotonated at any of the alpha positions (C1, C3, or C5). Deprotonation at C3 or C5 will not lead to an intramolecular reaction, because the resulting ring size would be too small to form:

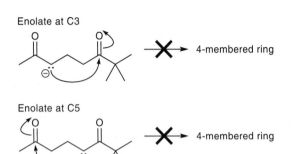

In contrast, deprotonation at C1 leads to an intramolecular aldol condensation that forms a six-membered ring:

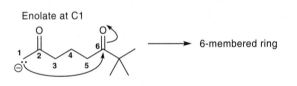

When drawing the product, we use the numbers to guide us. The reaction occurs between C1 (where an enolate is formed) and C6:

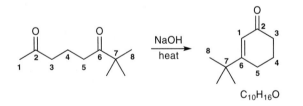

These numbers help us draw the correct ring size and they help us draw the substituent in the correct location (the *tert*-butyl group is connected to C6).

(b) In the first step, the starting material is deprotonated at C1. The resulting enolate ion then attacks the other carbonyl group in an intramolecular fashion, giving an alkoxide ion containing a (newly formed) six-membered ring. This alkoxide ion is then protonated by water to give a β-hydroxy ketone. This β-hydroxy ketone is then deprotonated at the alpha position to give an enolate ion, which then ejects hydroxide as a leaving group to give the product.

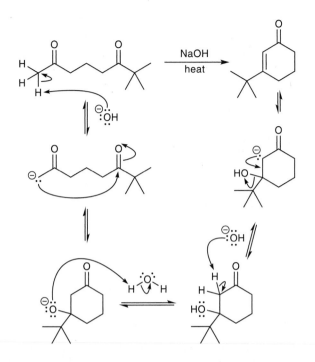

8.94 The starting material is diethyl malonate, and the desired product is a carboxylic acid, so we need to perform a malonic ester synthesis. Notice that the desired carboxylic acid has two alkyl groups (methyl groups) connected to the alpha position,

and we will need to install each of these alkyl groups individually. Our synthesis will have the following steps:

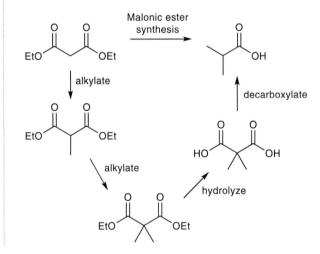

Below is a complete synthesis, showing all reagents for this malonic ester synthesis:

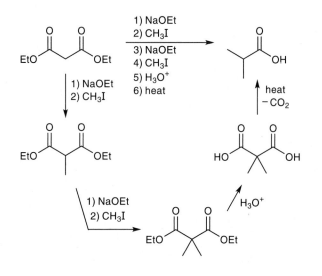

8.95 LDA is a strong, sterically hindered base, and it will irreversibly deprotonate the starting material (at the alpha position that bears protons) to give an ester enolate. This enolate then attacks the other ester group in an intramolecular fashion. The resulting tetrahedral intermediate than expels methoxide to reform the carbonyl group. The resulting product is indeed the final product of the reaction, however, under these strongly basic conditions (where LDA and methoxide ions are present), this product is deprotonated to give an enolate:

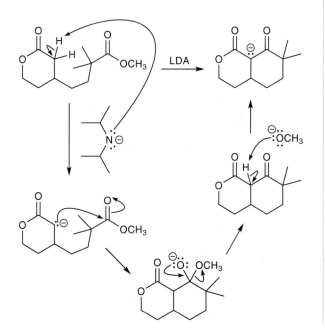

When the reaction is complete, aqueous acid (H_3O^+) is introduced into the reaction flask, thereby protonating the anion and regenerating the product:

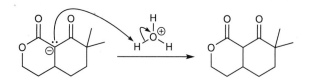

8.96 The starting material and the product have the same molecular formula (C_9H_{16}), so they are constitutional isomers. We have not learned a way to achieve this isomerization in one step, so we will have to devise a multistep synthesis. Notice that the starting material has a seven-membered ring, while the product has a six-membered ring. We have not learned a way to remove a carbon atom from a ring, so we will need to open the ring, and then close it back up again in a separate step.

To open the ring, we can use an ozonolysis process, which would give a compound with two carbonyl groups. This compound is ideally suited for an intramolecular aldol condensation, because that process will generate the desired carbon skeleton:

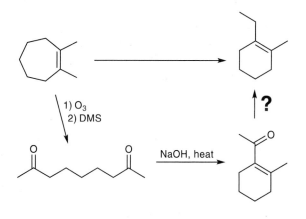

To complete the synthesis, we just need to remove the carbonyl group, and this can be achieved by converting the carbonyl group into a thioacetal, followed by desulfurization with Raney Nickel:

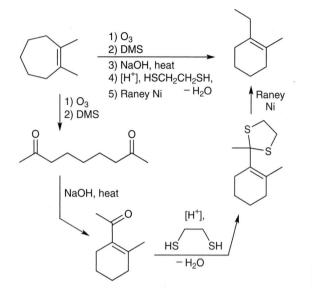

Note that, for removal of the carbonyl group, a Clemmensen reduction should be avoided here. That process employs HCl, which might add across the alkene.

8.97 The product is an imine, which can be made from the corresponding ketone, so this can be the last step of our synthesis:

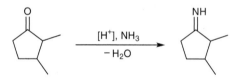

To complete the synthesis, we must bridge the gap between the starting ketone and the ketone shown above:

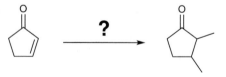

We must install methyl groups at the alpha and beta positions, and in the process, we must remove the double bond between the alpha and beta positions. All of these objectives can be achieved in just two steps: first the starting material is treated with lithium dimethyl cuprate (Me₂CuLi), and rather than working up the resulting enolate, we instead treat the resulting enolate with methyl iodide to give an alkylation reaction. The entire synthesis is shown here:

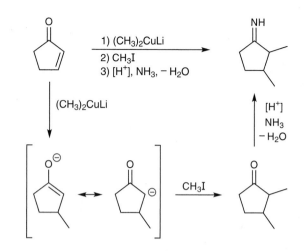

8.98 The starting material has only one aromatic ring, and the product has two aromatic rings, so we consider beginning our synthesis with an aldol condensation between two molecules of the starting ketone. This would give the following α,β-unsaturated ketone:

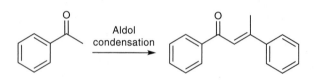

To complete the synthesis, we must convert this α,β-unsaturated ketone into the desired product:

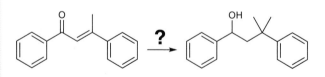

Notice that the product has one additional methyl group, which must be installed at the beta position. This can be achieved via a Michael addition,

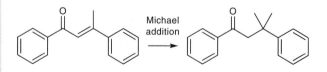

and the resulting compound can be converted into the desired product via reduction of the ketone. The complete synthesis is shown here.

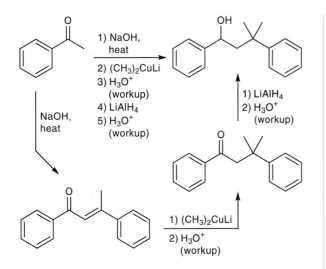

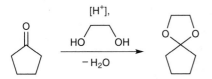

8.99 The product is an acetal, which can be made from the corresponding ketone, so this can be the last step of our synthesis:

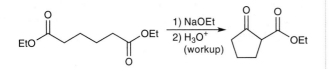

Now we must bridge the gap between the starting material and the ketone above:

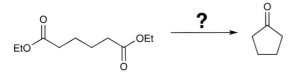

This process involves forming a ring. Notice that the starting material has two ester groups, and an intramolecular Claisen condensation would give the desired ring size:

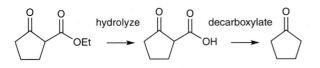

The product of the Claisen condensation is a beta keto ester, and it can be hydrolyzed and then decarboxylated to give the desired ketone:

The complete synthesis is shown here:

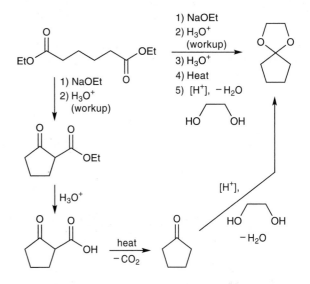

8.100 The starting material is diethyl malonate, which makes us think of the malonic ester synthesis (a process that generates a carboxylic acid). Indeed, the product is an ester, which can be made from a carboxylic acid via a Fischer esterification. This gives the following strategy:

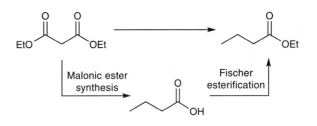

The complete synthesis is shown here:

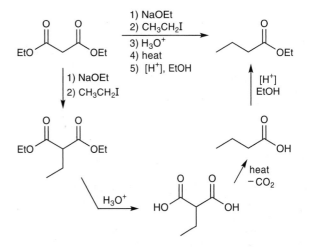

8.101 The starting material has a six-membered ring, while the product has a five-membered ring, and we have not learned any way to remove an atom from a ring. So we must open the ring somehow, and then close it back up in a separate step. Notice that the product is an α,β-unsaturated ketone, which can be made via an intramolecular aldol condensation, so this can be the last step of our synthesis:

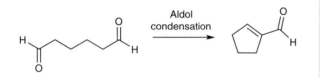

Now we must bridge the gap between the starting material and the dialdehyde shown above:

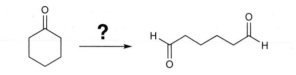

This requires opening the six-membered ring in the starting ketone. There are certainly many ways to do this. One method is to convert the ketone into an alkene (via the two steps shown below), followed by ozonolysis:

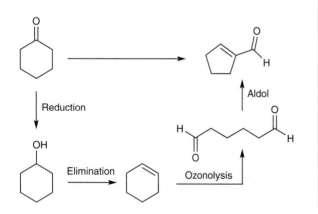

The complete synthesis is shown below:

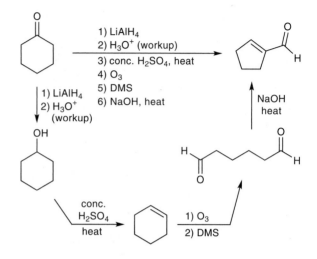

Alternatively, we can open the ring by treating the starting ketone with a peroxy acid, such as MCPBA, giving a Baeyer-Villiger oxidation. The resulting cyclic ester (called a lactone) can then be reduced to give a diol, followed by oxidation with PCC to give the desired dialdehyde:

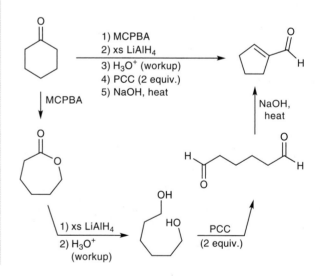

8.102 Comparing the starting material and the product, we see that we must install two alkyl groups (highlighted below):

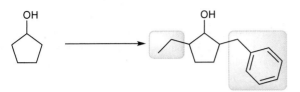

We have not learned a way to do this directly, but we have learned a way to install alkyl groups on either side of a ketone,

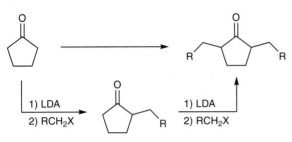

and we know how to interconvert alcohols and ketones:

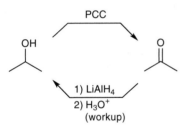

So our strategy is to oxidize the starting alcohol to give a ketone, and then to perform two successive alkylation reactions, and then finally to reduce the ketone back to an alcohol. The complete synthesis is shown here:

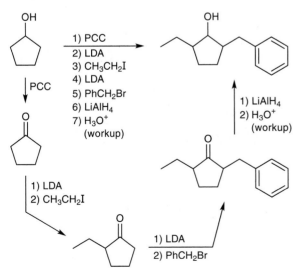

8.103 The desired product is an acetal, which can be made from the corresponding diol and ketone; this can be the last step of our synthesis:

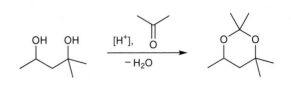

Now we must bridge the gap between the starting ketone and the diol above.

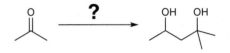

This can be accomplished in just two steps. First an aldol reaction will give the desired carbon skeleton, and then reduction of the ketone will give the diol that will be used in the last step of the synthesis:

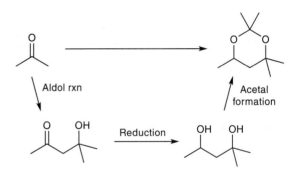

The complete synthesis is shown here:

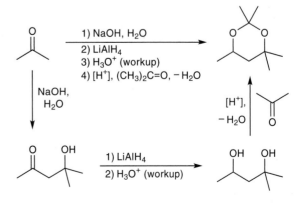

8.104 The desired product can be made via an alkylation reaction, where each of the starting materials shown below (aldehyde and alkyl bromide) has two carbon atoms:

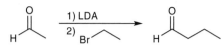

The aldehyde shown above can be made by treating ethanol with PCC; while ethyl bromide can be made by treating ethanol with PBr$_3$. The complete synthesis is summarized below:

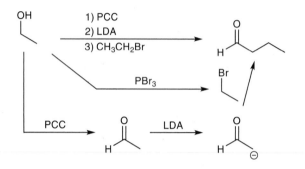

8.105

(a) Two molecules of the starting material react with each other, liberating water in the process, giving the following product:

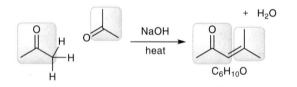

(b) The product has four different kinds of protons, labeled **a–d** below, so the ^{1}H NMR spectrum of the product will have four signals. Notice that the protons labeled "**c**" are not equivalent to the protons labeled "**d**" because they occupy different electronic environments ("**c**" protons are *cis* to the carbonyl group, while "**d**" protons are *trans* to the carbonyl group).

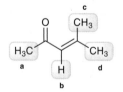

(c) The product has six different kinds of carbon atoms, labeled **a–f** below, so the ^{13}C NMR spectrum of the product will have six signals. Notice that the carbon atom labeled "**e**" is not equivalent to the carbon atom labeled "**f**" because those two methyl groups occupy different electronic environments (one is *cis* to the carbonyl group, while the other is *trans* to the carbonyl group).

8.106

(a) There are only two possible carboxylic acids that have the molecular formula C$_4$H$_8$O:

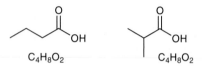

The first compound above will have four signals in its ^{1}H NMR spectrum and four signals in its ^{13}C NMR spectrum. In contrast, the second compound above will have only three signals in each spectrum. Therefore, we conclude that the first product shown above (butanoic acid) is the correct product, and it is being made with a malonic ester synthesis:

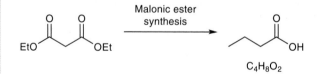

(b) The desired carboxylic acid has an ethyl group connected to the alpha position:

So we must use an ethyl halide (such as ethyl iodide or ethyl bromide) as the reagent in the second step of our malonic ester synthesis, shown here:

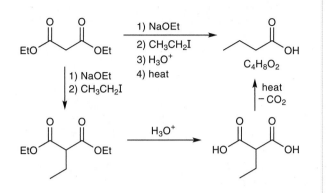

8.107

(a) Compound **A** is an aldehyde that has an alpha position bearing a proton (highlighted):

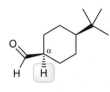

Upon treatment with a strong base, this proton can be removed to give an enolate, where the alpha position is no longer tetrahedral because it is sp^2 hybridized. This planar enolate can then be protonated from either face. Protonation can occur on the top face, to give compound **B**.

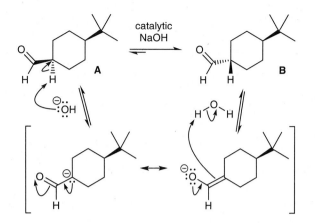

(b) Compound **B** can adopt a chair conformation in which both substituents occupy equatorial positions:

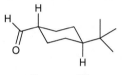

Compound **B**

In contrast, compound **A** cannot adopt such a chair conformation. Each chair conformation of compound **A** has one group in an axial position:

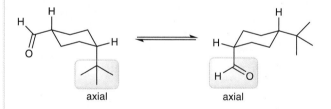

This renders compound **A** less stable than compound **B**. The base-catalyzed isomerization favors the more stable compound, which is compound **B**.

8.108 LDA is a strong, sterically hindered base, and it will irreversibly deprotonate the alpha position of the starting ketone. The resulting enolate can then attack the diester (attacking one of the carbonyl groups) to give a tetrahedral intermediate that ejects ethoxide as a leaving group to regenerate the carbonyl group. The resulting compound is then deprotonated to give an enolate, which then attacks the surviving ester group. The resulting tetrahedral intermediate then ejects ethoxide as a leaving group to regenerate the carbonyl group. And finally, there is an acidic proton (alpha to three different carbonyl groups) that will be removed under these conditions to give an enolate:

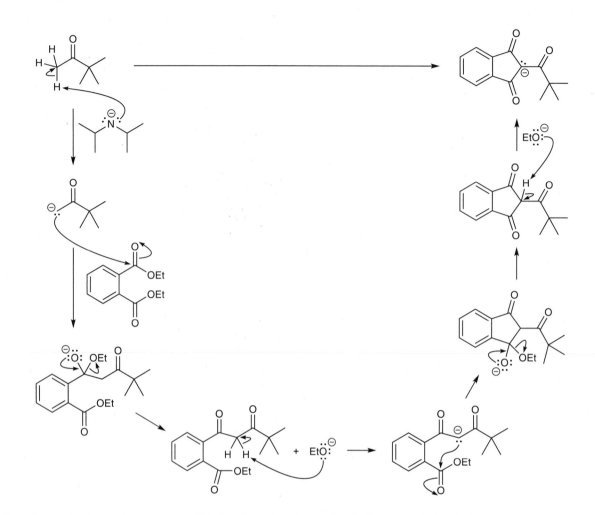

Once the reaction is complete, aqueous acid is introduced into the reaction flask to protonate the enolate and generate the uncharged, final product:

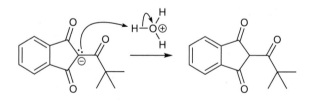

8.109 The first step of this process is a Claisen condensation, which is achieved with a strong base, followed by aqueous acidic workup, to give a beta-keto ester. Notice that we use NaOEt as the base because the starting ester is an ethyl ester. If it had been a methyl ester, we would have used NaOMe to avoid converting one type of ester into another. The next step of this process is alkylation of the beta-keto ester, which can be achieved by treating it with NaOEt, followed by methyl iodide. The next step is another alkylation, which can be achieved by treating with NaOEt, followed by benzyl iodide. Finally, the last step involves removal of the ester group entirely, which can be accomplished via acid-catalyzed hydrolysis of the ester, followed by decarboxylation to give the product:

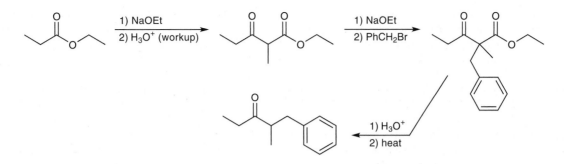

8.110 The first part of this synthesis is an iodoform reaction, which converts the methyl ketone into a carboxylic acid. The resulting compound has a carbonyl group that is beta to the carboxylic acid group, so this compound will undergo decarboxylation upon heating to give the aldehyde shown. And finally, this aldehyde is alkylated upon treatment with LDA followed by methyl iodide, to install a methyl group at the alpha position:

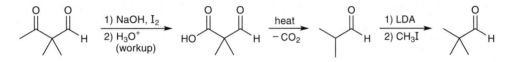

8.111 The starting material has a cyano group, which is hydrolyzed upon treatment with aqueous acid to give a carboxylic acid. The resulting compound is a beta-keto acid. Under the conditions of its formation, the beta-keto acid will undergo decarboxylation to give 3-pentanone, which has the molecular formula $C_5H_{10}O$:

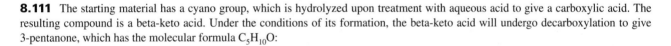

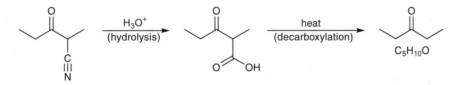

This ketone ($C_5H_{10}O$) is converted into an ester upon treatment with a peroxy acid, such as MCPBA, and the ester undergoes a Claisen condensation upon treatment with NaOEt followed by aqueous acidic workup to give the final product ($C_8H_{14}O_3$), The complete synthesis is shown here:

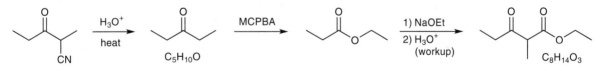

8.112

(a) The starting material has an ester, which is hydrolyzed to give a carboxylic acid (compound **A**) upon treatment with aqueous acid. Compound **A** is a beta-keto acid, which undergoes decarboxylation upon heating to give compound **B**. Finally, compound **B** is reduced upon treatment with lithium aluminum hydride, followed by aqueous acidic workup, to give an alcohol (compound **C**):

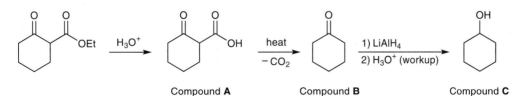

(b) Compound **A** has a carboxylic acid group, so the IR spectrum of compound **A** is expected to have a very broad signal from 2200 to 3600 cm^{-1} (covering about one third of the spectrum). In contrast, compound **B** does not have a carboxylic acid group, so the IR spectrum of compound **B** will not have a broad signal from 2200 to 3600 cm^{-1}. Therefore, the conversion of **A** to **B** can be monitored by looking at the signal between 2200 and 3600 cm^{-1}. This signal will disappear as the reaction proceeds.

   IR spectroscopy can also be used to monitor the conversion of compound **B** to compound **C**. Compound **B** is a ketone and will produce a signal at approximately 1720 cm^{-1}, while compound **C** is an alcohol and will have a broad signal between 3200 and 3600 cm^{-1}. So the conversion of **B** to **C** can be monitored by the disappearance of the signal at 1720 cm^{-1} and the appearance of a broad signal from 3200 to 3600 cm^{-1}.

(c) Compound **A** has seven carbon atoms, each of which occupies a unique electronic environment, so the ^{13}C NMR spectrum of compound **A** will have seven signals. In contrast, compound **B** has only six carbon atoms AND compound **B** has symmetry as well, so the ^{13}C NMR spectrum of compound **B** will have only four signals.

(d) Each compound (**B** and **C**) will have a ^{13}C NMR spectrum with only four signals. In compound **B**, one of these four signals (for the carbonyl group) will appear near 200 ppm. In contrast, compound **C** is an alcohol, not a ketone, so the ^{13}C NMR spectrum of compound **C** will lack a signal so far downfield. Instead, one of the four signals in the ^{13}C NMR spectrum of compound **C** will appear between 50 and 100 ppm (characteristic of a carbon atom next to an OH group).

8.113   LDA is a strong, sterically hindered base, which irreversibly deprotonates the alpha position to give an enolate of the thioester. This enolate then functions as a nucleophile and attacks the carbonyl group, giving an alkoxide ion. This alkoxide ion then functions as a nucleophile and attacks the carbonyl group of the thioester in an intramolecular fashion. The resulting tetrahedral intermediate then ejects a leaving group (CH$_3$S$^-$) to give the product.

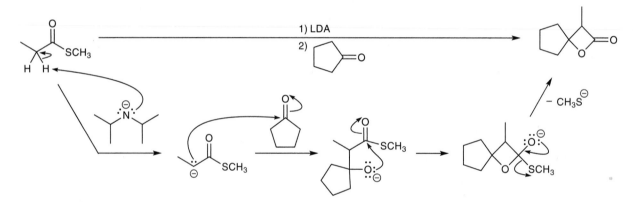

8.114   This compound has an anhydride group, which is hydrolyzed upon treatment with aqueous acid. A mechanism for this hydrolysis is shown below. First the anhydride is attacked by water to give a tetrahedral intermediate. This intermediate then reforms a carbonyl group to expel a carboxylate ion as the leaving group. And finally, proton transfer steps give a compound with one ketone group and two carboxylic acid groups:

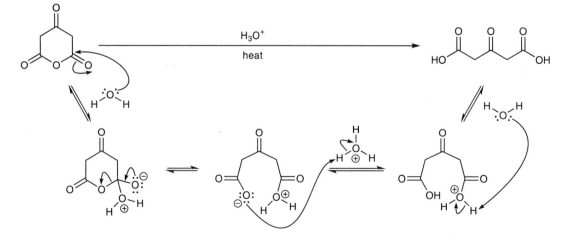

In the mechanism shown above, the anhydride was attacked directly by water, because anhydrides are sufficiently electrophilic to be attacked without first being protonated. But it is okay if you drew the anhydride being protonated first (as shown below), since acid is present (it just makes the mechanism a bit longer than necessary):

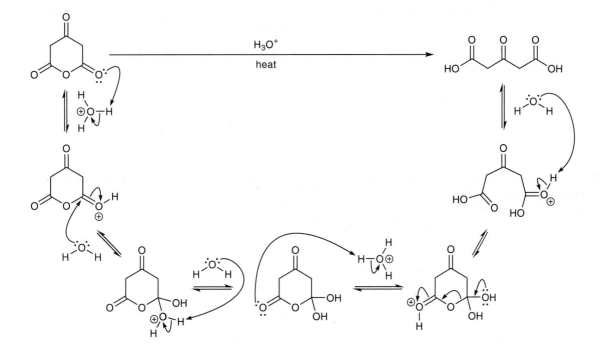

Then, heating causes decarboxylation, followed by tautomerization. Tautomerization is occurring under acidic conditions, so we draw a mechanism for tautomerization showing acid-catalyzed conditions:

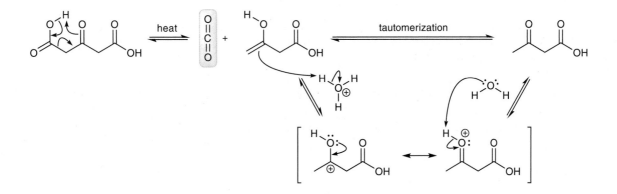

Continued heating once again causes decarboxylation, followed by tautomerization:

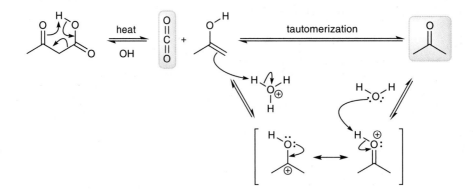

8.115 When treated with sodium ethoxide, diethyl malonate is deprotonated to give an enolate ion (exactly like the first step of the malonic ester synthesis). Then, when the epoxide (an electrophile) is introduced into the reaction flask, the enolate functions as a nucleophile and attacks the epoxide (much like the second step of the malonic ester synthesis, except that the electrophile is an epoxide rather than an alkyl halide). The resulting alkoxide ion can function as a nucleophile and attack one of the ester groups in an intramolecular fashion (thereby providing the necessary five-membered ring). This generates a tetrahedral intermediate, which can then expel a leaving group (ethoxide) to give a compound that has both a lactone (a cyclic ester) as well as another ester group. Under the basic condition of its formation, this compound is deprotonated to give an enolate ion:

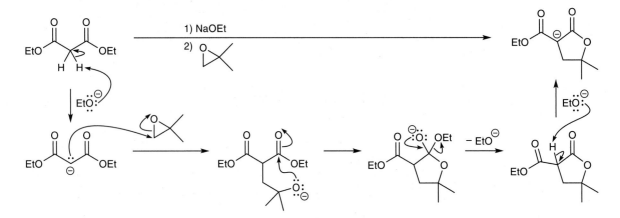

Next, aqueous acid is introduced into the reaction flask. Aqueous acid plays several roles. First, the enolate is protonate, as shown here:

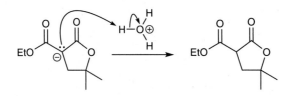

Next, the ester group is hydrolyzed, which occurs via several steps. First, the ester group is protonated, making it more electrophilic, and then the protonated ester group is attacked by water, generating a tetrahedral intermediate. This intermediate then undergoes two successive proton transfer steps, followed by loss of water. The resulting oxonium ion is then deprotonated by water to give a carboxylic acid:

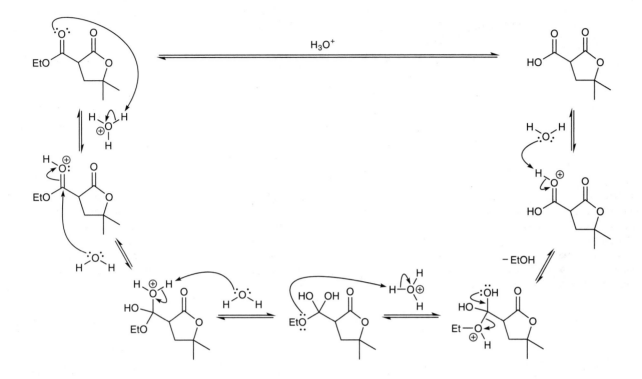

Notice the proximity of the ester group to the carboxylic acid group (the ester carbonyl is beta to the carboxylic acid group). When heated, this compound can undergo decarboxylation as shown below, to give an enol. Under the conditions of its formation, this enol quickly tautomerizes to give the product, which is a cyclic ester (a lactone):

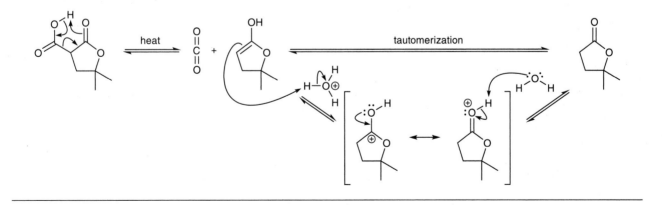

CHAPTER 9

9.2 In order to perform a Gabriel synthesis, we must identify the alkyl halide that should be used. This is the only choice that we need to make when planning a Gabriel synthesis. To do this, we just draw a halogen in place of the NH$_2$ group, as follows:

Below is the complete Gabriel synthesis in which this primary alkyl bromide is used to make the desired primary amine:

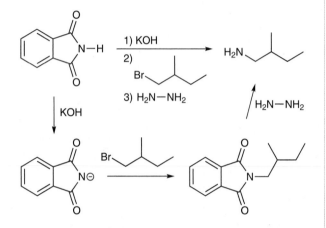

9.3 In order to perform a Gabriel synthesis, we must identify the alkyl halide that should be used. This is the only choice that we need to make when planning a Gabriel synthesis. To do this, we just draw a halogen in place of the NH$_2$ group, as follows:

This alkyl bromide is called benzyl bromide. Below is the complete Gabriel synthesis in which benzyl bromide is used to make the desired amine (benzyl amine):

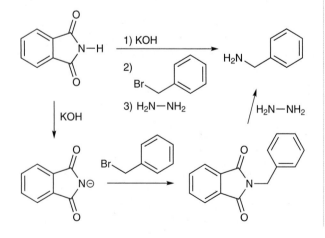

9.4 In order to perform a Gabriel synthesis, we must identify the alkyl halide that should be used. This is the only choice that we need to make when planning a Gabriel synthesis. To do this, we just draw a halogen in place of the NH$_2$ group, as follows:

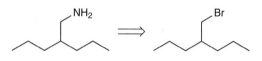

Below is the complete Gabriel synthesis in which this primary alkyl bromide is used to make the desired primary amine:

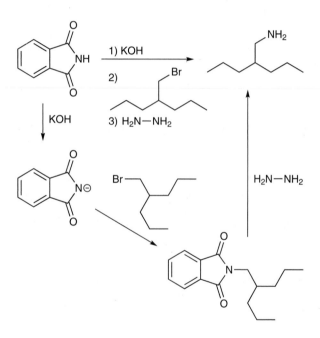

9.5 In order to perform a Gabriel synthesis, we must identify the alkyl halide that should be used. This is the only choice that we need to make when planning a Gabriel synthesis. To do this, we just draw a halogen in place of the NH$_2$ group, as follows:

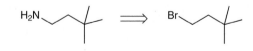

Below is the complete Gabriel synthesis in which this primary alkyl bromide is used to make the desired primary amine:

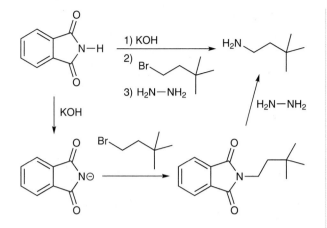

9.7 We begin by drawing the alkyl bromide that would be required to make the desired amine via a Gabriel synthesis:

This compound has the leaving group connected to an sp^2 hybridized carbon atom, and S_N2 reactions are not observed at sp^2 hybridized centers. Therefore, the desired amine cannot be made via a Gabriel synthesis.

9.8 We begin by drawing the alkyl bromide that would be required to make the desired amine via a Gabriel synthesis:

This alkyl bromide is both primary and benzylic and can serve as an excellent substrate in an S_N2 reaction. Therefore, the desired amine can be made via a Gabriel synthesis, shown below:

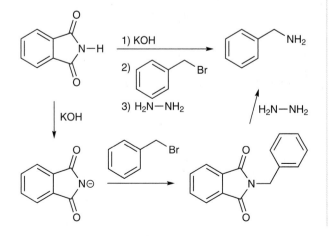

9.9 We begin by drawing the alkyl bromide that would be required to make the desired amine via a Gabriel synthesis:

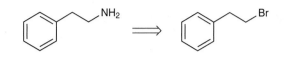

This alkyl bromide is primary and can serve as a substrate in an S_N2 reaction. Therefore, the desired amine can be made via a Gabriel synthesis, shown below:

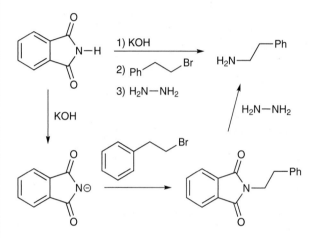

9.10 We begin by drawing the alkyl bromide that would be required to make the desired amine via a Gabriel synthesis:

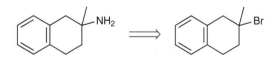

This alkyl bromide is tertiary, and S_N2 reactions do not occur with tertiary substrates (because of steric effects). Therefore, the desired amine cannot be made via a Gabriel synthesis.

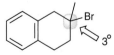

9.12 This problem involves the conversion of a ketone into a secondary amine, which should alert us to the possibility of a reductive amination. The ketone is first converted into an imine, which is then reduced to give the desired secondary amine:

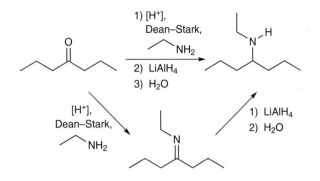

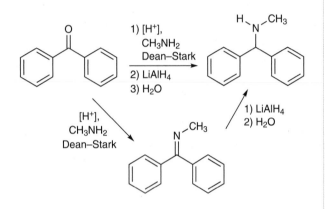

9.13 This problem involves the conversion of a ketone into a secondary amine, which should alert us to the possibility of a reductive amination. The ketone is first converted into an imine, which is then reduced to give the desired secondary amine:

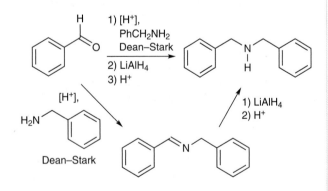

9.14 This problem involves the conversion of an aldehyde into a secondary amine, which should alert us to the possibility of a reductive amination. The ketone is first converted into an imine, which is then reduced to give the desired secondary amine:

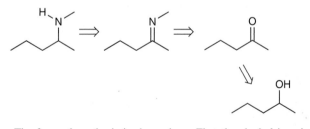

9.15 This problem involves the conversion of a ketone into a secondary amine, which should alert us to the possibility of a reductive amination. The ketone is first converted into an imine in an intramolecular fashion (giving a cyclic imine), which is then reduced to give the desired secondary amine:

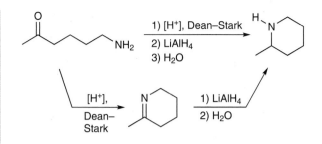

9.16 This problem involves the conversion of a primary amine into a secondary amine, which should alert us to the possibility of a reductive amination. The primary amine is first converted into an imine, which is then reduced to give the desired secondary amine:

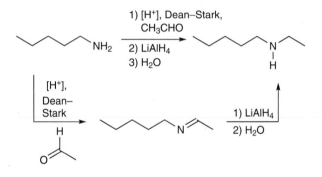

9.18 The following is a retrosynthetic strategy for making the desired amine from the starting alcohol:

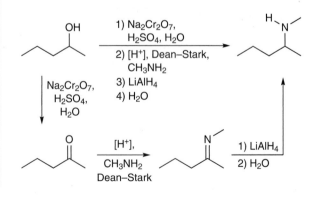

The forward synthesis is shown here. First the alcohol is oxidized to give a ketone, which is then converted into an imine. Finally, the imine is reduced to give the desired product:

9.19 The following is a retrosynthetic strategy for making the desired amine from the starting alkene:

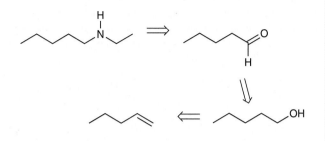

The forward synthesis is shown here. First the alkene is converted into a primary alcohol via hydroboration–oxidation to give *anti*-Markovnikov addition of H_2O. Then, the alcohol is oxidized with PCC to give an aldehyde. This aldehyde is then converted into an imine, which is reduced to give the desired product:

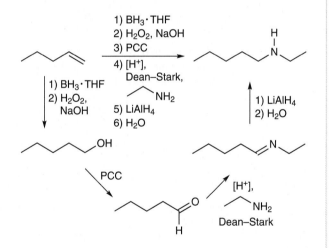

9.20 The following is a retrosynthetic strategy for making the desired amine from the starting acetal:

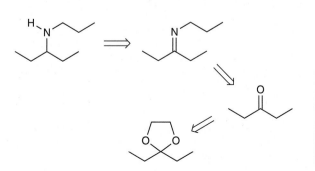

The forward synthesis is shown here. First the acetal is treated with aqueous acid to give a ketone, which is then converted into an imine. Finally, the imine is reduced to give the desired product:

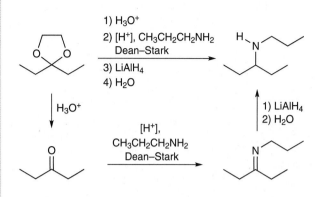

9.21 The following is a retrosynthetic strategy for making the desired amine from the starting alkene:

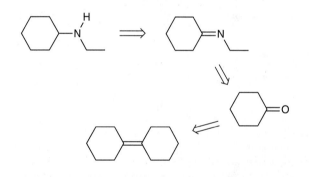

The forward synthesis is shown here. First the alkene undergoes ozonolysis to give two equivalents of a ketone. The ketone is then converted into an imine, and finally, the imine is reduced to give the desired product:

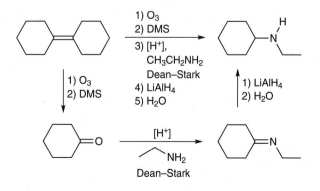

9.22 The following is a retrosynthetic strategy for making the desired amine from the starting acid halide:

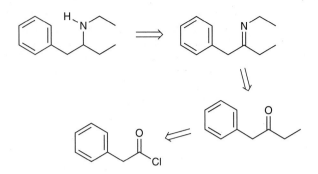

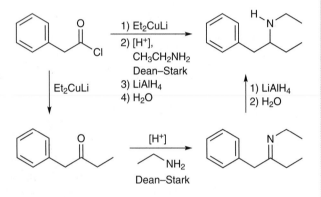

The forward synthesis is shown here. First the acid chloride is converted into a ketone upon treatment with a lithium dialkyl cuprate. This ketone is then converted into an imine, and finally, the imine is reduced to give the desired product:

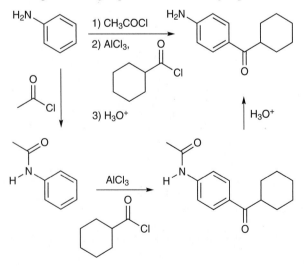

9.24 The desired transformation is a Friedel–Crafts acylation reaction at the *para* position, relative to an *ortho–para* director. Unfortunately, an acid halide (the reagent for a Friedel–Crafts acylation) will react with the amino group. We can circumvent this problem by first converting the amino group into an amide group. Then, we can perform the desired Friedel–Crafts acylation, followed by converting the amide group back into an amino group:

9.25 If we try direct chlorination of the starting material, we will obtain a trichlorinated product, because the *ortho* positions and *para* position are all activated, and the ring is very reactive. To achieve monochlorination rather than polychlorination, we can first convert the amino group into an amide group, thereby making the ring less activated. Chlorination of this compound will install a chlorine atom in the *para* position, and finally, the amide group can be hydrolyzed to convert it back into an amino group:

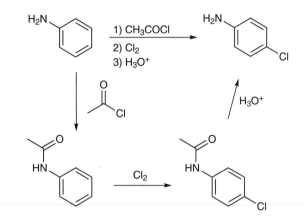

9.27 The starting material is a primary amine, and the reagents indicate an *in situ* preparation of nitrous acid (HONO). Under these conditions, a primary amine is converted into a diazonium salt:

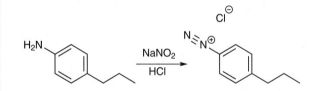

9.28 The starting material is a primary amine, and the reagents indicate an *in situ* preparation of nitrous acid (HONO). Under these conditions, a primary amine is converted into a diazonium salt:

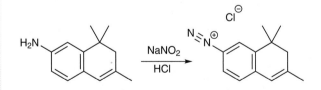

9.29 The starting material is a secondary amine, and the reagents indicate an *in situ* preparation of nitrous acid (HONO). Under these conditions, a secondary amine is converted into a nitrosamine:

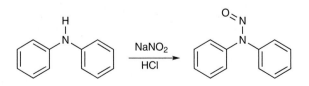

9.30 The starting material is a primary amine, and the reagents indicate an *in situ* preparation of nitrous acid (HONO). Under these conditions, a primary amine is converted into a diazonium salt:

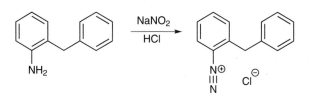

9.32 The starting material is a disubstituted aromatic ring where one of the substituents is an amino group. This amino group must be replaced with a bromine atom. This transformation can be achieved by first converting the starting material into a diazonium salt, and then using a Sandmeyer reaction to convert the diazonium salt into the desired product.

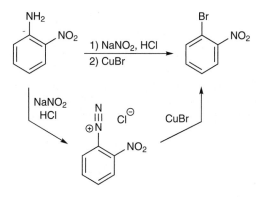

9.33 The starting material is a disubstituted aromatic ring where one of the substituents is an amino group. This amino group must be replaced with a cyano group. This transformation can be achieved by first converting the starting material into a diazonium salt, and then using a Sandmeyer reaction to convert the diazonium salt into the desired product.

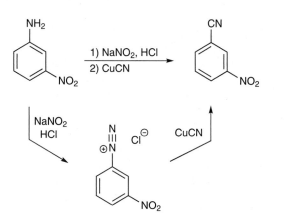

9.34 The starting material is a trisubstituted aromatic ring where one of the substituents is an amino group. This amino group must be replaced with a chlorine atom. This transformation can be achieved by first converting the starting material into a diazonium salt, and then using a Sandmeyer reaction to convert the diazonium salt into the desired product.

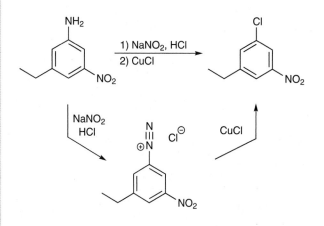

9.35 The starting material is a trisubstituted aromatic ring where one of the substituents is an amino group. This amino group must be replaced with a bromine atom. This transformation can be achieved by first converting the starting material into a diazonium salt, and then using a Sandmeyer reaction to convert the diazonium salt into the desired product.

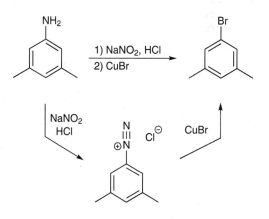

9.36 The starting material is an aromatic compound where one of the substituents is an amino group. This amino group must be replaced with a cyano group. This transformation can be achieved by first converting the starting material into a diazonium salt, and then using a Sandmeyer reaction to convert the diazonium salt into the desired product.

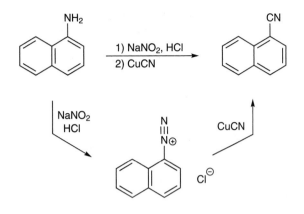

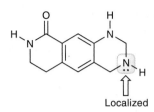

Localized

9.37 This compound has three nitrogen atoms. Let's begin by assessing the basicity of the nitrogen atom in the upper left corner of the structure. The lone pair on this nitrogen atom is delocalized by resonance, as shown here:

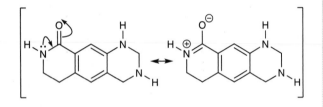

Since this lone pair is delocalized, it is less available to function as a base. Protonation at this position is unfavorable, because it would cause a loss of resonance. Therefore, this nitrogen atom is not strongly basic.

Now let's consider the nitrogen atom in the upper right corner of the structure. The lone pair on this nitrogen atom is also delocalized by resonance, as shown here:

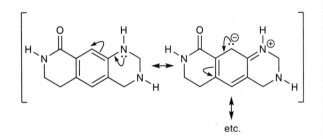

etc.

Since this lone pair is also delocalized, it is also less available to function as a base. Indeed, this nitrogen atom is also not strongly basic.

Finally, the remaining nitrogen atom (highlighted below) has a lone pair that is localized (not participating in resonance). As a result, the lone pair on this nitrogen atom is available to function as a base. Indeed, this nitrogen atom is the most basic site in the compound. That is, if this compound were treated with an acid, this nitrogen atom (highlighted) would be selectively protonated.

9.38 The starting material has two functional groups: a ketone and a primary amine. When treated with catalytic acid, under conditions where water is removed, these two functional groups can react with each other, in an intramolecular fashion, to give a cyclic imine. To draw this imine, it is helpful to place numbers on the compound, and then use those numbers to keep track of the atoms, like this:

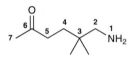

Notice that the numbers do not need to be in accordance with IUPAC rules (in this case, IUPAC rules would dictate that the numbers be assigned from left to right, rather than right to left). The numbers assigned here are not IUPAC numbers, but rather, they are tools that we can use to help us draw the product of an intramolecular reaction. Once we are done drawing that product, we can erase the numbers, which is why they don't need to follow IUPAC rules.

There is one other thing that we can do to help us draw the product of the first step of this process. Specifically, we will redraw the starting material so that the reactive centers (the electrophilic ketone and the nucleophilic amine) are in close proximity:

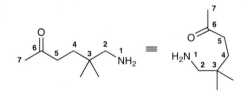

Now we are ready to draw the imine product. Notice that a C=N bond is formed between the positions numbered 1 and 6, thereby forming a six-membered ring:

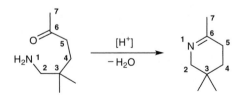

Finally, the last step of the synthesis is reduction of the imine (followed by aqueous workup) to give the secondary amine shown below:

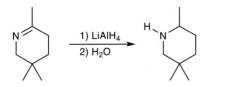

9.39 The starting material is a primary amine, and the desired product is a secondary amine, which requires the installation of a single alkyl group (in this case, we must install a cyclopentyl group). This type of transformation cannot be achieved via an S_N2 reaction (by treating the starting primary amine with cyclopentyl bromide). There are at least two reasons why this approach would be inefficient: (1) treating an amine with an alkyl halide generally results in exhaustive alkylation to give a quaternary ammonium ion, and that is not the desired outcome in this case, and (2) an S_N2 reaction is less efficient on a secondary substrate (such as cyclopentyl bromide), and in such a case, the competing E2 reaction is likely to dominate. For these two reasons, we cannot directly alkylate the starting material to give the desired product.

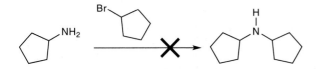

Instead, reductive amination can be used to achieve the desired transformation. First, the starting material is treated with cyclopentanone, with acid catalysis and removal of water, to give an imine. Then, the imine is reduced with lithium aluminum hydride, followed by aqueous workup, to give the product, which is a secondary amine.

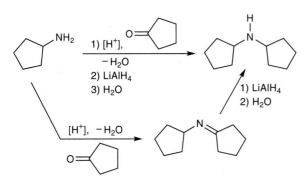

9.40 The starting material is phthalimide, and the product is a primary amine, so this is a Gabriel synthesis. In a Gabriel synthesis, phthalimide is first treated with KOH, followed by an alkyl halide, thereby installing an alkyl group. In this case, we need to install an *n*-propyl group, so we treat the starting material with KOH, followed by *n*-propyl bromide:

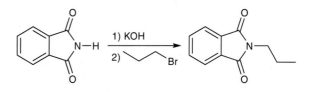

And then hydrazine is used to release the primary amine, as shown here:

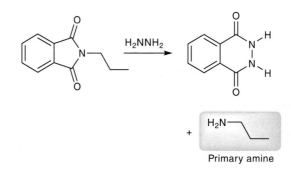

9.41 A Gabriel synthesis begins with phthalimide as the starting material, which is first treated with KOH, followed by an alkyl halide. To determine which alkyl halide to use, we simply redraw the desired amine where the amino group is replaced with a halide:

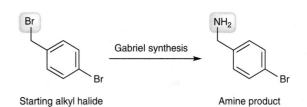

Starting alkyl halide   Amine product

The starting alkyl halide shown above can be used to make the desired amine. Below is a complete Gabriel synthesis of the desired amine:

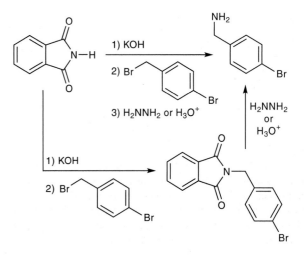

9.42 Compounds **A** and **B** each have a nitrogen atom with a lone pair that is delocalized by resonance:

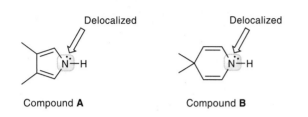

But there is another factor to consider when comparing these compounds. Note that compound **A** is aromatic, where the lone pair is contributing toward the six π electrons of the aromatic system:

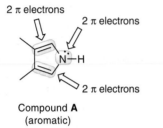

Compound **A**
(aromatic)

As a result, compound **A** will be a very poor base, because the lone pair is not available to function as a base (protonation would cause a loss of aromaticity). In contrast, compound **B** is not aromatic, because the ring lacks a continuous system of overlapping *p* orbitals (the ring has an *sp³* hybridized carbon atom, highlighted below):

Compound **B**
(not aromatic)

Therefore, the lone pair in compound **B** is more available to function as a base. Compound **B** is a stronger base than compound **A**.

9.43 The starting material is cyclic, and the product is acyclic, so our synthesis must involve a step that opens the ring. One way to open a ring is with ozonolysis, which requires that we first install a double bond in the ring. This can be accomplished by treating the starting alcohol with concentrated sulfuric acid, to give cyclohexene, which is then opened with ozonolysis:

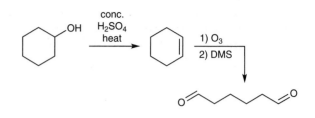

To complete the synthesis, we must convert the dialdehyde above into the desired product:

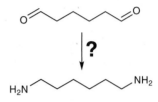

Notice that each carbonyl group must be reduced and converted into an amine. This can be achieved via reductive amination. The complete synthesis is shown below:

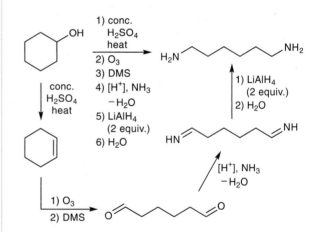

9.44 This problem requires that we replace one halogen for another:

Notice that in both the starting material and the product, the halogen is connected to an *sp²* hybridized carbon atom. Recall that S_N1 and S_N2 reactions generally do not occur at *sp²* hybridized centers, so we have not learned a direct way to achieve this type of transformation. Therefore, we will need to develop a multistep approach for replacing the chlorine atom with a bromine atom.

Let's work backwards. In this chapter, we have seen a new way to install a bromine atom on an aromatic ring. Specifically, we can treat an aryl diazonium ion with CuBr, and this can be the last step of our synthesis:

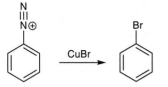

The diazonium salt above can be made from aniline, so the following two steps can be the last two steps of our synthesis:

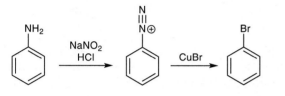

To complete the synthesis, we must convert the starting material (chlorobenzene) into aniline. Indeed, we have seen that this can be accomplished with elimination-addition:

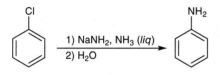

The complete synthesis is shown here:

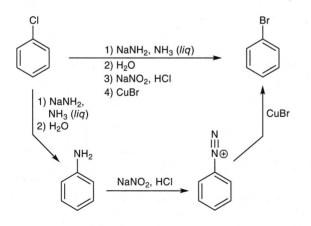

9.45 Let's work backwards. We have only seen one way to install a cyano group on an aromatic ring (by treating an aryl diazonium ion with CuCN), so this can be the last step of our synthesis:

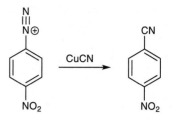

The diazonium salt above can be made from the corresponding aniline, so the following two steps can be the last two steps of our synthesis:

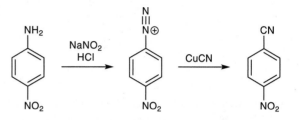

To complete the synthesis, we must convert benzene into *para*-nitroaniline:

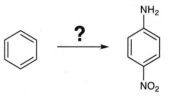

Recall that amino groups are *ortho-para* directors, while nitro groups are *meta*-directors. So we will need to install the amino group first, and then use the directing effects of the amino group to install the nitro group:

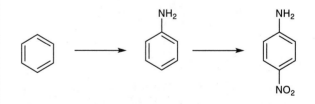

We have not seen a direct way to install an amino group on an aromatic ring, but we have seen a way to install an amino group in two steps: (1) install a chlorine atom on the ring, and then (2) use elimination-addition to replace the chlorine atom with an amino group:

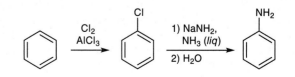

Our synthesis is now almost complete. We just need a way to convert aniline into *para*-nitroaniline:

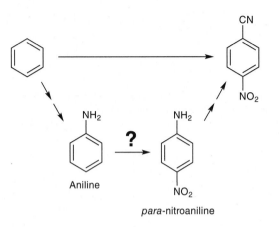

Aniline

?

para-nitroaniline

We have seen that direct nitration of aniline is generally ineffective. It is better to first acylate the amino group (thereby converting it from a strong activator to a moderate activator), then perform the nitration reaction, and then remove the acyl group:

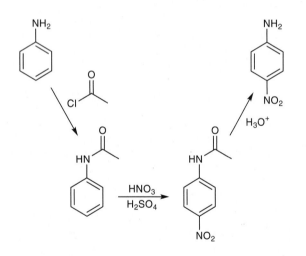

The complete synthesis is shown below. Chlorination of benzene gives chlorobenzene, which is then converted into aniline via an elimination-reduction process. Then, in order to install the nitro group, we first acylate the amino group, then install the nitro group, and then remove the acyl group. Finally, we convert the amino group into a diazonium ion, which we treat with CuCN to give the desired product:

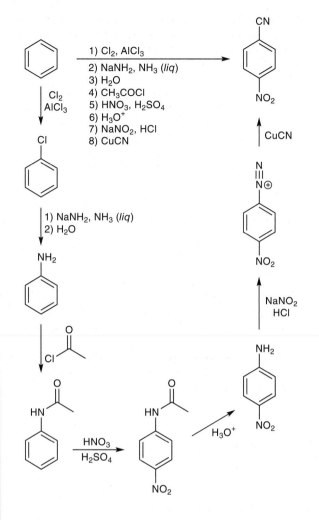

1) Cl_2, $AlCl_3$
2) $NaNH_2$, NH_3 (*liq*)
3) H_2O
4) CH_3COCl
5) HNO_3, H_2SO_4
6) H_3O^+
7) $NaNO_2$, HCl
8) CuCN

1) $NaNH_2$, NH_3 (*liq*)
2) H_2O

HNO_3 / H_2SO_4

H_3O^+

CuCN

$NaNO_2$ HCl

9.46 The product is a nitrosamine, which can be made by treating the corresponding secondary amine with $NaNO_2$ and HCl, so this can be the last step of our synthesis:

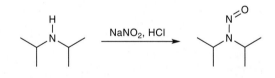

NaNO₂, HCl

To complete the synthesis, we must convert the starting material (a primary amine) into a secondary amine. This can be accomplished via reductive amination. The complete synthesis is shown here:

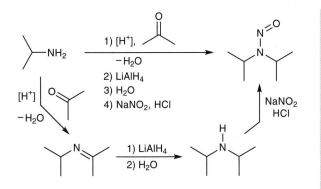

9.47 The product is a secondary amine, which can be made from the reaction between an aldehyde and a primary amine (using a reductive amination), and this can be the last part of our synthesis:

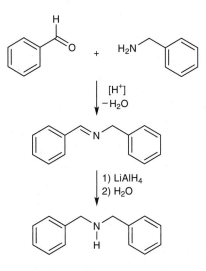

To complete the synthesis, we need to determine a way to convert the starting material into the aldehyde shown above (benzaldehyde). Benzaldehyde can be made from the starting material in just two steps: (1) S_N2 with hydroxide, followed by (2) oxidation with PCC:

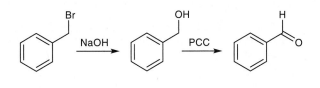

The complete synthesis is shown here:

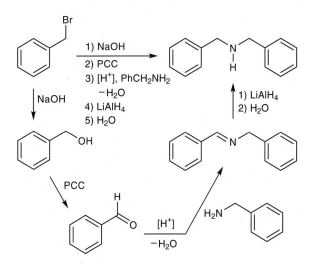

9.48 The desired product is a *meta*-disubstituted aromatic ring, where both substituents are bromine atoms. Recall that a bromine substituent is an *ortho-para* director, and this creates an obstacle for us. Specifically, after installing one bromine atom, the second bromine atom will be directed to the wrong location (the second bromine atom will be directed to the *ortho* or *para* positions, rather than being directed to the *meta* position).

In this chapter, we have seen another way to install a bromine atom on an aromatic ring. We have seen that an aryl diazonium ion can be treated with CuBr to install a bromine atom, and this can be the last step of our synthesis:

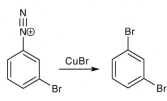

The diazonium ion above can be made from the corresponding aniline, so the last two steps of our synthesis can be as follows:

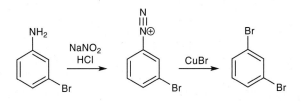

In the problem statement, we saw how an amino group can be made from a nitro group, so the substituted aniline above can be made from the corresponding nitro-substituted aromatic ring:

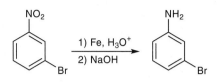

To complete the synthesis, we must convert the starting material (benzene) into the nitro-substituted compound shown above, and this can be accomplished in two steps: (1) nitration, followed by (2) bromination:

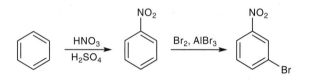

Nitration of the ring gives nitrobenzene, followed by bromination, which would occur at the *meta* position, as desired.

The complete synthesis is shown here:

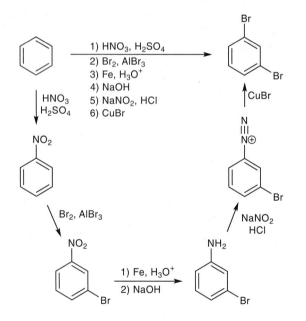

9.49

(a) A Michael reaction is a 1,4-addition, in which a nucleophile and a proton are installed in a 1,4-fashion across a conjugated π system, followed by tautomerization:

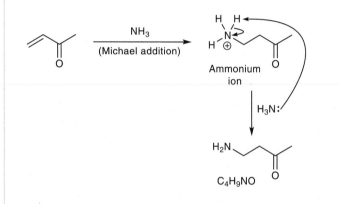

In this case, the attacking nucleophile is ammonia, and we expect the resulting ammonium ion to lose a proton to give the following product, which has the molecular formula C_4H_9NO:

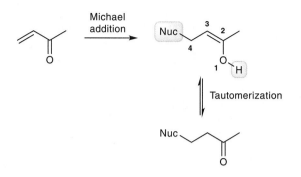

(b) The product has a primary amino group, so we expect signals at 3350 and 3450 cm^{-1} in the IR spectrum of the product. The starting material lacks this functional group, so its IR spectrum will not have signals at 3350 and 3450 cm^{-1}. Moreover, the starting material has a conjugated C=C bond, which should produce a signal at approximately 1600 cm^{-1}, while the product lacks a C=C bond, so the IR spectrum of the product will not have a signal at 1600 cm^{-1}.

In addition, the starting material has a conjugated C=O bond (giving a signal at approximately 1680 cm^{-1}), while the product has a C=O bond that is not conjugated (giving a signal at approximately 1720 cm^{-1}).

(c) The starting material has four different kinds of protons (labeled **a–d** below), so the starting material will have four signals in its ^{1}H NMR spectrum:

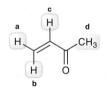

Similarly, the product has four different kinds of protons (labeled **a–d** below), so the product will also have four signals in its ^{1}H NMR spectrum:

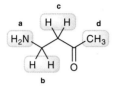

Since each spectrum will have four signals, we cannot differentiate these compounds based on the number of signals in each spectrum. However, we can differentiate these compounds based on the chemical shifts, multiplicities, and integration values of the signals in each spectrum.

In each spectrum, there will be a singlet with an integration of 3H at approximately 1.9 ppm, corresponding with the methyl group (highlighted in each structure below):

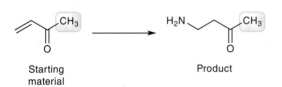

Starting material   Product

So we will focus our attention on the other three signals in each spectrum. For the starting material, all three signals (**a–c**) should appear between 4 and 6 ppm, corresponding with vinylic protons. Each of these signals will be a doublet of doublets, and each of these signals will have an integration value of 1H. In contrast, the ^{1}H NMR spectrum of the product will have more signals near 2 ppm, with most signals (all except for the methyl signal) having a relative integration value of 2H, rather than 1H.

(d) The starting material has a C═C bond, so the ^{13}C NMR spectrum of the starting material should have two signals between 100 and 150 ppm. The product will lack such signals in its ^{13}C NMR spectrum. Also, the ^{13}C NMR spectrum of the product will have a signal between 50 and 100 ppm, corresponding to the carbon atom that is adjacent to the nitrogen atom (the ^{13}C NMR spectrum of the starting material will not have a signal between 50 and 100 ppm).

(e) As described in the problem statement, ammonia functions as a nucleophile and attacks the beta position of the starting material. The resulting intermediate can then undergo an intramolecular proton transfer (or two, successive, intermolecular proton transfers) to give an enol, which then tautomerizes to the ketone. Since ammonia is a base, we have drawn the base-catalyzed mechanism for tautomerization (rather than the acid-catalyzed mechanism).

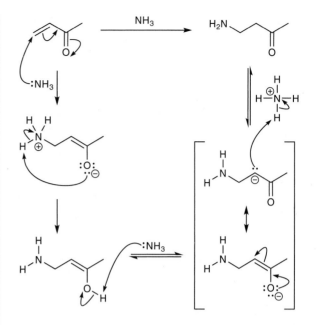

9.50

(a) Compound **A** has no N—H bonds, so the IR spectrum of compound **A** will have no signals above 3000 cm^{-1}. Compound **B** has one N—H bond, so the IR spectrum of compound **B** will have one signal at approximately 3400 cm^{-1}. Compound **C** has two N—H bonds, so the IR spectrum of compound **C** will have two signals near 3400 cm^{-1}.

(b) Compound **C** has four different kinds of protons (labeled **a–d** below), so compound **C** will have four signals in its ^{1}H NMR spectrum:

In contrast, compound **D** has only three different kinds of protons (labeled **a–c** below), so compound **D** will have only three signals in its ^{1}H NMR spectrum:

(c) Compound **C** has three different kinds of carbon atoms (labeled **a–c** below), so compound **C** will have three signals in its ^{13}C NMR spectrum:

In contrast, compound **D** has only two different kinds of carbon atoms (labeled **a** and **b** below), so compound **D** will have only two signals in its ^{13}C NMR spectrum:

(d) Only primary amines (RNH$_2$) will react with NaNO$_2$ and HCl to give a diazonium ion. Compound **A** is a tertiary amine (R$_3$N), and compound **B** is a secondary amine (R$_2$NH). Since compounds **A** and **B** are not primary amines, they are not converted into diazonium ions upon treatment with NaNO$_2$ and HCl. Compounds **C** and **D** are primary amines, so each of these compounds will react with NaNO$_2$ and HCl to give a diazonium ion:

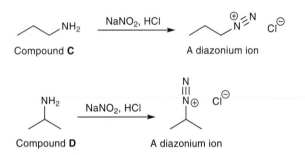

(e) A tertiary amine will react with only one equivalent of an alkyl halide when treated with excess alkyl halide. Compound **A** is a tertiary amine, so it will react with only one equivalent of methyl iodide when treated with excess methyl iodide to give a quaternary ammonium salt:

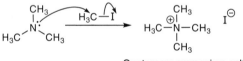

Quaternary ammonium salt

9.51 A primary amine is first converted into a secondary amine via reductive amination:

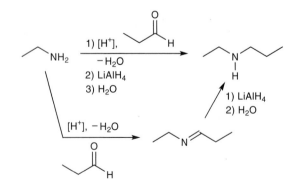

And then, this secondary amine is converted into an enamine upon treatment with a ketone under acidic conditions, with removal of water:

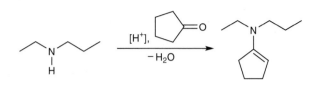

9.52 Compound **A** is benzyl amine, which is formed via a Gabriel synthesis, as shown here:

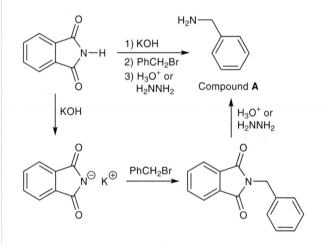

Compound **A** is a primary amine, and it is converted into a secondary amine (compound **B**) via reductive amination, as shown here:

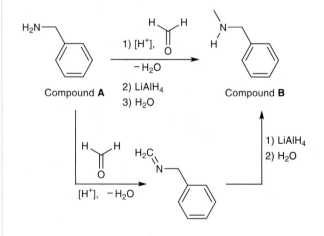

Compound **B** is a secondary amine, which is converted into a nitrosamine upon treatment with NaNO₂ and HCl:

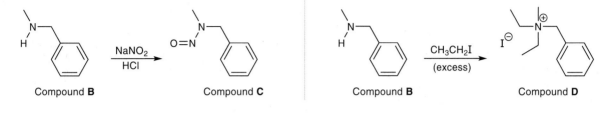

Compound **B**   Compound **C**

Compound **B** is a secondary amine, which is converted into a quaternary ammonium ion upon treatment with excess ethyl iodide:

Compound **B**   Compound **D**

9.53 Recall that a primary amine will react with NaNO₂ and HCl to give a diazonium ion:

Primary amine   A diazonium ion

Our starting material has two NH₂ groups, so we will consider each of these NH₂ groups being converted into a diazonium ion:

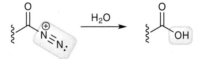

Note that an acyl diazonium ion has an excellent, built-in leaving group (nitrogen gas), so in the presence of water, we expect a nucleophilic acyl substitution reaction:

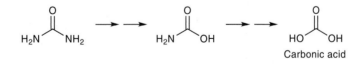

This ultimately happens to each NH₂ group, giving carbonic acid:

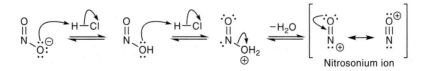

Carbonic acid

Now let's draw all of the mechanistic steps for this process. This is going to be a long mechanism, and we will break it down into parts. Let's begin with the formation of the reactive intermediate, a nitrosonium ion. When NaNO₂ and HCl are mixed together, an equilibrium is established in which some nitrosonium ion is present, as shown here:

Nitrosonium ion

In the presence of this reactive nitrosonium intermediate, one of the NH$_2$ groups in the starting material is first converted into a diazonium ion, as shown below. The lone pair on the nitrogen atom attacks the nitrosonium ion, and the resulting intermediate is deprotonated. Three more successive proton transfer steps, followed by loss of water, gives a diazonium ion:

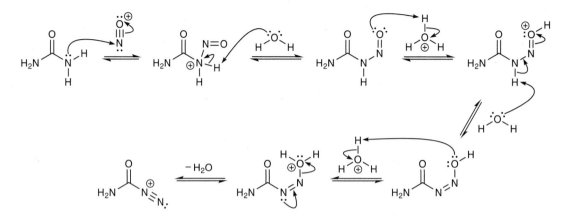

The diazonium ion has an excellent, built-in leaving group, so before we convert the other NH$_2$ group to a diazonium ion, we will first draw a nucleophilic acyl substitution process, which replaces the diazonium ion with an OH group. As shown below, water attacks the carbonyl group, giving a tetrahedral intermediate, which then collapses to reform the carbonyl group, by expelling nitrogen gas as a leaving group, followed by a proton transfer step:

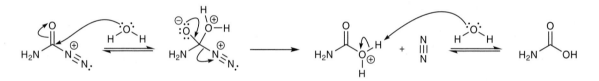

Now we turn our attention to the other NH$_2$ group, which will follow the same fate as the previous NH$_2$ group. In the presence of the reactive nitrosonium intermediate, the remaining NH$_2$ group in the starting material is first converted into a diazonium ion, as shown below. The lone pair on the nitrogen atom attacks the nitrosonium ion, and the resulting intermediate is deprotonated. Three more successive proton transfer steps, followed by loss of water, gives a diazonium ion:

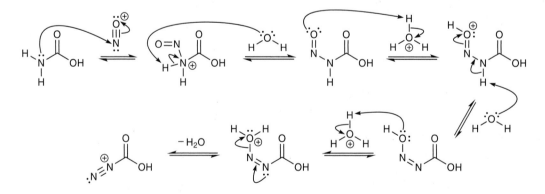

Once again, this diazonium ion has an excellent, built-in leaving group, so we draw a nucleophilic acyl substitution process, which replaces the diazonium ion with an OH group. As shown below, water attacks the carbonyl group, giving a tetrahedral intermediate, which then collapses to reform the carbonyl group, by expelling nitrogen gas as a leaving group, followed by a proton transfer step to give carbonic acid:

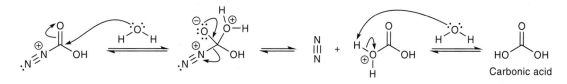

Carbonic acid undergoes spontaneous degradation in the presence of water to give CO_2. In catalytic acid, one of the OH groups can be protonated to give an excellent leaving group, which can then be expelled to give a cation, which is then deprotonated by water to give carbon dioxide:

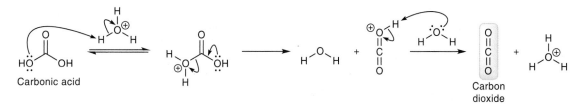

Alternatively, we can draw a slightly different mechanism for the degradation of carbonic acid. This alternate mechanism relies on the fact that carbonic acid exists in equilibrium with its conjugate base, so some of the conjugate base is present. This conjugate base can expel a hydroxide ion to give carbon dioxide directly:

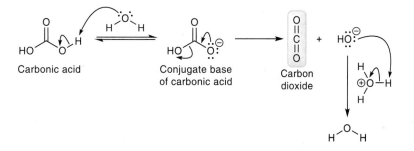

In general, we should avoid expelling a hydroxide ion in acidic conditions, but this case is exceptional, because the step that liberates hydroxide also liberates CO_2 which is a gas that bubbles out of solution. This has the effect of pushing the reaction to completion, despite the formation of a hydroxide ion.

9.54 Recall that a primary amine will react with $NaNO_2$ and HCl to give a diazonium ion:

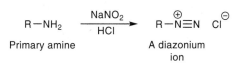

Our starting material is a primary amine, and under these conditions, it will be converted into a diazonium ion:

Now let's draw all of the mechanistic steps for this process. Let's begin with the formation of the reactive intermediate, a nitrosonium ion. When $NaNO_2$ and HCl are mixed together, an equilibrium is established in which some nitrosonium ion is present, as shown here:

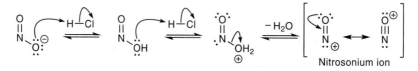

In the presence of this reactive nitrosonium intermediate, the NH_2 group in the starting material is converted into a diazonium ion, as shown below. The lone pair on the nitrogen atom attacks the nitrosonium ion, and the resulting intermediate is deprotonated. Three more successive proton transfer steps, followed by loss of water, gives a diazonium ion:

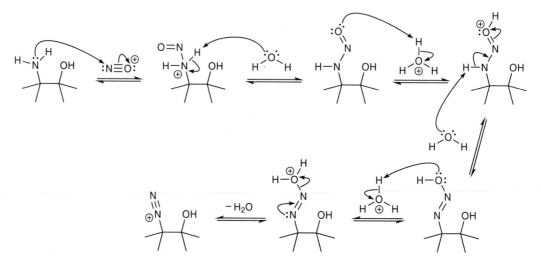

This diazonium ion has an excellent, built-in leaving group (nitrogen gas), which can leave by itself to give a tertiary carbocation (much as we saw in S_N1 reactions). In general, tertiary carbocations don't rearrange, unless doing so will produce a resonance-stabilized carbocation. Indeed, in this case, a methyl shift will generate a resonance-stabilized cation, shown below, which is then deprotonated by water to give the product:

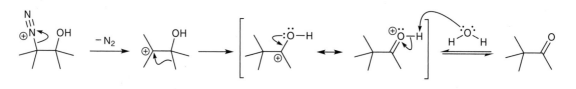

CHAPTER 10

10.1 This is a Diels–Alder reaction between 1,3-butadiene and a monosubstituted ethylene, so the product will have one chiral center. Since the starting materials are achiral, the product is a racemic mixture:

center. Since the starting materials are achiral, the product is a racemic mixture:

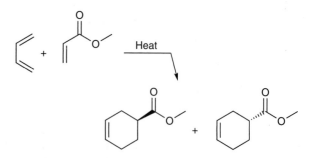

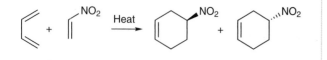

10.2 This is a Diels–Alder reaction between 1,3-butadiene and a monosubstituted ethylene, so the product will have one chiral

10.3 This is a Diels–Alder reaction between 1,3-butadiene and a monosubstituted ethylene, so the product will have one chiral center. Since the starting materials are achiral, the product is a racemic mixture:

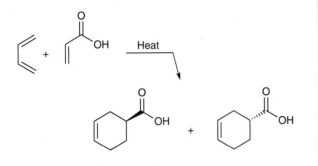

10.4 This is a Diels–Alder reaction between 1,3-butadiene and a monosubstituted ethylene, so the product will have one chiral center. Since the starting materials are achiral, the product is a racemic mixture:

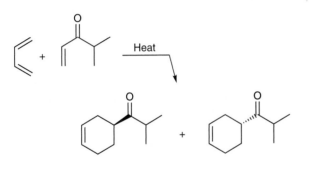

10.5 This is a Diels–Alder reaction between 1,3-butadiene and a *cis*-disubstituted ethylene. Notice that the configuration of the dienophile is preserved in the product, so the product is *cis*-disubstituted. Since the product has a plane of symmetry, it is a *meso* compound and has no enantiomers:

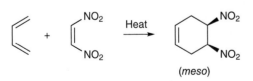

10.6 This is a Diels–Alder reaction between 1,3-butadiene and a *trans*-disubstituted ethylene. Notice that the configuration of the dienophile is preserved in the product, so the product is *trans*-disubstituted. A racemic mixture of enantiomers is expected, as shown:

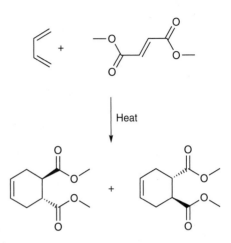

10.7 This is a Diels–Alder reaction between 1,3-butadiene and a *cis*-disubstituted ethylene. Notice that the configuration of the dienophile is preserved in the product, so the product is *cis*-disubstituted. Since the product has a plane of symmetry, it is a *meso* compound and has no enantiomers:

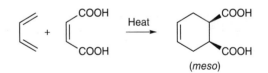

10.8 This is a Diels–Alder reaction between 1,3-butadiene and a *trans*-disubstituted ethylene. Notice that the configuration of the dienophile is preserved in the product, so the product is *trans*-disubstituted. A racemic mixture of enantiomers is expected, as shown:

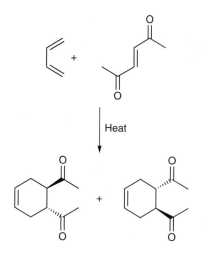

10.9 This is a Diels–Alder reaction between 1,3-butadiene and a disubstituted alkyne. Notice that the product has no chiral centers, and only one product is obtained:

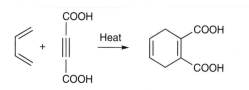

10.11 This is a Diels–Alder reaction involving cyclopentadiene, so we expect the product to have a bicyclic skeleton. The dienophile is *cis*-disubstituted, so the product must be *cis*-disubstituted. Notice that both substituents occupy *endo* positions. Since the product has a plane of symmetry, it is a *meso* compound and has no enantiomers:

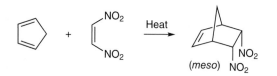

10.12 This is a Diels–Alder reaction involving cyclopentadiene, so we expect the product to have a bicyclic skeleton. The dienophile is *trans*-disubstituted, so the product must be *trans*-disubstituted. A mixture of enantiomers is obtained:

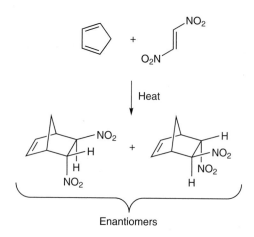

Enantiomers

10.13 This is a Diels–Alder reaction involving cyclopentadiene, so we expect the product to have a bicyclic skeleton. The dienophile is a monosubstituted ethylene, so we expect the following pair of enantiomers. Notice that the substituent occupies an *endo* position in each enantiomer.

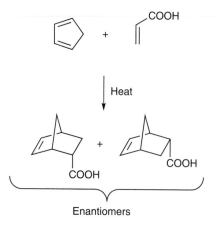

Enantiomers

10.14 This is a Diels–Alder reaction involving cyclopentadiene, so we expect the product to have a bicyclic skeleton. The dienophile is *cis*-disubstituted, so the product must be *cis*-disubstituted. Notice that both substituents occupy *endo* positions. Since the product has a plane of symmetry, it is a *meso* compound and has no enantiomers:

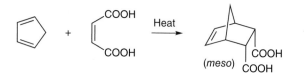

10.15 This is a Diels–Alder reaction involving cyclopentadiene, so we expect the product to have a bicyclic skeleton. The dienophile is *trans*-disubstituted, so the product must be *trans*-disubstituted. A mixture of enantiomers is obtained:

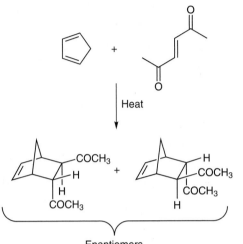

Enantiomers

10.16 This is a Diels–Alder reaction involving cyclopentadiene, so we expect the product to have a bicyclic skeleton. The dienophile is a monosubstituted ethylene, so we expect the following pair of enantiomers. Notice that the substituent occupies an *endo* position in each enantiomer.

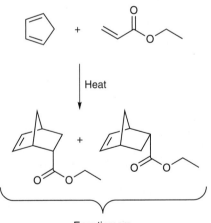

Enantiomers

10.17 Recall that a Diels–Alder reaction generates a cyclohexene derivative:

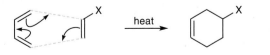

In order to draw the starting materials for a Diels–Alder reaction, it can be useful to imagine the reverse process, called a retro-Diels–Alder reaction. Indeed, retro-Diels–Alder reactions can occur at high temperature, so we will consider the compounds that would be generated if our desired product were to undergo a retro-Diels–Alder reaction at high temperature:

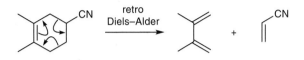

This theoretical retro-Diels–Alder reaction enables us to identify a diene and a dienophile that can be used to prepare the desired product via a forward Diels–Alder reaction:

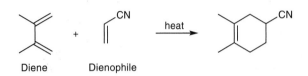

Diene   Dienophile

10.18 In order to draw the starting materials for a Diels–Alder reaction, it can be useful to imagine the reverse process, called a retro-Diels–Alder reaction. Indeed, retro-Diels–Alder reactions can occur at high temperature, so we will consider the compounds that would be generated if our desired product were to undergo a retro-Diels–Alder reaction at high temperature:

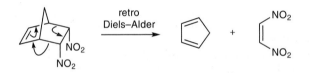

This theoretical retro-Diels–Alder reaction enables us to identify a diene and a dienophile that can be used to prepare the desired product via a forward Diels–Alder reaction:

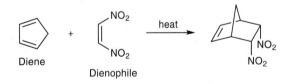

Diene   Dienophile

Notice that the two nitro groups must be *cis* to each other in the starting dienophile in order to be *cis* to each other in the product, because the configuration of the dienophile is preserved during a Diels–Alder reaction:

Dienophile

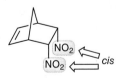

Product

10.19 In order to draw the starting materials for a Diels–Alder reaction, it can be useful to imagine the reverse process, called a retro-Diels–Alder reaction. Indeed, retro-Diels–Alder reactions can occur at high temperature, so we will consider the compounds that would be generated if our desired product were to undergo a retro-Diels–Alder reaction at high temperature. In this case, we can imagine two possible sets of starting materials, depending on how we push the curved arrows during the theoretical retro-Diels–Alder reaction. Both options are shown here:

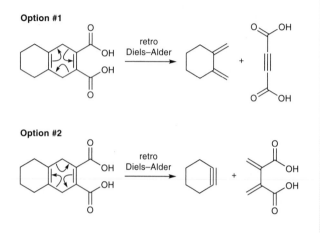

Of the two options, option #1 is preferred, because the dienophile has electron-withdrawing groups. Furthermore, option #2 is unrealistic, because it requires a dienophile that has a triple bond incorporated within a six-membered ring. Such a compound would be extremely high in energy (if it exists at all), because triple bonds have linear geometry.

The theoretical retro-Diels–Alder reaction shown above as option #1 enables us to identify a diene and a dienophile that can be used to prepare the desired product via a forward Diels–Alder reaction:

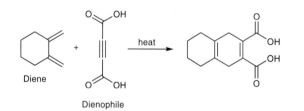

10.20 In a Diels–Alder reaction, the diene reacts with the dienophile in a concerted process. We draw the curved arrows for this process, and these curved arrows guide us when drawing the product.

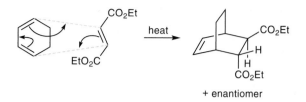

Notice that the starting dienophile has a *trans* configuration, and this *trans* configuration is retained in the product (the ester groups are *trans* to each other in the product as well).

Also notice that both starting materials are achiral, so the product (which contains chiral centers) must be optically inactive (a racemic mixture of enantiomers).

10.21 In a Diels–Alder reaction, the diene reacts with the dienophile in a concerted process. We draw the curved arrows for this process, and these curved arrows guide us when drawing the product. Note that the aldehyde groups are on a planar alkene (in the product), so there are no endo/exo positions in this case.

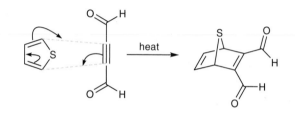

10.22 In order to draw the starting materials for a Diels–Alder reaction, it can be useful to imagine the reverse process, called a retro-Diels–Alder reaction. Indeed, retro-Diels–Alder reactions can occur at high temperature, so we will consider the compounds that would be generated if our desired product were to undergo a retro-Diels–Alder reaction at high temperature:

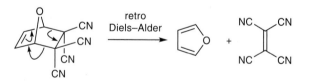

This theoretical retro-Diels–Alder reaction enables us to identify a diene and a dienophile that can be used to prepare the desired product via a forward Diels–Alder reaction:

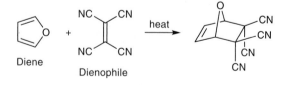

10.23 The structure of compound **A** contains both a diene and a dienophile,

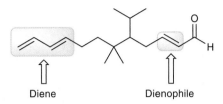

so compound **A** can undergo an intramolecular Diels–Alder reaction. We begin by assigning numbers to the parent chain of compound **A** (these number will help us keep track of all atoms during the reaction). Then, we use those numbers to help us redraw compound **A** so that the diene and the dienophile are near each other and aligned to react with each other:

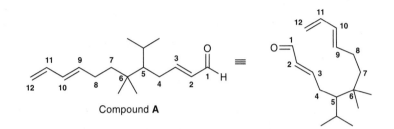

Compound **A**

Next, we draw the curved arrows for this process, and these curved arrows guide us when drawing the product:

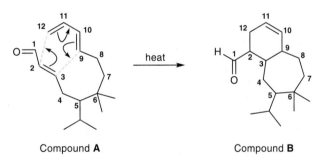

Compound **A**   Compound **B**

10.24 In order to draw the starting materials for a Diels–Alder reaction, it can be useful to imagine the reverse process, called a retro-Diels–Alder reaction. Indeed, retro-Diels–Alder reactions can occur at high temperature, so we will consider the compounds that would be generated if our desired product were to undergo a retro-Diels–Alder reaction at high temperature:

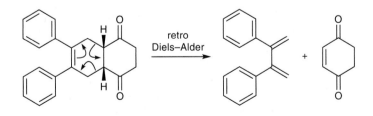

This theoretical retro-Diels–Alder reaction enables us to identify a diene and a dienophile that can be used to prepare the desired product via a forward Diels–Alder reaction:

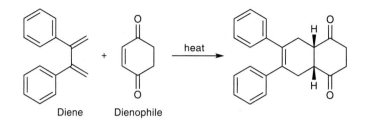

Diene   Dienophile

10.25   In a Diels–Alder reaction, the diene reacts with the dienophile in a concerted process. We draw the curved arrows for this process, and these curved arrows guide us when drawing the product.

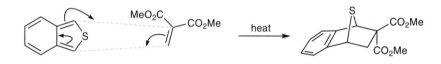

10.26   The problem statement indicates that the starting triene reacts with two equivalents of maleic anhydride (a dienophile). This suggests that the product is formed via two successive Diels–Alder reactions. We begin by drawing the curved arrows for the first Diels–Alder reaction, and these curved arrows guide us in drawing the product of the first Diels–Alder reaction:

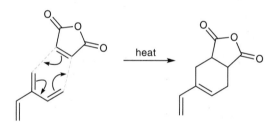

Notice that the resulting compound has a diene in its structure (bottom left of structure), and this diene can react with a second equivalent of maleic anhydride in a second Diels–Alder reaction. We draw the curved arrows for this Diels–Alder reaction, and once again, the curved arrows will guide us to draw the final product (which has 14 carbon atoms):

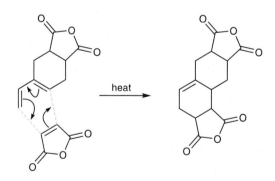

10.27 The desired product is an acetal, which can be made from the corresponding diol:

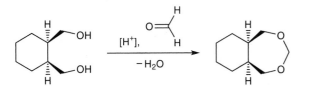

The reaction above can be the last step of our synthesis. To complete the synthesis, we need to convert the starting material into the diol above:

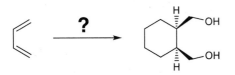

Notice that the starting material is a diene, and the product has a six-membered ring, which is strongly suggestive of a Diels–Alder reaction (Diels–Alder reactions generate a six-membered ring that contains a C=C bond, which can then be reduced via hydrogenation):

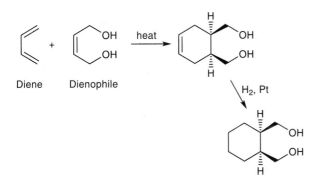

However, the Diels–Alder reaction shown above is not likely to be efficient, because the dienophile does not have electron-withdrawing groups connected directly to the C=C bond. This process will be much more efficient if we use a dienophile with electron-withdrawing groups that can be later converted into the desired CH_2OH groups. For example, aldehyde groups are electron-withdrawing groups, so the following Diels–Alder reaction will be more efficient:

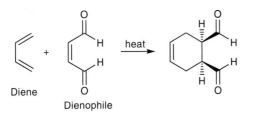

After the Diels–Alder reaction, the aldehyde groups can be reduced to give the desired CH_2OH groups, and the C=C bond can be reduced as well:

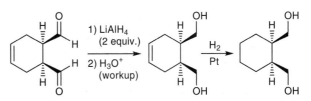

The complete synthesis is shown below:

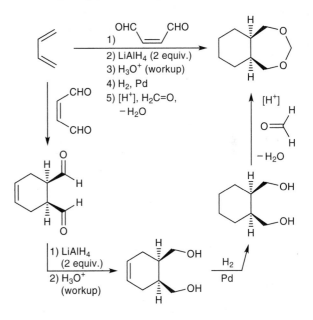

10.28 The desired product is a cyclohexene derivative, which can be made via a Diels–Alder reaction:

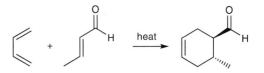

The reaction above can be the last step of our synthesis. To complete the synthesis, we need to make both the diene and the dienophile from acetaldehyde (as our only source of carbon atoms). Let's begin by converting acetaldehyde into the necessary dienophile, since that can be accomplished with just one reaction (an aldol condensation):

Aldol condensation

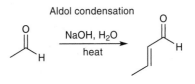

Next, we turn attention to the diene, and we consider how 1,3-butadiene can be made from acetaldehyde. There are certainly many ways to do this. One such way is shown in the synthesis below. Acetaldehyde is first reduced to give an alcohol, which is then treated with PBr$_3$ to convert the alcohol to an alkyl bromide. Insertion of magnesium gives a Grignard reagent, which will attack acetaldehyde (followed by aqueous acidic workup) to give 2-butanol. Treatment with concentrated sulfuric acid then gives an elimination reaction, generating *trans*-2-butene. This alkene is then treated with molecular bromine to give a dibromide, which is then treated with two equivalents of *tert*-butoxide (a sterically hindered base) to give two successive, elimination reactions, giving 1,3-butadiene. Then, our synthesis is complete when we treat 1,3-butadiene with the dienophile (the aldol condensation product) to give the desired substituted cyclohexene derivative:

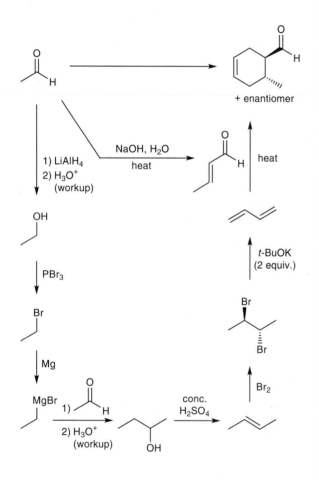

10.29 An alkene will react with molecular bromine (Br$_2$) to give an *anti*-addition of Br and Br across the pi bond, generating a *trans*-dibromide. Treating the *trans*-dibromide with two equivalents of *tert*-butoxide (a sterically hindered base) gives two successive, elimination reactions to afford 1,3-cyclohexadiene:

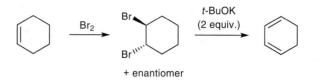

In the reaction scheme in the problem statement, the final step converts 1,3-cyclohexadiene into a substituted cyclohexene derivative, so this is a Diels–Alder reaction. To draw the necessary dienophile, it can be useful to imagine the reverse process, called a retro-Diels–Alder reaction. Indeed, retro-Diels–Alder reactions can occur at high temperature, so we will consider the compounds that would be generated if our desired product were to undergo a retro-Diels–Alder reaction at high temperature:

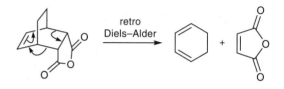

This theoretical retro-Diels–Alder reaction enables us to identify the dienophile that can be used to prepare the desired product via a forward Diels–Alder reaction:

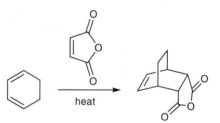

10.30 The first step of the process converts 1,3-butadiene into a substituted cyclohexene derivative, so this is a Diels–Alder reaction. To draw the necessary dienophile, it can be useful to imagine the reverse process, called a retro-Diels–Alder reaction. Indeed, retro-Diels–Alder reactions can occur at high temperature, so we will consider the compounds that would be generated if our desired product were to undergo a retro-Diels–Alder reaction at high temperature:

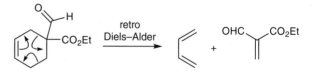

This theoretical retro-Diels–Alder reaction enables us to identify the dienophile that can be used to prepare the desired product via a forward Diels–Alder reaction:

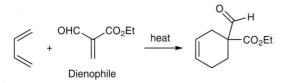

The next step of the process is a reduction of the C=C bond, which can be accomplished with molecular hydrogen (H₂) in the presence of a suitable metal catalyst:

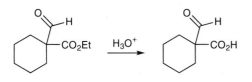

Next, this compound is treated with aqueous acid, which will hydrolyze the ester group to give a carboxylic acid:

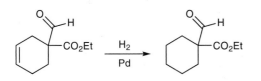

Notice that the resulting compound has a carbonyl group that is beta to the carboxylic acid group, and therefore, this compound can undergo decarboxylation upon heating to give a compound with only seven carbon atoms (rather than eight carbon atoms):

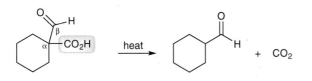

Finally, this aldehyde is converted into the corresponding acetal upon treatment with ethylene glycol in catalytic acid:

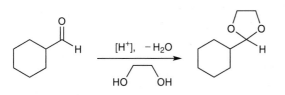

10.31

(a) We begin by drawing all significant resonance structures of the diene, utilizing the pattern of a lone pair adjacent to a π bond:

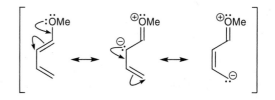

Notice that two locations on the diene are electron-rich (δ−), shown here in the following resonance hybrid:

Now let's draw all significant resonance structures of the dienophile (carbonyl resonance and allylic carbocation):

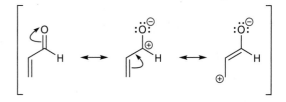

Notice that two locations are electron-poor (δ+), shown here in the following resonance hybrid:

After drawing the resonance structures of the diene and of the dienophile, we are now able to suggest an explanation for the observed regioselectivity. When the diene and the dienophile react with each other, an electron-rich carbon atom (δ−) of the diene will pair up with an electron-poor (δ+) carbon atom of the dienophile, as shown here.

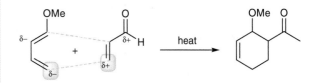

When opposite charges are aligned in the transition state, the effect is to lower the energy of the transition state, which favors formation of the product above.

(b) We begin by drawing all significant resonance structures of the diene:

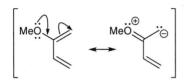

Notice that one of the locations of the diene is electron-rich (δ−), shown here in the following resonance hybrid:

Now let's draw all significant resonance structures of the dienophile:

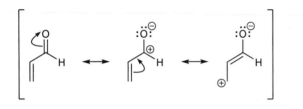

Notice that two locations are electron-poor (δ+), shown here in the following resonance hybrid:

When the diene and the dienophile react with each other, the electron-rich carbon atom (δ−) of the diene will pair up with an electron-poor (δ+) carbon atom of the dienophile, giving the following product:

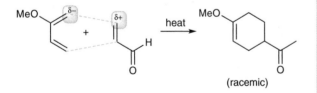

(racemic)

Note that both starting materials are achiral, so the product must be formed as a racemic mixture.

(c) We begin by drawing all significant resonance structures of the diene:

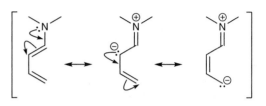

Notice that two locations are electron-rich (δ−), shown here in the following resonance hybrid:

Now let's draw all significant resonance structures of the dienophile:

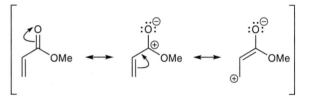

Notice that two locations are electron-poor (δ+), shown here in the following resonance hybrid:

When the diene and the dienophile react with each other, an electron-rich carbon atom (δ−) of the diene will pair up with an electron-poor (δ+) carbon atom of the dienophile, giving the following product.

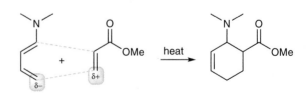

(d) We begin by drawing all significant resonance structures of the diene:

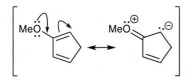

Notice that one of the locations is electron-rich (δ−), shown here in the following resonance hybrid:

Now let's draw significant resonance structures of the dienophile:

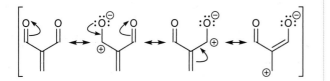

Notice that three locations are electron-poor (δ+), shown here in the following resonance hybrid:

When the diene and the dienophile react with each other, the electron-rich carbon atom (δ−) of the diene will pair up with an electron-poor (δ+) carbon atom of the dienophile, giving the following product:

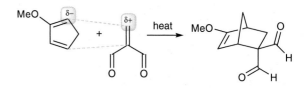

INDEX